STRINGS '93

PROCEEDINGS OF THE CONFERENCE
STRINGS '93

May 24–29, 1993

Berkeley, USA

Editors

M. B. Halpern & G. Rivlis
Lawrence Berkeley Laboratory

A. Sevrin
CERN
and
Vrije Universiteit Brussel

World Scientific
Singapore • New Jersey • London • Hong Kong

Published by

World Scientific Publishing Co. Pte. Ltd.
P O Box 128, Farrer Road, Singapore 9128
USA office: Suite 1B, 1060 Main Street, River Edge, NJ 07661
UK office: 57 Shelton Street, Covent Garden, London WC2H 9HE

STRINGS '93

ISBN 981-02-2187-8

Printed in Singapore by Uto-Print

Preface

These are the proceedings of the international conference Strings '93, which was held at the Lawrence Berkeley Laboratory in Berkeley, California from May 24 through May 29, 1993. This conference is the most recent one in a series of international meetings devoted to string theory. Earlier meetings in the series include those held at Maryland ('87 and '88), Texas A&M ('89 and '90) and Stony Brook ('91).

For six days, over 200 people attended 52 lectures covering subjects including superstring phenomenology, properties of string compactifications, conformal field theory, 2-D gravity, integrable systems, black hole physics, QCD in two dimensions, etc.

Though the conference was organized on a shoestring budget, it would not have been possible at all without the generous financial support of the Physics Division of the Lawrence Berkeley Laboratory, through the U.S. Department of Energy, and the further support of the National Science Foundation and of the Provost of the University of California at Berkeley.

We would like to thank the Physics Division of the Lawrence Berkeley Laboratory and in particular the Division director Bob Cahn, both for the financial support and the use of the LBL facilities. Special thanks are also due to Betty Moura and Luanne Neumann for donating their excellent secretarial skills.

The editors,

Martin B. Halpern
Gil Rivlis
Alexander Sevrin

STRINGS '93

May 24-29, 1993
Berkeley, California, USA

The Main Auditorium
Building 50
Lawrence Berkeley Laboratory

Time	m	t	w	t	f	s
9.00 · 10.00	**Strominger** — Quantum Mechanics and Black Holes	**Reshetikin** — Quantum Groups and Abelian Categories in Physics	**Dine** — Issues in string Phenomenology	**Zwiebach** — Directions in String Field Theory	**Ambjørn** — Strings for c>1 and Quantum Gravity for d>2	**Polchinski** — Exact String Solution for a 4-d Black Hole Throat
10.00 · 10.50	**Jones** — Subfactors, Loop Groups and Conformal Field Theory	**Ooguri** — Mirror Transformation in String Theory	**Witten** — Topology Change and the Landau-Ginzburg Correspondence	**Miwa** — Vertex Operators in Solvable Lattice Models	**Sonnenschein** — Twisted G/H Topological Theories and 2D Gravity	10.00 10.30 **Pope** — W-String Theories
10.50 · 11.20	Coffee	Coffee	Coffee	Coffee	Coffee	10.30 11.00 **Tye** — Fractional Superstrings
11.20 · 12.10	**van Nieuwenhuizen** — Loop Calculations in Conformal Field Theory	**Bern** — New QCD Results from Superstrings	**Schwarz** — Does String Theory have a Duality Symmetry Relating Weak and Strong Coupling?	**Gervais** — Going from Conformal to Light-Cone Gauge in W Gravity	11.20 11.50 **Cohn Fukuma Rabin**	11.00 12.00 **Migdal** — Induced QCD: Achievements, Problems and Perspectives
12.10 · 1.00	**Distler** — Strings in Three Dimensions	**Kaplunovsky** — Scenarios for Superstring Unification	**Verlinde** — Quantum Black Hole Evaporation	**Halpern** — Solving the Ward Identities of Irrational CFT	11.50 12.20 **Siegel** — Manifest Duality in Low-Energy Superstrings · 12.20 12.50 **Porrati** — Schwinger Instability in Open Strings	
1.00 · 2.30	Lunch	Lunch	Lunch	Lunch	Lunch	
2.30 · 3.00	**Horowitz** — Exact 3D Black Holes in String Theory	**Lerche** — N=2 Superconformal Structure of Topological (W)-Gravity	**Schoutens** — Quantum S-Matrix for 2D Black Holes	**Greene** — Mirror Manifolds and Space-Time Topology Change	**Kiritsis** — Exact Duality Symmetries of Strings in Curved Backgrounds	
3.00 · 3.30	**Taylor** — Two-Dimensional QCD and Strings	**Sevrin** — Gauged WZW Models, Extended Virasoro Algebras and 2D Gravity	**Rivlis** — Topological 2D Gravity and the Degeneration Equation	**LeClair** — Affine Lie Algebras and Correlation Functions in Massive Field Theory	**Van Proeyen** — Regularization and the BV Formalism	
3.30 · 3.55	Coffee	Coffee	Coffee	Coffee	Coffee	
3.55 · 4.25	**Berkovits Freed Schmidhuber**	**Jevicki** — Nontrivial Backgrounds in String Theory and Matrix Models	**Douglas** — Exact Results for Two Dimensional Yang-Mills Theory	**Bilal Chaudhuri Sonoda**	**Crescimanno Morrison Myers**	
4.25 · 4.55	**Saleur** — Mass-Less Flows in sine-Gordon and Minimal Models	**Bars** — Strings and QCD are equivalent in 2D Old and New Connections	**Banquet** at 6.30	**Giveon** — Axial-Vector Duality as Gauge Symmetry and Topology Change in String Theory	**Minahan** — Equivalence of Two Dimensional QCD and the c=1 Matrix Model	

Proceedings of the Conference

STRINGS '93

Lawrence Berkeley Laboratory
Berkeley, California, USA
May 24 – 29, 1993

Editors

M. B. Halpern
G. Rivlis
A. Sevrin

Organizers

M. B. Halpern
G. Rivlis
A. Sevrin
S. Mandelstam
K. Bardakçi
O. Alvarez

Table of Contents

1. Black Holes

EXACTLY SOLVABLE QUANTUM-CORRECTED
2D DILATON-GRAVITY THEORIES

ADEL BILAL[*]

Joseph Henry Laboratories, Princeton University
Princeton, NJ 08544, USA

ABSTRACT

We review the basic aspects of the exactly solvable quantum-corrected dilaton gravity theories in two dimensions, emphasizing the role of conformal invariance.

1. Introduction

Our starting point is the classical action for dilaton gravity in two dimensions as written by Callan, Giddings, Harvey and Strominger (CGHS) [1]

$$S_{\rm cl} = \frac{1}{2\pi} \int {\rm d}^2 x \sqrt{-g} \left[e^{-2\phi} \left(R + 4(\nabla\phi)^2 + 4\lambda^2 \right) - \tfrac{1}{2} \sum_{i=1}^{N} (\nabla f_i)^2 \right] . \qquad (1.1)$$

Here ϕ is the dilaton field with $G \equiv e^{\phi}$ playing the role of the gravitational coupling constant, λ^2 is referred to as cosmological constant and the f_i are N massless conformally coupled matter fields. This action admits classical (non-radiating) static black hole solutions, depending on one parameter m, the black hole mass. The $m = 0$ solution where $\phi = -\lambda\sigma$ is called the linear dilaton vacuum (LDV).

The goal is to quantize the theory described by this action $S_{\rm cl}$. If the number of matter fields is different from 24, $N \neq 24$, one has to include various contributions to the conformal anomaly. Thus we add[†]

$$S_{\rm anom} = -\frac{\kappa}{8\pi} \int {\rm d}^2 x \sqrt{-g} R \frac{1}{\nabla^2} R \qquad (1.2)$$

We will keep κ as a parameter to be determined later on. Note that $S_{\rm anom}$ is $\mathcal{O}(e^{2\phi}) \equiv \mathcal{O}(G^2)$ with respect to the gravitational part of $S_{\rm cl}$ and may be thought of as the one-loop contribution of the matter fields.

---

[*] on leave of absence from Laboratoire de Physique Théorique de l'Ecole Normale Supérieure, 24 rue Lhomond, 75231 Paris Cedex 05, France (unité propre du CNRS)
e-mail: bilal@puhep1.princeton.edu

[†] A possible term $\mu^2 \int \sqrt{-g}$ is supposed to be fine-tuned to vanish.

2. Conformal invariance and transformation to free fields

Conformal gauge

Let us first choose conformal gauge, $g_{++} = g_{--} = 0$, $g_{+-} = -\frac{1}{2}e^{2\rho}$, where $\sigma^{\pm} = \tau \pm \sigma$. Then

$$S_{\mathrm{cl}} = \frac{1}{\pi} \int \mathrm{d}^2\sigma \left[e^{-2\phi}\left(2\partial_+\partial_-\rho - 4\partial_+\phi\partial_-\phi + \lambda^2 e^{2\rho} \right) + \frac{1}{2}\sum_{i=1}^{N} \partial_+ f_i \partial_- f_i \right],$$

$$S_{\mathrm{anom}} = -\frac{\kappa}{\pi} \int \mathrm{d}^2\sigma \partial_+\rho\partial_-\rho .$$

(2.1)

The equations of motion derived in conformal gauge must be supplemented by the g_{++} and g_{--} equations of motion as constraints, $T_{++} = T_{--} = 0$, where

$$T_{\pm\pm} = e^{-2\phi}\left(4\partial_\pm\phi\partial_\pm\rho - 2\partial_\pm^2\phi \right) - \kappa\left(\partial_\pm\rho\partial_\pm\rho - \partial_\pm^2\rho \right) + \frac{1}{2}\sum_{i=1}^{N}(\partial_\pm f_i)^2 .$$

(2.2)

Note that for $\kappa > 0$ the kinetic term of $S_{\mathrm{cl}} + S_{\mathrm{anom}}$ is degenerate at $e^{-2\phi} = e^{-2\phi_c} \equiv \kappa$. We expect something singular to happen when $\phi = \phi_c$.

Conformal invariance

Since we are dealing with a theory of gravity, we started with a diffeomorphism invariant theory. Then we fixed conformal gauge, leaving as symmetries the subgroup of conformal diffeomorphisms $\sigma^+ \to f^+(\sigma^+)$, $\sigma^- \to f^-(\sigma^-)$. Quantization should preserve these conformal symmetries. In particular, we need to ensure that the resulting theory is a $c_{\mathrm{tot}} = 0$ conformal theory. The latter is a necessary condition that relies only on the short distance properties of the quantum theory. They may be inferred even though the full quantum theory might not be known. For $\kappa > 0$, a non-trivial complication is the presence of the critical value of ϕ where we expect a singularity. Typically $\phi = \phi_c$ on some line. Although the presence of this boundary type line complicates the elaboration of a complete quantum theory, it should not affect the short-distance singularities of the propagators away from it. Hence we should be able to check whether or not a theory is conformally invariant away from this line. This is the approach taken here (see refs. [2] and [3]): we will display a class of theories that are conformally invariant, at least when we need not consider the line of singularity, or if it is absent as for $\kappa < 0$.

Transformation to free fields

The kinetic part of $S_{\mathrm{cl}} + S_{\mathrm{anom}}$ can be written as

$$S_{\mathrm{kin}} = \frac{1}{\pi} \int \mathrm{d}^2\sigma e^{-2\phi}\left(-4\partial_+\phi\partial_-\phi + 2\partial_+\rho\partial_-\phi + 2\partial_+\phi\partial_-\rho - \kappa\partial_+\rho\partial_-\rho \right) + S_{\mathrm{matter}} .$$

(2.3)

Here, we assume $\kappa > 0$, while things work similar for $\kappa < 0$. Let now [2][‡]

$$\omega = e^{-\phi}/\sqrt{\kappa} \ , \quad \chi = \rho + \omega^2 \ . \tag{2.4}$$

Then

$$S_{\text{kin}} = \frac{1}{\pi} \int d^2\sigma \left[-\kappa \partial_+ \chi \partial_- \chi + 4\kappa(\omega^2 - 1)\partial_+ \omega \partial_- \omega \right] + S_{\text{matter}} \tag{2.5}$$

is diagonalized. We can bring the ω-kinetic term into a standard form by a further (local) field redefinition:

$$\Omega = \omega\sqrt{\omega^2 - 1} - \log\left(\omega + \sqrt{\omega^2 - 1}\right) + \tfrac{1}{2}\left(1 - \log \tfrac{\kappa}{4}\right) \ \Rightarrow \ \partial\Omega = 2\sqrt{\omega^2 - 1}\partial\omega \ , \tag{2.6}$$

so that finally

$$S_{\text{kin}} = \frac{1}{\pi} \int d^2\sigma \left[-\kappa \partial_+ \chi \partial_- \chi + \kappa \partial_+ \Omega \partial_- \Omega + \frac{1}{2}\sum_{i=1}^{N} \partial_+ f_i \partial_- f_i \right] \ . \tag{2.7}$$

Note that χ and Ω have opposite signature, but it now seems that the kinetic term can no longer become singular. What has happened to the singularity (for $\kappa > 0$) at $\phi = \phi_c$? Of course, it has been hidden in the transformation from ϕ to Ω. This transformation is not one to one (for $\kappa > 0$) and we have a singularity when $d\Omega/d\phi = 0$ which precisely happens at $\phi = \phi_c$.

Not only the kinetic part of the action is very simple when written in terms of the new fields χ and Ω, but also the stress tensor:

$$T_{\pm\pm} = -\kappa(\partial_\pm \chi)^2 + \kappa\partial_\pm^2 \chi + \kappa(\partial_\pm \Omega)^2 + \frac{1}{2}\sum_{i=1}^{N}(\partial_\pm f_i)^2 \ . \tag{2.8}$$

These forms of the kinetic part of the action and of the stress tensor are valid for both signs of κ, only the precise form of the field transformations are different.

For $\kappa < 0$ quantization is straightforward. For $\kappa > 0$, we disregard the subtleties connected with the presence of a singular line $\phi = \phi_c$ for the moment and proceed with a "naive" quantization. The kinetic part of the action then shows that χ and Ω have standard massless free field propagators, and it is straightforward to compute the short distance expansion of the stress tensor with itself. We find that it generates a (continuum) Virasoro algebra with central charge

$$c = 1 + 1 - 12\kappa + N - 26 = N - 24 - 12\kappa \tag{2.9}$$

which vanishes precisely if $\kappa = \frac{N-24}{12}$. The field Ω contributes 1 to the central charge. χ contributes $1 - 12\kappa$ since the χ part of $T_{\pm\pm}$ has the Feigin-Fuchs form with background charge $\sim \sqrt{\kappa}$. Furthermore, the matter fields just contribute the usual N, while, although not written explicitly, we also have ghosts from the conformal gauge fixing, and they contribute -26, as always.

---

‡ Here we rescale χ and Ω by a factor 2 with respect to ref. [2], and Ω also is shifted by a constant.

Thus, with $\kappa = (N - 24)/12$ we expect to have a conformal field theory with vanishing central charge. One might worry however, that when we transformed our fields from ϕ and ρ to χ and Ω, a complicated Jacobian would appear in the functional integral, turning χ and Ω into interacting fields. This is not so, as discussed in ref. [4]. Indeed, the initial "measure" for ϕ and ρ is precisely such that together with this Jacobian one obtains a standard free field measure for χ and Ω.

3. $(1,1)$-operators and exact solutions

So far we only discussed the kinetic part of the action. There is also the "interaction" part of the action which contains the cosmological constant $\sim \lambda^2 e^{-2\phi} e^{2\rho}$. This term behaves like a perturbation of our conformal theory. If it is a marginal operator, however, it will preserve the conformal invariance. A necessary condition for a marginal operator is that it has conformal dimension $(1,1)$. The cosmological constant operator has indeed dimension $(1,1)$ classically, i.e. if we do a Poisson bracket computation with $T_{\pm\pm}$, but this is no longer true at the quantum level. It is easy to see that the only operators (with no derivatives) of definite conformal dimension (Δ, Δ) are $\lambda^{2\Delta} : e^{\alpha\chi + \beta\Omega} :$ with $\Delta = \frac{\alpha}{2} + \frac{\alpha^2 - \beta^2}{\kappa}$ Any of these operators with $\Delta = 1$ (if truely marginal) will lead to a conformal theory. However, in the weak coupling limit, $e^{2\phi} \to 0$, we want to recover the classical dilaton-gravity action $S_{\rm cl}$. Hence, we must take $\alpha = 2$ and $\beta = -2$. It is easy to see that this operator is indeed marginal. Then the full action reads:

$$S = \frac{1}{\pi} \int \mathrm{d}^2\sigma \left[-\kappa\partial_+\chi\partial_-\chi + \kappa\partial_+\Omega\partial_-\Omega + \lambda^2 e^{2(\chi - \Omega)} + \frac{1}{2}\sum_{i=1}^{N} \partial_+ f_i \partial_- f_i \right] . \qquad (3.1)$$

Note that this action differs from the classical dilaton gravity action only by higher order corrections, i.e. terms that are $\mathcal{O}(G^2) = \mathcal{O}(e^{2\phi})$ with respect to $S_{\rm cl}$.

The equations of motion that follow from the action (3.1) are very simple. First of all, the N matter fields are just free fields, $\partial_+\partial_- f_i = 0$ with solution $f_i = f_i^+(\sigma^+) + f_i^-(\sigma^-)$. For χ and Ω we have:

$$\partial_+\partial_-(\chi - \Omega) = 0 \quad , \qquad \partial_+\partial_-(\chi + \Omega) = -\frac{2}{\kappa}\lambda^2 e^{2(\chi - \Omega)} . \qquad (3.2)$$

The general solution reads

$$2(\chi - \Omega) = f^+(\sigma^+) + f^-(\sigma^-) \equiv \log\partial_+\alpha^+(\sigma^+) + \log\partial_-\alpha^-(\sigma^-) ,$$
$$\frac{\kappa}{2}(\chi + \Omega) = -\lambda^2\alpha^+(\sigma^+)\alpha^-(\sigma^-) + \beta^+(\sigma^+) + \beta^-(\sigma^-) . \qquad (3.3)$$

This solution is just as simple as the one for the classical dilaton-gravity (i.e. the solution of the equations of motion derived from $S_{\rm cl}$ alone) which reads $2(\rho - \phi) = \log\partial_+\alpha^+(\sigma^+) + \log\partial_-\alpha^-(\sigma^-)$ and $e^{-2\phi} = -\lambda^2\alpha^+(\sigma^+)\alpha^-(\sigma^-) + \beta^+(\sigma^+) + \beta^-(\sigma^-)$. Let us also give the stress tensor when evaluated on the solutions (3.3):

$$T_{\pm\pm} = -\partial_\pm f^\pm \partial_\pm \beta^\pm + \partial_\pm^2 \beta^\pm + \frac{\kappa}{4}\partial_\pm^2 f^\pm , \qquad (3.4)$$

where $f^\pm \equiv \log\partial_\pm\alpha^\pm$.

4. The RST variant

There are a few variations of the preceding formalism. Here, we shall only discuss the RST variant. So far we started with a given kinetic part of the action and improved the cosmological constant operator by higher order corrections until it had dimension $(1,1)$. Russo, Susskind and Thorlacius (RST) [5], motivated by the search of simple and exactly solvable equations of motion , followed a slightly different route. Their procedure amounts to keeping the cosmological constant operator fixed and modifying the kinetic part of the action and hence also the stress tensor until the old cosmological constant operator has dimension $(1,1)$ with respect to the new stress tensor. It turned out that this was very easy to achieve. All one needs is to add

$$\delta S_{\mathrm{RST}} = -\frac{\kappa}{4\pi} \int d^2\sigma \sqrt{-g}\phi R = \frac{1}{\pi} \int d^2\sigma\, \kappa\, \partial_+\phi\partial_-\rho \, , \qquad (4.1)$$

where the second expression is valid in conformal gauge only. In this RST-model the field transformations to the χ, Ω-fields are simplest:

$$\Omega = \frac{e^{-2\phi}}{\kappa} + \frac{\phi}{2} \, , \qquad \chi = \frac{e^{-2\phi}}{\kappa} - \frac{\phi}{2} + \rho \, . \qquad (4.2)$$

When written in terms of these fields, the action and stress tensor take on exactly the same form (3.1) and (2.8) as before. Note also, that for $\kappa > 0$ the modified kinetic action is degenerate at $e^{-2\phi} = e^{-2\phi_c} \equiv \frac{\kappa}{4}$. This is precisely the value of ϕ where $d\Omega/d\phi = 0$. Although the precise value of ϕ_c is shifted, the qualitative feature of a singular line for $\kappa > 0$ is present as well.

REFERENCES

1. C. Callan, S. Giddings, J. Harvey and A. Strominger, Phys. Rev. **D45** (1992) R1005.

2. A. Bilal and C. Callan, Nucl. Phys. **B394** (1993) 73.

3. S. de Alwis, Phys. Lett. **B289** (1992) 278, Phys. Lett. **B300** (1993) 330.

4. A. Bilal, "Aspects of exactly solvable quantum-corrected 2D dilaton gravity theories", Princeton University preprint PUPT-1411 (July 1993), hep-th@xxx/9307052.

5. J. Russo, L. Susskind and L. Thorlacius, Phys. Rev. **D46** (1992) 3444.

THREE-DIMENSIONAL BLACK HOLES IN STRING THEORY

GARY T. HOROWITZ

Physics Department, University of Califoronia
Santa Barbara, CA 93106-9350 USA

In order to understand strong gravitational effects in string theory, we need to find exact solutions describing gravitational phenomena such as black holes. A few years ago, Witten[1] found a two dimensional black hole using a gauged Wess-Zumino-Witten (WZW) model based on the group $SL(2, R)$. What I would like to do here is describe another exact black hole solution which is extremely simple. In fact, it may be the simplest example of a black hole in string theory. This black hole is three dimensional and is also based on the $SL(2, R)$ WZW model. But in this case, there is no gauging required. This work was done in collaboration with my student Dean Welch[2], and related work was done independently by Kaloper[3]. We were all strongly influenced by Banados, Teitelboim, and Zanelli's observation[4] that black holes exist in three dimensional general relativity with a negative cosmological constant.

We start with the question: When does the group $SL(2, R)$ describe a black hole? One can parameterize $SL(2, R)$ by

$$g = \begin{pmatrix} x_1 + x_2 & x_3 + x_0 \\ x_3 - x_0 & x_1 - x_2 \end{pmatrix} \tag{1}$$

The statement that this matrix has unit determinant yields the equation

$$-x_0^2 - x_1^2 + x_2^2 + x_3^2 = -1 \tag{2}$$

The group invariant metric can be obtained by viewing this as a surface in the four dimensional flat space of signature $(- - + +)$

$$ds^2 = -dx_0^2 - dx_1^2 + dx_2^2 + dx_3^2 \tag{3}$$

The resulting space has Lorentz signature and constant negative curvature: It is just three dimensional anti-de Sitter space. This space is clearly invariant under $SO(2,2)$. (Since $SO(2,2)$ is locally the product of $SL(2, R)$ with itself, this is also the symmetry of the $SL(2, R)$ WZW model.) The six independent Killing vectors consist of two rotations (in the $(0, 1)$ and $(2, 3)$ planes) and four boosts. A convenient way to parameterize the surface is to choose two commuting Killing vectors and let two of the coordinates be the parameters along these symmetry directions. If t and φ are parameters along the two rotations, the metric takes the familiar form

$$ds^2 = -\left(1 + r^2\right) dt^2 + \left(1 + r^2\right)^{-1} dr^2 + r^2 d\varphi^2 \tag{4}$$

To obtain the black hole, we let $\hat{t}$ and $\hat{\varphi}$ be parameters along the boosts in the (0,3) and (1,2) planes. Explicitly, set

$$
\begin{aligned}
x_1 &= \hat{r}\cosh\hat{\varphi} & x_0 &= \sqrt{1-\hat{r}^2}\,\cosh\hat{t} \\
x_2 &= \hat{r}\sinh\hat{\varphi} & x_3 &= \sqrt{1-\hat{r}^2}\,\sinh\hat{t}
\end{aligned}
\tag{5}
$$

Notice that $\hat{r}^2 > 0$ only covers the region $x_1^2 - x_2^2 > 0$. Since this is not the entire space, it might be more natural to use the radial coordinate $\rho = \hat{r}^2$ which takes both positive and negative values. However, we will continue to use $\hat{r}$ to maintain agreement with ref [4]. In terms of these coordinates, anti-de Sitter space becomes

$$
ds^2 = \left(1 - \hat{r}^2\right)d\hat{t}^2 + \left(\hat{r}^2 - 1\right)^{-1}d\hat{r}^2 + \hat{r}^2 d\hat{\varphi}^2
\tag{6}
$$

Since $\hat{t}$ and $\hat{\varphi}$ are both parameters along a boost, they can take any real value. The key point is that we can now identify $\hat{\varphi} = \hat{\varphi} + 2\pi$. After we do so, (6) describes a black hole. The surfaces of constant $\hat{t}$ and $\hat{r} < 1$ are now compact trapped surfaces, i.e., both outgoing and ingoing light rays converge. The event horizon is at $\hat{r} = 1$. The region $\hat{r} = 0$ is not a curvature singularity, since the curvature is constant. However it does correspond to geodesic incompleteness because the boost symmetry has a fixed point there. In terms of $SL(2,R)$, translations in $\hat{\varphi}$ correspond to the axial symmetry

$$
\delta g = \epsilon \left[\begin{pmatrix} 1 & 0 \\ 0 & -1 \end{pmatrix} g + g \begin{pmatrix} 1 & 0 \\ 0 & -1 \end{pmatrix} \right]
\tag{7}
$$

while translations of $\hat{t}$ correspond to the vector symmetry

$$
\delta g = \epsilon \left[\begin{pmatrix} 1 & 0 \\ 0 & -1 \end{pmatrix} g - g \begin{pmatrix} 1 & 0 \\ 0 & -1 \end{pmatrix} \right]
\tag{8}
$$

By gauging one of these symmetries, one obtains Witten's two dimensional black hole. By making a discrete identification of the axial symmetry (7), one obtains the three dimensional black hole.

There are several generalizations of the black hole (6). First, one can identify $\hat{\varphi}$ with a period other than 2π. This turns out to change the mass M of the black hole. Second, one can identify a general linear combination of $\hat{t}$ and $\hat{\varphi}$. This turns out to add angular momentum J to the black hole. Finally, one can simply rescale the metric. This corresponds to changing the level k of the WZW model. Putting these together, one obtains a three parameter family of solutions:

$$
ds^2 = \left(M - \frac{r^2}{k}\right)dt^2 - Jdtd\varphi + r^2 d\varphi^2 + \left(\frac{r^2}{k} - M + \frac{J^2}{4r^2}\right)^{-1}dr^2
\tag{9}
$$

where φ is identified with $\varphi + 2\pi$. Several things should be noted about this metric. Since it is obtained from identifications of anti-de Sitter space, its curvature is still constant. In fact, $R_{\mu\nu} = -\frac{2}{k}g_{\mu\nu}$. When $0 < J^2 < M^2 k$, there are two horizons at the zeros $r_\pm$ of g_{rr}^{-1}. In this case, g_{tt} becomes positive outside the event horizon. This is analogous to what happens in four dimensions and means that the spacetime has an ergosphere: Observers near the horizon cannot stay at rest with respect to infinity but must rotate with the black hole. All solutions approach anti-de Sitter space asymptotically. ¿From (4) and (9), it is clear that anti-de Sitter space is recovered everywhere not when $M = 0$, but when $M = -1$ (and $J = 0$). The zero mass solution corresponds to the limit that the event horizon shrinks to zero size.

We now briefly consider this solution from the standpoint of the low energy string action. In three dimensions, this action is

$$S = \int d^3x \sqrt{-g}\, e^{-2\phi} \left[\frac{4}{k} + R + 4(\nabla\phi)^2 - \frac{1}{12} H_{\mu\nu\rho} H^{\mu\nu\rho} \right] \tag{10}$$

It is easy to verify that the black hole metric (9) is an extremum of this action with $\phi = 0$ and $H_{\mu\nu\rho} = \frac{2}{\sqrt{k}}\epsilon_{\mu\nu\rho}$. This is what one would expect from the WZW model. The antisymmetric tensor plays the role of an effective cosmological constant in three dimensions.

Since φ is a spacelike symmetry with compact orbits, one can construct the dual of this solution using the low energy duality transformation[5]

$$\tilde{g}_{\varphi\varphi} = 1/g_{\varphi\varphi}, \qquad \tilde{g}_{\varphi\alpha} = B_{\varphi\alpha}/g_{\varphi\varphi}$$
$$\tilde{g}_{\alpha\beta} = g_{\alpha\beta} - (g_{\varphi\alpha}g_{\varphi\beta} - B_{\varphi\alpha}B_{\varphi\beta})/g_{\varphi\varphi}$$
$$\tilde{B}_{\varphi\alpha} = g_{\varphi\alpha}/g_{\varphi\varphi}, \qquad \tilde{B}_{\alpha\beta} = B_{\alpha\beta} - 2g_{\varphi[\alpha}B_{\beta]\varphi}/g_{\varphi\varphi}$$
$$\tilde{\phi} = \phi - \frac{1}{2}\ln g_{\varphi\varphi} \tag{11}$$

where α, β denote the t, r directions. (For a more general discussion of duality of WZW models see ref [6].) The result, after a simple coordinate transformation, is the three dimensional black string[7]

$$\tilde{ds}^2 = -\left(1 - \frac{\mathcal{M}}{\hat{r}}\right) d\tilde{t}^2 + \left(1 - \frac{Q^2}{\mathcal{M}\hat{r}}\right) d\hat{x}^2 + \left(1 - \frac{\mathcal{M}}{\hat{r}}\right)^{-1}\left(1 - \frac{Q^2}{\mathcal{M}\hat{r}}\right)^{-1} \frac{k\, d\hat{r}^2}{4\hat{r}^2}$$
$$\tilde{\phi} = -\frac{1}{2}\ln \hat{r}\sqrt{k}, \qquad \tilde{H}_{\hat{x}\hat{r}\hat{t}} = \frac{Q}{\hat{r}^2} \tag{12}$$

where $\mathcal{M} = r_+^2/\sqrt{k}$ and $Q = -J/2$. (This was also noticed by Ali and Kumar[8].)

This dual solution describes an equivalent conformal field theory[9]. Perhaps the most interesting aspect of this equivalence is that the asymptotic behavior has changed. While the black hole is asymptotically anti-de Sitter, its dual (12) is asymptotically flat! In addition, $\tilde{H}_{\mu\nu\rho}$ vanishes asymptotically in the dual description. The "cost" of this improved asymptotic behavior is that the dilaton $\tilde{\phi}$ is no longer constant, but decreases linearly with proper distance. The asymptotic regions are mapped into each other under duality, but their behavior is different since the orbits of the φ translation symmetry are becoming very long, $g_{\varphi\varphi} = r^2$. Since the dual descriptions are equivalent, there is a sense in which strings do not see a three dimensional negative cosmological constant. More precisely, the effect of the cosmological constant is equivalent to a linear dilaton. Could some generalization of this effect may play a role in resolving the cosmological constant problem in four dimensions?

Acknowledgments

This work was supported in part by NSF Grants PHY-8904035 and PHY-9008502.

References

1. E. Witten, *Phys. Rev.* **D44**, (1991), 314. 2. G. Horowitz and D. Welch, *Phys. Rev. Lett.* **71**, (1993), 328.

3. N. Kaloper, Alberta-THY-8-93, hep-th/9303007.

4. M. Banados, C. Teitelboim, and J. Zanelli, *Phys. Rev. Lett.* **69**, (1992), 1849; M. Banados, M. Henneaux, C. Teitelboim, and J. Zanelli, to appear in *Phys. Rev. D*

5. T. Buscher, *Phys. Lett.* **B201**, (1988), 466; *Phys. Lett.* **B194**, (1987), 59.

6. E. Kiritsis, CERN-TH.6797/93, hep-th/9302033.

7. J. Horne and G. Horowitz, *Nucl. Phys.* **B368**, (1992), 444.

8. A. Ali and A. Kumar, IP-BBSR-92-80, hep-th/9303032.

9. M. Roček and E. Verlinde, *Nucl. Phys.* **B373**, (1992), 630.

MONOPOLE CONFORMAL FIELD THEORY
AND STRING BLACK HOLES

JOSEPH POLCHINSKI

Institute for Theoretical Physics
University of California
Santa Barbara, CA 93106

1. Introduction

During the spring we had a program at the ITP on nonperturbative string theory, which turned out to be largely concerned with black holes. I had resolved that I would not get caught up in this, that I would think about supersymmetry breaking in string theory. I would like to take a few minutes to explain how this developed into the work described in the title.

In order to produce a string theory with a large ratio between the string scale and the supersymmetry breaking scale, it might be useful to start with a string theory which has a small *tree-level* breaking of supersymmetry. This could combine with the usual nonperturbative breakings, such as gaugino condensation, to produce a stable minimum with good properties. Unfortunately, a small tree-level breaking seems very hard to achieve. In the first attempt, the breaking came from an antisymmetric tensor field background,[1] but this turned out to be quantized in string units.[2] There are various additional no-go results: the small breaking cannot come from a small continuous parameter,[3] and various schemes with small but discrete breaking do not work.[4]

A simple way to produce the desired result would be to take the compactified dimensions to be a conformal field theory (CFT) with a central charge slightly off from the usual value of $(c_R, c_L) = (9, 22)$ (for the heterotic string).[5] To see why this breaks supersymmetry, note that we can produce a good string background by giving the four-dimensional fields a compensating central charge shift, for example with a linear dilaton. So the central charge shift in the internal theory corresponds to a dilaton potential in four dimensions.* Since there exist discrete series of CFT's with central charges

---

*Note that the linear dilaton here is of no interest in itself—the goal in the end is

arbitrarily close to integers, we can make the tree level dilaton potential as small as we like.

A simple series of such CFT's is obtained by taking two of the compact dimensions to be a two-sphere rather than Ricci-flat, with a magnetic monopole field

$$F_{\mu\nu} = Q\epsilon_{\mu\nu} \tag{1}$$

in the two dimensions, where the Dirac quantization makes $2Q$ an integer. The relevant terms in the low energy heterotic string Lagrangian are

$$R - \frac{\alpha'}{4} F_{\mu\nu} F^{\mu\nu} = \frac{2}{r^2} - \frac{\alpha' Q^2}{2r^4}. \tag{2}$$

This is stationary when

$$r^2 = Q^2 \alpha'/2, \tag{3}$$

the desire of the manifold to contract being balanced by the desire of the magnetic field to expand. This is a tree level string solution, at least for Q^2 large so that the α' expansion is good. The central charge is shifted from its Ricci-flat value by $\delta c = -3/Q^2$ and so can be arbitrarily small.

There is an obvious problem here, that as the dilaton potential is made small by taking Q large, the size of the internal manifold also grows large. The supersymmetric mass splittings, of order the square root of the dilaton potential, scale the same way as the Kaluza-Klein masses, as Q^{-1}. This is the same problem that has afflicted all previous attempts.[1,2,4,5] Now, one could try to reduce the size of the internal manifold by dividing by a large discrete group, but it seems that the best that one can do is to use a large $Z(n)$ subgroup of $SU(2)$ rotations. This can reduce the large two-sphere to a large one-dimensional manifold, but that is still unsatisfactory. So this does not lead to a large hierarchy between the compactification and tree-level supersymmetry breaking scales.[†] There does not seem to be any general

---

a static solution, after all perturbative and nonperturbative contibutions to the dilaton potential are included.

[†]There have been recent attempts to deal with this problem by assuming that in fact the compactification scale is not far above the weak scale.[6] We do not understand how this can evade the basic dimensional problem[7] that the gauge couplings in this higher-dimensional theory grow as a power of the energy, so the theory is strongly coupled just above the compactification scale. Certainly any attempt to use perturbative string theory in such a background is self-inconsistent, but more generally we do not see how this can be consistent with the idea of gauge-gravitational unification, since the gravitational coupling is still tiny when the gauge couplings diverge.

14

principle in CFT which would forbid such a hierarchy, but I have asked many conformal field theorists (both rational ones and irrational) for examples, without success.

So it quickly became obvious that this was not going to fly. But then Andy Strominger mentioned that if one had the exact CFT corresponding to the monopole compactification, one could build a four-dimensional string black hole. And so...

2. The Monopole CFT

In this section I will find the exact CFT corresponding to the monopole background. In the next I apply this to some exact solutions, and finally will speculate about solutions which are not known exactly. We will see that the exact solution has some surprises at small values of the monopole charge. This is based on ref. 8, with Giddings and Strominger.

The world-sheet action for the monopole CFT is

$$
S = \frac{1}{4\pi} \int d^2z\, r_2^2 G^{S_2}_{\mu\nu} \partial_z X^\mu \partial_{\bar z} X^\nu + \frac{1}{2\pi} \int d^2z \left\{ \bar\lambda_R(\partial_{\bar z} - i\omega_\mu \partial_{\bar z} X^\mu)\lambda_R \right.
$$
$$
\left. \bar\lambda_L(\partial_z - i2Q A^M_\mu \partial_z X^\mu)\lambda_L + i2Q F^M_{\mu\nu} \psi^\mu_R \psi^\nu_R \bar\lambda_L \lambda_L \right\} . \tag{4}
$$

The conventions are $\alpha' = 2$ and $d^2z = 2d\sigma d\tau$. Here $G^{S^2}_{\mu\nu}$ is the unit round metric on the two sphere, λ_L is the current algebra fermion, λ_R is the super-symmetric fermion in tangent space,

$$
\lambda_R = (e^1_\mu + ie^2_\mu)\psi^\mu_R , \tag{5}
$$

ω_μ is the spin connection for vectors on the two sphere, and A^M_μ is the gauge field of a magnetic monopole of unit charge. In the 'northern' coordinate patch the monopole potential is

$$
A^{M(N)}_\phi = \frac{1 - \cos\theta}{2}, \tag{6}
$$

and in the 'southern' patch it is

$$
A^{M(S)}_\phi = A^{M(N)}_\phi + ie^{-i\phi}\partial_\phi e^{i\phi} = -\frac{1 + \cos\theta}{2}. \tag{7}
$$

The current algebra fermion λ_L has charge $e = 1$. The action is well defined for Q half-integer, but we will find below that Q must in fact be an integer.

This action has a great deal of symmetry. In addition to $SU(2)$ rotation invariance, there is one non-anomalous fermion rotation

$$\delta\lambda_L = i\epsilon\lambda_L, \qquad \delta\lambda_R = iQ\epsilon\lambda_R. \tag{8}$$

Further, the $(1,0)$ heterotic supersymmetry is promoted to $(2,0)$ because the $U(1)$ acts nontrivially on the supercurrent. With this much symmetry, there are a number of ways to determine the exact CFT. We will use a bit of guesswork: the $SU(2)$ symmetry with central charge equal to 3 minus the curvature effect suggests a connection to the $SU(2)$ WZNW model. Of course the WZNW model would have an extra $SU(2)$ and no supersymmetry, so it's not quite the same.

In order to make the connection to the WZNW model clearer we will bosonize, introducing a third embedding field X^3 which is periodic with period 2π. The bosonized action is of the form

$$S = \frac{1}{4\pi} \int d^2z \left(G^{(3)}_{MN} + B^{(3)}_{MN} \right) \partial_z X^M \partial_{\bar{z}} X^N. \tag{9}$$

Rather than give explicit forms for the three-dimensional metric and torsion, we can describe them. The two-dimensional metric $G_{\mu\nu}$ is that of a sphere of some radius r_2. The one-dimensional metric G_{33} is that of a circle of some radius r_1. The off-diagonal metric $G_{\mu 3}$ is a monopole field of charge $2Q + 2$, while the antisymmetric tensor $B_{\mu 3}$ is a monopole field of charge $2Q - 2$. These are proportional respectively to the sum and difference of the gauge and spin connections, the latter being a monopole field of charge 1.

This space is locally $S_2 \times S_1$, with winding number $2Q + 2$ and torsion $\frac{1}{8\pi^2} \int H = Q - 1$. The $SU(2)$ group manifold is locally $S_2 \times S_1$ with winding number 1 (the Hopf fibration). So if the latter has torsion $k = (2Q+2)(Q-1)$ and we mod out a $Z(2Q + 2)$ acting on the S_1, the result is topologically equivalent to the monopole CFT. Not surprisingly the actions also agree, with appropriate radii

$$r_2 = (Q+1)r_1 = r_{SU(2)} = \alpha'|Q^2 - 1|. \tag{10}$$

The monopole CFT is $SU(2)/Z(2Q + 2)_R$, modding on a discrete subgroup of the right-moving $SU(2)$. At $Q = 0$ the level is negative, equivalent to a world-sheet parity transformation, and one mods by $Z(2)_L$ instead. The exact central charge is

$$c = 3k/(k + 2). \tag{11}$$

One can work out various details of the equivalence. There is indeed a $(2,0)$ world-sheet supersymmetry, the familiar twist on the $SU(2)$ current algebra. Wavefunctions for charged fields in this background must be appropriate monopole harmonics. Carrying out the bosonization, one finds that the vertex operators are current algebra fields times exponentials of the right-moving boson,

$$e^{i\Delta J_3 X_R^3/(Q+1)},\qquad (12)$$

consistent with the identification of the manifold as a coset. One can also come to the same spectrum from a different route. Twisting an $SU(2)$ current algebra and imposing level matching, one finds that a modular invariant spectrum is obtained keeping all fields with

$$J_3 = F_R + Q F_L, \qquad \Delta J_3 = (Q-1)(F_R - F_L) \qquad (13)$$

with $2F_{R,L}$ being integers. Here J_3 is the spin of the current algebra field and ΔJ_3 the shift in the exponential. This spectrum agrees with that from bosonization. Another way to think about this is as a modular-invariant heterosis of an $SU(2)$ current algebra with an $N = 2$ minimal model.

Comparing the exact result with the semiclassical, the surprise is the existence of a nontrivial solution for $Q = 0$ and the absence of one for $|Q| = 1$, the central charge vanishing in the latter case. Semiclassically the curvature is balanced by the magnetic field. At $Q = 0$ it is balanced by the spin connection instead. At $|Q| = 1$ the spin and gauge connections offset and the solution collapses. These are outside the validity of the α' expansion, so one needs the exact solution. We will have more to say about these special cases later.

3. Exact Solutions

What does this have to do with black holes? We can make a four-dimensional spacetime by combining the monopole CFT with a linear dilaton tx CFT, with energy-momentum tensor

$$\begin{aligned}
T &= \frac{1}{2}\left(\partial_z t \partial_z t + \psi_t \partial_z \psi_t\right) - \frac{1}{2}\left(\partial_z x \partial_z x + \psi_x \partial_z \psi_x\right) - \alpha \partial_z^2 x \\
\tilde{T} &= \frac{1}{2}\bar{\partial}_z t \bar{\partial}_z t - \frac{1}{2}\bar{\partial}_z x \bar{\partial}_z x - \alpha \bar{\partial}_z^2 x.
\end{aligned} \qquad (14)$$

The total four-dimensional central charge

$$\left(3 + 12\alpha^2 + \frac{3k}{k+2},\ 2 + 12\alpha^2 + \frac{3k}{k+2}\right) \tag{15}$$

must equal (6,5), so $\alpha = 1/\sqrt{2k+4} = 1/2Q$. Together with a central charge $(9,21)$ internal CFT, this is a good heterotic string background (one unit of left-moving central charge has already been used in the monopole CFT).

This background is the extremal limit of a four-dimensional magnetic monopole-black hole solution found in the α' expansion by Garfinkle, Horowitz, and Strominger.[9] In that limit the horizon is at the end of an infinite throat; the throat is given by the product CFT that we have found. That is, one takes the limit by keeping an eye on the middle of the throat as the horizon runs off to infinity in one direction and the asymptotic region to infinity in the other. A different way to take this limit is to keep one's eye on the horizon. Then the limiting background is a black hole at the end of a semi-infinite throat. Again this is a simple product of two exact CFT's, the monopole theory and the two dimensional black hole.[10] A third limit, where one keeps one's eye fixed on the mouth between the throat and the asymptotically flat region and ends up with an asymptotic region and a semi-infinite throat, does not correspond to a simple product CFT and is not known exactly. Also the non-extremal black holes are not known in terms of exact CFT's. We will return to these points in the next section.

With the exact solution one can check a number of physical properties. The large-Q solutions, where the α' expansion is valid, are not spacetime supersymmetric,[9] and with the exact solution one can verify that this is true for all Q. One would need a weight $(0, \frac{1}{8})$ Ramond field in the spectrum of the monopole CFT, and there is none. Without spacetime supersymmetry there is the possibility of an instability. From the exact solution one finds that the solution is indeed unstable if the magnetic field is in an unbroken non-Abelian group, from the lowest partial wave of the charged gauge bosons. It is stable if the non-Abelian symmetry is broken at a sufficiently high scale. This is the same as has been found in other monopole backgrounds.[11]

An interesting check on the construction is an index calculation, suggested by Tom Banks. A chiral fermion of charge e in a monopole field of charge Q should have $2Qe$ zero modes, and this is what is found from the exact spectrum. This allows us to determine the value of Q directly from the conformal field theory, assuring that the sigma-model value has not been in some way renormalized.

18

The $Q = 0$ solution corresponds to a 'neutral remnant,' a neutral black hole with an infinite throat. As far as I know, such an object has not been found before, and may play a role in resolving the black hole information paradox.

4. Speculations

The exact solutions found here describe only the horizon plus throat region of the extremal black hole. As we have noted, the non-extremal solutions, and the mouth region of the extremal solution, are much more complicated and not exactly known. Within the exact solution one can find the marginal perturbation corresponding to the widening of the mouth toward the asymptotic region, but the methods do not exist to integrate this and find the full solution.

For this reason, we cannot be sure that the $Q = 0$ solution, or for that matter $Q = 2$, $Q = 3$, or any specific small value of Q, can be connected onto the asymptotic spacetime. The arguments are only indirect: the existence of a solution in the α' expansion for large Q, counting of parameters which indicates that it is consistent for such solutions to exist, and the argument that, although we cannot integrate the mouth-widening perturbation, it is hard to see what it could integrate to except an asymptotically flat exterior.

There is one other argument, which also has interesting implications for the strange case $|Q| = 1$. Consider the mouth region of the extremal solutions in the regime where they are known, large Q. As a function of the radial coordinate x, the radius r of the transverse two-sphere starts large in the asymptotic region and goes asymptotically to the WZNW fixed-point value down the throat. This resembles qualitatively the WZNW renormalization-group flow. The resemblance is not entirely accidental. These solutions have a radial dilaton gradient, so that under a world-sheet Weyl transformation $\delta g_{ab} = 2\delta\omega g_{ab}$ the world-sheet field x transforms as

$$\delta x = 2\delta\omega G^{xx}\partial_x\Phi. \tag{16}$$

When the dilaton gradient is large enough, this classical effect overwhelms any quantum effect and the radial coordinate indeed acts as a world-sheet scale. There is a certain limit where this approximation becomes exact—if we take the central charge to infinity by giving the dilaton a large radial gradient. For the real string theory this is an uncontrolled approximation, but it does indicate the existence of asymptotically flat solutions for all Q.

It is interesting to assume that this '$x \sim$scale' idea is valid and to apply it to $Q = 1$. At $Q = 1$ the torsion vanishes by cancellation between the right and left connections. If we start at large r (weak coupling) and flow, the sigma model without torsion develops a mass gap and is trivial in the infrared. The picture then is that in the mouth region the theory becomes massive and down the throat we have only the two-dimensional xt theory—effectively the angular directions have collapsed to zero radius. Note that this is not the same as saying that there is no throat at all, which is how one might naively have interpreted the triviality of the angular CFT. In particular, there are still massless fermion zero modes travelling down the infinite throat, as required by the index argument given at the end of the previous section. There is a nice interplay here between index arguments, chiral symmetry breaking, and instanton effects, as explained in detail in ref. 8. The angular CFT actually has four states, all mapping to the unit operator in the low energy theory but differing in their massive degrees of freedom.

There are some possible general lessons to draw here. First, spacetime duality is perhaps less general than might have been expected. Here we have a case where a dimension contracts to zero radius in a real physical sense—all angular excitations become arbitrarily massive down the throat. Second, this solution suggests a general mechanism by which timelike singularities may be eliminated in string theory, by replacement with an infinite-length zero-radius throat.

Finally, the renormalization group picture gives a general method for generating approximate string backgrounds: choose your favorite renormalization group flow, and combine with a radial coordinate with dilaton gradient. One interesting flow is in the $O(2)$ sigma model with θ-parameter. This has a fixed point at $\theta = \pi$.[12] The flow to this fixed point is a string black hole with curvature balanced by axion hair.[13] However, it has two instabilities, a classical one due to a tachyonic mode of the axion, and a quantum mechanical instability to axion fluctuations.

5. Acknowledgements

I would like to thank Tom Banks, Mike Douglas, Steve Giddings, Jeff Harvey, Steve Shenker, and Andy Strominger for discussions and collaboration. This work was supported in part by NSF grants PHY-90-09850 and PHY-91-57463.

6. References

1. M. Dine, R. Rohm, N. Seiberg, and E. Witten, *Phys. Lett.* **B156** (1985) 55.

2. R. Rohm and E. Witten, *Ann. Phys. (NY)* **170** (1986) 454.

3. M. Dine and N. Seiberg, *Nucl. Phys.* **B301** (1988) 357;
T. Banks and L. Dixon, *Nucl. Phys.* **B307** (1988) 93.

4. I. Antoniadis, C. Bachas, D. Lewellen and T. Tomaras, *Phys. Lett.* **B207** (1988) 441;
C. Kounnas and M. Porrati, *Nucl. Phys.* **B310** (1988) 355;
S. Ferrara, C. Kounnas, M. Porrati and F. Zwirner, *Nucl. Phys.* **B318** (1989) 75.

5. S. P. del Alwis, J. Polchinski, and R. Schimmrigk, *Phys. Lett.* **B218** (1989) 449.

6. I. Antoniadis, *Phys. Lett.* **246B** (1990) 377.

7. V. Kaplunovsky, *Phys. Rev. Lett.* **55** (1985) 299.

8. S. B. Giddings, J. Polchinski, and A. Strominger, *Four Dimensional Black Holes in String Theory*, hep-th/9305083, to appear in *Phys. Rev. D.*

9. D. Garfinkle, G. Horowitz, and A. Strominger, *Phys. Rev.* **D43** (1991) 3140, erratum *Phys. Rev.* **D45** (1992) 3888.

10. E. Witten, *Phys. Rev.* **D44** (1991) 314;
G. Mandal, A Sengupta, and S. Wadia, *Mod. Phys. Lett.* **A6** (1991) 1685.

11. R.A. Brandt and F. Neri, *Nucl. Phys.* **B161** (1979) 253;
K. Lee, V.P. Nair, and E.J. Weinberg, *Phys. Rev. Lett.* **68** (92) 1100;
Phys. Rev. **D45** (1992) 2751.

12. F. D. M. Haldane, *Phys. Lett.* **A93** (1983) 464.

13. S. B. Giddings, J. A. Harvey, J. Polchinski, S. H. Shenker, and A. Strominger, *Hairy Black Holes in String Theory*, hep-th/9309152, preprint NSF-ITP-93-58;
A somewhat different application of this flow appeared in I. Kogan, *Mod. Phys. Lett.* **A6** (1991) 3297.

Black Hole Evaporation and Quantum Gravity.

Kareljan Schoutens, Herman Verlinde

Joseph Henry Laboratories

Princeton University, Princeton, NJ 08544

and

Erik Verlinde

TH-Division, CERN

CH-1211 Geneva 23

and

Institute for Theoretical Physics

University of Utrecht

P.O. BOX 80.006, 3508 TA Utrecht

Abstract

In this note we consider some consequences of quantum gravity on the process of black hole evaporation. In particular, we will explain the suggestion by 't Hooft that quantum gravitational interactions effectively exclude simultaneous measurements of the Hawking radiation and of the matter falling into the black hole. The complementarity of these measurements is supported by the fact that the commutators between the corresponding observables can be shown to grow uncontrollably large. The only assumption that is needed to obtain this result is that the creation and annihilation modes of the in-falling and out-going matter act in the same Hilbert space. We further illustrate this phenomenon in the context of two-dimensional dilaton gravity.

1. Introduction.

The black hole evaporation phenomenon can be viewed as a consequence of the fact that the horizon of a black hole on the one hand forms a surface of infinite red-shift, while on the other hand it represents a perfectly regular part of space-time. Any outgoing wave that reaches an asymptotic observer with a finite frequency corresponds to an exponentially high frequency mode near the horizon, and the reasonable assumption that these ultra-high energy modes are in their ground state was used by Hawking to show that the asymptotic observer will see thermal radiation [1]. Soon after this discovery, Hawking made the remarkable suggestion that the resulting black hole evaporation process will inevitably lead to a fundamental loss of quantum coherence. The mechanism by which the quantum radiation is emitted indeed appears to be insensitive to the detailed history of the black hole, and thus it seems hard to imagine how one can prevent that information gets forever lost to an outside observer.

The only possibility for maintaining quantum coherence, it seems, would be if quantum gravity somehow leads to non-local effects that gradually bring out this information in the form of subtle correlations in the out-going radiation. Superficially, however, the evaporation process for large black holes involves only physics at low energies. After all, the process entirely takes place within space-time regions where both the curvature and the energy-momentum flux of the radiation remain small everywhere. It is therefore often argued that any strong quantum gravitational effects take place either too late or too far behind the horizon to provide a possible mechanism for a complete information transport to the outside world.

This last reasoning however is incomplete. Namely, as has been emphasized by several authors (see e.g. [2] and [3]), it does not take into account the important fact that the exponential red-shift effect associated with the black hole horizon leads to a breakdown of the usual separation of length scales. In a certain way, this red-shift effectively works as a magnifying glass that makes the consequences of the short distance, or rather, high energy physics near the horizon visible at larger scales to an asymptotic observer. Direct examination of Hawking's original derivation (or any later one) of the black hole emission spectrum indeed shows that one inevitably needs to make reference to particle waves that have arbitrarily high frequency near the horizon as measured in the reference frame of the in-falling matter. While this point has been noted by many authors, it is usually put aside with the argument that, since there are no physical ultra-high energy particles running along the horizon, one does not need to know anything about their physics except how to describe their local vacuum state. However, after one realizes that the frequencies involved here are so large that the corresponding energies exceed any macroscopic mass by an exponentially growing factor, it becomes in fact far from obvious that quantum gravitational effects remain small. At these frequencies it may indeed no longer be appropriate to think of particles as being superimposed on top of some classical background geometry.

In this respect it is important to note that, while often we think of quantum gravity as relevant only to the physics at sub-planckian distances, at these ultra-high frequencies quantum gravitational effects can in principle take macroscopic proportions. Only if as an asymptotic observer we would restrict our observations to extremely low frequencies, such that they remain reasonably small when propagated back to the horizon, we can to a very good approximation work with the classical geometry. If instead we measure out-going particles of moderate frequencies at infinity, then the history of these modes must involve geometries that can be very different from the classical one.

As has been emphasized by 't Hooft, this fact may lead to large deviations from our semi-classical intuition. In particular, he has suggested that, when one wants to simultaneously consider the observations of an asymptotic observer and those of an in-falling observer, such measurements will in general involve observables whose commutators will grow uncontrollably large. Hence these simultaneous measurements are essentially forbidden: they are *complementary* in the usual sense of quantum mechanics. If this suggestion is true, then it will obviously have very important consequences.

An apparent weakness in the argumentation of 't Hooft, however, is that this complementarity between the in-falling and asymptotic observers appears to follow almost directly from the starting assumption that the black hole evaporation process should be describable in terms of an S-matrix. This assumption by itself immediately turns the black hole horizon into an ultra-strong coupling regime, because any S-matrix element involving a generic *out*-state will be very singular in that region. This very fact, on the other hand, is often used as an argument to show why the S-matrix assumption must be wrong, as it appears to contradict the fact that horizon should be regular to an in-falling observer. It is therefore important to establish whether the same non-trivial quantum gravitational effects can be derived in a more general framework that is not based on this S-matrix assumption.

As a first small step in this direction, two of us [6] and independently Susskind and collaborators [7] presented several arguments suggesting that the idea of complementarity can not be disproven without making reference to Planckian physics. In this note, however, we would like to go further and indicate how one can actually *derive* the existence of these large commutators in the standard set-up chosen by Hawking, essentially without making any further unnecessary assumptions. The key new ingredient in our discussion will be that we will take into account the quantum mechanical nature of the matter forming the black hole. We will make this step by simply replacing in Hawking's original formulas the classical in-falling stress-energy by the corresponding quantum mechanical operator. One of the surprising consequences of this procedure is that it automatically includes an important part of the gravitational back reaction and it also gives some new insight into the question of energy conservation.

Another very important new effect is that the asymptotic coordinate system, since it is dynamically determined in terms of the in-falling matter, becomes *operator valued*. In a Heisenberg picture, this will imply that the observables that measure the asymptotic radiation will *not commute* with the observables associated with the in-falling matter. Normally, if we draw a Cauchy surface in a space-time diagram, like that in figure 1, we expect that all operators on this surface that are spacelike separated commute with each other. This assumption is in fact essential in the standard argumentation that the asymptotic Hilbert space of *out*-modes is incomplete. However, as we will explicitly show in this note, the commutators between *out*-modes ϕ_{out} and in-going modes ϕ_{in} in figure 1 are non-zero, and in fact will grow extremely large. It is clear that this result will require a drastic revision of the standard semi-classical picture of the evaporation process.

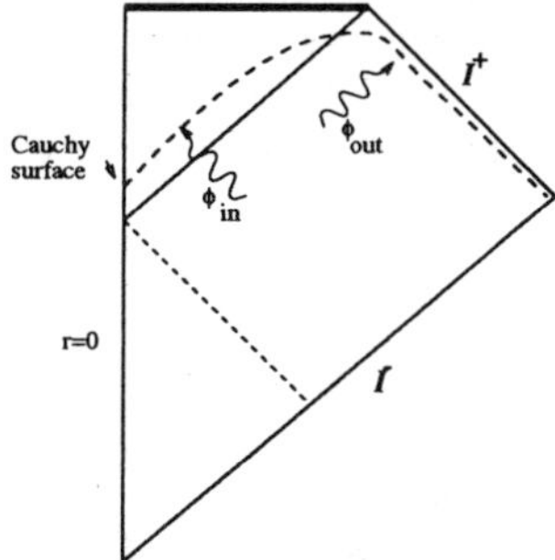

Fig 1. We will show in this note that, due to the fact that the asymptotic coordinate system is dynamically determined in terms of the in-falling matter, the commutator between the out-going modes ϕ_{out} and in-falling modes ϕ_{in} will grow uncontrollably large.

In the first part of this note we will present this calculation for the s-wave sector of the $3 + 1$-dimensional problem. After reviewing Hawking's original derivation we will argue on the basis of energy conservation that this derivation, as it stands, must be corrected already after a very short time. We will then summarize the derivation of the commutator between the *in* and *out* fields, and briefly discuss some implications of this result. In the second half of these notes we will illustrate this phenomenon and some possible physical consequences more explicitly in the context of two-dimensional dilaton gravity. This second part will also contain as a new result an exact quantum construction of the physical Hilbert space for an arbitrary number of massless matter fields. In this model the dynamical nature of the asymptotic coordinate system follows naturally from the gravitational dressing of the matter fields.

2. Black Hole Evaporation in the s-Wave Sector

In this section we begin with a short summary of Hawking's original derivation of the thermal spectrum of the out-going radiation. To simplify the formulas we restrict our attention to the s-wave sector of a massless field ϕ. We introduce two null-coordinates u and v such that at a large distance $r \to \infty$ we have $u \to t + r$ and $v \to t - r$.

2.1. HAWKING'S DERIVATION.

In his pioneering paper [1], Hawking did not not refer to the local vacuum state near the horizon, but instead he tried to establish a direct relation between the out-going state at future null infinity $\mathcal{I}^+$ directly to the in-state at past null infinity $\mathcal{I}^-$. Specifically, he imagined sending a small test-particle backwards in time from future null infinity $\mathcal{I}^+$ and letting it propagate all the way through to $\mathcal{I}^-$ (see fig 2). To relate the form of this signal in the two asymptotic regions, he then used the free wave equation on the *fixed* background geometry of the collapsing black hole, while ignoring the effects due to gravitational back reaction.

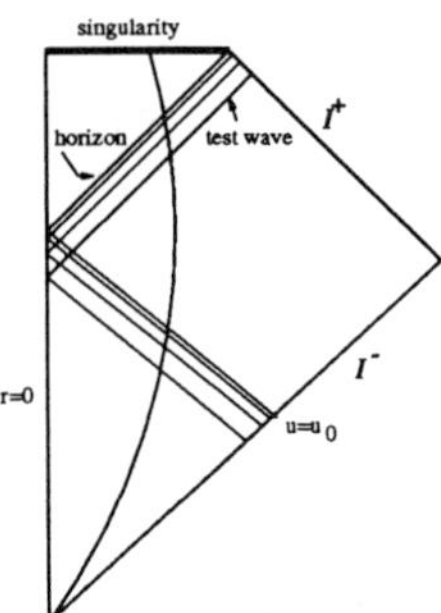

Fig 2. Following Hawking, we construct the out-going state at $\mathcal{I}^+$ by sending back a test wave and letting it propagate all the way to $\mathcal{I}^-$. Note that the frequency of this test wave diverges near the event horizon.

From the condition that the field is regular at the origin $r = 0$ one deduces that the outgoing s-wave ϕ_{out} of the massless scalar field ϕ and the corresponding in-coming wave ϕ_{in} are related by a reparametrization

$$\phi_{in}(u) = \phi_{out}(v(u)). \tag{2.1}$$

The diffeomorphism $v(u)$ typically takes the form

$$v(u) = u - 4M \log[(u_0 - u)/4M], \tag{2.2}$$

where M denotes the black hole mass and u_0 the critical in-going time, *i.e.* the location of the in-going light-ray that later will coincide with the black hole horizon. Thus an outgoing s-wave with a given frequency ω translates to an in-signal

$$e^{i\omega v(u)} = e^{i\omega u}\left(\frac{u_0 - u}{4M}\right)^{-i4M\omega}\theta(u_0 - u). \tag{2.3}$$

that decomposes as a linear superposition of incoming waves with very different frequencies. The out-going modes b_ω (*i.e.* the Fourier coefficients of ϕ_{out}) are therefore related to the in-coming modes a_ξ via a non-trivial Bogoljubov transformation of the form

$$b_\omega = \sum_\xi \alpha_{\omega\xi} a_\xi + \sum_\xi \beta_{\omega\xi} a_\xi^\dagger$$

$$b_\omega^\dagger = \sum_\xi \beta_{\omega\xi}^* a_\xi + \sum_\xi \alpha_{\omega\xi}^* a_\xi^\dagger. \tag{2.4}$$

The coefficients $\alpha_{\omega\xi}$ and $\beta_{\omega\xi}$ of this transformation are the Fourier transform of the function that appears on the right-hand side of (2.3). Up to some irrelevant phase the Bogoljubov coefficients take the form

$$\alpha_{\omega\xi} = \sqrt{\frac{\omega}{\xi}}\, e^{i(\xi-\omega)u_0} e^{\pi M\omega}\Gamma(1 - i4M\omega)$$

$$\beta_{\omega\xi} = \sqrt{\frac{\omega}{\xi}}\, e^{i(\xi+\omega)u_0} e^{-\pi M\omega}\Gamma(1 + i4M\omega) \tag{2.5}$$

At this level of approximation, these coefficients are c-numbers.

Once this relation between the *out*-modes and *in*-modes is established, one can for any given *in*-state find the corresponding *out*-state. In general, this *out*-state can be represented by a density matrix ρ_{out}, such that

$$\mathrm{tr}(\rho_{out}\mathcal{O}(b^\dagger, b)) = \langle in|\mathcal{O}(\beta^* a + \alpha^* a^\dagger, \alpha\, a + \beta\, a^\dagger)|in\rangle. \tag{2.6}$$

for all operators $\mathcal{O}$ that are constructed from the b-modes. When this map is applied to the in-vacuum $|0\rangle$ one obtains the Hawking state ρ_H. The explicit form of this Hawking state can be computed explicitly in terms of the coefficients (2.5), and describes a constant flux of thermal radiation at the Hawking temperature $T_H = \frac{1}{8\pi M}$. Notice that the Hawking state also depends on the time u_0 at which the black hole horizon forms, but that this dependence only shows up in the expectation values of operators $\mathcal{O}$ carrying non-zero energy.

Since the transformation (2.4) is not invertible, the map $|in\rangle \to \rho_{out}$ defined in (2.6) maps *pure in*-states into *mixed out*-states. Indeed, in-coming waves that have support for $u > u_0$ cannot be re-constructed out of the outgoing s-waves, and thus the out-going creation operators $b_\omega^\dagger$ generate only a subspace $\mathcal{H}_{out}$ of the complete Hilbert space $\mathcal{H}$ generated by the $a_\xi^\dagger$-modes. As long as the black hole is still there this is not so surprising, because the total *out* Hilbert space also contains a sector describing the matter that has fallen into the black hole. The important question is however whether the outgoing radiation, after the black hole has completely evaporated, is still described by a mixed state or whether eventually quantum coherence is restored. Clearly, this last possibility can only occur if the gravitational back reaction leads to a drastic modification of the above semi-classical picture.

2.2. ULTRA-HIGH FREQUENCIES.

In the above derivation of the Hawking state we have ignored the gravitational back reaction of the test wave on the geometry. This is commonly believed to be a good approximation, at least for for macroscopic black holes, since one only considers space-time regions in which both the curvature as well as the expectation value of the stress-tensor are small compared to the Planck scale. This argument, however, is in our opinion at best sufficient to show that the Hawking state correctly describes the global features of the out-going radiation, such as the average energy-flux. To study the actual information content of the radiation, on the other hand, one has to know the *entire* quantum state and in this case the behaviour of one *single* expectation value provides an insufficient criterion for the reliability of the approximation.

Because the out-going radiation has an (approximately) thermal spectrum with temperature $T_H = \frac{1}{8\pi M}$, the typical frequency ω of an out-going mode b_ω is of the order

$$\omega \sim \frac{1}{M}$$

in Planck units. So, as long as M is much larger than the Planck mass, each Hawking particle carries only a very small fraction of the total mass of the black hole. On the other hand, from (2.3) we see that the corresponding in-coming modes are very rapidly oscillating near $u \to u_0$. This implies that the calculation of the Hawking state at late time v_1 requires one to consider very high in-coming frequencies, typically of the order of

$$\xi \sim \frac{e^{(v_1-u_0)/4M}}{M}. \tag{2.7}$$

Hence, already after a very short time of the order of $M\log(M)$ after the formation of the black hole, we need to consider frequencies ξ that are much larger than the mass M of the black hole itself! Note that even for a macroscopic solar-mass black hole this is already after a fraction of a second!

It is now immediately clear that, due to the appearance of these large frequencies, the result (2.4)-(2.5) for the relation between the *in* and *out*-modes can not be taken literally. Namely, suppose that we would trust the transformation (2.4) as an accurate approximation of the exact operator identification. Then, physically, it describes how a particle, while propagating in the (fixed) background geometry of a collapsing star, changes its energy from its initial value ξ to a final value ω. In fact, the quantum mechanical amplitude for this process is given by one of the Boboljubov coefficients

$$\langle 0|b_\omega a_\xi^\dagger|0\rangle = \alpha_{\omega\xi}. \tag{2.8}$$

In general the initial energy ξ is much larger than the final energy ω, because a particle will lose almost all its energy while trying to escape from near the horizon out to infinity. However, suppose that we do include the back reaction, then the total energy of the particle-black hole system should be conserved of course. We must therefore conclude that all the energy that is lost by the particle has gone into the black hole. So, if in the final state the mass of the black hole is equal to M, the initial black hole mass must have been

$$M_{initial} = M + \omega - \xi,$$

just like in any other physical scattering process. Now it is clear that for real physical particles the gravitational back reaction will surely lead to significant corrections in the amplitude (2.8) as soon as the difference $\xi-\omega$ is of the order of M. In this regime the linear transformation (2.4) between the *in* and *out*-modes no longer gives an accurate description of particle-black hole scattering, and must therefore be corrected. In our opinion this also means that to correctly describe the Hawking process one should use a modified version of the Bogoljubov transformation that includes gravitational corrections.

While it should be evident from this that the formula (2.4), when interpreted as an operator identification, needs to be corrected, it is perhaps not so clear that these corrections will also lead to real modifications in the calculation of the Hawking state. One could indeed argue that in this calculation one doesn't quite need the relation (2.4) to its full strength, since one is only interested in propagating an *in*-state that looks like the *vacuum* near $u = u_0$. In particular, it would appear therefore that one can use the (approximate) Lorentz invariance of this local vacuum state to drastically reduce the problem of the exploding frequencies. However, this mechanism is clearly insufficient to completely eliminate the problem, because this would essentially require that Lorentz invariance is an *exact* symmetry of the *complete in*-state. Since the *in*-state also describes the in-falling matter forming the black hole, this is clearly not the case. We will return to this point later.

3. Gravitational Back Reaction.

This brings us to the question: how important are the gravitational corrections? In particular, can the consequences be large enough to avoid that information gets lost inside the black hole? Unfortunately, there is not yet enough known about quantum gravity to definitely answer this question. We can however, with some reasonable assumptions, try to investigate what the possibilities are.

3.1. SOME GENERAL OBSERVATIONS.

To begin with, let us recall another often mentioned argument for why, in spite of the ultra-high energies involved, the quantum gravitational effect should still be negligible in the calculation of the Hawking state. Namely, one could argue that in this calculation it is unnecessary, or even wrong, to trace the history of the out-going test particle all the way back to $\mathcal{I}^-$, since intuitively the Hawking particles emitted by the black hole begin their lives as virtual particles produced by pair creation out of the vacuum near the horizon. Since this pair production takes place after the black hole has already been formed and settled into a "steady state", it appears that the out-going radiation never had a chance to interact with the in-falling matter. In other words, the test-wave used in Hawking's derivation nor the virtual particles near the horizon are physically detectable, and should according to this argument have no measurable effects on the back ground geometry or anything else.

To test the validity of this argument, let us assume that the black hole evaporation process satisfies some very basic physical rules. Specifically, we will assume that the basic starting points that were used in Hawking's original derivation remain valid also after including the gravitational back reaction. These physical principles can be formulated as follows:

(1) *There exists a quantum-mechanical time-evolution.*
In quantum mechanics we can describe the time evolution of a system using either a Schrödinger or a Heisenberg picture. If we use the latter, operators are time-dependent. The existence of a time-evolution implies that all the operators at a given time t_1 must be expressible in terms of the operators at any given earlier time $t_0 < t_1$. We will not necessarily require that this time evolution is strictly unitary: in principle we will allow that certain operators are 'destroyed' e.g. by the black hole singularity, or transported into some other universe. However, we do want to *exclude* the possibility that operators are created out of nothing during the time evolution, since this would render the quantum mechanical time evolution unpredictable in a clearly unacceptable way. In other words, we will require that at any given time all observable operators must have a past.

(2) *The initial asymptotic data can be described in terms of a free field Fock space.*
This is a standard postulate of the LSZ asymptotic theory, and assumes that in the remote past all particles are well-separated from each other. In this limit all (gravitational) interactions between the asymptotic particles can be ignored. Of course, free field theory will no longer be adequate to describe the black hole formation and evaporation process, but we will assume that our first postulate is satisfied at all times. In particular, this means that *in principle* all physical observables at later times are expressible in terms of these initial free fields ϕ_{in}.

(3) *Energy is conserved.*
Asymptotically, the space-time metric reduces at all times to the flat Minkowski metric, and thus we can define a conserved energy operator E that keeps track of the energy carried in or out of the system by the asymptotic in and out-going particles. By our assumption (2) there exists a unique vacuum state and all other states have positive energy.

It is important to note that none of the above three physical assumptions is (obviously) contradicted by the semi-classical picture of the evaporation process. We will see, however, that they will give very strong restrictions on the possible scenarios. In particular, it is clear that the first two requirements, while not as strong, will become essentially equivalent to an S-matrix assumption if we would add CPT invariance as an additional postulate.[*] It should therefore be mentioned that there are perhaps ways in which one can imagine weakening one or more of these assumptions, but we are not aware of any accurately formulated, acceptable alternatives.

One of the immediate consequences of the first physical requirement (1) is that any observable that can be used to distinguish asymptotic states (*i.e.* the asymptotic modes b_ω) must be expressible in terms of operators defined at earlier times. Hence, although the Hawking radiation may perhaps be thought of as arising from pair creation near the

---

[*] For more detailed discussions of CPT invariance in relation with black hole evaporation see [8].

32

horizon, this does *not* imply that the corresponding creation and annihilation modes b_ω and $b_\omega^\dagger$ are pair created out of nothing. If this were true this pair creation process itself would already violate a basic principle of quantum mechanics. Thus the above mentioned argument, that one should not propagate the particle waves all the way back to $\mathcal{I}^-$, is clearly inaccurate, at least when we interpret this propagation as the quantum mechanical time evolution of a Heisenberg operator. In quantum mechanics, to evolve states forward in time, we need to know how to evolve observables backwards in time.

The first two physical assumptions combined imply that the out-going modes $(b_\omega, b_\omega^\dagger)$ can be expressed in terms of the in-coming modes $(a_\omega, a_\omega^\dagger)$. Given the form of the black hole geometry, this indeed seems a reasonable conclusion. It is furthermore the same assumption that underlies Hawking's derivation, which suggests that this relation between the b- and the a-modes will in a suitable approximation take the form of a Bogoljubov transformation as in (2.4).

3.2. ENERGY CONSERVATION

In most semi-classical models of black hole formation and evaporation the collapsing matter is assumed to be in some semi-classical (or coherent) state. It then appears to be a sensible procedure to replace the operators associated with it by classical c-number quantities. However, the above argument indicates that this may no longer be a good approximation when the energy that is gained or lost by the test-particle is comparable to the mass M of the black hole. The main problem indeed with treating the black hole geometry and the in-falling matter as c-numbers is that it makes energy conservation totally obscure. All quantum processes take place in a time dependent classical background, which becomes an (inexhaustible) source of energy. This causes the problem of the diverging frequencies.

In the following, we will therefore try to develop a formulation in which we treat (part of) the in-falling matter quantum mechanically. To do this exactly would of course be difficult, but as a reasonable first approximation we will use Hawking's formulas as a starting point, while replacing all quantities that depend on the background geometry by *operator-valued* quantities. In other words, we will continue to work with the relation (2.4) between the *in-* and *out*-modes, but with the Bogoljubov coefficients replaced by suitable quantum operators, that act on the Hilbert space of the in-falling matter. Equation (2.4) will thus no longer be a linear relation but a highly *non-linear operator identification*, expressing the *out*-modes in terms of the *in*-modes.

It will become clear that in this framework one automatically includes an important part of the interaction between the out-going modes and the in-falling matter. Moreover, it will enable us to formulate energy conservation as a meaningful and precise requirement. Namely, we may imagine taking the commutator of the energy operator E with the operator valued Bogoljubov transformation (2.4), and require that we get the same answer on the left- and the right-hand-side. The physical interpretation of this requirement is that all energy that the test-wave gains in propagating back to $\mathcal{I}^-$, should come from the matter forming the black hole. We thus deduce that the coefficients $\alpha_{\omega\xi}$ and $\beta_{\omega\xi}$ must carry energy $E = \xi - \omega$ and $E = \xi + \omega$, respectively.

From the classical expression (2.5) of the coefficients we see that the energy balance can indeed be restored while keeping essentially the same expression, by giving parameter u_0 a non-trivial commutation relation with E as follows

$$[E, u_0] = i. \tag{3.1}$$

so that

$$[E, \alpha_{\omega\xi}] = i\frac{d}{du_0}\alpha_{\omega\xi} = (\xi - \omega)\alpha_{\omega\xi} \tag{3.2}$$

and similarly for $\beta_{\omega\xi}$. Although the precise meaning of this observation can become fully clear only in a more complete quantum theory of black holes, it certainly suggests that in understanding the quantum back reaction of the Hawking radiation one should take into account the operator character of u_0. To illustrate this point, we will now discuss one particularly important consequence of this new insight.

3.3. Gravitational shift interactions.

A crucial assumption in Hawking's derivation is that the incoming particles described by $\phi_{in}(u)$ with $u > u_0$ and the outgoing particles described $\phi_{out}(v)$ form independent sectors of the Hilbert space, and that the corresponding field operators commute with each other. The underlying classical intuition is that the fields $\phi_{in}(u)$ with $u > u_0$ will propagate into the region behind the black hole horizon, and thus become unobservable from the out-side. However, this intuition ignores the important fact that the in-falling particles in fact *do* interact with the out-going radiation, because they slightly change the black hole geometry. In the spherically symmetric theory, this change in the geometry is represented by a small shift in the black hole mass M of and the time u_0 at which the black hole horizon was formed.

Consider a spherical shell of matter with energy δM that falls in to the black hole at some late time u_1. The Schwarzschild radius will then increase slightly with an amount

$2\delta M$, and the time u_0 will also change very slightly. A simple calculation shows that[†]

$$\delta u_0 = -c\,\delta M\,e^{-(u_1-u_0)/4M} \tag{3.3}$$

At first it seems reasonable to ignore this effect as long as the change δM is much smaller than M. However, in view of our preceding discussion on energy conservation, it may be a good idea to study this point somewhat closer.

The exponential u-dependence that occurs in the formula (3.3) is typical of black holes and has to do with the diverging red-shift. This time it helped in our favour because it exponentially suppressed the effect on u_0 of the in-going matter. But in other physical quantities it is easy to get exponentially growing factors that enhance physical effects that seemed to be unimportant at first. For example, the variation in u_0, although very small, has an enormous effect on the wave-function $\phi_{out}(v)$ of an out-going particle. By combining (2.1), (2.2) and (3.3) one easily verifies that as a result of the in-falling shell, the outgoing particle-wave is delayed by an amount that grows rapidly as a function of v

$$\phi_{out}(v) \to \phi_{out}(v - 4M\log(1 - c\frac{\delta M}{4M}e^{(v-u_1)/4M})). \tag{3.4}$$

Notice that even for a very small perturbation δM the argument of the field ϕ_{out} goes to infinity after a finite time $v_{lim} - v_1 \sim -4M\log(\delta M/M)$. The physical interpretation of this fact is that a matter-particle that is on its way to reach the asymptotic observer at some time $v > v_{lim}$ will, as a result of the additional in-falling shell, cross the event-horizon and be trapped inside the black-hole horizon[‡]. This implies that the asymptotic wave-function of an *individual particle* is *very sensitive* to the gravitational back-reaction. To see what this means for the *collective* state of the outgoing radiation is clearly a much more subtle matter. In fact, it can be shown that the transformation (3.4) is an approximate symmetry of the Hawking state, and this is undoubtedly the reason why it is usually not taken in to account. However, as we will show in the remainder of this section, the fact that the gravitational back reaction is important for individual particles is sufficient to substantially change the usual semi-classical picture.

3.4. The exchange algebra between *in* and *out*-fields.

How does one take the effect (3.4) into account? At this point we come back to our idea that the parameter u_0 is not just as a classical number but should be treated as a quantum

---

[†] Here, c is a constant of order one and happens to be equal to $4e$

[‡] This same calculation is often used to show that "white holes" are unstable under small perturbations.

operator. To make this more concrete, let us divide up the in-falling matter in a classical piece plus a small quantum part that is described in terms of a quantum field $\phi_{in}(u)$. Of course, u_0 is mainly determined by the classical in-falling matter, but in addition it has a small quantum piece. Using (3.3) we find

$$u_0 = u_0^{cl} - c \int_{u_0^{cl}}^{\infty} du \, e^{(u_0^{cl} - u)/4M} T_{in}(u) \tag{3.5}$$

where $T_{in}(u)$ denotes the stress-energy tensor of the $\phi_{in}(u)$ with support $u > u_0^{cl}$. From here on the calculation is simple, straightforward and unavoidable. The only additional ingredient we need is the same fundamental relation (2.1), that formed the starting point of Hawking's derivation. The only difference is now that in the reparametrization (2.2) we include the seemingly negligible quantum contribution. Our goal is to calculate the algebra of the outgoing field $\phi_{out}(v)$ for late times with the in-coming field $\phi_{in}(u)$ for $u > u_0^{cl}$. First we compute

$$[u_0, \phi_{in}(u)] = -ic \, \exp((u_0^{cl} - u)/4M) \partial_u \phi_{in}(u), \tag{3.6}$$

or equivalently

$$e^{i\xi u_0} \phi_{in}(u) e^{-i\xi u_0} = \phi_{in}(u - 4M \log(1 - \frac{c}{4M}\xi e^{(u_0^{cl} - u)/4M})), \tag{3.7}$$

where we simply used the fact that the stress-tensor generates coordinate transformations.

We can now compute the exchange algebra between the *in* and *out*-fields by combining (2.1), (2.2) and (3.7). One finds the following result

$$\phi_{out}(v)\phi_{in}(u) = \exp(ic \, e^{(v-u)/4M} \partial_v \partial_u) \phi_{in}(u)\phi_{out}(v), \tag{3.8}$$

which is valid at for $u > u_0^{cl}$. This exchange algebra is the quantum implementation of the gravitational back-reaction (3.4) of the in-falling matter on the out-going radiation. An equivalent physical interpretation of this non-local algebra is that it represents a gravitational shock-wave interaction between the incoming and out-going matter waves [2].

We want to emphasize that to derive this result we did not need any assumptions other than those already made in the usual derivation of Hawking evaporation. The only extra ingredient that we took into account is the small quantum contribution to u_0. In this sense the relation (3.8) appears to be unavoidable and independent of which scenario one happens to believe in.

3.5. SOME PHYSICAL CONSEQUENCES.

Of course, much work is needed to analyze the precise physical consequences of this algebra, but at this point it is clear that the presence of these large commutators implies that the standard semi-classical picture of the black hole evaporation process needs to be drastically revised. In particular, it tells us that, due to the quantum uncertainty principle, we should be very careful in making simultaneous statements about the in-falling and out-going fields. Mathematically, the Hilbert space of the scalar-fields on a Cauchy surface as drawn in figure 1 does *not* decompose into a simple tensor product of a Hilbert space inside the black hole and one out-side. Instead, in view the exponentially non-local nature of the commutator between the *in* and *out*-fields, it is clear that the *out* Hilbert space is not even approximately independent of the Hilbert space of the in-falling matter!

It should be pointed out, however, that the exchange algebra (3.8) in itself can not be responsible for the complete transfer of information to the out-region, since it only depends on a single quantum number u_0 of the in-going matter. In fact, as we will show explicitly in section 4.4, the above algebra is perfectly consistent with Hawkings analysis, and can even be used to rederive his results in this new setting. On the other hand, the algebra (3.8) has the important consequence that it introduces large non-local quantum effects, that may bring other (strong coupling) effects into the picture. In relation to this it is important to note that, in deriving the algebra (3.8) for $\phi_{in}(u)$ we excluded the *in*-region $u \leq u_0^{cl}$, that is mapped in (2.1) on the asymptotic *out*-region. In this region the algebra of *in* and *out*-fields contains an additional term, which describes the direct propagation of the modes. In the leading semi-classical limit

$$[\phi_{out}(v), \phi_{in}(u)] = i\theta(v - v(u)) \tag{3.9}$$

where $v(u)$ denotes the diffeomorphism given in (2.2) that maps the out-interval onto the interval $u < u_0^{cl}$. This direct interaction between *in* and *out* modes is clearly capable of transporting detailed information. The question that should be studied further is whether the (yet to be found) complete interaction between the *in* and *out* modes that combines the exchange relation (3.8) and the direct propagation (3.9) is sufficient to transport *all* the in-going quantum numbers back out-wards into the Hawking radiation.

To illustrate why this is *in principle* possible, let us make the following simple observation. We want to know whether the out-going observables cover the complete collection of incoming fields. To begin with, it seems reasonable to assume that via the direct propagation (3.9) the out-going operators contain at least all incoming operators inside the region $u < u_0^{cl}$. This is nothing new and in complete accordance with the usual semi-classical reasoning. The important new consequence of the algebra (3.8), however, is that it appears to hand as an additional peace of information to the outside observer the quantum operator

$$P_{in} = \int_{u_0^{cl}}^{\infty} du\, e^{(u_0^{cl} - u)/4M} T_{in}(u)$$

that acts on the incoming fields at $u > u_0^{cl}$. At first sight this may not seem such a big deal, since this is just one single quantum number. It may therefore come perhaps as a surprise that, if both these seemingly innocent assumptions are indeed correct, then this immediately implies that the out-going modes must carry *all* in-going information. Namely, one should not think of P_{in} as an operator measuring just one quantum number (in fact, it has no eigen states), but rather as an operator that *generates* a certain type of translations on the interval $u > u_0^{cl}$. It then becomes immediately evident that the collection of all *in*-fields in the interval $u < u_0^{cl}$ *together* with the above operator P_{in} form a complete basis of operators for the incoming Hilbert space!

Finally, we note that the algebra (3.8) has the surprising property that it is *symmetric* between the *in* and *out*-fields, although the derivation certainly *looked* asymmetric. This suggests that a complete description of the black hole formation and evaporation process, that properly incorporates this interaction, should also be symmetric between the *in* and *out* fields.

4. Quantum Back Reaction in 2d Dilaton Gravity

In this section we shall illustrate the above discussion by focusing on the specific example of $d = 2$ dilaton gravity [9]-[10]. It is generally agreed that the physics of this model is closely analogous to what happens in four dimensions and, indeed, advocates on all sides of the black hole paradox have skillfully used the model to strengthen their case. We cheerfully joined this tradition in [12]-[13], where we proposed a specific S-matrix description of quantum dilaton gravity. Here we will briefly review this approach, and show how it naturally leads to the same type of non-local interactions between the *in* and *out*-fields as discussed above for the s-wave reduction of the four-dimensional theory.

4.1. FREE FIELD FORMALISM

Below we briefly review the analysis presented in [12]-[13]. We will actually be a little bit more general by allowing a general number N of matter fields flavors and by considering a more general class of boundary conditions [14]. Our intention in this section is to illustrate the points made in the previous section in this specific situation. In particular, it will become clear that the basic principles (1), (2) and (3) proposed in section 3.1 are indeed realized in this simple model, and that their implications can be understood.

The starting point of our analysis is the following action for dilaton gravity coupled to N massless scalar fields in the conformal gauge [9] [14]

$$S = \frac{1}{\pi} \int d^2x \left[(2e^{-2\phi}(2\partial_u\partial_v\rho - 4\partial_u\phi\partial_v\phi + \lambda^2 e^{2\rho}) + \tfrac{1}{2}\sum_{i=1}^{N} \partial_u f_i \partial_v f_i \right.$$

$$\left. + \tfrac{N}{12}\phi\,\partial_u\partial_v(\rho - \phi) - \tfrac{N-24}{6}\partial_u(\rho - \phi)\partial_v(\rho - \phi) \right] . \tag{4.1}$$

This action contains a specific combination of 1-loop correction terms, that was motivated in [14]. Introducing

$$\Omega(u,v) = e^{-2\phi} + \tfrac{N}{24}\phi \,, \qquad \hat{\rho} = \rho - \phi \,, \tag{4.2}$$

the equations of motion of this action can be written as

$$\partial_u\partial_v\,\hat{\rho} = 0$$

$$\partial_u\partial_v\Omega + \lambda^2 e^{2\hat{\rho}} = 0 \,. \tag{4.3}$$

The stress-energy tensor takes the form

$$T_{uu} = -2\partial_u\Omega\partial_u\hat{\rho} + \partial_u^2\Omega - 2\kappa(\partial_u\hat{\rho}\partial_u\hat{\rho} - \partial_u^2\hat{\rho}) + \tfrac{1}{2}\sum_i \partial_u f_i \partial_u f_i \,, \tag{4.4}$$

where we defined $\kappa = \frac{N-24}{12}$.

The first of the field-equations (4.3) allows us to introduce free field variables according to

$$e^{2\hat{\rho}} = \partial_u X^+(u)\,\partial_v X^-(v) \,. \tag{4.5}$$

The general solution of the second equation in (4.3) is then given by

$$\Omega = -\lambda^2 X^+(u)X^-(v) + \omega^+(u) + \omega^-(v) \,. \tag{4.6}$$

We define variables $P_\pm$ according to

$$\partial_u\omega^+(u) = P_+\partial_u X^+ - \frac{\kappa}{2}\frac{\partial_u^2 X^+}{\partial_u X^+}, \quad \partial_v\omega^-(v) = -P_-\partial_v X^- - \frac{\kappa}{2}\frac{\partial_v^2 X^-}{\partial_v X^-} \,. \tag{4.7}$$

These definitions are such that at the quantum level the following commutation relations are valid

$$[\partial X^+(u_1), \partial P_+(u_2)] = 2\pi i \, \delta'(u_1 - u_2) \, ,$$

$$[\partial f_i(u_1), \partial f_j(u_2)] = 2\pi i \, \delta_{ij} \, \delta'(u_1 - u_2) \, . \tag{4.8}$$

The total stress-energy tensor, which takes the form

$$T_{uu} = \partial_u P_+ \partial_u X^+ + \frac{\kappa}{2}\partial_u^2 \log(\partial_u X^+) + \tfrac{1}{2}\partial_u f_i \partial_u f_i \tag{4.9}$$

(normal ordering is implied), satisfies a Virasoro algebra of central charge $c = 2 - 12\kappa + N = 26$.

The semiclassical theory defined by these equations has been analyzed recently in [14]. Here we shall briefly sketch a full quantum treatment of the theory, slightly generalizing the work of [12]-[13], where we focussed on the special case $N = 24$.

4.2. PHYSICAL OPERATORS AND GRAVITATIONAL DRESSING

In the quantum theory, states that are annihilated by the total stress tensor (4.9) are called physical states. Physical states that are at the same time descendants with respect to the total stress-tensor are called spurious physical states; it can be shown that such states decouple from the theory. For the physical spectrum we are thus interested in the space of physical states modulo spurious physical states.

The analysis of the physical state condition in quantum dilaton gravity is closely analogous to a similar analysis in light-cone gauge critical string theory. Following this lead, one discovers that the following oscillators create physical states (called DDF states)

$$\alpha_i(\omega) = \int du \, (\lambda X^+(u))^{i\omega} \partial_u f_i(u) \, . \tag{4.10}$$

Furthermore, there are additional physical operators given by

$$\alpha^-(\omega) = \int du \left[: (\lambda X^+)^{1+i\omega} \partial_u P_+(u) : \right.$$

$$\left. + \frac{\lambda\hat{\kappa}}{2}(1 + i\omega)(\lambda X^+)^{i\omega} \partial_u \log(\partial_u X^+) \right] , \tag{4.11}$$

with $\hat{\kappa} = \kappa + 1$. These operators are found to satisfy the following algebra

$$[\alpha_i(\omega_1), \alpha_j(\omega_2)] = \delta_{ij} \, \omega_1 \, \delta(\omega_1 + \omega_2)$$

$$[\alpha^-(\omega_1), \alpha_i(\omega_2)] = -\lambda \, \omega_2 \, \alpha_i(\omega_1 + \omega_2) \, . \tag{4.12}$$

A most remarkable relation is the following

$$[\tilde{\alpha}^-(\omega_1), \tilde{\alpha}^-(\omega_2)] = \lambda\,(\omega_1 - \omega_2)\tilde{\alpha}^-(\omega_1 + \omega_2)\,, \tag{4.13}$$

where $\tilde{\alpha}^-(\omega) = \alpha^-(\omega) - \frac{\lambda}{2} : \alpha_i \alpha_i : (\omega)$. This relation shows that the difference of the generator $\alpha^-(\omega)$ and the light-cone Virasoro generator $\frac{\lambda}{2} : \alpha_i \alpha_i : (\omega)$ generates a *centerless* Virasoro algebra. This immediately suggests that states created by the $\alpha^-(\omega)$ are equivalent (modulo spurious physical states) to states that can be created by using the oscillators $\alpha_i(\omega)$ only. A more careful analysis confirms that the states created by using only the oscillators $\alpha_i(\omega)$ indeed form a complete basis of all physical states. (This statement is the equivalent of the so-called no-ghost theorem in critical string theory.) In [13] we proved this for $N = 24$ ($\kappa = 0$) by using a finite volume regularization. Since this proof only used the algebraic relations satisfied by the oscillators $\alpha_i(\omega)$ and $\alpha^-(\omega)$, it can be generalized immediately to the general case $N \neq 24$ ($\kappa \neq 0$).

We have thus obtained a very convenient basis for a description of the space of physical states from the point of view of an observer at $\mathcal{I}_R^-$, who uses $\tau_+ = \lambda^{-1}\log(\lambda X^+)$ as time variable. A similar basis, in terms of oscillators $\beta_i(\omega)$ appropriate to an asymptotic *out*-observer at $\mathcal{I}_R^+$ can be constructed by interchanging the role of the coordinates X^+ and X^-. It is appropriate to call the oscillators α_i 'dressed oscillators', since they correspond to 'bare' oscillators of the matter fields f_i that are dressed with dilaton gravity excitations (represented by X^+, X^-) that describe the back-reaction of the matter excitation on the dilaton gravity system. Our formalism thus incorporates this back reaction in an explicit, algebraic way that avoids the semi-classical notion of a background dilaton gravity configuration. In the next subsections we will show how this algebraic treatment of the back-raction problem can be used to derive Hawking radiation in the *out*-state and to investigate the quantum gravitational corrections to the semi-classical result.

4.3. BOUNDARY CONDITION.

In the formalism for quantum dilaton gravity that we described sofar, the left and right-moving degrees of freedom are completely independent: incoming left-moving signals propagate freely to $\mathcal{I}_L^+$ and outgoing right-movers have their origin at $\mathcal{I}_L^-$. We are, however, interested in a quantum mechanical theory that describes the evolution of initial left-moving data on $\mathcal{I}_R^-$ to right-moving signals on $\mathcal{I}_R^+$, in analogy with the 3+1-dimensional theory. We shall now describe how such a theory can be obtained by specifying a boundary condition in the strong coupling regime.

Let us first use the classical theory to motivate the specific boundary condition we will choose. Classically we can pick a large but constant value of the dilaton field, and require the line where this value is attained to be the boundary of the two-dimensional world.[*] If we choose a physical coordinate system (x^+, x^-) in which the rescaled metric $d\tilde{s}^2 = e^{-2\phi} ds^2$ (which is flat everywhere) takes the form $d\tilde{s}^2 = dx^+ dx^-$, it can be shown that the resulting equation of motion for the boundary $(\mathbf{x}^+, \mathbf{x}^-)$ reads as follows

$$-\frac{m}{2}\sqrt{\partial_{\mp}\mathbf{x}^{\pm}} \pm \lambda^2 \mathbf{x}^{\pm} + p_{\mp}(\mathbf{x}^{\mp}) = 0 \;, \tag{4.14}$$

where

$$p_{\pm}(\mathbf{x}^{\pm}) = \pm \int\limits_{\mathbf{x}^{\pm}}^{\pm\infty} dx^{\pm}\, T_{\pm\pm} \tag{4.15}$$

denotes the integrals of the in- and out-going momentum flux. Note that the $x^{\pm}$ coordinates are related to the standard asymptotic coordinates r and t via $x^{\pm} = \pm\lambda^{-1}\exp(\lambda(r \pm t))$.

The total system of matter and boundary only describes a well-defined dynamical system if one restricts to field configurations below a certain critical energy flux. As an example, we consider the classical boundary equation (4.14) when the incoming wave is a shock wave located at $x^+ = q^+$, with amplitude p_+

$$T_{++}(x^+) = p_+\delta(x^+ - q^+) \;. \tag{4.16}$$

As long as the total energy $E = p_+ q^+$ carried by the pulse is smaller than $\frac{m^2}{4\lambda^2}$, the boundary trajectory is time-like everywhere and given by

$$(\lambda^2 \mathbf{x}^- - p_+)\mathbf{x}^+ = -\frac{m^2}{4\lambda^2} \tag{4.17}$$

for $\mathbf{x}^+ < q^+$ and

$$\mathbf{x}^-(\lambda^2 \mathbf{x}^+ + p_-) = -\frac{m^2}{4\lambda^2} \tag{4.18}$$

with

$$p_- = \frac{\lambda^2 p_+ q^+}{\frac{m^2}{4\lambda^2 q^+} - p_+} \tag{4.19}$$

for $\mathbf{x}^+ > q^+$. The typical form of this boundary trajectory is depicted in fig 3a. Note that the mirror point behaves as a particle with negative rest mass: when the shock wave hits it, it does not bounce back to the left but in the opposite direction to the right.

---

[*]A boundary condition of this type was first introduced by Russo et al in [11].

In case $p_+q^+ > \frac{m^2}{4\lambda^2}$ then the solution to the equation (4.14) cannot be time-like every where, but turns space-like for $\mathbf{x}^+ > q^+$ (see fig 3b). So in this regime it is no longer classically consistent to treat the boundary as a reflecting mirror. Obviously, this situation precisely corresponds to the formation of a black hole. This example suggest that the following inequality

$$p_+(x^+) < \frac{m^2}{4\lambda^2 x^+} \qquad (4.20)$$

for all x^+, with $p_+(x^+)$ as defined in (4.15), is a necessary and possibly sufficient criterion for the incoming energy flux to ensure that the classical boundary remains timelike. Note that this inequality does not imply any specific *local* upper bound on the energy flux T_{++}.

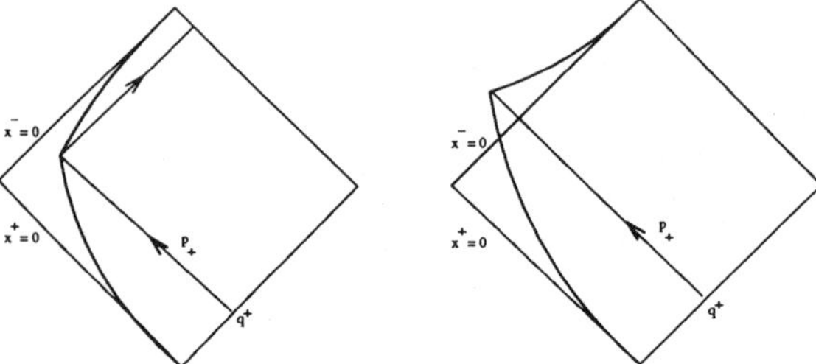

Fig 3a and 3b. Schematic depiction of the classical boundary trajectory for a sub-critical (left) and a super-critical (right) shock wave.

Let us now describe how the above boundary condition translates to the quantum theory described in the previous subsection. We choose the (u, v)-coordinate system in such a way that the boundary becomes identified with the line $u = v$, and denote the parameter along this boundary by s. We require that the dilaton field takes a large and constant value along the boundary, such that

$$\partial_s \Omega = 0 \ . \qquad (4.21)$$

In terms of the X and P-fields this condition reads

$$\partial_s X^+ (\lambda^2 X^- - P_+) + \partial_s X^- (\lambda^2 X^+ + P^-) + \frac{\kappa}{2}\partial_s \log(\partial_s X^+ \partial_s X^-) = 0. \qquad (4.22)$$

The above boundary condition is coordinate invariant and this allows us to impose the additional constraint that the gravitational and matter components of the energy momentum flux each separately reflect off the boundary. So (4.22) is supplemented with the condition

$$T^g_{uu} = T^g_{vv} \qquad (4.23)$$

where (compare with (4.9))

$$T^g_{uu} = \partial_u P_+ \partial_u X^+ + \frac{\kappa}{2}\partial_u^2 \log(\partial_u X^+)$$
$$T^g_{vv} = -\partial_v P_- \partial_v X^- + \frac{\kappa}{2}\partial_v^2 \log(\partial_v X^-) \ . \tag{4.24}$$

The two equations (4.22) and (4.23) combined specify the precise reflection condition that relates the incoming canonical variables (X^+, P_+) to the outgoing canonical variables (X^-, P_-).

We would like to make manifest that this relation defines a canonical transformation. To this end, we should write a generating functional $S[X^+, X^-]$ of the coordinate fields, such that the momenta $P_\pm$ defined by

$$P_\pm = \frac{\delta S[X]}{\delta \partial_s X^\pm} \tag{4.25}$$

identically solve the boundary equations (4.22) and (4.23). This condition results in a set of functional equations for $S[X]$ that can be solved explicitly. The form of the solution is unique, once we fix the constant value of Ω along the boundary. If we set $\Omega(X^+, X^-) = \frac{m^2}{4\lambda^2}$, then the generating functional $S[X]$ takes the following form

$$S[X] = m \int ds \sqrt{\partial_s X^+ \partial_s X^-} - \lambda^2 \int ds\, X^+ \partial_s X^-$$
$$+ \frac{\kappa}{2} \int ds \log(\partial_s X^+)\partial_s \log(\partial_s X^-) \ . \tag{4.26}$$

Using (4.25) we then obtain the following relations between the in-going variables (X^+, P_+) and the outgoing variables (X^-, P_-)[†]

$$\pm \frac{m}{2}\sqrt{\partial_s X^+ \partial_s X^-} + \partial_s X^\pm (\lambda^2 X^\mp \mp P_\pm) + \frac{\kappa}{2}\partial_s \log(\partial_s X^\mp) = 0 \ . \tag{4.27}$$

As was shown in [14], the relations (4.27) can be used to derive the following semi-classical equation of motion for the coordinates $(\mathbf{x}^+, \mathbf{x}^-)$ of the boundary curve

$$-\frac{m}{2}\sqrt{\partial_\mp \mathbf{x}^\pm} \pm \lambda^2 \mathbf{x}^\pm + p_\mp(\mathbf{x}^\mp) \pm \frac{\kappa}{2}\partial_\mp \log \partial_\mp \mathbf{x}^\pm = 0 \ , \tag{4.28}$$

where the physical fields $p_\pm(\mathbf{x}^\pm)$ are identified with the integrals (4.15) of the respective components of the matter stress tensor.

[†]This equation in fact requires a suitable normal ordering prescription. In the semi-classical limit such corrections will not be important.

The equation (4.28) can be compared to the equation (4.14) which can be derived in classical dilaton gravity. This comparison shows that in general the classical boundary trajectory receives quantum corrections. At the quantum level, the dynamics of the boundary curve is affected by the fact that the stress-energy flux $T_{\pm\pm}$ does not vanish in the physical vacuum defined by an asymptotic observer. In particular, this implies that, in order to have a *time-like* vacuum boundary curve, the parameter m has to satisfy the inequality

$$m^2 > 8\kappa\lambda^2 \ . \tag{4.29}$$

We refer to [14] for further discussion of this point.

Before we continue, we would like to reflect on the profound consequences of the equations (4.27). After the boundary condition has been implemented, the operators (X^+, P_+) and (X^-, P_-) all act in the same Hilbert space and have non-trivial commutation relations. In particular, the out-going coordinates X^- no longer commute with the in-going coordinates X^+. In the special case $\kappa = 0$, one can explicitly compute this commutator, which takes the form (again modulo normal ordering effects)

$$[X^-(u_1), X^+(u_2)] = \frac{2\pi i}{\lambda^2} \, e^{\frac{-\lambda^2}{m}\theta(u_{12})\int_{u_2}^{u_1} du \sqrt{\partial_u X^+ \partial_u X^-}} \ . \tag{4.30}$$

The function on the right-hand side is equal to $2\pi i\lambda^{-2}$ for $u_2 > u_1$ and dies off exponentially with the physical distance for $u_1 > u_2$. In the limit $m = 0$, which was considered in detail in [13], the above algebra reduces to the standard free field commutation relation

$$[X^-(u_1), X^+(u_2)] = \frac{2\pi i}{\lambda^2} \, \theta(-u_{12}) \ . \tag{4.31}$$

In the next subsection we will show that this non-trivial commutator between the physical coordinates X^+ and X^- leads to an exchange algebra between physical *in-* and *out*-fields very similar to the one discussed in section 3. We will further demonstrate that, among other things, it directly implies the existence of Hawking radiation.

4.4. THE EXCHANGE ALGEBRA AND HAWKING RADIATION.

In this subsection we focus on the special case of $N = 24$ matter fields, so that $\kappa = 0$. Let us introduce the following physical operators

$$\hat{A}_i(p_+) = \int du \, e^{-ip_+ X^+(u)} \partial_u f_i(u) \tag{4.32}$$

that create in-coming particles with a definite Kruskal-momentum p_+. These operators can be expressed as linear combinations of the energy eigenmodes $\alpha_i(\omega)$. For example, for $p_+ > 0$ we have

$$\widehat{A}_i(p_+) = -i \int d\omega\, e^{-\frac{\pi}{2}\omega}\Gamma(-i\omega)\left(\frac{p_+}{\lambda}\right)^{i\omega}\alpha_i(\omega).\tag{4.33}$$

This equation, which can also be read as the definition of $\widehat{A}_i(p_+)$, shows that these Kruskal modes are in fact somewhat singular operators, because they contain $\alpha_i(\omega)$ modes of arbitrarily high frequency. In the following we will mostly ignore this singularity, as it will not affect the main conclusions.

Using the operators $\widehat{A}_i(p^+)$ we can build a more or less realistic incoming state in which the matter is localized in a finite time interval $x_0^+ < x^+ < x_0^+ + \Delta x^+$ and carries a large total energy $E \pm \Delta E$. This *in*-state may be represented as a sum of eigenstates of the total Kruskal-momentum with eigenvalues concentrated around $P_+ = E/x_0^+$, and each of these eigenstate can be constructed by acting with a string of $\widehat{A}_i(p^+)$-operators on the vacuum. Thus a typical incoming state is a linear combination of states of the form

$$|\psi\rangle = \prod_n \widehat{A}_{i_n}(p_n^+)|0\rangle.\tag{4.34}$$

In a Heisenberg picture the *out*-state is given by this same expression, but to interpret this state physically we have to know how to write it in terms of the *out*-fields.

The idea is now to analyse this state by acting with the out-going modes and to try to commute this modes through the $\widehat{A}$-operators. For the outgoing modes we choose to work in a coordinate representation

$$f_i^{(out)}(x^-) = \int \frac{d\omega}{\omega}(\lambda x^-)^{i\omega}\beta_i(\omega)\tag{4.35}$$

where x^- is a c-number and $\beta_i(\omega)$ the operator (compare with (4.10)

$$\beta_i(\omega) = \int dv(-\lambda X^-(v))^{-i\omega}\partial_v f_i(v).\tag{4.36}$$

From the commutation relation (4.30) between the X^+ and X^--fields it is straightforward to derive the exchange properties of the *in* and *out* fields (see [13]). The details of this algebra depend on the particular boundary condition that is chosen in the strong coupling region, but there is one particular universal feature that is independent of this choice. Namely, for any choice of the parameter m, the two coordinate fields X^+ and X^- have the following commutator in the region $u_1 < u_1$

$$[X^+, X^-] = \frac{2\pi i}{\lambda^2}.\tag{4.37}$$

46

This shows that the *in*-fields $f^{(in)}$, since they depend X^+, are capable of shifting the argument X^- of the *out*-fields $f^{(out)}$. This is exactly the same gravitational effect we discussed before in section 3, and corresponds physically to the fact that the *in*-coming fields interact gravitationally with the *out*-fields even before they disappear into the black hole and/or reflect off the boundary. We would now like to study the physical effect of this interaction.

Since this interaction essentially takes place in the region $u_1 < u_2$, we will now simplify the discussion and for the moment work with the simplified commutation relation (4.37). We thus find the following exchange algebra of $\hat{A}_i(p_+)$ and $f_j^{(out)}(x^-)$:

$$f_j^{(out)}(x^-)\,\hat{A}_i(p_+) = \hat{A}_i(p_+)\,f_j^{(out)}(x^- - \frac{p_+}{\lambda^2}), \qquad (4.38)$$

We explicitly see that the *in*-mode shifts the *out*-fields by an amount proportional to the incoming Kruskal-momentum p_+. Note that this interaction is essentially the same as that derived in the *s*-wave reduced Einstein theory with $1/4M$ replaced by λ^{-2}.

If the coordinate x^- were a normal Minkowski coordinate, ranging from $-\infty$ to $+\infty$, a constant shift in x^- would have had no physical effect whatsoever: it could simply be absorbed by shifting the Minkowski vacuum. In our case, however, it is crucial that $x^\pm$ parametrizes only a Rindler wedge $\pm x^\pm > 0$ and that the vacuum of the f-fields is defined accordingly in terms of the Rindler type modes $\alpha(\omega)$ and $\beta(\omega)$. We will now show that the resulting distortion of the out-modes is responsible for the production of Hawking radiation.

To determine how the out-going vacuum state is affected by the coordinate shift (4.38), let us rewrite the exchange algebra in terms of the β-modes. One finds

$$\beta_j(\omega)\hat{A}_i(p_+) = \hat{A}_i(p_+) \int d\xi\, \mathrm{B}_{\omega\xi}(p_+)\beta_j(-\xi) \qquad (4.39)$$

with

$$\mathrm{B}_{\omega\xi}(p_+) = \left(\frac{p_+}{\lambda}\right)^{-i(\xi+\omega)} \frac{\Gamma(1-i\omega)\Gamma(i(\xi+\omega))}{\Gamma(1+i\xi)}. \qquad (4.40)$$

The linear combination of β-modes on the right-hand-side contains both creation- and annihilation- operators. Consistency of the algebra (4.39) further requires that these combinations again satisfy canonical commutation relations, so we see that exchanging a $\beta_j(\omega)$-oscillator with $\hat{A}_i(p_+)$ leads to a Bogoljubov-transformation. Note that the transformed modes occurring on the right-hand side of (4.39) do not form a complete basis of all β-modes, since they cover only the interval $x^- < -\lambda^{-2}p_+$. The exchange property (4.39) can therefore in general only be used in one direction.

Finally, we can act on the state $|\psi\rangle$ given in (4.34) with the β_j-modes and repeatedly use (4.39) until we can act on the vacuum. These manipulations are of course the direct quantum counterpart of the standard semi-classical calculation. The repeated use of the exchange-algebra describes the propagation of the out-going particles through the in-falling matter, while taking into account the gravitational interaction between the two. Now it is clear from (4.38) that in this procedure only the total momentum P_+ plays a role, so the exchange relation between the β-modes and the product $\prod_a \widehat{A}_{i_a}(p_a)$ is again of the same form (4.39). In this way we find that, just as in the semi-classical calculation, the asymptotic out-state is given by the Bogoljubov transform (4.39)-(4.40) of the vacuum. At this point we can simply refer to the standard analysis [1] to conclude that, in this approximation, the *out*-state describes a constant out-going flux of thermal Hawking radiation.

It is important to note that above we have of course dealt with an idealized situation. In the first place we assumed that $|\psi\rangle$ is an exact eigen state of P_+, and as noted before, such states in fact do not exist. If instead we would consider a state for which the in-coming energy is bounded, we should find that the resulting black hole radiate for only a finite amount of time. This can in fact be very directly seen in the above algebraic derivation. The exchange algebra we derived for dilaton gravity is namely *energy preserving* in the precise sense that the total energy carried by the *in*-modes plus that carried by the *out*-modes is the same on both sides of the equation. This tells us that every time a out-going particle is emitted, the in-falling matter loses a proportional amount of energy, while it also gets shifted. This process can continue for a long time, until most of the in-coming energy has been radiated away. We believe that by this time new (strong coupling) effects will start to take place, while also the direct reflection off the $r = 0$ boundary will again become important. These effects will produce corrections to the Hawking spectrum, that in principle should be capable of transporting all information to the out-side world. The detailed physics of these processes depends on the boundary conditions provided by the strong coupling physics. A specific proposal for these boundary conditions for dilaton gravity has been worked out in detail in [13].

5. Concluding Remarks.

To summarize, we have shown in this note that gravitational interactions lead to non-trivial commutators between the observables that measure the Hawking radiation and the matter falling into the black hole. To derive this commutator algebra (3.8) we did not need to make any new assumptions other than those already made in Hawking's original derivation of the black hole evaporation effect [1]. A very important consequence of this result is that it calls into question the standard assertion that the Hilbert space of the scalar-fields on a Cauchy surface as drawn in figure 1 decomposes into a simple tensor product of a Hilbert space inside the black hole and one out-side. Instead, our result supports the physical picture that there is a certain *complementarity* between the physical realities as seen by an asymptotic observer and by an in-falling observer. Indeed, for the latter, the in-falling matter will simply propagate freely without any perturbation, but he (or she) will not see the out-going radiation. For the asymptotic observer, on the other hand, the Hawking radiation is physically real, while the in-falling matter will appear to evaporate completely before it falls into the hole. With our result, the apparent discrepancy between these observations can be explained by the fact that the two observers use different *non-commuting* sets of observables to assign physical meaning to the same quantum state. It is clear that this insight will have important repercussions for the information paradox (see also section 3.5).

As a further comment, we note that our derivation of the commutator between the in-falling and out-going modes can of course be generalized to other observers, by properly taking into account the precise *operator* relation between the physical coordinates used by these observers. In this way one will only find strong coupling effects when the relation between these coordinate systems becomes very singular, such as in the case worked out in section 3. Our reasoning does therefore not single out the horizon as a singular part of space-time. It are only the *observables* that the asymptotic observer uses to distinguish different physical situations that become singular, since his (or her) coordinate system degenerates there. The coordinate system of the in-falling observer, on the other hand, is regular, so she (or he) will see nothing special happening near the horizon.

Finally, we expect that our result may also be of importance in relation with the entropy of black holes. In [15] it was noted that a naive free field calculation of the one-loop correction to the black hole entropy gives an infinite answer. This infinity arises due to the diverging contribution of states that are packed arbitrarily close to the horizon. Our result suggests a possible remedy of this problem, because it shows that in the coordinate system appropriate for this calculation the in- and out-going fields no longer commute when they come very close to the horizon. It is tempting to speculate that this will effectively reduce the number of allowed states, and thereby eliminate the diverging contribution in the entropy calculation. We leave this problem for future study.

Acknowledgements.

We would like to thank U. Danielson and C. Stephens for stimulating discussions. The research of K.S. is supported by DOE-Grant DE-AC02-76ER-03072, and that of H.V. is supported by NSF Grant PHY90-21984 and the David and Lucille Packard Foundation. The research of E.V. is partly supported by an Alfred P. Sloan Fellowship, and by a Fellowship of the Royal Dutch Academy of Sciences.

References

[1] S.W. Hawking, *Commun. Math. Phys.* **43** 199 (1975) 199.

[2] G. 't Hooft, *Nucl. Phys.* **B335** (1990) 138, and *"Unitarity of the Black Hole S-Matrix,"* Utrecht preprint THU-93/04.

[3] T. Jacobson, *"Black Hole Radiation in the Presence of a Short-Distance Cutoff,"* Maryland preprint UMDGR-93-32.

[4] C. R. Stephens, G. 't Hooft and B. F. Whiting, *"Black Hole Evaporation without Information Loss,"* Utrecht preprint THU-93/20.

[5] See e.g. N. Birrell and P. Davies, *Quantum Fields in Curved Space* (Cambridge, 1982) and references therein.

[6] E. Verlinde and H.Verlinde, unpublished.

[7] L. Susskind, L. Thorlacius and J. Uglum, *Phys. Rev.* **D48** (1993), 3743; L. Susskind and L. Thorlacius, *"Gedanken Experiments involving Black Holes,"* Stanford preprint SU-ITP-93-19.

[8] D. Page, *Phys. Rev. Lett.* **44** (1988) 301; R. M. Wald, *Phys. Rev.* **D 21** (1980) 2742; A. Strominger, *Phys. Rev.* **D 48** (1993), 5769.

[9] C. Callan, S. Giddings, J. Harvey, and A. Strominger, *Phys. Rev.* **D45** (1992) 1005.

[10] H. Verlinde, *"Black Holes and Strings in Two Dimensions"*, in the proceeding of the Sixth Marcel Grossman Meeting, World Scientific (1992).

[11] J. Russo, L. Susskind and L. Thorlacius, *Phys. Rev* **D 47** (1993) 533.

[12] E. Verlinde and H.Verlinde, *Nucl. Phys.* **B 406** (1993) 43

[13] K. Schoutens, E. Verlinde and H. Verlinde, *Phys. Rev.* **D 48**(1993) 2690.

[14] T-D. Chung and H. Verlinde, *"Dynamical Moving Mirrors and Black Holes,"* Princeton preprint, PUPT-1430.

[15] G. 't Hooft, *Nucl. Phys.* **B 256** (1985), 727.

2. Conformal Field Theory and Integrable Systems

HANDLE OPERATORS IN R.C.F.T.

M. CRESCIMANNO

Center for Theoretical Physics, M.I.T., 77 Mass. Ave.
Cambridge, MA.   02139-4307

ABSTRACT

For the series associated to a group or coset R.C.F.T. there is a simple universal form for the inverse of the handle operator in the ring of fusions. These formulae may be easily understood from the quantization of the associated Chern-Simons theory.

1. Introduction

Among the data that defines a rational conformal field theory (RCFT), the fusion algebra plays a central role[1]. It restricts the form of the modular representations and fixes a convenient basis for discussing the dimension of the space of blocks on higher genus ($g > 0$). Let roman letters label the integrable representations of a group or coset RCFT and $\mathcal{O}_j$ be the operators of fusion

$$\mathcal{O}_i \times \mathcal{O}_j = \mathrm{N}_{ij}^k \mathcal{O}_k \ . \tag{1.1}$$

As described in Ref.[2,3], the dimension of the space of conformal blocks in genus $g > 0$ is given by

$$\dim \mathcal{H}^g = Tr(K^{g-1}) \ , \tag{1.2}$$

where K is a matrix in the space of integrable representations and is a particular linear combination of the operators $\mathcal{O}_j$

$$K = \sum_l Tr(\mathcal{O}_l) \ \mathcal{O}_l. \tag{1.3}$$

This is a very general characterization of K. For a gaussian model (i.e. one composed of any number of non-interacting massless bosons) K is the unit matrix scaled by the number of integrable representations (a general property of K^{-1} is $Tr(K^{-1}) = 1$.) For the case of group or coset RCFT, K has no obvious classical group-theoretic interpretation. That is, for example, given G_k and forming K of Eq.(1.3) for this theory it is clear that K depends on the level k and on the group theory of G in a complicated way.

The purpose of this note is to demonstrate that the inverse matrix, K^{-1} does admit a simple interpretation in terms of classical group-theoretic ideas. Indeed, we show that K^{-1} for a given group or coset RCFT is a ratio

$$K^{-1} = w^2/vol(k) \ , \tag{1.4}$$

where w^2 is a fixed linear combination of representations (depending only on the group theory of G) and $vol(k)$ is the naive volume of the moduli space of flat G-connections over the torus and is a simple combinatorial factor that depends on the

54

level k. For example, for $SU(2)_k$ one finds

$$K^{-1} = \frac{1}{2(k+2)}[3\mathcal{O}_1 - \mathcal{O}_3] \tag{1.5}$$

for all level k. Here the subscripts refer to the dimensions of the representations.

2. Chern-Simons and the Handle Operator

Perhaps the simplest and most revealing way of understanding formulae of this type is via the quantization of Chern-Simons (CS) theory[4]. Here, we sketch derivation of Eq.(1.4). For more details see Ref.[5,6,7]. Recall that the Hilbert space of CS theory is naturally isomorphic to the space of conformal blocks of a CFT, and that one convenient way to construct the Hilbert space of CS theory is by quantizing the space, $\mathcal{M}$ of flat gauge connections over a given surface. The moduli space $\mathcal{M}$ is not quite a manifold; in general it contains a set of measure zero where there are cusps, self-intersections, disconnected points, and other pathologies. Fortunately, we will only concern ourselves here with the moduli space of flat gauge connections over the torus and this, for the classical Lie groups, has only cusp-type singularities. Thus, for that case it is enough to study the quantization of the covering space $\hat{\mathcal{M}}$ of the moduli space $\mathcal{M} = \hat{\mathcal{M}}/W$, where the group W is essentially a realization of the Weyl group. As usual, the quantization on $\hat{\mathcal{M}}$ proceeds via the symplectic form $\Omega = (k+c)Tr$ where c is the quadratic casimir of the adjoint representation and Tr is the matrix through which the inner products in the Cartan subalgebra are taken (for the simply laced Lie algebras it is just the Cartan matrix; for the non-simply laced case it is an appropriate symmetrization of the Cartan matrix.) Call $\hat{\mathcal{H}}$ the quantum Hilbert space that results from the quantization of covering space $\hat{\mathcal{M}}$. The Hilbert space $\hat{\mathcal{H}}$ is isomorphic to the space of conformal blocks of a gaussian model , and admits an action of the covering group W. Call that group action $W_{\mathcal{H}}$. As discussed in the literature, the Hilbert space $\mathcal{H}$ of the CS theory is found by modding $\hat{\mathcal{H}}$ by this group action

$$
\begin{array}{ccc}
 & \Omega & \\
\hat{\mathcal{M}} & \longrightarrow & \hat{\mathcal{H}} \\
W \downarrow & & \downarrow W_{\mathcal{H}} \\
\mathcal{M} & & \mathcal{H}
\end{array}
$$

The "handle-squashing" operator K^{-1} is then just the push-forward of the handle-squashing operator in the $\hat{\mathcal{H}}$ theory. Since the $\hat{\mathcal{H}}$ theory is gaussian, we know that on $\hat{\mathcal{H}}$ the K^{-1} is the unit matrix divided by $\dim\hat{\mathcal{H}}$. For example, in G_k CS theory the K^{-1} of the Hilbert space associated to the cover of moduli space over the torus is

$$K_{\hat{\mathcal{H}}}^{-1} = \left|\frac{\Lambda_w}{(k+c)\Lambda_r}\right|^{-1} \mathbf{1} \, , \tag{2.1}$$

where Λ_w is the co-root lattice and Λ_r is the root lattice of G.

Now, just as all the points in $\mathcal{M}$ are invariant under the action of W, all the states of $\mathcal{H}$ form a covariant multiplets under the action of $W_{\mathcal{H}}$. As described in the literature, gauge invariance of the entire partition function requires that the the states in $\psi \in \mathcal{H}$ are alternating states, that is, they satisfy $w\psi = det(w)\psi \quad \forall w \in W_{\mathcal{H}}$. There is a fundamental alternating operator Γ which "projects" all the states in $\hat{\mathcal{H}}$ onto the states in $\mathcal{H}$

$$\Gamma = \frac{1}{\sqrt{|W|}} \sum_{w \in W} det(w) \Pi_i B_i^{(\omega_i, w\rho)} \tag{2.2}$$

where $\rho = \frac{1}{2}\Sigma_{\alpha>0}\alpha$ and the inner product in the exponent of the raising operators B_i are in terms of the co-root basis, $\{\omega_i\}$. $|W|$ is the order of the Weyl group. To compute modular representations, fusions, etc. in the ring of operators on $\mathcal{H}$ one simply conjugates the corresponding quantity in the $\hat{\mathcal{H}}$ by the operator Γ. Thus, by pushing-forward Eq.(2.1) we find the handle-squashing operator for the case of G_k to be $K^{-1} = \frac{|W|}{dim\hat{\mathcal{H}}}\Gamma^\dagger \mathbf{1}\Gamma$ which we may write (using the consequence of symmetry $\Gamma^\dagger = -\Gamma$) as

$$K_G^{-1} = -\frac{1}{vol(G_k)}\Gamma^2 \;, \tag{2.3}$$

where

$$vol(G_k) = \frac{\left|\frac{\Lambda_w}{(k+c)\Lambda_r}\right|}{|W|} \tag{2.4}$$

may be thought of as the naive symplectic volume of the $\mathcal{M}$ and Γ^2 is, of course, invariant under the action of the Weyl group and, for a given G, is a fixed linear combination of representations and is thus independent of the level.

For cosets "without fixed points" there is also a simple formula for the handle-squashing operator K^{-1} of the form Eq.(2.3). For a simple coset G_k/H_k, where the bonus currents act without fixed points

$$K_{G/H}^{-1} = |Z|^2 (K_G^{-1} \otimes K_H^{-1})|_{G/H} \tag{2.5}$$

where $Z = Z_G \cap H$ is the common center and $|Z|$ is the number of elements it has. Since the center action may be thought of as acting via $Z \times Z$ as a further identification in $\hat{\mathcal{M}}_G \times \hat{\mathcal{M}}_H$ and since it acts freely, we see that collecting the factors we may represent $K_{G/H}^{-1}$ again as a ratio of a particular group theoretic part and a naive symplectic volume of the coset's moduli space of flat connections $vol(G_k/H_k) = vol(G_k)vol(H_k)/|Z|^2$.

3. Examples and Summary

Using the above ideas we now list a few examples of these explicit handle operator formulae

$$K_{U(1)}^{-1} = \frac{1}{2k}\mathcal{O}_1$$

$$K_{SU(2)}^{-1} = \frac{1}{2(k+2)}(3\mathcal{O}_1 - \mathcal{O}_3)$$

$$K_{SU(3)}^{-1} = \frac{1}{3(k+3)^2}(9\mathcal{O}_1 - 6\mathcal{O}_8 + 3[\mathcal{O}_{10} + \mathcal{O}_{\bar{10}}] - \mathcal{O}_{27})$$

$$K^{-1}_{SO(5)} = \frac{1}{4(k+3)^2}(22\mathcal{O}_{(0,0)} - 4\mathcal{O}_{(1,0)} - 7\mathcal{O}_{(0,2)} - 2\mathcal{O}_{(2,0)} + 6\mathcal{O}_{(1,2)} - 3\mathcal{O}_{(3,0)} - 3\mathcal{O}_{(0,4)} + \mathcal{O}_{(2,2)})$$

$$K^{-1}_{SU(2)/U(1)} = \frac{1}{k(k+2)}(3\mathcal{O}_1 \otimes \mathcal{O}_1 - \mathcal{O}_3 \otimes \mathcal{O}_1)$$

Note that for low level k the operators in the above expressions are to be understood as being related to the integral representations modulo the action of the 'Weyl reflection about the k-line'[5].

We have described in general for group and simple coset type RCFT's how formulae for K^{-1} of this type arise. The formulae have a natural interpretation in CS theory as a ratio of a group-theoretic part and a naive volume of moduli space.

There is an intriguing mathematical connection between these ideas in the RCFT context and $N = 2$ theories. Also it seems possible to explore properties of the space of blocks in higher genus with some of these techniques. Finally, these formulae may be useful for better understanding the classical ($k \to \infty$) limit of CS theory.

4. Acknowledgements

We wish to thank S. Axelrod, K. Bardakci, S. A. Hotes, H. J. Schnitzer and I. M. Singer for conversations, and the organizers of this conference. This work was supported in part by funds provided by the U.S. Department of Energy (D.O.E.) under contract #DE-AC02-76ER03069, and by the Division of Applied Mathematics of the U.S. Department of Energy under contract #DE-FG02-88ER25066.

5. References

1. E. Verlinde, *Nucl. Phys.* **B300** (1988) 360.
2. H. Verlinde and E. Verlinde, "Conformal Field Theory and Geometric Quantizatio," published in **Trieste Superstrings** (1989), 422.
3. R. Bott, *Surveys in Diff. Geom.* **1** (1991) 1.
4. E. Witten, *Commun. Math. Phys.* **121** (1989) 351.
5. M. Crescimanno and S. A. Hotes, *Nucl. Phys.* **B372** (1992) 683.
6. M. Crescimanno, *Nucl. Phys.* **B393** (1993) 361.
7. M. Crescimanno, *Mod. Phys. Lett.* **A8** (1993) 1877.

The WZWN model at two loops and induced chiral actions.

B. de Wit
Institute for Theoretical Physics
Utrecht University
Princetonplein 5, 3508 TA Utrecht
The Netherlands

M.T. Grisaru
Department of Physics
Brandeis University
Waltham, MA 02254, USA

P. van Nieuwenhuizen
Institute for Theoretical Physics
State University of New York at Stony Brook
Stony Brook, NY 11794-3840, USA

Abstract

In these proceedings, we report on some work we have done concerning a field theoretic approach to a model which is usually treated with the methods of conformal field theory. This is the WZWN model [1, 2, 3] in which we compute the 2-loop 2-point function (and the relevant 1-loop n-point functions). Since the WZWN model turns out to be the induced action for the coupling of chiral matter to a chiral gauge field we also review the issue of loop calculations in chiral nonlocal field theories. A detailed version of our work has appeared in *Nuclear Physics B* [4]. Here we shall provide some background material, not all of which appears in the published article.

1 Perturbative field theory in the WZNW model

As is well known, a certain class of two-dimensional nonlocal actions (of which the Polyakov induced gravity action is the prototype) arises when one couples an external gauge field to some classically gauge-invariant matter system and one integrates out the latter. The gauge-field induced actions arise because the quantization of the matter-systems generates anomalies. An alternative description of the resulting gauge-field dynamics is by means of a WZWN action.

Unlike the theory based on nonlocal induced actions the classically equivalent WZNW model is a renormalizable quantum field theory. It is not the first time that this model has been treated by perturbative methods. The calculation of ultraviolet divergences, with the aim of determining critical points and anomalous dimensions, was performed by Witten [3] and Bos [5]. The latter used dimensional regularization and discussed the proper treatment of the epsilon tensor and evanescent counter-terms. In addition there is the work of Leutwyler and Shifman [6], who recovered the level shift $k \to k + \tilde{h}$ that one encounters in conformal field theory [7]. We should stress that calculations of this type are not intended as an alternative for conformal field theory. They do test the extent to which conventional perturbative field theory can reproduce the results of conformal field theory. On the other hand, there are aspects and features that at present can only be studied in perturbation theory.

Our calculations are based on a covariant background field action, which can be conveniently treated by dimensional regularization, after which the effective action of the WZNW model is extracted by a chiral trucation. The approach is rather insensitive to the short-distance singularities of the theory, but long-distance singularities appear, related to the fact that the quantum fields are not primary fields (in spite of the fact that we calculate one-particle irreducible correlation functions of primary operators, namely, of currents). We treat these infrared singularities by using the so-called R^* scheme [8], which is extremely convenient. The divergences then induce logarithmic terms associated with the arbitrariness of the infrared subtractions, and those terms are present in the final answers.

The covariant background field action, which is related to the gauged WZNW action, reads

$$\mathcal{L} = -i\epsilon^{\mu\nu}\, \vec{\pi} \cdot (\partial_\mu \vec{J}_\nu - \partial_\nu \vec{J}_\mu - \vec{J}_\mu \times \vec{J}_\nu)$$
$$-\tfrac{1}{2}(D_\mu\vec{\pi})^2 - \tfrac{1}{6} i\,\epsilon^{\mu\nu}\, \vec{\pi} \cdot (D_\mu\vec{\pi} \times D_\nu\vec{\pi})$$
$$+\tfrac{1}{24}(D_\mu\vec{\pi} \times \vec{\pi})^2 + \cdots , \tag{1.1}$$

where $D_\mu\vec{\pi} = \partial_\mu\vec{\pi} - \vec{J}_\mu \times \vec{\pi}$. Here $\vec{\pi}$ are the quantum fields, whereas J_μ contains the background fields according to $J_+ = \partial_+ g\, g^{-1}$ and $J_- = \partial_- \bar{g}\, \bar{g}^{-1}$, see section 3. The chiral truncation in the final results consists of setting $J_- = 0$.

At one-loop we determined the two- and three-point functions. The results indicate that the effective action (i.e. the one-particle irreducible graphs) constitutes a WZNW action with level $k + \tilde{h}$, up to infrared-induced logarithms. This can for instance be verified from the result for the (covariant) two-point function,

$$\Pi^{ab}_{\mu\nu}(p) = \frac{\tilde{h}}{4\pi}\, g^{ab} \left(\delta_{\mu\nu} - \tfrac{p_\mu p_\nu}{p^2}\right)\left[2 - \ln \tfrac{p^2}{\mu^2}\right]. \tag{1.2}$$

As we were not aware of any direct evidence according to which the coupling constant of the *effective* action should be quantized, thus forcing the level shift to be a pure one-loop effect, we computed the two-loop contribution to the two-point function [4]. As it turns out the calculation is independent of the precise prescription for the epsilon tensors, provided one includes possible evanescent counterterms. The result of the calculation is

$$\Pi^{ab}_{\mu\nu}(p) = \frac{\tilde{h}}{4\pi}\, g^{ab} \left(\delta_{\mu\nu} - \tfrac{p_\mu p_\nu}{p^2}\right)\left[\tfrac{1}{12}\frac{\tilde{h}}{k} \ln^2 \tfrac{p^2}{\mu^2}\right], \tag{1.3}$$

Up to infrared-induced logarithms, the two-loop contribution thus vanishes. We regard this as evidence that the effective action receives only one-loop corrections leading to a level shift $k \to k + \tilde{h}$.

The precise treatment of the infrared divergences remains problematic in this perturbative approach. We would hope that these divergences cancel when summing to all orders of perturbation theory, something that is beyond our present ability to check. It is difficult to see how one could use renormalization group and resummation methods to improve matters here. Moreover there seem no alternative methods available for evaluating the *finite* terms in the effective action. The ultraviolet divergences, on the other hand, are well controlled in our calculational scheme as a result of gauge invariance and global power counting.

In section 2 we review properties of the classical WZWN model. In section 3 we discuss the chiral induced action for spin 1 gauge fields. It is classically equivalent to the WZWN model as we discuss at the end of section 3, but it is formulated in terms of chiral gauge fields A_+ and it is nonlocal. In section 4 we discuss loop calculations in this and other nonlocal field theories and discuss various open problems. In section 5 we construct the covariant gauge-invariant background-field dependent action of (1.1) which we use for the loop calculations in the WZWN model. In section 6 we discuss the results for the 1-loop and 2-loop calculations based on this action.

2   The classical WZNW action.

We start by summarizing a number of well-known features of the WZNW action and establish our notation. In section 3 we derive its relation with the action induced by chiral matter coupled to gauge fields associated with a group G.

The WZNW action is constructed from two terms, written as an integral over a closed two-dimensional surface ∂B and as an integral over the enclosed volume B, respectively,

$$
\begin{aligned}
I[g] &= \frac{1}{16\pi\chi} \int_{\partial B} d^2x \, Tr \, \partial_\mu g \, g^{-1} \, \partial^\mu g \, g^{-1} \,, \\
\Gamma[g] &= \frac{1}{24\pi\chi} \int_B d^3x \, \epsilon^{\mu\nu\rho} \, Tr \, \partial_\mu g \, g^{-1} \, \partial_\nu g \, g^{-1} \, \partial_\rho g \, g^{-1} \,,
\end{aligned}
\tag{2.1}
$$

where the fields $g(x)$ take their values in the group G. For a compact group the Euclidean action $I[g]$ is thus negative. The normalization of $\Gamma[g]$ is such that, for a compact group, it is well defined modulo an integer times 2π. We give an elementary proof in appendix A. Furthermore we note that $I[g^{-1}] = I[g]$ and $\Gamma[g^{-1}] = -\Gamma[g]$.

We wrote the group elements and the corresponding trace in an irreducible representation of G with generators T_a. Structure constants are defined by $[T_a, T_b] = f_{ab}{}^c T_c$. The index χ of this representation is given by $Tr T_a T_b = -\chi\, g_{ab}$, where g_{ab} denotes the Cartan-Killing metric, defined by $f_{ac}{}^d f_{bd}{}^c = -\tilde{h}\, g_{ab}$ with $\tilde{h}$ the dual Coxeter number. The dual Coxeter number $\tilde{h}$ is given by $\tilde{h}^{-1} = \theta_m M^{mn}\theta_n$, where $\vec{\theta}$ is the longest root and M^{mn} is the matrix inverse of $M_{mn} \equiv \sum_{\text{all roots}} a_m a_n$. Clearly M_{mn} is proportional to the Killing metric $f_{mp}{}^q f_{nq}{}^p$ when the indices m and n lie in the Cartan subalgebra. For SU(2), with roots $\pm a$, one finds $M = 2a^2$ and $\theta^2 = \frac{1}{2}$, hence $\tilde{h} = 2$. For $SU(n)$ we have $\tilde{h} = n$ and, for the fundamental representation $\chi = \frac{1}{2}$; for $SO(n)$ with $n > 4$, these numbers are $n - 2$ and 1, respectively. In the adjoint representation we always have $\chi = \tilde{h}$. We will repeatedly make use of the group theory result $f_{ad}{}^e f_{be}{}^f f_{cf}{}^d = \frac{1}{2}\tilde{h} f_{abc}$, which follows from writing the Jacobi identities in terms of matrices $(M_a)_b{}^c = f_{ba}{}^c$ as $[M_a, M_b] = f_{ab}{}^c M_c$, and then using that $Tr M_a M_b M_c = \frac{1}{2} Tr M_a[M_b, M_c]] = -\frac{1}{2}\tilde{h} f_{abc}$.

A central role is played by the following multiplication rules [10]

$$
\begin{aligned}
I[gh] &= I[g] + I[h] + \frac{1}{8\pi\chi} \int_{\partial B} d^2x\, Tr\, \partial_\mu h\, h^{-1}\, g^{-1}\partial^\mu g\,, \\
\Gamma[gh] &= \Gamma[g] + \Gamma[h] + \frac{1}{8\pi\chi} \int_{\partial B} d^2x\, \epsilon^{\mu\nu}\, Tr\, \partial_\mu h\, h^{-1}\, g^{-1}\partial_\nu g\,,
\end{aligned}
\qquad (2.2)
$$

Parametrizing $g = \exp\pi$ with $\pi = \pi^a T_a$, and defining $g(t) = \exp t\pi$, we find from $I[g] = \int_0^1 dt \frac{d}{dt} I[g(t)]$, upon using $\frac{d}{dt}[\{\partial_\mu g(t)\}g^{-1}(t)] = \exp t\pi(\partial_\mu\pi)\exp -t\pi$, that the integrand of $I[g]$ is proportional to $\int_0^1 dt(\partial_\mu\pi)(\exp -t\pi)\partial^\mu(\exp t\pi)$. Similarly, the integrand of $\Gamma[g]$ is proportional to $\partial_\mu\pi(\partial_\nu \exp -t\pi)(\partial_\rho \exp t\pi)$, which, being a total derivative, allows one to cast $\Gamma[g]$ as a two-dimensional integral. The relation between the definition of $\epsilon^{\mu\nu\rho}$ and $\epsilon^{\mu\nu}$ is given by $\int d^3x \epsilon^{\mu\nu\rho}\frac{\partial}{\partial x^\rho} F(x) = \int d^2x \epsilon^{\mu\nu} F(x)$.

62

In this way, both $I[g]$ and $\Gamma[g]$ become proportional to $\int_0^1 \partial_\mu \pi (\exp -t\pi)(\partial_\nu \exp t\pi)dt$, and one finds [see the first reference in [5], eq (8)].

$$
\begin{aligned}
I[g] &= -\frac{1}{8\pi} \int d^2x \sum_{n=1}^{\infty} \frac{1}{(2n)!} \, \partial_\mu \vec{\pi} \cdot (\vec{\pi} \times (\vec{\pi} \times \cdots \partial^\mu \vec{\pi})) \\
&= -\frac{1}{16\pi} \int d^2x \left\{ (\partial_\mu \vec{\pi})^2 - \tfrac{1}{12}(\vec{\pi} \times \partial_\mu \vec{\pi})^2 + \mathcal{O}(\pi^6) \right\}, \\
\Gamma[g] &= \frac{1}{8\pi} \int d^2x \sum_{n=1}^{\infty} \frac{1}{(2n+1)!} \, \varepsilon^{\mu\nu} \partial_\mu \vec{\pi} \cdot (\vec{\pi} \times (\vec{\pi} \times \cdots \partial_\nu \vec{\pi})) \\
&= -\frac{1}{48\pi} \int d^2x \left\{ \varepsilon^{\mu\nu} \vec{\pi} \cdot (\partial_\mu \vec{\pi} \times \partial_\nu \vec{\pi}) + \mathcal{O}(\pi^5) \right\},
\end{aligned}
\tag{2.3}
$$

where we used the notation

$$
\vec{x} \cdot \vec{y} \equiv g_{ab}\, x^a y^b, \qquad (\vec{x} \times \vec{y})^a \equiv f_{bc}{}^a\, x^b y^c. \tag{2.4}
$$

Note that the sum in $I[g]$ extends over even powers in $\vec{\pi}$, while $\Gamma[g]$ contains the terms odd in π. Observe also that $I[g]$ and $\Gamma[g]$ are both real.

For the Euclidean WZNW action we choose the metric $\eta^{\mu\nu} = \mathrm{diag}\,(+,+)$, the Levi-Cività symbol with $\varepsilon^{12} = 1$ and define

$$
S_W^{\pm}[g] = k\,(I[g] \pm i\Gamma[g]). \tag{2.5}
$$

Here k is an integer coupling constant such that the action is defined modulo an integer times $2\pi i$ and $\exp S_W^{\pm}$ is unambiguous. Introducing complex coordinates,

$$
x^{\pm} = \frac{\lambda}{\sqrt{2}}\,(x^1 \pm ix^2), \qquad \partial_{\pm} = \frac{1}{\lambda\sqrt{2}}\,(\partial_1 \mp i\partial_2) \tag{2.6}
$$

(so that $\partial_\mu \partial^\mu = 2\lambda^2\, \partial_+ \partial_-$), where λ is an arbitrary normalization factor to facilitate comparison with corresponding expressions in the literature*, the action becomes

$$
\begin{aligned}
S_W^{\pm}[g] = {}& \frac{k\lambda^2}{8\pi\chi} \int_{\partial B} d^2x\, Tr\, \partial_+ gg^{-1} \partial_- gg^{-1} \\
&+ \frac{k\lambda^2}{8\pi\chi} \int_B d^3x\, Tr\, \partial_3 gg^{-1} (\partial_+ gg^{-1} \partial_- gg^{-1} - \partial_- gg^{-1} \partial_+ gg^{-1})
\end{aligned}
\tag{2.7}
$$

---

*In Minkowski space one defines

$$
x^{\pm} = \frac{\lambda}{\sqrt{2}}\,(x \pm t), \qquad \partial_{\pm} = \frac{1}{\lambda\sqrt{2}}\,(\partial_x \pm \partial_t), \qquad S_W^{\pm}[g] = k\,(I[g] \pm \Gamma[g]).
$$

where $d^3x = dx^1dx^2dx^3$. It satisfies the Polyakov-Wiegmann formula [10]

$$S_W^{\pm}[gh] = S_W^{\pm}[g] + S_W^{\pm}[h] + \frac{k\,\lambda^2}{4\pi\chi}\int d^2x\; Tr\;\partial_{\pm}h\,h^{-1}\,g^{-1}\partial_{\mp}g\,. \tag{2.8}$$

Combining (2.3) and (2.6) gives

$$S_W^{\pm}[g] = \frac{k\,\lambda^2}{4\pi}\int d^2x\;\left\{-\tfrac{1}{2}\partial_+\vec{\pi}\cdot\partial_-\vec{\pi}\mp\tfrac{1}{6}\vec{\pi}\cdot(\partial_+\vec{\pi}\times\partial_-\vec{\pi})+\tfrac{1}{24}(\vec{\pi}\times\partial_+\vec{\pi})\cdot(\vec{\pi}\times\partial_-\vec{\pi})+\mathcal{O}(\pi^5)\right\}. \tag{2.9}$$

Let us define $A_+ = \partial_+g\,g^{-1} = (A_1 - iA_2)/(\lambda\sqrt{2})$. Under *left* multiplication of g with an x-dependent element of the group near the identity, A_+ transforms as

$$\delta\vec{A}_+ = \partial_+\vec{\eta} + \vec{\eta}\times\vec{A}_+\,, \tag{2.10}$$

where $\eta^aT_a = \delta g\,g^{-1}$. Using (2.8) shows at once that the action changes under this variation as follows

$$\begin{aligned}
\delta S_W^+[g] &= -\frac{k\,\lambda^2}{4\pi\chi}\int d^2x\; Tr\;\delta g\,g^{-1}\partial_-(\partial_+g\,g^{-1}) \\
&= \frac{k\,\lambda^2}{4\pi}\int d^2x\;\vec{\eta}(x)\cdot\partial_-\vec{A}_+(x)\,.
\end{aligned} \tag{2.11}$$

The first equality shows the well-known result that A_+ is holomorphic by virtue of the field equations (applying the same procedure with respect to right multiplication yields the conjugate result that $g^{-1}\partial_-g$ is anti-holomorphic), while the second equality takes the form of an anomalous variation, as we discuss below. Corresponding results can be obtained for $S_W^-[g]$ and $A_- = \partial_-g\,g^{-1}$.

3   The induced action.

The WZWN model is the induced action which one obtains if one couples a chiral matter system to a chiral gauge field, and integrates out the matter. If the matter couples to the gauge field as $\mathcal{L}(int) = -\frac{1}{\pi\chi}Tr\,J(x)A_+(x)$ (with $x^{\pm}$ as defined in (2.6) but with $\lambda = \sqrt{2}$) and the current satisfies a Kac-Moody algebra

$$J_a(x)J_b(y) = \frac{-\tfrac{1}{2}kg_{ab}}{(x^- - y^-)^2} + \frac{f_{ab}{}^c J_c(y)}{x^- - y^-} + \dots \tag{3.1}$$

one can evaluate the induced action

$$\exp S_{ind}{}^+[A_+] = <0|\exp -\frac{1}{\pi\chi}\int d^2x\, Tr\, J(x)A_+(x)|0> \tag{3.2}$$

by using that the vacuum expectation values of normal-ordered products (the regular terms in the operator-product expansion) vanish. Expanding the exponent, and using $<0|J(x)|0> = 0$, one finds terms with 2,3, ... A-fields. These terms can be computed by using the following identity

$$\frac{1}{(x^- - y^-)^n} = \pi \frac{(-)^{n-1}}{(n-1)!}\frac{(\partial_{x^-})^{n-1}}{\partial_{x^+}}\delta^2(x-y), \quad x^\pm = x^1 \pm ix^2 \tag{3.3}$$

This identity can be proved by integrating y over the plane minus a small disk around x. Multiplying both sides with ∂_{x^+}, one obtains on the left-hand side the derivative of the theta function $\theta[x^- - y^-)(x^+ - y^+) - \epsilon^2]$ which is proportional to the delta function.

The kinetic term for the A_+ field is contained in

$$\frac{1}{2}\frac{1}{\pi^2}\int d^2x\, A_+{}^a(x)\int d^2y\, A_+{}^b(y)\frac{-\frac{1}{2}kg_{ab}}{(x^- - y^-)^2} \tag{3.4}$$

and using the identity for $n = 2$, $(x^- - y^-)^2 = -\pi\frac{\partial_{x^-}}{\partial_{x^+}}\delta^2(x-y)$, one obtains

$$S_{ind}[A_+] = \frac{k}{4\pi}\int A_+{}^a\frac{\partial_-}{\partial_+}A_{+a}d^2x + \ldots \tag{3.5}$$

The term with three A_+ fields is due to the correlation function of three currents

$$<J_a(x)J_b(y)J_c(z)> \sim f_{abc}\frac{1}{x^- - y^-}\frac{1}{(y^- - z^-)^2} + f_{acb}\frac{1}{x^- - z^-}\frac{1}{(y^- - z^-)^2} \tag{3.6}$$

Using the identity in (3.3), we find a result proportional to

$$\int d^2x\, d^2y\, d^2z\, f_{abc}\left\{\left[(\partial_+^x)^{-1}\delta^2(x-y)\right]\left[\partial_-^y/\partial_+^y\delta^2(y-z)\right] - y \leftrightarrow z\right\}A^a(x)A^b(y)A^c(z) \tag{3.7}$$

One may partially integrate with the inverses of derivatives, although the Leibniz rule does not hold: $\partial_+{}^{-1}(fg) \neq (\partial_+^{-1}f)g + f(\partial_+^{-1})g$. One can, however, still manipulate the $\partial_\pm^{-1}$ derivatives as in the Leibniz rule by a trick: rewrite $(\partial_+^{-1}f)g =$

$(\partial_+^{-1}f)(\partial_+/\partial_+)g$ and partially integrate with the ordinary derivative ∂_+. In the example above one can first partially integrate $(\partial_+^x)^{-1}$ and then perform the x-integration. Then $[\partial_-/\partial_+\delta^2(y-z)]A^c(z)$ is written with this trick as $[\partial_-\delta^2(y-z)]\partial_+^{-1}A^c(z)$ and the final partial integration with ∂_- can be performed. The result is a term proportional to $A^a[(\partial_+^{-1}A^b),(\partial_-/\partial_+A^c)]$. In this way one may obtain the general form of the induced action [11]. It is given by

$$
\begin{aligned}
S_{ind}^+[A_+] &= \frac{k\,\lambda^2}{4\pi\chi}\sum_{n=0}^{\infty}\frac{1}{2+n}Tr\int d^2x\, A_+\Big[\frac{1}{\partial_+}A_+,\cdots\Big[\frac{1}{\partial_+}A_+,\frac{\partial_-}{\partial_+}A_+\Big]\cdots\Big]_{n\ times} \\
&= -\frac{k\,\lambda^2}{4\pi}\int d^2x\, \vec{A}_+\cdot\Big[\frac{1}{2}\frac{\partial_-}{\partial_+}\vec{A}_+ + \frac{1}{3}\frac{1}{\partial_+}\Big(\vec{A}_+\times\frac{\partial_-}{\partial_+}\vec{A}_+\Big) \\
&\qquad +\frac{1}{4}\frac{1}{\partial_+}\Big(\vec{A}_+\times\frac{1}{\partial_+}\Big(\vec{A}_+\times\frac{\partial_-}{\partial_+}\vec{A}_+\Big)\Big) + \cdots\Big]
\end{aligned}
\tag{3.8}
$$

and likewise for $S_{ind}^-[A_-]$. This result follows also from Feynman diagram calculations of the type considered in section 4.

Because of the chiral anomaly the induced action satisfies an anomalous Ward identity; under gauge transformations $\delta\vec{A}_\pm = \partial_\pm\vec{\eta} + \vec{\eta}\times\vec{A}_\pm$ it will transform as[†]

$$
\delta S_{ind}^\pm[A_\pm] = \frac{k}{2\pi}\int d^2x\, \vec{\eta}(x)\cdot\partial_\mp\vec{A}_\pm(x)\,.
\tag{3.9}
$$

The right-hand side originates from the anomaly of the underlying chiral theory, and is obtained by using $\delta A_+ = \partial_+\eta$ in the first term in (3.8). This anomaly is the unique (local) solution of the Wess-Zumino consistency equations. As there is only one (gauge) field it is not possible to construct gauge-invariant actions in terms of this field, so that the induced action is uniquely determined by the anomalous Ward identity.

An example of a matter system which realizes (3.2) is a chiral spinor ψ_-. The action

$$
\mathcal{L} = \frac{1}{2\pi\chi}Tr\psi_-(\partial_+\psi_- - [A_+,\psi_-]) \equiv -\frac{1}{2\pi}\psi_-^{\ a}\partial_+\psi_-^{\ b}g_{ab} + \frac{1}{\pi}A_+^{\ a}J_a
\tag{3.10}
$$

[†]As is well known, this equation is equivalent to the statement that the two-dimensional vector potential $(A_\pm,u_\mp)$, with $u_\mp = -(4\pi/k\lambda^2)\,\delta S_{ind}^\pm[A_\pm]/\delta A_\pm$, is curvature-free.

is locally invariant under $\delta\psi_- = [\eta, \psi_-]$ and $\delta A_+ = \partial_+\eta + [\eta, A_+]$. The ψ propagator is obtained from the path integral as $< \psi_-{}^a(x)\psi_-{}^b(y) >= \pi g^{ab}\left(\frac{\partial}{\partial x^+}\right)^{-1}\delta^2(x-y)$ which is equal to $g^{ab}(x^- - y^-)^{-1}$ according to the identity in (3.3). Then the current $J_a = -\frac{1}{2}f_{abc}\psi_-{}^b\psi_-{}^c$ satisfies a Kac-Moody with central charge $k = \tilde{h}$.

Another matter system whose induced covariant action is the sum of the two chiral nonlocal induced actions $S_{ind}{}^\pm[A_\pm]$ in (3.2) and a local A_+A_- term, is the WZWN model itself. Indeed, consider $S_W{}^+[h]$. (Similar consideration hold of course for $S_W{}^-[h]$). As well-known, it is invariant under two "semi-local" transformations $h(x^+, x^-) \to g_1(x^+)h(x^+, x^-)g_2(x^-)$. This follows from (2.11), and from rewriting (2.11) into a form with

$$Tr g^{-1}\delta g \partial_+(g^{-1}\partial_- g) \qquad (3.11)$$

There are thus <u>two</u> Noether currents

$$j_- = -\tfrac{k}{2}h^{-1}\partial_- h \; ; \; j_+ = \tfrac{k}{2}\partial_+ h h^{-1} \qquad (3.12)$$

If one chooses to gauge the vector symmetry $h \to \gamma h \gamma^{-1}$ with local $\gamma = \gamma(x^+, x^-)$, one may use the order-by-order in k Noether method to obtain a gauge invariant action [11]

$$\Gamma[h, A_+, A_-] = S_W{}^+[h]$$
$$+ \; \tfrac{1}{\pi \chi}\int d^2x Tr[j_- A_+ + j_+ A_- - \tfrac{k}{2}A_+ A_- + \tfrac{k}{2}A_- h A_+ h^{-1}] \qquad (3.13)$$

where $A_\pm$ are the components of the gauge field. Parametrizing $A_+ = \partial_+ g g^{-1}$ and $A_- = \partial_- g'((g')^{-1})$ one finds from the Polyakov-Wiegmann identity that

$$\Gamma[h, A_+, A_-] = S_W{}^+[g'^{-1}hg] - S_W{}^+[g'^{-1}g] \qquad (3.14)$$

From this result we can immediately check the vector gauge invariance

$$h \to \gamma h \gamma^{-1} \; , \; g \to \gamma g \; , \; g' \to \gamma g' \qquad (3.15)$$

corresponding to $\delta A_+ = \partial_+\gamma + [\gamma, A]$ and $\delta A_- = \partial_-\gamma + [\gamma, A_-]$. The axial vector transformations

$$h \to \gamma^{-1}h\gamma^{-1} \; , \; g' \to \gamma^{-1}g' \; , \; g \to \gamma g$$

$$\delta A_+ = \partial_+ \gamma + [\gamma, A_+], \qquad \delta A_- = -\partial_- \gamma - [\gamma, A_-] \qquad (3.16)$$

are clearly not symmetries. Integrating out h, one obtains indeed the covariant induced action of section 2 provided one uses a regularization scheme which maintains the vector gauge invariance. A general method to construct consistent regulators which maintain certain symmetries, has been worked out in ref. [12] and it would be interesting to apply this method to this case and obtain explicitly the regulator for the WZWN action which maintains vector invariance. Using the non-anomalous vector symmetry $\delta A_- = \partial_- \gamma + [\gamma, A_-]$ to gauge A_- to zero, we retrieve the action $S_{ind}{}^+[A_+]$.

As we now show, the WZNW action is related to the induced action $S_{ind}^{\pm}[A_\pm]$ obtained from coupling a chiral system to a chiral gauge field. It thus depends on the gauge fields A_+^a or A_-^a in a form which is purely two-dimensional, though nonlocal. The correspondence with the WZNW actions follows from the fact that $S_{ind}^{\pm}[A_\pm]$ with $A_\pm = \partial_\pm g \, g^{-1}$ are solutions to the same anomalous Ward identity (3.9) as the WZNW action, see (2.11). As this solution is unique (see below (3.9)), we conclude that

$$S_{ind}^+[A_+(g)] = S_W^+[g], \qquad S_{ind}^-[A_-(\bar{g})] = S_W^-[\bar{g}], \qquad (3.17)$$

with $A_+(g) = \partial_+ g \, g^{-1}$ and $A_-(\bar{g}) = \partial_- \bar{g} \, \bar{g}^{-1}$ and $g(x)$ and $\bar{g}(x)$ two independent group-valued fields.

Combining these results with (2.8) gives

$$
\begin{aligned}
S_{ind}^+[A_+] + S_{ind}^-[A_-] &= S_W^+[g] + S_W^-[\bar{g}] \\
&= S_W^+[\bar{g}^{-1}g] + \frac{k\lambda^2}{4\pi\chi} \int d^2x \, Tr \, \partial_+ g \, g^{-1} \, \partial_- \bar{g} \, \bar{g}^{-1} \\
&= S_W^+[\bar{g}^{-1}g] - \frac{k\lambda^2}{4\pi} \int d^2x \, \vec{A}_+ \cdot \vec{A}_-, \qquad (3.18)
\end{aligned}
$$

which shows the well-known result that the sum of the two chiral induced actions yields a covariant action up to a local term quadratic in A. This term is related to the ultraviolet subtraction one has to perform in order to preserve gauge invariance of the covariant (nonchiral) theory. For example, using Pauli-Villars regularization, the mass term $MTr\psi_+\psi_-$ is gauge-invariant, and the fermion loop with one external A_+ and one external A_- leads in the limit of M tending to infinity to the local A_+A_- term [9].

4 Loop calculations in nonlocal chiral field theories.

Given the action in (3.8) one can derive propagators for the A fields, and then construct Feynman diagrams in momentum space, just as in ordinary field theories [9]. Also the effective action for spin 2 fields h and spin 3 fields b [13] can be obtained in this way from the chiral nonlocal induced actions. Corresponding calculations in x-space become extremely complicated due to the fact that the inverse derivatives ∂_+^{-1} do not satisfy the Leibniz rule, and tricks like $\int(\partial_+^{-1}(fg))hk = -\int fg\partial_+^{-1}(hk)$ leave one with composite objects $\partial_+^{-1}(gh)$. In the cases where both approaches have been used (1 loop calculations for spin 1, spin 2 and spin 3), the results agree [4, 13, 14, 15, 16, 17]. However, even in momentum space, there are a number of problems that have to be faced [14], whose resolution is not entirely clear to us.

To perform momentum-space calculations one takes the naive Fourier transform of the vertices and propagators, ignoring their nonlocal and singular nature. In particular, one freely integrates by parts $1/\partial_+$ without worrying about possible boundary effects. A more careful approach to this issue is clearly desirable.

In x space the techniques that have been used avoid the problem of singular long-distance behavior by working on a compact Riemann surface. In momentum space the long-distance problem reappears in the form of infrared divergences from higher-order zeroes of momentum denominators. One regulates them by writing higher-order poles as derivatives of first-order poles. This leads to very simple integrals, but it may convert infrared divergences into ultraviolet ones. One could have added small masses instead, but the resulting integrals would have become very complicated, and also the procedure would have regulated, not removed, infrared divergences; whereas if we work on a compact Riemann surface there should be no infrared divergences at all. At the one-loop level the procedure for removing the infrared divergences by writing higher-order poles as derivatives of first-order poles, seems reasonable, but perhaps at higher loops one will have to worry about overlaps of such divergences and some form of the R^* subtraction procedure [8] will have to be adopted.

The corrections one computes are ultraviolet finite, but at intermediate stages one encounters divergent one-loop integrals, and these one regulates with a factor $\exp(-\epsilon k^2)$, where k is the loop momentum [9]. For example, for the self energy of the field $A_+{}^a$ one finds [4]

$$\Pi^{ab}(p) = \int d^2k \left[(\frac{k_-}{k_+})^2 - \frac{(k+p)_-}{(k+p)_+} \frac{k_-}{k_+} \right] \frac{k_+}{k_-} \frac{(k+p)_+}{(k+p)_-} \delta^{ab} \qquad (4.1)$$

In local field theories the x-space propagator is regulated, for example, by adding a short-distance cutoff $\theta(|x - y|^2 - \epsilon^2)$, and although the Fourier transform of the product of the x-space propagator times the θ function gives a convolution one might nevertheless try to identify the effect of the θ factor with a momentum-space exponential cutoff for each propagator. This would have given less desirable, and probably incorrect answers, for the following reason. A self-energy diagram would get two exponential factors corresponding to the two propagators, namely $\exp(-\epsilon k^2)$ and $\exp -\epsilon[(k + p)^2]$, while the seagull graph would get only one factor, $\exp(-\epsilon k^2)$, and the logarithmic divergences one encounters would no longer cancel. One might also think of regulating the nonlocal, and actually singular vertices instead of, or in addition to, the propagators; but besides raising additional questions, the actual calculations would become rather unwieldy. Using a single factor, $\exp(-\epsilon k^2)$, for the whole one-loop integral does lead to finite (i.e. ϵ-independent) results. However, it is not clear to us how this regularization should be implemented at higher loops where questions of overlapping divergences have to be faced. For local field theories the BPHZ subtraction procedure disentangles them and is consistent with unitarity, but here we are dealing with nonlocal divergences and the whole procedure has to be re-examined.

Some of the integrals one already encounters at the one-loop level for spin 2 [17, 14] and spin 3 [14, 15] problems are actually linearly or quadratically divergent and as we have said though the divergences ultimately cancel, they sometimes lead to finite routing-dependent contributions. It is possible to find simple recipes how to pick the routing of momenta in one-loop graphs in such a way that one ends up with finite results (which, in addition, agree with results obtained from x-space): for a self-energy graph containing two identical internal lines one symmetrizes with respect to the two routings [equivalently, by using once the regulator $\exp(-\epsilon k^2)$ and once the regulator $\exp[-\epsilon(k+p)^2]$ and taking the average]; while a self-energy graph containing two different lines (for example a spin 2 and field h and a spin 3 field b) give the same total result whether one routes the k momentum through the h line and the $p+k$ momentum through the b line or vice versa. For seagull graphs one uses the same momentum k in the regulator as in the loop, which seems reasonable but is not *a priori* obvious. (Already in the case of pure gravity a different choice for the h seagull would have given a different answer from what one expects on the basis of general arguments and explicit x-space calculations.) However, if one goes beyond the two-point function, and certainly at higher loops, the issue of routing dependence, or proper choice of momentum in the regulator, will have to be resolved in a more convincing manner. Based on our limited experience it may well be that the correct rule is to average over possible momentum choices for lines corresponding to identical particles, whereas for distinct particles no averaging is necessary. Such a rule might emerge from a closer examination of the nonlocal kinetic and interaction terms in the original Lagrangian.

The evaluation of momentum space Feynman integrals such as in (4.1) with chiral momenta k_+ and k_- can be done by using symmetric integration according to which momentum integrals such as $\int d^2k \, k_{\pm}^n$ should vanish for any positive or negative n. This eliminates, in particular, tadpoles. (Seagull graphs are not zero because the four-point couplings are nonlocal). One breaks the integration region, typically, into $k^2 \leq p^2$ and $k^2 \geq^2$ and expands $(p_+ + k_+)^{-1}$ into powers of k_+/p_+ and p_+/k_+, respectively. This "chiral regularization scheme" is more suitable for loop calculations than dimensional regularization, because the latter has well-known problems with $\epsilon^{\mu\nu}$. However, one may wonder whether the results for loop calculations in nonlocal (chiral) field theories (such as the induced action for spin 1, spin 2 and spin 3) depend on the regularization scheme chosen.

In local field theories different regularization schemes give results which differ by local counterterms, and either these local terms have no physical significance, or if symmetries are present, Ward identities may be imposed to fix their value. But for nonlocal field theories different regularization schemes will in general lead to different nonlocal counterterms (note that our integrals contain overall factors like p_-^3/p_+), and one may wonder whether this leads to such freedom in the effective action that nothing definite remains, unless Ward identities can be used to remove any ambiguity. Usually, one has only (anomalous) symmetries left in the induced action, and we could require that our regularization scheme, or loop-momentum routing, respect the coresponding anomalous Ward identities for the effecive action. The problem with this approach is that in the usual derivation of Ward identities there is always a moment when one assumes that the Jacobian for a symmetry change of variables is unity, and in our case with an anomalous symmetry of the induced action, it is not clear that the Jacobian which arises in the Ward identity for this system is unity.

Ultimately, regularization-scheme and routing ambiguities might be resolved by imposing "integrability" (or "renormalizability"). There are indications [13, 16, 19, 20, 21] that the full effective chiral action computed to all orders in c, can be obtained from the *leading* part of the induced action, $cS_{ind}^{(0)}(h, b)$, by replacing c with $Z_c c$, h with $Z_h h$, and b with $Z_b b$, where the Z factors are of the form $1 + a_1/c + a_2/c^2 +$

.... [Ordinary renormalizability in a *local* field theory only states that the (local) quantum action, not the effective action, is of this form; whereas here one would find that after including all contributions from all $S_{ind}^{(n)}$, $n = 0, 1, \ldots$, the total final result is expressible in terms of a renormalized $S_{ind}^{(0)}$.] If this is the case, then the effective action satisfies another Ward identity, namely the (rescaled) Ward identity for the induced action $S_{ind}^{(0)}$. (Comparison with the original Ward identity would then give an equation for the Jacobian mentioned above.)

There are, of course, also rigid nonanomalous symmetries of the induced action for Yang-Mills theory or pure gravity: the $R(1)$ or $Sl(2, R)$ symmetries, whose parameters are those parts of $\epsilon(x^+, x^-)$ for which $\partial_- \epsilon = 0$ or $\partial_-^3 \epsilon = 0$. Similar statements hold for the $c \to \infty$ part of the induced action of W_3 gravity. One could use the fact that these symmetries are also symmetries of the effective action if renormalizability holds, and this might provide another means of fixing the ambiguities inside the Polyakov scheme (or any other regularization scheme; equivalently, they might provide an Adler-Rosenberg-type procedure for determining the answers in a regularization-free approach).

Incidentally, for the two-point function, different regularization schemes which preserve rigid Lorentz invariance ("matching of $+$ and $-$ indices") all give results proportional to the original kinetic actions, since they are the only structures allowed, but for higher-point functions, new structures could be generated, which are not present in the original ($c \to \infty$ part) of the action, and hence "renormalizability" is a strong requirement.

Another possible approach to restricting the freedom generated by using arbitrary regularization schemes is to impose unitarity. In local field theories the presence of ambiguous or undetermined local terms does not affect unitarity, but here the ambiguities show up in nonlocal terms and arbitrary choices might conflict with unitarity constraints. In calculations with purely $k_\pm$ denominator factors, unitarity constraints have been forgotten because one ignores the presence of $i\epsilon$ factors, not only in the propagators, but also in the nonlocal interaction vertices. Starting from a matter-coupled system, one find $i\epsilon$'s in various places of the induced action. If one then goes on, and considers loops with gauge fields, one tends to ignore these original $i\epsilon$'s, but may put in new $i\epsilon$'s which make the path integral well defined. That is Polyakov's approach [9]. However, it may well be that a better understanding of the unitarity restrictions that follow from the original matter-coupled action would help with some of the issues we have discussed.

In calculations, one doesn't see essential differences between the Yang-Mills, pure-gravity and W_3 case. Yet, the original matter-coupled theory is renormalizable in the case of Yang-Mills and pure gravity and nonrenormalizable in the case of W_3 gravity (in the latter case there is a coupling constant with the wrong dimension). A directly observable consequence of this nonrenormalizability in the W_3 case is that the momentum integrals are *a priori* more divergent. The fact that in the end all the divergences cancel seems a bit miraculous, and no doubt hints at underlying special properties of the theory such as, perhaps, the integrability we described above. One would hope that these special properties would also take care of the routing and regularization ambiguities and ultimately explain the recipes we have used. Also the fact that the results obtained with x-space methods based on conformal field theory techniques, fully agree with the results obtained from momentum-space methods, seems to indicate that both methods are equivalent.

5 Quantization and the background field method.

In the background field formulation of the WZNW action $S_W^+[g]$ (see e.g. [3, 5]) one replaces the group element by the product of two group elements, say $g \to h\,g$, where g is the background field and h is expanded with respect to the real quantum fields π^a according to $h = \exp \pi^a T_a$. The background field only appears through the current $J_+ = \partial_+ g\, g^{-1}$, as follows from application of (2.8),

$$S_W^+[h\,g] - S_W^+[g] = S_W^+[h] + \frac{k\,\lambda^2}{4\pi\chi} \int d^2x \; Tr \; J_+\, h^{-1}\partial_- h\,. \tag{5.1}$$

Both sides of this equation are invariant under

$$
\begin{aligned}
g(x) \;&\rightarrow\; U(x^+)\, g(x)\,, \\
h(x) \;&\rightarrow\; U(x^+)\, h(x)\, U^{-1}(x^+)\,, \\
J_+(x) \;&\rightarrow\; \partial_+ U(x^+)\, U^{-1}(x^+) + U(x^+)\, J_+(x)\, U^{-1}(x^+)\,.
\end{aligned}
\tag{5.2}
$$

Using the expression for $h^{-1}\partial_- h$ in terms of the quantum fields π^a and the equalities (2.9) we deduce the following form of the action,

$$S_W^+[h\,g] - S_W^+[g] \;=\; \frac{k\,\lambda^2}{4\pi} \int d^2x \left\{ -\vec{J}_+ \cdot \partial_- \vec{\pi} - \sum_{n=2}^{\infty} \frac{1}{n!}\, \partial_- \vec{\pi} \cdot (\vec{\pi} \times (\vec{\pi} \times \cdots D_+\vec{\pi})) \right\}, \tag{5.3}$$

with

$$D_+\vec{\pi} = \partial_+\vec{\pi} - \vec{J}_+ \times \vec{\pi}\,. \tag{5.4}$$

A similar expression can be derived for $S_W^-[\bar{g}]$ using the background field splitting $\bar{g} \rightarrow \bar{g}\, h^{-1}$. One obtains then an action involving $J_- = \partial_- \bar{g}\, \bar{g}^{-1}$,

$$
\begin{aligned}
S_W^-[h^{-1}\bar{g}] - S_W^-[\bar{g}] \;&=\; S_W^+[h] - \frac{k\,\lambda^2}{4\pi\chi} \int d^2x \; Tr \; J_-\, \partial_+ h\, h^{-1} \\
&=\; \frac{k\,\lambda^2}{4\pi} \int d^2x \left\{ \vec{J}_- \cdot \partial_+\vec{\pi} - \sum_{n=2}^{\infty} \frac{1}{n!}\, D_-\vec{\pi} \cdot (\vec{\pi} \times (\vec{\pi} \times \cdots \partial_+\vec{\pi})) \right\},
\end{aligned}
\tag{5.5}
$$

which is invariant under

$$h(x) \rightarrow U(x^-)\, h(x)\, U^{-1}(x^-)\,, \qquad \bar{g}(x) \rightarrow U(x^-)\, \bar{g}(x)\,. \tag{5.6}$$

Note that with the exception of the term linear in $\vec{\pi}$, which is separately invariant, (5.3) and (5.5) follow from covariantizing $S_W^+[h]$ in (2.9).

However, we are interested in neither the former chiral action, nor the latter. Rather, we prefer a covariant action containing both J_+ and J_- with a local gauge invariance, since this will simplify our quantum calculations considerably as we shall discuss later. It is not hard to find an action involving g, h and $\bar{g}$ with a local gauge invariance, namely the gauged WZNW model in (3.14) described by

$$S[\vec{\pi}, \vec{J}] = S_W^+[\bar{g}^{-1} h g] - S_W^+[\bar{g}^{-1} g] \,, \tag{5.7}$$

which is manifestly invariant under

$$
\begin{aligned}
h(x) &\rightarrow U(x)\, h(x)\, U^{-1}(x) \,, \\
g(x) &\rightarrow U(x)\, g(x) \,, \\
\bar{g}(x) &\rightarrow U(x)\, \bar{g}(x) \,.
\end{aligned}
\tag{5.8}
$$

The latter two tranformation laws induce the standard gauge transformations for the currents

$$J_\pm(x) \rightarrow \partial_\pm U(x)\, U^{-1}(x) + U(x)\, J_\pm(x)\, U^{-1}(x) \,. \tag{5.9}$$

Moreover, for $\bar{g} = 1$ we recover (5.3) and for $g = 1$ we recover (5.5) (using $S_W^-[\bar{g}] = S_W^+[\bar{g}^{-1}]$).

The covariant action can be evaluated by fully covariantizing the actions (5.3) and (5.5). The result is

$$
\begin{aligned}
S[\vec{\pi}, \vec{J}] \;=\; & S_W^+[h] - \frac{k\,\lambda^2}{4\pi\chi} \int d^2x \, Tr\left[\partial_+ h\, h^{-1} J_- - J_+\, h^{-1}\, \partial_- h + J_+\, h^{-1}\, [J_-,\, h]\right] \\
=\; & -\frac{k}{8\pi} \int d^2x \; i\varepsilon^{\mu\nu}\, \vec{\pi} \cdot (\partial_\mu \vec{J}_\nu - \partial_\nu \vec{J}_\mu - \vec{J}_\mu \times \vec{J}_\nu) \\
& + \frac{k}{8\pi} \int d^2x \, \sum_{n=1}^{\infty} \Big\{ -\frac{1}{(2n)!}\, D_\mu \vec{\pi} \cdot (\vec{\pi} \times (\vec{\pi} \times \cdots D^\mu \vec{\pi})) \\
& \qquad\qquad + \frac{i}{(2n+1)!}\, \varepsilon^{\mu\nu}\, D_\mu \vec{\pi} \cdot (\vec{\pi} \times (\vec{\pi} \times (\vec{\pi} \times \cdots D_\nu \vec{\pi}))) \Big\} \,,
\end{aligned}
\tag{5.10}
$$

where the first term under the summation sign contains $2n$ and the second $2n + 1$ fields $\vec{\pi}$, and the covariant derivative is given by

$$D_\mu \vec{\pi} = \partial_\mu \vec{\pi} - \vec{J}_\mu \times \vec{\pi} \,. \tag{5.11}$$

Note again that the term linear in $\vec{\pi}$ does not follow from covariantizing the action (2.3), but is separately gauge invariant.

The last form of the action forms the basis of our subsequent calculations. We write down the Lagrangians for the terms of second, third and fourth order in the quantum fields $\vec{\pi}$. Dropping an overall factor $k/8\pi$, these Lagrangians are

$$\mathcal{L}^{(2)} = -\tfrac{1}{2}(\partial_\mu \pi^a)^2 - f^{abc} J_\mu^a \, \partial^\mu \pi^b \, \pi^c - \tfrac{1}{2} f^{abe} f^{cde} J_\mu^a \, J^{c\mu} \, \pi^b \pi^d \,, \tag{5.12}$$

$$\mathcal{L}^{(3)} = i\varepsilon^{\mu\nu}\Big\{ -\tfrac{1}{6} f^{abc} \, \pi^a \, \partial_\mu \pi^b \, \partial_\nu \pi^c - \tfrac{1}{3} f^{abc} f^{cde} J_\mu^a \, \pi^b \, \pi^d \, \partial_\nu \pi^e$$

$$+\tfrac{1}{6} f^{acf} f^{feg} f^{bdg} \, J_\mu^a \, J_\nu^b \, \pi^c \, \pi^d \, \pi^e \Big\} \,, \tag{5.13}$$

$$\mathcal{L}^{(4)} = \tfrac{1}{24} f^{abe} f^{cde} \partial_\mu \pi^a \, \pi^b \, \partial^\mu \pi^c \, \pi^d - \tfrac{1}{12} f^{abf} f^{cfg} f^{dge} \, J_\mu^e \, \partial^\mu \pi^a \, \pi^b \, \pi^c \, \pi^d$$

$$+\tfrac{1}{24} f^{eag} f^{gbh} f^{hci} f^{idf} \, J_\mu^e \, J^{f\mu} \, \pi^a \, \pi^b \, \pi^c \, \pi^d \,. \tag{5.14}$$

6 Two-Loop calculations in the WZWN model.

We shall now compute one-particle irreducible contributions to the current correlation functions of the gauged WZWN model. These are diagrams with external J-lines, and internal π lines. We shall use conventional dimensional regularization techniques (even though sometimes using contour integration methods for the complexified variables p_+ and p_- is much simpler. [4]). Although individual diagrams are ultraviolet divergent by power counting, the gauge invariance of the covariant action ensures the ultraviolet finiteness of the results. To spell this out in more detail: according to the BPHZ scheme extended to take also infrared divergences into account, if one subtracts all lower-loop infrared and ultraviolet divergences, the n-loop divergences must be local. Hence, $p_\mu p_\nu / p^2$ divergences are forbidden. Gauge invariance states that the 2-point function is transversal; in particular, divergences must be proportional to $\delta_{\mu\nu} - p_\mu p_\nu / p^2$. This implies that there can be no divergences to $\delta_{\mu\nu}$ either, hence the complete 2-point function is finite.

However, there are some subtleties. First of all, although all 1-loop graphs with external J lines are finite, 1-loop n-point functions with two external <u>quantum</u> fields need not be finite, and when they are part of a 2-loop diagram with two external sources, one needs counter terms for the divergent subgraphs. As we shall argue, no such counter terms are needed if one treats the $\epsilon^{\mu\nu}$ tensors as n-dimensional objects, but if one treats the $\epsilon^{\mu\nu}$ tensors as 2-dimensional objects, the counter terms are of "evanescent type", i.e., they are proportional to operators which vanish for $d = 2$. Namely, for the π - π selfenergy one needs

$$\Delta\mathcal{L} = -\frac{\tilde{h}}{32\pi\epsilon}\hat{\delta}^{\mu\nu}\,\partial_\mu\pi\partial_\nu\pi \tag{6.1}$$

where the indices of $\hat{\delta}^{\mu\nu}$ take only value in the $d - 2$ extra dimensions. No such counter terms are needed for the one-loop $J\pi\pi$ 3-point function since the index J_μ of the external current is 2-dimensional.

The treatment of the ϵ tensor has a long history which we summarize here. Naive continuation away from two dimensions for products of the epsilon tensor,

$$\epsilon_{\mu\nu}\epsilon^{\rho\sigma} = f(\epsilon)(\delta^\rho_\mu\delta^\sigma_\nu - \delta^\rho_\nu\delta^\sigma_\mu), \tag{6.2}$$

with $f(0) = 1$ and δ^ν_μ the d-dimensional Kronecker symbol, becomes inconsistent at dimensions different from two when products of more than two epsilon tensors are considered [5]. Furthermore its consequences are in disagreement with general results, both for the WZWN model away from the critical point [5] and for two-dimensional sigma models with torsion [18]. The generally accepted rule is to keep the epsilon tensors in two dimensions, so that the tensors δ^ν_μ on the right-hand side of (6.2) should be replaced by two-dimensional Kronecker symbols and $f(\epsilon) = 1$.

So, why are there no counterterms needed when one treats $\epsilon^{\mu\nu}$ as n-dimensional? The one-loop counter terms with external π fields are proportional to the Lagrangian $\mathcal{L}(\pi, J)$ because of background gauge invariance and manifest rotational invariance in d dimensions. The set of all counter terms with precisely two π fields can thus be absorbed into the original Lagrangian by multiplicative renormalization of all π fields (but not the J fields). However, it is well-known that in the path integral a rescaling of quantum fields has no effect since one integrates over these quantum fields. Therefore, the counter terms corresponding to this wave-function renormalization do not contribute. If one does include the counter terms for the 2-point and 3-point 1-loop subgraphs in the 2-loop 2-point function, one finds by actual calculation that these contributions indeed sum up to zero, confirming this general argument. If, on the other hand, one treats $\epsilon^{\mu\nu}$ as 2-dimensional, then one loses manifest rotational invariance in d dimensions, and the counter terms have no longer the same form as the original Lagrangian. The difference of the n-dimensional and 2-dimensional counter term will therefore contribute, and it is given by (6.1). Clearly such counter terms give only a finite contribution to the final 2-loop result (because $\hat{\delta}^{\mu\nu}$ is effectively proportional to ϵ). Of course, one should also take into account that all 2-loop graphs involving (two) ϵ-tensors will give different finite contributions depending on whether one treats ϵ as n-dimensional as 2-dimensional. Remarkably, if one treats $\epsilon^{\mu\nu}$ consistently everywhere as either n-dimensional or 2-dimensional, the result is the same!

Another problem concerns the infrared divergences. For these in ordinary field theory on R_2 there is no cure: the best we can do is separate them from what we believe to be the relevant contribution. Conceivably, summing the infrared-induced logarithms in the final answer to all orders of perturbation theory would lead to their complete cancellation (for example, like $\ln \frac{p^2}{\mu^2}(1 - \ln p^2/\mu^2)^2$ tends to zero as μ^2 tends to zero), but this is only a hope, and we have no solid support for this conjecture. So, how then do we subtract the infrared divergences? We use the infrared subtraction methods of Chetyrkin, Tkachov, Simirnov, Kataev and others [8]. It amounts at the one-loop level to the replacement

$$\frac{1}{k^2} \rightarrow \frac{1}{k^2} + \frac{\pi}{\epsilon}\delta^2(k) \tag{6.3}$$

At higher loops the rules needed are quoted in [4]. To remove contributions proportional to $\ln 4\pi$, the Euler constant (and at higher loops the Riemann function $\zeta(2)$) we add for each loop an additional factor $\Gamma(1 - \epsilon)/(4\pi)^{-\epsilon}$. Of course, the removal of infrared divergences should not spoil the gauge invariance, and it can be checked that the subtracted terms separately are also gauge invariant. For example, the evanescent counter terms give a contribution at the 2-loop level from which one must subtract the infrared divergent part, and the final result from the evanescent counter term is both transverse (gauge invariant) and finite (due to the infrared subtraction). We now summarize the results of the 1-loop and 2-loop corrections to the 2-point function.

For the one-loop 2-point function, the sum of the selfenergy graph and the seagull graph yields

$$\Pi_{\mu\nu}^{ab}(p) = \frac{\tilde{h}}{4\pi} g^{ab} \frac{[\Gamma(1-\epsilon)]^3 \Gamma(1+\epsilon)}{\epsilon \Gamma(2-2\epsilon)} \left[\frac{p^2}{\mu^2}\right]^{-\epsilon} \left(\delta_{\mu\nu} - \frac{p_\mu p_\nu}{p^2}\right), \tag{6.4}$$

where μ is the mass parameter required by dimensional regularization. The pole at $\epsilon = 0$ represents an infrared singularity and is removed by the infrared counterterms discussed above. It amounts to adding to this result the expression

$$\Delta_{IR}\Pi_{\mu\nu}^{ab}(p) = \tilde{h} g^{ab} \int \frac{d^2k}{(2\pi)^2} \left[\frac{\pi}{\epsilon}\delta^{(2)}(p-k)\frac{k_\mu k_\nu}{k^2} - \frac{\pi}{\epsilon}\delta^{(2)}(k)\delta_{\mu\nu}\right]. \tag{6.5}$$

After this subtraction we obtain

$$\Pi_{\mu\nu}^{ab}(p) = \frac{\tilde{h}}{4\pi} g^{ab}(2 - \ln\frac{p^2}{\mu^2})\,(\delta_{\mu\nu} - \frac{p_\mu p_\nu}{p^2}). \tag{6.6}$$

which is the result mentioned in (1.2).

The two-loop contributions to the two-point function with two external J fields, can be divided into a set of diagrams involving no $\epsilon^{\mu\nu}$ tensor, and a set with precisely two $\epsilon^{\mu\nu}$ tensors. The first set consist of 5 diagrams, while the latter set contains 8 diagrams (of which the last two contain an evanescent counter term, hence they only contribute if $\epsilon^{\mu\nu}$ is treated as a 2-dimensional object). In the table we record separately the results for the terms proportional to $\delta_{\mu\nu}$ and those proportional to $p_\mu p_\nu/p^2$, and in each case we begin by recording the contribution from 2-dimensional ϵ tensors first, and then the n-dimensional corrections are indicated in square brackets. In VII we record the contributions of the two evanescent counter terms. Of course, in the set of five diagrams there are no corrections since these diagrams do not depend on $\epsilon^{\mu\nu}$, but in the set of 8 diagrams, the corrections for the first, second and fifth diagrams involve extra factors $1 - \epsilon$ and $1 - 2\epsilon$. The fourth diagram is the same in both cases, but the contribution from the third diagram is written first for the case of two-dimensional ϵ tensors, and then below it the contribution due to n-dimensional ϵ tensors is recorded. Finally, it is clear from VII that the last two diagrams only contribute if one uses two-dimensional ϵ tensors.

In the last two columns of the table we expand these contributions in powers of ϵ, and multiply by an overall factor ϵ^2 to simplify the notation. As the reader may note, all divergences and also all finite parts sum up to zero, both in the $\delta_{\mu\nu}$ sector and in the $p_\mu p_\nu / p^2$ sector, and both for two-dimensional and for n-dimensional ϵ tensors. Also the set of 5 diagrams is separately gauge invariant, as are the contributions from the evanescent counter terms.

A Normalization of the WZWN term.

The normalization of the topological term $\Gamma[g]$ can easily be checked for the case that the group G is SU(2). In the 2×2 representation with $T_a = \frac{1}{2} i \sigma_a$ and σ_a the Pauli matrices, one has $f_{ab}{}^c = -\epsilon_{abc}$. Hence $f_{ap}{}^q f_{bq}{}^p = -2$, and since this is by definition $-\tilde{h} g_{ab}$ where $\tilde{h}$ is the dual Coxeter number, ($\tilde{h} = 2$ for SU(2)) we find $g_{ab} = \delta_{ab}$. From $Tr\, T_a T_b = -g_{ab}\chi$ we then find the well-known result that $\chi = 1/2$ in the fundamental representation of SU(2).

We parametrize a given group element of SU(2) by Euler angles, $g = \exp T_3 \psi \exp T_1 \theta \exp T_3 \varphi$. This yields

$$g = \begin{pmatrix} e^{\frac{1}{2}i(\psi+\varphi)} \cos \theta/2 & i \exp^{\frac{1}{2}i(\psi-\varphi)} \sin \theta/2 \\ ie^{-\frac{1}{2}i(\psi-\varphi)} \sin \theta/2 & e^{-\frac{1}{2}i(\psi+\varphi)} \cos \theta/2 \end{pmatrix} \tag{A.1}$$

(It is easy to check this result for small ψ, φ and θ). We take $0 \leq \theta \leq \pi$ because $\cos \theta/2$ and $\sin \theta/2$ are then positive, and to obtain all possible phases of g_{11}, we need $-2\pi \leq \psi + \varphi \leq 2\pi$. Similarly, we obtain all possible phases of g_{21} if $-2\pi \leq$

$\psi - \varphi \leq 2\pi$. In the ψ, φ plane this corresponds to a figure of the shape of a diamond, but one obtains the same surface if one uses the rectangular region $0 \leq \varphi \leq 2\pi$ and $0 \leq \psi \leq 4\pi$. Hence, the ψ, φ, θ domain which covers SU(2) just once is given by $0 \leq \theta \leq \pi, 0 \leq \psi \leq 4\pi, 0 \leq \varphi \leq 2\pi$.

From the explicit representation of g given above, one can easily compute the group vielbeins $g^{-1}\partial_\mu g$. One finds

$$g^{-1}\partial_\psi g \;=\; T_3$$
$$g^{-1}\partial_\theta g \;=\; T_1 \cos \varphi + T_2 \sin \varphi$$
$$g^{-1}\partial_\psi g \;=\; T_3 \cos \theta - \sin \theta (T_2 \cos \varphi - T_1 \sin \varphi) \tag{A.2}$$

From this result it is straightforward to obtain

$$\epsilon^{\mu\nu\rho} Tr g^{-1}\partial_\mu g g^{-1}\partial_\nu g g^{-1}\partial_\rho g$$
$$= 3 Tr g^{-1}\partial_\psi g (g^{-1}\partial_\theta g g^{-1}\partial_\varphi g - g^{-1}\partial_\psi g g^{-1}\partial_\theta g)$$
$$= \tfrac{3}{2} \sin \theta. \tag{A.3}$$

Next we show that SU(2) equals S_3. This follows easily from the fact that $g = a_0 + i\vec{a} \cdot \vec{\sigma}$ is unitary if $g^{-1} = (a_0 - i\vec{a} \cdot \vec{\sigma})(a_0{}^2 + \vec{a} \cdot \vec{a})^{-1}$ equals $g^\dagger = (a_0^* - i\vec{a}^* \cdot \vec{\sigma})$. This yields that $g = (a_0 + i\vec{a} \cdot \vec{\sigma}) \exp i\delta$ with real a_0 and $\vec{a}$, and $\det g = 1$ if $\exp i\delta = 1$ and $a_0{}^2 + a_k{}^2 = 1$. Hence $g = a_0 + i\vec{a} \cdot \vec{\sigma}$ with real a_0 and a_k satisfying $a_0{}^2 + a_k{}^2 = 1$.

We find then for (2.1), using $\chi = \tfrac{1}{2}$, $\epsilon^{\mu\nu\rho} = \epsilon^{\psi\theta\varphi}$ and $B = SU(2)$

$$\Gamma[g] = \tfrac{1}{12\pi} \int_0^{4\pi} d\psi \int_0^{\pi} d\theta \int_0^{2\pi} d\varphi \tfrac{3}{2} \sin \theta = 2\pi \tag{A.4}$$

This demonstrates that (2.1) has the correct normalization for the fundamental representation of SU(2). For other representations of SU(2), one gets an integer multiple of this result, since $Tr T_a T_b$ in an arbitrary representation is an integer times $Tr T_a T_b$ in the fundamental representation. (The carrier space is obtaining by tensoring and reducing.) For other groups it is a theorem that one can continuously deform a three-sphere S_3 in the group manifold such that at the end all points of this S_3 are of the form $\exp x^a T_a (a = 1, 3)$ with T_a forming an SU(2) subalgebra. For different embeddings of this SU(2) subgroup into the group G, the trace $Tr T_a T_b$ can be different by an integer factor, and in general the minimum value of $\Gamma[g]$ for which $\exp i\Gamma[g]$ is still unambiguous, is obtained by taking the T_a to be the generators which correspond to the largest root ($E_\theta, E_{-\theta}$ and $[E_\theta, E_{-\theta}]$). For example, for $G = SU(3)$, these generators correspond to the isospin SU(2) subgroup, while "the other SU(2) subgroup" gives a result that is a factor 4 larger. We thank A. Van Proeyen for discussions.

References

[1] J. Wess and B. Zumino, Phys. Lett. **37B** (1971) 95.

[2] S.P. Novikov, Sov. Math. Doklady **24** (1981) 222; Usp. Math. Nauk. **37** (1982) 3.

[3] E. Witten, Nucl. Phys. **B223** (1983) 422; Commun. Math. Phys. **92** (1984) 455.

[4] B. de Wit, M.T. Grisaru and P. van Nieuwenhuizen, *Nucl. Phys. B* **408** (1993) 229.

[5] M. Bos, Phys. Lett. **B189** (1987) 435; Ann. Phys. (N.Y.) **181** (1988) 177.

[6] H. Leutwyler and M. Shifman, Int. J. Mod. Phys. **A7** (1992) 795.

[7] A. Tseytlin, *Nucl. Phys. B* **399** (1993) 601, *Nucl. Phys. B* **411** (1994) 509; I. Bars and K. Sfetsos, *Phys. Rev. D* **48** (1993) 844; K. Sfetsos and A.A. Tseytlin, *Phys. Rev. D* to appear.

[8] K.G. Chetyrkin and F.V. Tkachov, Phys. Lett. **114B** (1982) 340;
V.A. Smirnov and K.G. Chetyrkin, Theor. Math. Phys. **63** (1985) 462;
K.G. Chetyrkin and V.A. Smirnov, "The R^* operation", Moscow State University Nuclear Physics Institute, preprint 89-3/80, Moscow 1989.
K.G. Chetyrkin, A.L. Kataev and F.V. Tkachov, Nucl. Phys. **B174** (1980) 345;
K.G. Chetyrkin and F.V. Tkachov, Nucl. Phys. **B192** (1981) 59.
M.T. Grisaru, D.I. Kazakov and D. Zanon, Nucl. Phys. **B287** (1987) 189.

[9] A.M. Polyakov, in *Fields, Strings and Critical Phenomena*, Les Houches 1988, E. Brézin and J. Zinn-Justin (eds.), North-Holland, 1990.

[10] A.M. Polyakov and P.B. Wiegmann, Phys. Lett. **131B** (1983) 121; **141B** (1984) 223.

[11] H. Ooguri, K. Schoutens, A. Sevrin and P. van Nieuwenhuizen, Commun. Math. Phys. **145** (1992) 515.

[12] A. Diaz, A. Van Proeyen, W. Troost and P. van Nieuwenhuizen, *Int.J. Mod. Phys. A* **4**, 3959 (1989)
M. Hatsuda, A. Van Proeyen, W. Troost and P. van Nieuwenhuizen *Nucl. phys. B* **335**, 166 (1990).

[13] K. Schoutens, A. Sevrin and P. van Nieuwenhuizen, Nucl. Phys. **B364** (1991) 584; **B371** (1992) 315.

[14] M.T. Grisaru and P. van Nieuwenhuizen, Int. J. Mod. Phys. **A7** (1992) 5891.

[15] G.W. Delius, M.T. Grisaru and P. van Nieuwenhuizen, Nucl. Phys. **B389** (1993) 25.

[16] K. Schoutens, A. Sevrin and P. van Nieuwenhuizen, in *Strings & Symmetries 1991*, proc. of the Stony Brook conference, eds N. Berkovits *et al.*, World Scientific, 1992.

[17] K.A. Meisner and J. Pavelchik, Mod. Phys. Lett. **A5** (1990) 763.

[18] S. Ketov, Nucl. Phys. **B294** (1987) 813;
C. Hull and P.K. Townsend, Phys. Lett. **191B** (1987) 115;
R.R. Metsaev and A.A. Tseytlin, Phys. Lett. **191B** (1987) 354;
D. Zanon, Phys. Lett. **191B** (1987) 363;
D.R.T. Jones, Phys. Lett. **192B** (1987) 391.

[19] A. Sevrin, K. Thielemans and W. Troost, preprint LBL-33738, KUL-TF-93/09.

[20] J. de Boer and J. Goeree, *The effective action of W_3 gravity to all orders*, Nucl. Phys. **B**, in print.

[21] A. Sevrin, R. Siebelink and W. Troost, Leuven preprint, in preparation.

Graph	$\delta_{\mu\nu}$	$p_\mu p_\nu/p^2$	$\delta_{\mu\nu}$	$p_\mu p_\nu/p^2$
1	$5/24$	0	$5/24$	0
2	$(-5/12)\{(1-2\epsilon)p^{2\epsilon}\}^{-1}$	$5/12\{(1-2\epsilon)p^{2\epsilon}\}^{-1} - \frac{5}{12}$	$\frac{-5}{12} - \frac{5}{6}\epsilon - \frac{5}{3}\epsilon^2$	$\frac{5}{6}\epsilon + \frac{5}{3}\epsilon^2$
3 + 4	$\frac{1}{6}\{(1-2\epsilon)p^{2\epsilon}\}^{-1} - \frac{1}{12}$	$-\frac{1}{6}\{(1-2\epsilon)p^{2\epsilon}\}^{-1} + \frac{1}{6}$	$\frac{1}{12} + \frac{1}{3}\epsilon + \frac{2}{3}\epsilon^2$	$-\frac{1}{3}\epsilon - \frac{2}{3}\epsilon^2$
5	$\frac{1}{8}\{(1-2\epsilon)^2 p^{4\epsilon}\}^{-1}$	$-\frac{1}{8}\{(1-2\epsilon)^2 p^{4\epsilon}\}^{-1} + \frac{1}{8}$	$\frac{1}{8} + \frac{1}{2}\epsilon + \frac{3}{2}\epsilon^2$	$-\frac{1}{2}\epsilon - \frac{3}{2}\epsilon^2$
I	$\dfrac{[1-2\epsilon]}{4(1-2\epsilon)(1-3\epsilon)p^{4\epsilon}}$	$[1-2\epsilon]^{-1}(\frac{1}{2}\frac{1}{(1-2\epsilon)p^{2\epsilon}} -\frac{1}{2}(1-\epsilon)\{(1-2\epsilon)(1-3\epsilon)p^{4\epsilon}\}^{-1})$	$\frac{1}{4} + \frac{5}{4}\epsilon + \frac{19}{4}\epsilon^2$ $[\frac{1}{4} + \frac{3}{4}\epsilon + \frac{9}{4}\epsilon^2]$	$-\epsilon - 5\epsilon^2$ $[-\epsilon - 3\epsilon^2]$
II	$[1-2\epsilon]\frac{1}{6}\{\frac{3}{4}\frac{1}{1-3\epsilon}\frac{1}{p^{4\epsilon}} - \frac{3}{2}\frac{1}{(1-2\epsilon)}\frac{1}{p^{2\epsilon}}\} + (2-\epsilon)(1-2\epsilon)p^{2\epsilon}\}^{-1} - 1 - \frac{1}{6}\{(1-13\epsilon)p^{4\epsilon}\}^{-1}$	$-\frac{1}{6}\{\frac{-1}{(1-3\epsilon)p^{4\epsilon}} + \frac{2-\epsilon}{(1-2\epsilon)p^{2\epsilon}} - 1\}$	$-\frac{1}{8} - \frac{1}{8}\epsilon - \frac{3}{8}\epsilon^2$ $[-\frac{1}{8} + \frac{1}{8}\epsilon - \frac{1}{8}\epsilon^2]$	$\frac{1}{2}\epsilon^2$
III	$\dfrac{-2+4\epsilon^2}{4(1-2\epsilon)(1+\epsilon)(1-3\epsilon)(2-3\epsilon)p^{4\epsilon}} - [\frac{1}{4}\frac{1}{(1-3\epsilon)p^{4\epsilon}}]$	$\frac{3}{8}\frac{1}{(1-2\epsilon)(1-3\epsilon)p^{4\epsilon}} - \frac{1}{4}\frac{1}{(1-2\epsilon)p^{2\epsilon}} - \frac{1}{8}\frac{1}{(1-\epsilon)(1-2\epsilon)} + [\frac{3}{8}\frac{1-\epsilon/3}{1-3\epsilon}\frac{1}{p^{2\epsilon}} - \frac{1}{4p^{2\epsilon}} - \frac{1}{8}]$	$\frac{1}{4} - \frac{11}{8}\epsilon - \frac{85}{16}\epsilon^2$ $[-\frac{1}{4} - \frac{9}{4}\epsilon - \frac{3}{4}\epsilon^2]$	$\epsilon + \frac{21}{4}\epsilon^2$ $[\epsilon + 3\epsilon^2]$
IV	$\frac{1}{8}\left(\frac{1}{1-3\epsilon} - \frac{1}{(1-2\epsilon)^2}\right)\frac{1}{p^{4\epsilon}}$	$-\frac{1}{8}\left(\frac{1+\epsilon}{1-3\epsilon} - \frac{1}{(1-2\epsilon)^2}\right)\frac{1}{p^{4\epsilon}}$	$\frac{1}{8}\epsilon - \frac{11}{16}\epsilon^2$ $[-\frac{1}{8}\epsilon - \frac{3}{8}\epsilon^2]$	0
V	$\dfrac{[(1-\epsilon)(1-2\epsilon)]}{8(1-\epsilon)(1-2\epsilon)}$	0	$\begin{cases} \frac{1}{8} + \frac{3}{8}\epsilon + \frac{7}{8}\epsilon^2 \\ [\frac{1}{8}] \end{cases}$	0
VI	0	0	0	0
VII	$\frac{1}{4}\epsilon^2 + [-\frac{1}{4}\epsilon^2]$	$-1/4\epsilon^2 + [-\frac{1}{4}\epsilon^2]$	$\frac{1}{4}\epsilon^2$ $[0]$	$-1/4\epsilon^2$ $[0]$

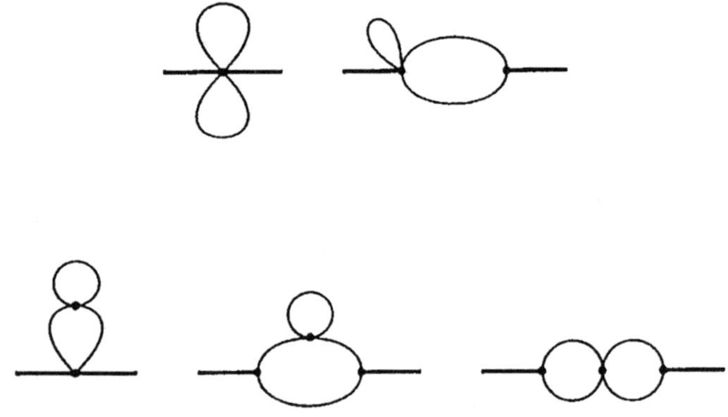

Figure 1: Two-loop diagrams without ϵ tensors.

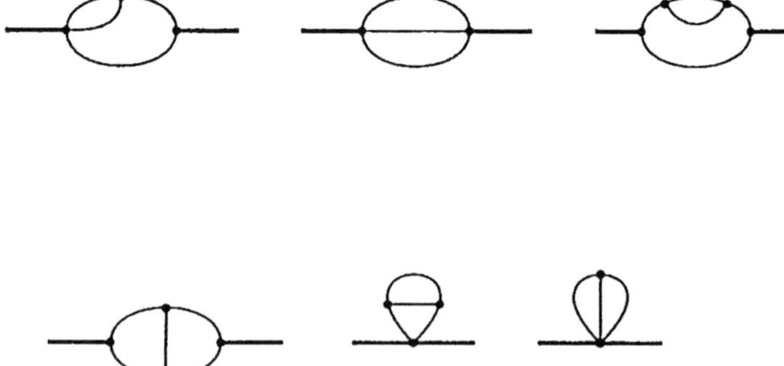

Figure 2: Two-loop diagrams with two ϵ tensors. To this set one should add the one-loop selfenergy graph and seagull graph with an insertion of the evanescent counter terms in (6.1).

Massless integrable quantum field theories and massless scattering in 1+1 dimensions

P. Fendley[†] and H. Saleur[♠*]

[†]Department of Physics, University of Southern California
Los Angeles CA 90089

[♠]Department of Physics and Department of Mathematics
University of Southern California
Los Angeles CA 90089

These lecture notes provide an elementary introduction to the study of massless integrable quantum field theory in 1+1 dimensions using "massless scattering". Some previously unpublished results are also presented, including a non-perturbative study of Virasoro conserved quantities.

*Lectures presented at Strings 93, May 1993
and at the Trieste Summer School on High Energy Physics and Cosmology, July 1993.*

* Packard Fellow

1. Introduction

A massless quantum field theory has no gap in the excitation spectrum. This can be seen, for example, in the Laplace representation of Green's functions. This of course does not imply scale invariance; in general properties will interpolate between those of the two different conformal field theories describing the UV and IR fixed points. A standard (although a little marginal) example of such theory is the $O(3)$ non-linear sigma model with topological angle $\Theta = \pi$, which has central charges $c_{UV} = 2$ and $c_{IR} = 1$.

In 1+1 dimensions, many massless theories are integrable. Such theories include well-known statistical-mechanical models like the continuum limit of the XXZ spin chain and the Kondo problem. Many more are provided by appropriate perturbations of conformal field theories. Their study is interesting for several reasons. A few properties are accessible experimentally; see for example [1,2]. Features of academic interest include Green's functions with different anomalous dimensions in the UV and IR, the consequences for the topology of the space of relativistic quantum field theories, a better understanding of the second law of thermodynamics associated with renormalization group trajectories, and a way of understanding perturbations of IR fixed points by irrelevant operators.

In these lectures we will discuss the scattering theories associated with integrable massless quantum field theories. In a massless theory the excitations should consist of right-moving and left-moving particles with $p = \pm E$, where we set the speed of light to be 1. S-matrices describing the "scattering" of such particles were calculated long ago in [3] for the XXX model. Such objects do not make much sense in traditional S-matrix theory where one requires the existence of in and out states; it is difficult for instance to imagine a physical process that would lead to scattering between two particles moving in the same direction at the speed of light. Massless S-matrices in $1 + 1$ dimensions seem to make sense only in the context of integrable quantum field theories. In this case the scattering is completely elastic: momenta are conserved individually. We build states by acting with creation and annihilation operators on the ground state. These operators have non-trivial commutation properties (the Zamolodchikov-Faddeev algebra [4,5]) encoded in the S-matrix (suppressing internal indices describing the particles):

$$\mathcal{R}^+(\theta_1)\mathcal{R}^+(\theta_2) = S(\theta_1 - \theta_2)\mathcal{R}^+(\theta_2)\mathcal{R}^+(\theta_1), \tag{1.1}$$

where by convention $\mathcal{R}^+(\theta_i)\mathcal{R}^+(\theta_j)$ creates a plane wave with $x_i < x_j$. This formal definition of S, directly inspired by the Bethe ansatz equations, makes sense in both massive and massless cases. Another way of describing this definition is as a matching condition on two-particle wavefunctions.

Some care must be taken when defining massless integrable theories. For instance, the usual proof that an infinite number of conserved quantities implies factorized scattering relies on the possibility of separating wave packets in general [6], which is not possible in the massless case. Analyticity properties are also not completely clear. Since the particles are massless, they are either right- or left-moving. Because S-matrices for pure left-left or right-right scattering can be obtained by a limiting process from physical S-matrices acting on massive particles, one requires from S_{LL} and S_{RR} exactly the same properties as for physical S-matrices. As we will discuss, the case of S_{LR} is subtler. In general these properties can be derived using the definition based on Bethe ansatz wavefunctions, without reference to any in and out states.

After having given the caveats, we would like to explain why finding massless S-matrices is a worthy endeavor. Even if the S-matrix found has no physical meaning in its own right, many quantities calculated from it do. In these notes we will show how to calculate the free energy at non-zero temperature using the thermodynamic Bethe ansatz (TBA). Modular invariance relates this to the Casimir energy on a circle, giving a "c-function", which for unitary models shows the evolution of the number of degrees of freedom in the flow from ultraviolet to infrared. This calculation can also be modified to give some excited-state energies as well, which give conformal dimensions in the critical limit. We will also show how to obtain the ground-state energy at zero temperature in a background field, a result related to the chiral $U(1)$ anomaly in the critical limit. Finally, we present here some new results on obtaining higher-spin Virasoro conserved charges from the massless scattering.

These results in fact suggest that some aspects of a conformal field theory can be described by a theory of massless particles with no left-right scattering [7]. This can be seen as follows. In these massless but not scale-invariant theories, we have a mass scale M; $M = 0$ gives the ultraviolet fixed point while $M \to \infty$ gives the infared one. The momenta and energy of the particles are parametrized by

$$
\begin{aligned}
E = p = \frac{M}{2} e^{\theta} \qquad &\text{for right movers} \\
E = -p = \frac{M}{2} e^{-\theta} \qquad &\text{for left movers}
\end{aligned}
\tag{1.2}
$$

A Lorentz-invariant S-matrix element S_{LL} describing scattering of two left movers depends only on the ratio of the two momenta, so it depends only on $\theta_1 - \theta_2$ and not on M. The left-right scattering also depends only on the rapidity difference but does depend on M, because the only Lorentz invariant is $s = (p_1 + p_2)^2$. We can always rescale $M \to \infty$ by shifting the rapidities. The LL and RR S-matrices are independent of this shift (although S_{LR} is not), so they are characterized solely by properties of the infrared fixed point. In this sense one can think of the LL and RR S-matrices as being the S-matrices for the conformal field theory. This should not seem bizarre — many properties of four-dimensional field theories (even ones with massless particles like QED) are described by particle theories!

Another reason for studying massless S-matrices is that finding them is often an easier task than doing the full Bethe ansatz, a result of the constraints of an integrable theory. In the massive case, this has become a highly-developed art (see [8-10] for reviews), and many of these lessons can be applied to the massless case. The basic method is to guess the particle content based on the knowledge of the symmetries of the problem (and on the Lagrangian, if one is known), and then find the simplest S-matrix consistent with these symmetries as well as the criteria of factorizability, unitarity and crossing symmetry. In many cases, such an S-matrix is the correct one, as can be checked by a variety of methods. The symmetries, in particular affine quantum group symmetries, must however be analyzed carefully.

In addition to the XXX model, massless S-matrices have been found for a number of models. Continuum theories include the flow from the tricritical Ising model to the Ising model [11], the $O(3)$ sigma model at $\Theta = \pi$ and the $SU(2)_1$ principal chiral model [7], the flows between the minimal models [12], the Kondo problem [13], the "sausage" sigma model [14], and the Landau-Ginzburg flows to the $N=2$ minimal models [15]. Lattice models include integrable higher-spin XXX chains [16] and the Hubbard model [17,18].

The purpose of these lectures is to provide a pedagogical introduction, so we will skip many technical details. We have tried to make the sections reasonably independent of one another so that they can be read separately. In Section 2 we give a simple example of a massless field theory, the sine-Gordon model with imaginary potential. Section 3 contains a discussion of the usual Thirring model (for properties we discuss, Thirring and sine-Gordon can be used interchangeably) and its massless limit. Its main purpose is to introduce the physically odd idea of left-left or right-right scattering between massless particles from the Bethe ansatz view point. Section 4 gives a simple introduction to the thermodynamic Bethe ansatz carried out with the example of massless scattering.

Section 5 explores a little more how one can describe certain aspects of conformal field theories by massless particles with no left-right scattering. Previously unpublished results regarding the non-perturbative analysis of conserved quantities in conformal field theories are presented. Section 6 contains some comments on integrable theories with nontrivial left-right scattering. Section 7 is based on [12]. There we carry out explicitly the study of the sine-Gordon model with imaginary coupling and background field, the latter being introduced to get a simpler calculation. We show that even if massless scattering appears a little odd physically, it at least provides the proper analytic continuation of physical quantities in appropriate directions of the parameter space. Section 8 contains conclusions and questions of interest.

2. A simple example of massless field theory

A simple but generic example of integrable massless field theory is provided by the sine-Gordon model with imaginary "coupling" (i.e. prefactor of the cosine term). Recall the hamiltonian

$$H = \int ds \left[\frac{1}{2}\Pi^2 + \frac{1}{2}(\partial\phi)^2 + \lambda \cos\beta_{SG}\phi \right] \tag{2.1}$$

where we set $\beta_{UV}^2 = 8\pi\frac{t}{t+1}$ at the UV fixed point. This model is well-studied for λ real and known to be a massive field theory with a trivial fixed point in the IR. Taking λ imaginary does not look too physical at first. There are however several reasons for doing so. The massless flows between minimal models are obtained as reductions of this model, and are unitary even though (2.1) is not. In addition, (2.1) describes the flow of the $O(n)$ model to its low-temperature phase [19]; this covers interesting physical situations in condensed-matter physics such as self-avoiding polymers.

To see the difference in behavior between λ real and imaginary consider the large-t limit, where the cosine term is almost marginal and reliable perturbation theory at order $1/t$ can be carried out following the analysis of the XY model [20]. The RG equations are

$$\frac{d\lambda}{d\beta} = \left(2 - \frac{\beta^2}{4\pi}\right)\lambda, \tag{2.2}$$

and

$$\frac{1}{\beta^2}\frac{d\beta^2}{db} = -\pi^2\lambda^2. \tag{2.3}$$

From these one deduces, at first nontrivial order

$$\frac{d\beta^2}{db} = -\frac{\beta^4}{4\pi}\left[\left(\frac{8\pi}{\beta^2}-1\right)^2 - \frac{1}{t^2}\right].\tag{2.4}$$

For λ real, the initial derivative of β is negative, and the coupling flows to zero at large distance. For λ imaginary, the initial derivative is positive and the coupling increases monotonically to the IR fixed point with

$$\beta^2_{IR} = 8\pi\frac{t-1}{t}\tag{2.5}$$

Correspondingly both the UV and IR fixed points are Gaussian models with different radii of compactification, and we have a flow "within" $c = 1$.

The most interesting aspect concerns the evolution of the running central charge. The latter can be defined in several ways away from the fixed points, for instance using the two-point function of the stress energy tensor [21] or finite-size effects [22,23]. Qualitatively, these functions should behave in a similar way and are usually expected to describe the evolution of the number of degrees of freedom. By analogy with the second law of thermodynamics we expect that such a function should decrease when following a RG flow. It has been proven that the first type of c-function always decreases in unitary theories (the c-theorem [21]), and all known unitary examples of the second type also decrease. In our non-unitary problem, it is easy to compute them at first non trivial order in $1/t$ where they coincide. One finds then [12]

$$c = 1 + \frac{12}{t^3}\frac{e^{4(b-b_0)/t}\bigl(1 - e^{4(b-b_0)/t}\bigr)}{(1 + e^{4(b-b_0)/t})^3},\tag{2.6}$$

that has a roaming behavior. It indeed has $c = 1$ in the UV and IR but goes up and down in between, reaching a pair of extrema with values

$$c_\pm = 1 \pm \frac{2}{\sqrt{3}}\frac{1}{t^3}.\tag{2.7}$$

Of course the fact that c can increase and does not obey a nonunitary version of the c-theorem is not surprising and is an obvious consequence of the imaginary coupling. One might hope to slightly modify the perturbation to obtain an exact flow which would stop at the value c_+. This does not seem impossible in view of [24], which describes flows that exactly interpolate between minimal models and flows that go "very near" them.

The fact that c is not monotonic sheds some doubt on the reliability of the running c-function as a measure of the number of degrees of freedom. Let us emphasize that (2.6) is indeed related to the absolute ground state of the theory — it *is* the "c_{eff}" supposed to qualitatively replace c in the non-unitary case.

3. Massless scattering at the conformal point

We consider the example of the massive Thirring model on a circle of length L with hamiltonian

$$H = \int dx \left[-i \left(\psi_1^+ \partial_x \psi_1 - \psi_2^+ \partial_x \psi_2 \right) + m_0 \left(\psi_1^+ \psi_2 + \psi_2^+ \psi_1 \right) + 2g_0 \psi_1^+ \psi_2^+ \psi_2 \psi_1 \right]. \qquad (3.1)$$

which is solvable by the Bethe ansatz [25,26]. The analogous calculation is done for the sine-Gordon model in [27]. As a result eigenenergies take the form

$$\mathcal{E} = \sum_i m_0 \cosh \xi_i, \qquad (3.2)$$

and momenta

$$P = \sum_i m_0 \sinh \xi_i, \qquad (3.3)$$

where the ξ_i are bare rapidities of pseudoparticles satisfying the Bethe ansatz equations

$$\exp(i m_0 L \sinh \xi_i) \prod_j \frac{\sinh(\xi_i - \xi_j + 2i\mu)}{\sinh(\xi_i - \xi_j - 2i\mu)} = 1, \qquad (3.4)$$

and $\cot \mu = -\frac{1}{2} g_0$. One recognizes in (3.4) the conditions for the wave function to be periodic, the phase shift being a combination of a free term and factorized scattering between pairs of pseudoparticles. Thus the elements of the product in (3.4) are the "bare" S-matrix elements for the pseudoparticles, which we denote as $S_0(\xi_i - \xi_j) \equiv \exp[i\phi_0(\xi_i - \xi_j)] \equiv \frac{1+i\Lambda}{1-i\Lambda}$. Equation (3.4) follows from the form of Bethe wave functions

$$\psi(x_1, \ldots, x_N | \xi_1 \ldots, \xi_N) = \exp(i m_0 \sum_i x_i \sinh \xi_i) \prod_{i<j} [1 + i\Lambda(\xi_i - \xi_j)\mathrm{sign}(\xi_i - \xi_j)]. \qquad (3.5)$$

This wave function is almost free, with only a phase shift when the coordinates are exchanged. Of course since we are dealing with fermions, the real wave function has still to be antisymmetrized so

$$\Psi(x_1, \ldots, x_N) = \sum_P \mathrm{sign}(P)\psi(x_1, \ldots, x_N | \xi_{P1}, \ldots, \xi_{PN}), \qquad (3.6)$$

(this makes sense because Λ is an odd function, or equivalently $S_0(\xi)S_0(-\xi) = 1$) and all rapidities must therefore be different.

The study of the quantum field theory requires building the ground state by filling the appropriate Dirac sea, and then finding the excitations and their scattering. We must mention a difficulty concerning the choice of UV cutoff. In the attractive regime $g_0 > 0$ essentially all cutoffs produce the same results, but the situation is different in the repulsive regime $g_0 < 0$. (This is unfortunately the regime where one truncates the sine-Gordon model to describe massive perturbations of Virasoro minimal models by the ϕ_{13} operator [28].) The rapidity cutoff of [29] leads to many more particles in the spectrum than the cutoff of [30], which is essentially a lattice cutoff [31]. The latter regularization reproduces the conjectures of [8] and seems in agreement with the expected physics of perturbed minimal models [32]. We avoid entering into technical details and give a "morally" correct discussion in the following.

Another difficulty arises in the eigenfunctions of (3.1) as a result of terms like (sign δ). Depending on the regularization, different relations between the bare coupling g_0 and the parameter μ are found. In [26] the relation $\mu = \frac{\pi + g_0}{2}$ is used instead. We shall carry out the discussion in terms of the variable μ.

The ground state is easily built by filling the sea with antipseudoparticles, i.e. filling the line $Im(\xi) = \pi$. There are various kind of excitations. We shall discuss only the case of solitons s and antisolitons a (these are the only particles with non-vanishing $U(1)$ charge). For instance a pair of solitons is simply obtained by making two holes in the ground state distribution. Antisolitons, or pairs of solitons and antisolitons are obtained similarly, with the addition of some strings [26] of pseudoparticles around $Im(\xi) = 0$. Of course the introduction of such holes induces a shift in the distribution of the ξ's, the so-called *backflow*. As a result the mass and rapidity of the solitons are renormalized, giving

$$E = m \cosh \theta, \ p = m \sinh \theta, \tag{3.7}$$

with

$$m = m_0 \frac{\gamma e^{\Lambda(1-\gamma)}}{\pi(\gamma - 1)} \tan \pi\gamma, \ \theta = \gamma\lambda, \tag{3.8}$$

with $\gamma \equiv \frac{\pi}{2\mu}$ and λ denotes the bare rapidity of the particles (s, a) (again there are some slight differences between authors for this formula). As shown in [26] the S-matrix of solitons can be extracted from the Bethe ansatz equations. The S-matrix elements are defined as follows. Suppose first we have only one soliton, with bare rapidity λ_1. Then the total phase shift collected by the wave function when the argument of the soliton goes around the circle is

$$\phi_1 = m_0 L \sinh \lambda_1 + \sum_j \phi_0(\lambda_1 - \bar{\xi}_j) \tag{3.9}$$

where the sum is taken over all pseudoparticles in the sea and $\bar{\xi}$ indicates shifted (with respect to the ground state) rapidities due to the backflow. Suppose then we have two solitons with bare rapidities λ_1 and λ_2. Then the total phase shift of the wave function when the argument of the first soliton again goes around the circle is

$$\phi_2 = m_0 L \sinh \lambda_1 + \sum_j \phi_0(\lambda_1 - \tilde{\xi}_j). \tag{3.10}$$

where $\tilde{\xi}$ are shifted rapidities. One then defines the S-matrix element by $\ln S = i(\phi_2 - \phi_1)$. Complete computation shows that it depends only on the difference of the rapidities. Moreover one also checks that for more particles, the phase shifts simply add and the scattering can be decomposed as a succession of two-particle ones. The resulting S-matrix, therefore a solution of the Yang-Baxter equation, has the well-known matrix elements

$$a = Z(\theta) \sinh\left(\frac{i\pi - \theta}{t}\right), \quad b = Z(\theta) \sinh\left(\frac{\theta}{t}\right), \quad c = Z(\theta) \sinh\left(\frac{i\pi}{t}\right) \tag{3.11}$$

where a corresponds to $ss \to ss$ scattering, b to $sa \to sa$ and c to $sa \to as$. We give the the expression for normalization factor Z in sect. 7. The symmetry under $s \leftrightarrow a$ gives the remainder of the elements. We have parametrized

$$\gamma \equiv \frac{\pi}{2\mu} \equiv \frac{\pi}{t+1}. \tag{3.12}$$

The S-matrix (3.11) can be manipulated to exhibit $\hat{U}_q sl(2)$ symmetry [33] with

$$q = -\exp\left(-\frac{i\pi}{t}\right). \tag{3.13}$$

This is a dynamical symmetry; there is also a kinematical symmetry $U_{q_0} sl(2)$ with $q_0 = -\exp\left(-\frac{i\pi}{t+1}\right)$ following from the Bethe ansatz equations [34]. Notice the shift of the denominator.

In order to reach the deep UV limit, we let the mass m_0 go to zero. Particles with non-vanishing energy must have rapidities with very large modulus, of the order $\xi_0 = \ln(M/m_0) >> 1$, where M is a not-yet-defined parameter with the dimension of mass.

There are thus two regions of interest in which we set respectively $\xi = \xi_0 + \theta$ and $\xi = -\xi_0 + \theta$, θ remaining finite. The spectrum obviously splits into right and left excitations with

$$E_R \approx \frac{m_0}{2} e^{\xi_0} e^{\theta} \equiv \frac{M}{2} e^{\theta} = p_R; \qquad E_L \approx \frac{m_0}{2} e^{\xi_0} e^{-\theta} \equiv \frac{M}{2} e^{-\theta} = -p_L. \qquad (3.14)$$

and we have a *doubling* of species $a_{L,R}, s_{L,R}$ (see 1). For pseudoparticles of the same kind, the phase shifts are unchanged as $\xi_i - \xi_j = \theta_i - \theta_j$. For particles of different kinds however $\xi_i - \xi_j \approx \pm 2\xi_0 \to \pm\infty$ so the phase shifts become constants, independent of the rapidities. As a result, in the computation of the S-matrix, the LL and RR scattering are the *same* as the ones for corresponding massive particles computed above $S_{LL} = S_{RR} = S^1$, while the LR scattering becomes trivial.

The S-matrices in the massless limit can also be obtained by studying the XXZ spin chain, which is a lattice regularization of the massless Thirring model. This is a simple generalization of the work of [3] on the XXX chain.

It should be clear finally that the properties of a massless theory with trivial LR scattering are independent of the mass scale M. Indeed, changing M is equivalent to shifting ξ_0, and the analysis only depends on rapidity differences.

4. Thermodynamic Bethe ansatz

The technique we now discuss involves computing the free energy of an integrable lattice model (e.g. the XXZ model) or an integrable quantum field theory (e.g. the Thirring model) on an infinite line at finite temperature T. There are two approaches to this calculation. In the traditional "bare" approach, one finds the energy and entropy of the states of the model using the Bethe ansatz. The thermodynamic state is the state which minimizes the free energy. The limit $T \to 0$ gives the ground-state energy and the vicinity $T \approx 0$ the structure of low-lying excitations. In the massive Thirring model this approach starts with the bare equations (3.4). In the second approach one forgets the bare theory completely and studies instead the thermodynamics of a gas composed of the various "physical" excitations (like the soliton and antisoliton of the sine-Gordon model) scattering with their respective S-matrices. As in the first approach, one determines their energy and entropy and again minimizes the free energy. The first approach was used in works like

---

^{1} In some cases like in [7] there is an additional phase in the definition of S_{LL} and S_{RR}.

[35,30]. The second approach, pioneered in [36], is more recent, and is usually called the thermodynamic Bethe ansatz (TBA).

The second approach allows some convenient short cuts: instead of solving the theory one first establishes that it is integrable, conjectures the excitations and their S-matrix using intuition and symmetry arguments (with some care), and uses the TBA to derive various properties. When both approaches can be implemented, they of course give the same results; the quasiparticle excitations used in the second can be found from the Bethe ansatz equations by filling the Fermi or Dirac sea. However, there are examples (at least in the massive case) where a lattice model is not integrable but its continuum limit is an integrable quantum field theory. Usually the only way of defining the continuum model is by a perturbed conformal field theory, so the usual Bethe ansatz methods cannot be applied; the only recourse is to use the second approach. A classic example is the Ising model at $T = T_c$ in a magnetic field [9][2].

As a simple example we describe the TBA for a single type of massless particle, say right-moving, with energy and momentum parametrized as in (1.2). The scattering is described by a single S-matrix element S_{RR}. Quantizing a gas of such particles a circle of length L requires the momentum of the ith particle to obey

$$\exp\left(i\frac{Me^{\theta_i}}{2}L\right)\prod_{j\neq i}S_{RR}(\theta_i - \theta_j) = 1. \tag{4.1}$$

One can think of this intuitively as bringing the particle around the world through the other particles; one obtains a product of two-particle S-matrix elements because the scattering is factorizable. This is the renormalized equivalent of the bare relation (3.4).

Going to the $L \to \infty$ limit, we introduce the density of rapidities indeed occupied by particles $\rho(\theta)$ and the density of holes $\tilde{\rho}$. A hole is a state which is allowed by the quantization condition (4.1) but which is not occupied, so that the density of possible rapidities is $\rho(\theta) + \tilde{\rho}(\theta)$. Taking the derivative of the log of (4.1) yields

$$2\pi[\rho(\theta) + \tilde{\rho}(\theta)] = \frac{ML}{2}e^{\theta} + \int_{-\infty}^{\infty} \Phi(\theta - \theta')\rho(\theta'), \tag{4.2}$$

where

$$\Phi(\theta) = \frac{1}{i}\frac{d}{d\theta}\ln S(\theta).$$

---

[2] Observe however that there is another integrable lattice model based on E_8 that is integrable and has the same scaling limit as Ising in a magnetic field.

98

To determine which fraction of the levels is occupied we do the thermodynamics [35]. The energy is

$$\mathcal{E} = \int_{-\infty}^{\infty} \rho(\theta)\frac{M}{2}e^{\theta}\,d\theta,$$

and the entropy is

$$\mathcal{S} = \int_{-\infty}^{\infty} \left[(\rho + \tilde{\rho})\ln(\rho + \tilde{\rho}) - \rho\ln(\rho) - \tilde{\rho}\ln(\tilde{\rho})\right]d\theta.$$

The free energy per unit length $\mathcal{F} = (\mathcal{E} - T\mathcal{S})/L$ is found by minimizing it with respect to ρ. The variations of $\mathcal{E}$ and $\mathcal{S}$ are

$$\delta\mathcal{E} = \int_{-\infty}^{\infty} \delta\rho\,\frac{M}{2}e^{\theta}\,d\theta$$

$$\delta\mathcal{S} = \int_{-\infty}^{\infty} \left[(\delta\rho + \delta\tilde{\rho})\ln(\rho + \tilde{\rho}) - \delta\rho\ln(\rho) - \delta\tilde{\rho}\ln(\tilde{\rho})\right]d\theta.$$

It is convenient to parametrize

$$\frac{\rho(\theta)}{\tilde{\rho}(\theta)} \equiv \exp\left(-\frac{\epsilon}{T}\right), \tag{4.3}$$

giving

$$\delta\mathcal{S} = \int_{-\infty}^{\infty} \left[\delta\rho\ln\left(1 + e^{\epsilon/T}\right) + \delta\tilde{\rho}\ln\left(1 + e^{-\epsilon/T}\right)\right]d\theta.$$

Using (4.2) allows us to find $\tilde{\rho}$ in terms of ρ. Denoting convolution by $\star$, this gives $2\pi(\delta\rho + \delta\tilde{\rho}) = \Phi \star \delta\rho$ so

$$\delta\mathcal{S} = \int_{-\infty}^{\infty} \left[\frac{\epsilon}{T} + \frac{\Phi}{2\pi}\star\ln\left(1 + e^{-\epsilon/T}\right)\right]\delta\rho\,d\theta.$$

Hence the extremum of $\mathcal{F}$ occurs for

$$\frac{M}{2}e^{\theta} = \epsilon + T\frac{\Phi}{2\pi}\star\ln\left(1 + e^{-\epsilon/T}\right). \tag{4.4}$$

and one has then, expressing $\tilde{\rho}$ from (4.2) and using (4.4)

$$\mathcal{F} = \mathcal{E}_0 - T^2\frac{M}{4\pi T}\int_{-\infty}^{\infty} e^{\theta}\ln\left(1 + e^{-\epsilon/T}\right)d\theta. \tag{4.5}$$

The ground state energy $\mathcal{E}_0$ cannot be obtained by this method since all the information we use is the structure of excitations above the ground state, so we set $\mathcal{E}_0 = 0$ for the rest of this section.

The limit $T \to 0$ of this system is interesting. We introduce the positive and negative parts of the pseudoenergy satisfying therefore

$$\frac{M}{2}e^\theta = \epsilon^+ + \epsilon^- - \frac{\Phi}{2\pi} \star \epsilon^-, \tag{4.6}$$

In this limit the solution is $\epsilon^- = 0$, $\epsilon^+ = \frac{M}{2}e^\theta$. It follows from this and (4.3) that $\rho \to 0$ as $T \to 0$, which is required because our TBA provides the structure of excitations over the ground state. In general $-\epsilon^-$ (resp. ϵ^+) gives the excitation energy for holes (resp. particles).

The knowledge of $\mathcal{F}$ leads to the determination of the central charge of the theory. We have $\mathcal{F} = -\frac{T}{L}\ln Z$, where Z is the partition function of the one dimensional quantum field theory at temperature T (in the following we refer to this point of view as "thermal"). In Euclidean formalism, this corresponds to a theory on a torus with finite size in time direction $R = 1/T$. By modular invariance, identical results should be obtained if one quantizes the theory with R as the space coordinate . For large L, $Z = e^{-E(R)L}$, where $E(R)$ is the ground-state (Casimir) energy with space a circle of length R. Thus $\mathcal{F} = E(R)/R$. In the following we refer to this as the "finite-size" point of view. Conformal invariance requires that at a fixed point this Casimir energy is $E(R) = -\frac{\pi c}{6R}$, where c is the central charge [37]. Going back to the thermal point of view, $\mathcal{F} = -\frac{\pi c T^2}{6}$ and the specific heat is $C = -\frac{\pi c T}{3}$.

Observe from (4.4) and (4.5) that the free energy does not depend on the mass scale M, because it can be rescaled by a shift in rapidities. By dimensional analysis one has therefore $\mathcal{F} = T^2 \times$ constant. This scale invariance is a manifestation of the fact that S_{LL} and S_{RR} are describing only conformal properties. With massive particles or with nontrivial left-right massless scattering, $\mathcal{F}$ does depend on M/T, giving a running central charge.

We can analytically find this central charge from (4.4). We take the derivative of (4.4) with respect to θ and solve for e^θ. Substituting this in (4.5), we have

$$
\begin{aligned}
\mathcal{F} &= -\frac{T}{2\pi}\int d\theta \left[\frac{d\epsilon}{d\theta}\ln(1 + e^{-\epsilon/T}) - \int d\theta' \ln(1 + e^{-\epsilon(\theta)/T})\Phi(\theta - \theta')\frac{d\epsilon}{d\theta'}\frac{1}{1 + e^{\epsilon(\theta')/T}}\right] \\
&= -\frac{T}{2\pi}\int d\theta \frac{d\epsilon}{d\theta}\left[\ln(1 + e^{-\epsilon/T}) + (\epsilon - \frac{M}{2}e^\theta)\frac{1}{1 + e^{\epsilon(\theta)/T}}\right] \\
&= -\mathcal{F} - \frac{T}{2\pi}\int d\theta \frac{d\epsilon}{d\theta}\left[\ln(1 + e^{-\epsilon/T}) + \frac{\epsilon}{1 + e^{\epsilon/T}}\right],
\end{aligned}
$$

where we use (4.4) again to get to the second line. We can replace the integral over θ with one over ϵ, giving an ordinary integral

$$\mathcal{F} = -\frac{T}{4\pi} \int_{\epsilon(-\infty)}^{\infty} d\epsilon \left[\ln(1 + e^{-\epsilon/T}) + \frac{\epsilon}{1 + e^{\epsilon(\theta)/T}} \right],$$

A change of variables gives

$$\mathcal{F} = -\frac{T}{2\pi} L\left(\frac{1}{1+x_0}\right),$$
(4.7)

where $L(x)$ is the Rogers dilogarithm function

$$L(x) = -\frac{1}{2} \int_0^x \left(\frac{\ln(1-y)}{y} + \frac{\ln y}{1-y} \right) dy,$$

and $x_0 \equiv \exp(\epsilon(-\infty)/T)$ is obtained from (4.4) as

$$\frac{1}{x_0} = \left(1 + \frac{1}{x_0} \right)^I,$$
(4.8)

with $I = \frac{1}{2\pi} \int \Phi$.

For example, when the S matrix is a constant, $\Phi = 0$, $x_0 = 1$ and

$$\mathcal{F} = -\frac{T\pi}{24},$$
(4.9)

where we used $L(1/2) = \frac{\pi^2}{12}$. Here we find $c_L = \frac{1}{4}$. In a left-right-symmetric quantum field theory, the right sector makes the same contribution, giving the total central charge $c = \frac{1}{2}$ required for free fermions.

For the nonunitary Lee-Yang S-matrix [38] one has $I = -1$. In that case $x_0 = \frac{1+\sqrt{5}}{2}$. Using $L\left(\frac{3-\sqrt{5}}{2}\right) = \frac{2}{5}\frac{\pi^2}{6}$ one finds $c_R = \frac{2}{10}$ and $c = \frac{2}{5}$ after left and right contributions have been collected. This is indeed the effective central charge for the Lee-Yang problem; it is not equal to the true central charge $c = -\frac{22}{5}$ because of the presence of an operator with negative dimension in the vacuum.

It is possible to do the same computation with the massless pair (a, s) scattering with (3.11). The computation is technically more complicated because the scattering is non-diagonal. We just refer the reader to references [39,32] for details. Simply observe that in the case $\mu = \frac{\pi}{2}$ the scattering becomes diagonal and because of the doubling of the number of species, the above calculation gives rise to $c = 1$ as expected.

Besides the central charge, some conformal dimensions can also be identified using the TBA. To do so one includes an imaginary chemical potential μ_b (not to be confused with

μ in section 3) for each species of particle b. As before, we minimize the corresponding free energy $\mathcal{G} = \mathcal{E} - TS - \sum_b \mu_b \mathcal{N}_b$, and find similar results with, in most cases, $\exp(-\epsilon_b/T)$ replaced by $\exp(-(\epsilon_b - \mu_b)/T)$. In the finite-size point of view, the introduction of a chemical potential amounts to considering the theory on a circle of length R with twisted boundary conditions. As is well known, the ground state energy in that case gives an effective central charge $c_{eff} = c - 24h$ where h is related to the twist. For RR scattering given by (3.11) for instance, with $\mu_s = -\mu_a = i\alpha\pi T/t$ one finds $h = \frac{\alpha^2}{4t(t+1)}$. Thus one recovers the dimensions of vertex operators in a Gaussian model.

The question of reconstructing the whole quantum field theory from a massless scattering theory with no left-right scattering is still open. For Virasoro minimal models two independent such theories are probably necessary, as there are two fundamental quantum groups, or two labels in the Kac table. A bit of progress in this direction is presented in the next section.

5. Integrable CFT and massless scattering: Virasoro conserved quantities

We showed in sect. 4 how one obtains the free energy in a massless integrable theory, and compared this result with conformal field theory predictions. In an integrable theory, the energy $\mathcal{E} \equiv \langle E \rangle$ is just the first of an infinite series of conserved charges. These conserved charges can often be expressed as suitably regularized powers of the energy-momentum tensor. Since we derive the particle densities of the thermodynamic state, we can calculate the expectation value of any quantity which can be expressed in terms of the particles. This suggests that the conserved charges should be related to expectation values $\langle E^n \rangle$. In this section we show that in the sine-Gordon model, this is indeed true, thus verifying non-perturbatively what had been shown classically and perturbatively in the quantum theory [40,41].

We start with free fermions with antiperiodic boundary conditions on a circle of length R in order to select the ground state. We are thinking about the system in the finite-size point of view discussed in the last section. Consider the quantity

$$
\begin{aligned}
\langle E^n \rangle_{gs} &= \frac{1}{2} \left(\frac{2\pi}{R}\right)^n \left\langle \sum_{j=-\infty}^{\infty} (j + 1/2)^n : \psi_{-j-1/2}\psi_{j+1/2} : \right\rangle_{gs} \\
&= (-)^n \frac{1}{2} \left(\frac{2\pi}{R}\right)^n \sum_{j=0}^{\infty} (j + 1/2)^n.
\end{aligned}
\tag{5.1}
$$

The sum can be evaluated by ζ-function regularization leading to

$$\langle E^{2k+1}\rangle_{gs} = \frac{1}{2}\left(\frac{2\pi}{R}\right)^{2k+1}\left(1 - \frac{1}{2^{2k+1}}\right)\zeta(-2k-1),\tag{5.2}$$

and

$$\langle E^{2k}\rangle_{gs} = 0,\tag{5.3}$$

where the last result follows from $\zeta(-2k) = 0$. For $k = 0$ one gets $\mathcal{E} = \langle E\rangle = -\frac{\pi}{6R}\frac{c}{2}$ with $c = 1/2$ ($c/2$ appears because we concentrate on one chirality).

We can compute the analogous quantity from the thermal point of view by putting the particles on a circle of large length L at temperature $T = 1/R$. The TBA analysis of sect. 4 gives

$$\begin{aligned}
\langle E^n\rangle_{TBA} &= \int_{-\infty}^{\infty} d\theta \left(\frac{Me^\theta}{2}\right)^n \rho(\theta) \\
&= \int_{-\infty}^{\infty} d\theta \left(\frac{Me^\theta}{2}\right)^n (\rho(\theta) + \tilde\rho(\theta))\frac{1}{1 + \exp(\epsilon/T)} \\
&= nLT \int_{-\infty}^{\infty} \frac{d\theta}{2\pi} \left(\frac{Me^\theta}{2}\right)^n \ln(1 + e^{-\epsilon/T}),
\end{aligned}\tag{5.4}$$

where we used the fact that $2\pi(\rho + \tilde\rho) = LT d\epsilon/d\theta$, which is proven by showing that they obey the same integral equation.

For free fermions, $\epsilon = Me^\theta/2$. The expectation value (5.4) is

$$\langle E^n\rangle_{TBA} = \frac{L}{R}\frac{n}{2\pi}R^{-n} \int_0^{\infty} x^{n-1}\ln(1 + e^{-x})dx,$$

or, restricting to $n = 2k + 1$

$$\langle E^{2k+1}\rangle_{TBA} = \frac{L}{R}\frac{(2k+1)!}{2\pi}R^{-2k-1}\left(1 - \frac{1}{2^{2k+1}}\right)\zeta(2k+2).\tag{5.5}$$

To compare (5.2) and (5.5) recall the identities [42]

$$\zeta(2k+2) = \frac{(2\pi)^{2k+2}}{2(2k+2)!}(-1)^k B_{2k+2}; \qquad \zeta(-2k-1) = -\frac{B_{2k+2}}{2k+2},$$

where B_n are Bernouilli numbers. Hence we find

$$\langle E^{2k+1}\rangle_{TBA} = (-1)^{k+1}\frac{L}{R}\langle E^{2k+1}\rangle_{gs}.\tag{5.6}$$

For $k = 0$ we recover $\mathcal{F}$ as derived in the previous section, using the fact that $\mathcal{E} = -\mathcal{F}$ for theories with no left-right scattering. More generally (5.6) follows from the relation of $\langle E^{2k+1} \rangle_{TBA}$ (resp. $\langle E^{2k+1} \rangle_{gs}$) to the energy-momentum tensor component T_{xx}^{k+1} (resp. T_{yy}^{k+1}) and $T_{xx} + T_{yy} = 0$ at the conformal point.

Notice that when n is even, the results are quite different, since $\langle E^{2k} \rangle_{TBA} \neq 0$ while (5.3) holds. The usual argument is that such even powers do not correspond to any local quantity in the quantum field theory, and therefore the two results cannot be compared as we did in (5.6).

The generalization of this computation to the case of nontrivial scattering is not straightforward, but it is useful. We can compute – at least numerically – the quantities $\langle E^{2k+1} \rangle_{TBA}$ from (5.4). In general the conformal field theory cannot easily be described by oscillators as in the free theory above, so we do not compute the equivalent of $\langle E^{2k+1} \rangle_{gs}$. What we can however compute using *only* the Virasoro algebra are quantities like

$$\left\langle \int : T^k : \right\rangle_h, \tag{5.7}$$

where the integral is over the period of the cylinder, and average is taken in a state of conformal weights (h, h), generalizing the ground state. The double dots indicate normal ordering on the cylinder. Recall that this normal ordering leaves room for non-vanishing expectation values. These can be computed by explicitly performing the subtraction of the divergent terms. The result coincides with the simpler zeta regularization. Such a quantity cannot be directly compared to $\langle E^{2k+1} \rangle_{TBA}$ because of a non-trivial renormalization factor. In the free fermion case for instance $\langle \int : T^2 : \rangle_{gs} = \left(\frac{2\pi}{R} \right)^3 \frac{49}{96} \zeta(-3)$ while $\langle E^3 \rangle_{gs} = \left(\frac{2\pi}{R} \right)^3 \frac{7}{16} \zeta(-3)$, because $: T^2 := \frac{3}{8} : \partial^2 \psi \partial \psi : - \frac{5}{24} : \partial^3 \psi \partial \psi :$. However this normalization factor occurs from short-distance singularities and therefore does not depend on boundary conditions. We can therefore compare ratios of the moments of $T, \partial T, \ldots$ and $\langle E^n \rangle$'s for different boundary conditions around the circle, or equivalently the choice of state in which the average (5.7) is taken. As explained in the previous section, in the thermodynamics this means taking imaginary chemical potentials for the particles.

A crucial point is that we must treat the theory as a non-minimal, non-unitary theory with central charge $c = 1 - \frac{6}{t(t+1)}$. We use this value of c, denoted as c_{min}, in the Virasoro algebra computations. Boundary conditions other than these twisted ones give rise to $c_{eff} = c_{min} - 24h$, where the conformal dimension h is computed with respect to the c_{min} ground state. The usual sine-Gordon ground state (with $c_{eff} = 1$) is then interpreted as arising from an operator of negative dimension. With these definitions one finds for instance, using the Virasoro algebra and ζ-function regularization,

$$\langle \int : T^2 : \rangle_h = \frac{1}{5760}(10c_{eff}^2 + 40c_{eff} + 4c_{min}), \tag{5.8}$$

$$\langle \int : T^3 : \rangle_h = -\frac{1}{(24)^3}\left(c_{eff}^3 + 12c_{eff}^2 + \frac{192}{5}c_{eff} + \frac{32}{7}c_{min} + \frac{6}{5}c_{min}c_{eff}\right), \tag{5.9}$$

$$\langle \int : T\partial^2 T : \rangle_h = -\frac{1}{(24)^3}\left(\frac{48}{5}c_{eff} + \frac{32}{7}c_{min}\right). \tag{5.10}$$

The c_{eff} result from the non-zero $\langle L_0 \rangle_h$ on the cylinder.

Our strategy has been simply to compute numerically the values of $\langle E^{2k+1} \rangle_{TBA}$ for various chemical potentials, denoting these by $\langle E^{2k+1} \rangle_\alpha$. For sine-Gordon the chemical potentials are $\mu_s = -\mu_a = i\alpha\pi T/t$; this results in $c_{eff} = 1 - 24h$, where $h = \frac{\alpha^2}{4t(t+1)}$. The general TBA equations with fugacities are written out in [32]. Since the numerics are crucial to obtaining our result, we describe the methods briefly. The multi-function generalization of (4.4) is of the form

$$\epsilon_a(\theta) = \nu_a(\theta) + \sum_b \int d\theta' \phi_{ab}(\theta - \theta')\ln(1 + \lambda_b e^{-\epsilon_b(\theta')/T})$$

To find ϵ_a numerically, we solve this iteratively. We guess the initial ϵ_a; $\epsilon_a = \nu_a$ usually works. Then one evaluates the right-hand side numerically by discretizing the integal; this gives the next guess for ϵ_a. Usually the iteration converges; occasionally one needs to use take a linear combination of $x(\text{guess})+(1-x)(\text{iteration})$ for the next guess. More elaborate methods to improve convergence are described in [43]. Once this procedure has obtained ϵ_a to the desired accuracy, the expression (5.4) for $\langle E^n \rangle$ can then be numerically evaluated. We note that this numerical procedure for solving non-linear integral equations is generally far simpler to implement that those for solving non-linear differential equations.

The numerical results are rather interesting. As before, we have $c_{min} = 1 - 6/t(t+1)$ and $c_{eff} = 1 - 6\alpha^2/t(t+1)$. We find

$$\langle E^3 \rangle_\alpha = f_t(10c_{eff}^2 + 40c_{eff} + 4c_{min})$$

$$\langle E^5 \rangle_\alpha = g_t(c_{eff}^3 + 12c_{eff}^2 + 40c_{eff} + 2c_{eff}c_{min} + \frac{16}{3}c_{min} + \frac{8}{21}c_{min}^2). \tag{5.11}$$

Moreover, we find that at least for $\langle E^3 \rangle$, the prefactor takes reasonably simple values; to excellent numerical accuracy we have

$$f_2 = \frac{7}{48} \quad f_3 = \frac{\pi^2}{70} \quad f_4 = \frac{1001}{7200} \quad f_5 = \frac{2\pi^2}{143}$$

This is a hint that these numbers can be derived analytically from the TBA, but we have tried and failed to do so. We can make the amusing observation that $\langle E^n \rangle$ for free fermions can be written in terms of polylogarithms, just like the dilogarithms written in the last section for $\langle E \rangle$. We also note that the free energy of the impurity in the Kondo problem (see [1]) can be written as a sum of these conserved quantities, another hint of interesting hidden structure.

Hence we can fit our numerical results to (5.8) for $n = 3$ and to a linear combination of (5.9) and (5.10) for $n = 5$. We see that (5.8) is indeed proportional to $\langle E^3 \rangle_{TBA}$ and that

$$\langle E^5 \rangle_\alpha \propto \langle \int : T^3 : + \left(\frac{c_{min} + 2}{12} \right) T \partial^2 T : \rangle_h. \tag{5.12}$$

After the numerical computation was completed we checked that (5.12) agrees with the conserved quantity at grade 5 in [44]. This is of course no surprise. Recall that in the classical sine-Gordon theory, the conserved quantities are precisely expressed as the sum of odd powers of the momenta: we simply check here that this result holds in the quantum theory as well. This is expected, but as far as we know, was checked only perturbatively so far [40,41]. Hence by massless scattering we recover not only the central charge and conformal weights of a conformal field theory, but also the conserved quantities which involve the Virasoro algebra itself, making the connection between the two points of view a little closer. One might wonder if there is an action of the Virasoro algebra on the massless particles.

6. Nontrivial left-right scattering

The massless Thirring model has no non-trivial LR scattering because L and R excitations are infinitely separated in the rapidity plane. On the other hand, the most interesting situations occur when the LR scattering is non-trivial. In that case, the theory is not scale invariant, and is described by two different conformal field theories in the UV and IR limits. To get such a situation in the Thirring model, we need to arrange for massless L and R excitations that both occur around the same region of rapidities. A way to do so is to choose a *purely imaginary* bare mass in (3.1) $m_0 = -i|m_0|$. Indeed the result (3.2) still holds, so

$$E = -i|m_0| \sum_i \cosh \xi_i. \tag{6.1}$$

If we restrict to the consideration of real energies we need $Im(\xi) = \pm\pi/2$. Then for $\xi = \pm i\frac{\pi}{2} + \nu$, $e = \pm|m_0|\sinh\nu$. One therefore expects the ground state to resemble figure 2 with the half lines

$$Re(\xi) < 0, \ Im(\xi) = \frac{\pi}{2}; \ Re(\xi) > 0, \ Im(\xi) = -\frac{\pi}{2}, \tag{6.2}$$

filled up. Actually, because the theory is not free, the determination of the ground state is slightly more delicate — the interaction between the various Bethe ansatz roots must be considered. The choice of the cutoff is also important, as well as the sign of g_0. One finds typically that the picture (6.2) is almost correct, up to some exponentially decaying density on the other side of the half-lines. This produces therefore the necessary massless excitations around a common region $\xi = 0$. More details of this approach will be presented in [45].

An imaginary mass in the Thirring model is like imaginary prefactor in front of the cosine term in sine-Gordon, so we recover the situation discussed in sect. 2. As explained there the appearance of imaginary numbers is more natural than may appear at first sight. Just as the massive minimal models perturbed by ϕ_{13} are related to the ordinary sine-Gordon model [28], the massless flow between minimal models [21,23] is related to the model with imaginary mass. The non-unitarity of the imaginary-mass model does not exclude unitarity for a subsector (the perturbed massless minimal model). For more details see [12].

It is easy to generalize the TBA of section 4 to models with a a non-trivial S_{LR}. This time, the running central charge depends nontrivially on M/T. Its UV and IR values can be easily found. As discussed in the introduction, we expect that the IR conformal field theory is characterized by only S_{LL} and S_{RR}, so its LR scattering should be trivial. One finds as before

$$c_{IR} = c_R + c_L = 2c_R = \frac{6}{\pi^2}L\left(\frac{1}{1+y_0}\right), \tag{6.3}$$

where y_0 is the solution of $1/y_0 = (1 + 1/y_0)^{I_1}$ with $I_1 = \frac{1}{2\pi}\int \Phi_{LL}$. In the UV coupling between left and right particles has to be considered leading to

$$c_{UV} = \frac{6}{\pi^2}\left[2L\left(\frac{1}{1+x_1}\right) - L\left(\frac{1}{1+x_0}\right)\right], \tag{6.4}$$

where $1/x_1 = (1 + 1/x_1)^{I_1+I_2}$ and $I_2 = \frac{1}{2\pi}\int \Phi_{LR}$.

A simple example of a massless integrable field theory is the flow from tricritical to critical Ising model. As discussed in [11] the spectrum consists of a right mover and a left mover, the Goldstino resulting from spontaneously-broken supersymmetry. Because the IR conformal field theory is a free fermion, S_{LL} and S_{RR} must be trivial. The left-right scattering is given by

$$S_{RL}(\theta_R - \theta_L) = -\tanh\left(\frac{\theta_R - \theta_L}{2} - i\frac{\pi}{4}\right). \tag{6.5}$$

The compatibility of left-right and right-left interchange of arguments of the wavefunction requires that

$$S_{LR}(\theta_L - \theta_R)S_{RL}(\theta_R - \theta_L) = 1, \tag{6.6}$$

so here

$$S_{LR}(\theta_L - \theta_R) = \tanh\left(\frac{\theta_L - \theta_R}{2} - i\frac{\pi}{4}\right). \tag{6.7}$$

In the IR limit (the Ising model) where $\theta_R - \theta_L \to \infty$ one checks that both matrix elements go to -1 as expected. With this S-matrix we have $I_1 = 0$ and $I_2 = 1/2$, so $x_0 = 1$ and $x_1 = \frac{\sqrt{5}-1}{2}$. Using values of dilogarithms given in sect. 4 we find $c_{IR} = 1/2$ and $c_{UV} = 7/10$ as desired.

Following (6.5) and (6.6) notice that

$$S_{RL}(\theta)S_{RL}(-\theta) = -1, \tag{6.8}$$

a result that must be carefully compared to the usual $S(\theta)S(-\theta) = 1$ for diagonal massive (or left-left or right-right) scattering.

7. Sine-Gordon model in a background field

In this section we discuss the sine-Gordon model in a background field coupled to the $U(1)$ soliton-number charge. In the traditional bare approach, this field would modify the Dirac or Fermi sea. In our approach, this makes it energetically favorable for physical particles to appear in the vacuum, even at zero temperature. In the sine-Gordon case with positive background field, only the negatively-charged particles appear in the vacuum. Their mutual scattering is diagonal, so the problem is technically easier than the finite-temperature TBA problem, where both kinds of particles appear in the thermodynamic state. We discuss both the cases λ real and λ imaginary, hence giving a (partially) non-perturbative treatment of the problem raised in sect. 2. For a more complete study see [12].

The sine-Gordon Hamiltonian with a constant external $U(1)$ gauge field A_μ is

$$H = \int dx \left[\frac{1}{2}\Pi^2 + \frac{1}{2}(\partial\varphi)^2 + \lambda \cos \beta_{SG}\varphi \right] - QA, \tag{7.1}$$

where

$$Q = \int j_0 dx = \frac{\beta_{SG}}{2\pi} \int_{-\infty}^{\infty} \frac{\partial\varphi}{\partial x} dx \tag{7.2}$$

In the ordinary case λ real, Q is the integer-valued soliton topological charge, normalized so that the soliton (antisoliton) has $Q = 1$ (-1). A is a constant with the dimension of a length. We then consider the corresponding specific vacuum energy $\mathcal{E}(A, \lambda)$ as a function of A. As before, we parametrize $\beta_{SG}^2 = 8\pi\frac{t}{t+1}$.

Before we turn to the scattering theory, it is worth looking at the action (7.1) from the perturbative (in λ) point of view. Dimensional arguments as well as explicit perturbative calculations show that the background field works as an infrared cutoff at scales $\sim A$ and therefore if $A \gg \lambda^{(1+t)/2}$ the theory is in the ultraviolet regime. As a leading $A \to \infty$ approximation we set $\lambda = 0$ in (7.1). This theory is the continuum limit of the XXZ model in a magnetic field, which has been studied in refs. [47-49]. As can easily be inferred from the action, it is a Gaussian model whose radius of compactification depends on A (amusing finite-size corrections occur in the related XXZ model in a field due to commensurability problems between R and the scale A). Redefining $\partial\varphi$ by a shift gives for the ground-state energy density

$$\mathcal{E}(A, 0) = -\frac{\beta_{SG}^2}{8\pi^2} A^2. \tag{7.3}$$

At any critical point, $\mathcal{E}(A, 0)$ is proportional to A^2, since there is no other scale in the problem. The coefficient is proportional to the chiral anomaly (found from the $J_L J_L$ OPE) [50].

For $\lambda \neq 0$ the scaling argument shows that $\mathcal{E}(A, \lambda)$ is a function of the dimensionless variable

$$\xi \equiv \lambda/A^{2/(1+t)}$$

and by parity has a perturbative expansion in ξ^2

$$\mathcal{E}_{\text{pert}}(\xi) = -\frac{A^2}{\pi} \sum_{l=0}^{\infty} k_{2l}\xi^{2l}. \tag{7.4}$$

One can use perturbed conformal field theory to derive

$$k_0 = \frac{t}{1+t}$$

$$k_2 = \frac{\pi^2}{4}\left(\frac{2t}{1+t}\right)^{2(t-1)/(t+1)}\frac{\Gamma\left(\dfrac{1-t}{1+t}\right)}{\Gamma\left(\dfrac{2t}{1+t}\right)}.$$

We expect the series (7.4) to have some finite radius of convergence ξ_0, defining therefore an analytic function $\mathcal{E}_{\text{pert}}(\xi)$ at $|\xi| < \xi_0$. The perturbation theory is the same for λ real or imaginary, so (7.4) holds all around $\xi = 0$. However, the scattering theory depends crucially on the nature of λ: for real λ the particles are massive, while for λ imaginary they are massless.

Consider first the unitary massive sine-Gordon model (λ real in (7.1)) and let m be the mass of the corresponding charged particle (soliton). As usual the on-mass-shell momenta (E, p) are parameterized in terms of rapidity θ

$$E = m\cosh\theta \; ; \qquad p = m\sinh\theta$$

In the field (7.1) every soliton (antisoliton) acquires additional energy $-A$ (A). It is clear that if $A \gg m$ the state without particles is no longer the ground state. The true vacuum contains a sea of positively-charged solitons which fill all possible states inside some (A-dependent) "Fermi interval" $-B < \theta < B$. The non-trivial scattering of the solitons certainly influences the structure of the ground-state sea. However, only one kind of particle is in the sea (this can be checked more completely [30,45]); for $A > 0$ this is the soliton. The solitons scatter diagonally among themselves, the two-particle amplitude being $a(\theta)$ from (3.11).

As in the finite-temperature case, we define the density of particles ρ and density of states $\rho + \tilde{\rho}$. To obtain the ground state energy we minimize

$$\mathcal{E}^{(Re)}(A) - \mathcal{E}^{(Re)}(0) = \int (m\cosh\theta - A)\rho(\theta)d\theta, \tag{7.5}$$

(the superscript Re is added to stress that currently we address the ordinary sine-Gordon model with real coupling λ) subject to the quantization

$$2\pi[\rho + \tilde{\rho}] = m\cosh\theta + \Phi \star \rho \tag{7.6}$$

where the kernel Φ follows from the soliton-soliton S-matrix and reads explicitly

$$\frac{\Phi(\theta)}{2\pi} = \frac{1}{2\pi i}\frac{d}{d\theta}\log a(\theta) = \int \frac{e^{i\omega\theta}\sinh\frac{\pi(t-1)\omega}{2}}{2\cosh\frac{\pi\omega}{2}\sinh\frac{\pi t\omega}{2}}\frac{d\omega}{2\pi}. \tag{7.7}$$

The equations (7.5) and (7.6) and the equations for B (given by minimizing the energy with respect to B) can be put in a more convenient form by defining the function $f(\theta)$ as

$$f(\theta) = A - m\cosh\theta + \int_{-B}^{B} d\theta'\,\Phi(\theta - \theta')f(\theta'), \tag{7.8}$$

where this equation is good only for $|\theta| < B$. Replacing $A - m\cosh\theta$ in (7.5) with this and using (7.6) one finds that

$$\mathcal{E}(A) = -\frac{m}{2\pi}\int_{-B}^{B} d\theta\ \cosh\theta\ f(\theta). \tag{7.9}$$

The boundary conditions $f(\pm B) = 0$ determine B.

We can understand the meaning of the function f as follows. Define ϵ^+ as the energy of particle excitations above the ground state, and ϵ^- as the energy of holes. By this definition, $\epsilon^+ \geq 0$ and $\epsilon^- \leq 0$. A variation of the energy is thus

$$\delta\mathcal{E}^{(Re)}(A) = \int (m\cosh\theta - A)\delta\rho(\theta)d\theta = \int \epsilon^+\delta\rho - \epsilon^-\delta\tilde{\rho}. \tag{7.10}$$

Using (7.6) to reexpress $\delta\tilde{\rho}$ as a function of $\delta\rho$ we find

$$\delta\mathcal{E}^{(Re)}(A) = \int \left(\epsilon^+ + \epsilon^- - \frac{\Phi}{2\pi}\star\epsilon^-\right)\delta\rho \tag{7.11}$$

so by comparing (7.10) and (7.11) we find

$$m\cosh\theta - A = \epsilon^+ + \epsilon^- - \frac{\Phi}{2\pi}\star\epsilon^-. \tag{7.12}$$

Using this in (7.5) gives

$$\mathcal{E}^{(Re)}(A) - \mathcal{E}^{(Re)}(0) = \frac{m}{2\pi}\int d\theta\left[\cosh\theta\epsilon^- + \epsilon^+\rho - \epsilon^-\tilde{\rho}\right]. \tag{7.13}$$

In the ground state $\rho(\theta) = 0$ when $\epsilon^+(\theta) > 0$ and $\tilde{\rho}(\theta) = 0$ when $\epsilon^-(\theta) < 0$. This means that the last two terms vanish, and we obtain (7.9), where $f = -\epsilon^-$.

From a formal point of view one may interpret the whole scattering approach and the resulting system (7.7)–(7.9) as a way of summing up the perturbative expansion (7.4) at real ξ and going beyond the radius of convergence *along λ real*. One would like a similar tool to sum up (7.4) at λ purely imaginary. To do so we make a guess inspired by the study of the massive Thirring model and section 6. We assume that in the sine-Gordon model with imaginary coupling or Thirring model with imaginary mass, the number of species doubles, so now we have a pair of left and right massless particles. We also assume that the LL (identical to RR) and the LR scattering are nontrivial. The energy spectrum (1.2) is gapless. Turn on the background field $A > 0$ as before. The positively charged particles are always excited in the ground state; we assume as in the massive case that they are the only particles contributing to the thermodynamics. Now the right- and left-movers fill respectively the semi-infinite Fermi intervals $-\infty < \theta < B$ and $-B < \theta < \infty$ with some Fermi boundary $B \sim \log A/M$. Again it is a straightforward Bethe ansatz exercise to derive the following system of integral equations

$$QA - \frac{Me^\theta}{2} = f_R(\theta) - \int_{-\infty}^{B} \Phi_{LL}(\theta - \theta')f_R(\theta')\frac{d\theta'}{2\pi} - \int_{-B}^{\infty} \Phi_{RL}(\theta - \theta')\rho_L(\theta')\frac{d\theta'}{2\pi};$$

$$QA - \frac{Me^{-\theta}}{2} = f_L(\theta) - \int_{-B}^{\infty} \Phi_{LL}(\theta - \theta')f_L(\theta')\frac{d\theta'}{2\pi} - \int_{-\infty}^{B} \Phi_{RL}(\theta - \theta')f_R(\theta')\frac{d\theta'}{2\pi}.$$

$$(7.14)$$

The positive functions $f_R(\theta)$ and $f_L(\theta)$ are defined in the Fermi intervals $\infty < \theta < B$ and $-B < \theta < \infty$ respectively, and are restricted by the boundary conditions

$$f_R(B) = f_L(-B) = 0 \tag{7.15}$$

Finally, the ground state energy, which we now call $\mathcal{E}^{(Im)}(A)$, is evaluated as follows

$$\mathcal{E}^{(Im)}(A) - \mathcal{E}^{(Im)}(0) = -\frac{M}{2\pi} \int_{-\infty}^{B} e^\theta\, f_R(\theta)d\theta, \tag{7.16}$$

where we have taken into account the obvious symmetry $f_R(\theta) = f_L(-\theta)$.

In eqn. (7.14) we introduced the $U(1)$ charges $\pm Q$ of the massless particles, which, with the knowledge assumed in these lectures, we cannot fix in advance. It is possible to carry out the computation with undetermined kernels, and fix them at the end by requiring the result to provide analytic continuation of the perturbative series to λ imaginary. To save time, let us give the answer and justify it. One has for $\Phi_{LL}(\theta)$ the *same* expression as

112

in the massive case, but with a *shift* $t \to t - 1$. For Φ_{RL} we have the same, with a further shift $\theta \to \theta + i\frac{\pi}{2}(t-1)$. This results in

$$
\begin{aligned}
\frac{\Phi_{LL}(\theta)}{2\pi} &= \frac{1}{2\pi i}\frac{d}{d\theta}\log a_{LL}(\theta) = \int \frac{e^{i\omega\theta}\sinh\frac{\pi(t-2)\omega}{2}}{2\cosh\frac{\pi\omega}{2}\sinh\frac{\pi(t-1)\omega}{2}}\frac{d\omega}{2\pi} \\
\frac{\Phi_{RL}(\theta)}{2\pi} &= \frac{1}{2\pi i}\frac{d}{d\theta}\log a_{RL}(\theta) = -\int \frac{e^{i\omega\theta}\sinh\frac{\pi\omega}{2}}{2\cosh\frac{\pi\omega}{2}\sinh\frac{\pi(t-1)\omega}{2}}\frac{d\omega}{2\pi}.
\end{aligned}
\tag{7.17}
$$

Although this new BA system (7.14)–(7.16) has a rather different form from that of eqns. (7.7)–(7.9), it is easy to relate the two in the UV region $A \to \infty$ where in both systems $B \to \infty$. In this limit the right and left Fermi intervals have a broad overlap at $-B < \theta < B$. Near say the right Fermi boundary $\theta \sim B$ (where the main contribution to (7.16) comes from) we can forget about the left one and solve (7.14) for $f_L(\theta)$ by the Fourier transform with $B \to \infty$. The resulting equation for $f_R(\theta)$ is

$$
rQA - \frac{Me^\theta}{2} = f_R(\theta) - \int_{-\infty}^{B}\Phi(\theta - \theta')f_R(\theta')\frac{d\theta'}{2\pi}
\tag{7.18}
$$

where in terms of the Fourier transforms

$$
\tilde{\Phi}(\omega) = \tilde{\Phi}_{LL}(\omega) + \frac{[\tilde{\Phi}_{RL}(\omega)]^2}{1 - \tilde{\Phi}_{LL}(\omega)} = \frac{\sinh\frac{\pi(t-1)\omega}{2}}{2\cosh\frac{\pi\omega}{2}\sinh\frac{t\omega}{2}}
\tag{7.19}
$$

(compare with eqn. (7.7)) and

$$
r = 1 + \frac{\tilde{\Phi}_{RL}(0)}{1 - \tilde{\Phi}_{LL}(0)} = \frac{t-1}{t}
$$

It coincides precisely with the corresponding limit $B \to \infty$ of eq. (7.8) provided

$$
\begin{aligned}
Q &= \frac{t}{t-1} \\
M &= m
\end{aligned}
\tag{7.20}
$$

We pause to discuss the logic. In the UV limit $\lambda = 0$ and we have a free boson after the shift $\partial\varphi \to \partial\varphi + A$. We can perturb this fixed point by λ real or λ purely imaginary. The first case is the usual sine-Gordon in the regime where there are only solitons in the scattering theory. The theory is massive, so the IR fixed point is the trivial one. The second case is like the sine-Gordon model with imaginary λ, where there are only L and R "solitons". As discussed in previous sections the theory is massless and the IR fixed point is now nontrivial. Using known results about flow between minimal models (for example the large-t expansion discussed in sect. 2), we expect the IR fixed point to exhibit the shift $t \to t - 1$. In the IR limit, the LR scattering becomes negligible, hence the form of LL and RR scattering above. For LR scattering we require that in the opposite UV limit, the two scattering theories look the same, which is natural since they hold in two regimes connected at the UV fixed point.

The expansion (7.4) is obtained by analyzing our two sets of background field energy equations using a generalized Weiner-Hopf technique [46,30]. This has been described in detail in [12], so we do not present the full calculation here. Instead, we will explain how to extract the relevant information from the kernels. The technique relies on the usual Weiner-Hopf trick of dividing the Fourier transforms of the kernels into a product of two pieces, the first of which has no poles or zeroes in the lower half plane and the second none in the upper half plane. Defining $1/K_+(\omega)K_-(\omega) \equiv 1 - \tilde{\Phi}(\omega)$, one finds that expressions of the form

$$\oint \frac{h(\omega)}{(\omega - i)^2} g(\omega) e^{2i\omega B} \frac{d\omega}{2\pi i}; \qquad g(\omega) \equiv \frac{K_+(\omega)}{K_-(\omega)} \tag{7.21}$$

occur regularly in the analysis; the contour covers the upper half plane. The function $h(\omega)$ is different depending on where we are in the analysis, but it is analytic in the upper half plane. First we discuss the massive case. The poles in the contour are at $\omega = i$ and at the zeros of $1 - \tilde{\Phi}$, which are at $\omega = 2in/(t+1)$. The pole at $\omega = i$ results in the bulk contribution $\mathcal{E}(0)$. Ignoring the bulk term, (7.21) can be written as a series in $\exp(-4B/(t+1))$. In particular, the boundary condition results in an equation

$$\frac{M}{A} e^B = const + \sum_n h_n g_n e^{-4nB/(t+1)}$$

where h_n and g_n are the residues of $h(\omega)/(\omega-i)^2$ and $g(\omega)$, respectively. The h_n themselves also obey an equation of this form, so for large A/M, we can write e^B and h_n each as a series in $(A/M)^{-4/(t+1)}$. The energy is also given by a term like (7.21), so it too must be a series in $(A/M)^{-4/(t+1)}$ as in (7.4):

$$\mathcal{E}^{Re}(A, M) = \mathcal{E}^{(Re)}(0) - \frac{A^2}{\pi} \sum_{n=0}^{\infty} k_n \left(\frac{M}{A} \right)^{4n/(t+1)}. \tag{7.22}$$

This gives the result (7.4) for ordinary sine-Gordon. For the imaginary coupling, we must first rewrite the equations (7.14) in Weiner-Hopf form. The general result is that

$$A \to Q(1 + \frac{\tilde{\Phi}_{LR}(0)}{1 - \tilde{\Phi}_{LL}(0)})A$$

$$\frac{1}{K_+(\omega)K_-(\omega)} = 1 - \tilde{\Phi}_{LL} - \frac{\tilde{\Phi}_{LR}^2}{1 - \tilde{\Phi}_{LL}}$$

$$g(\omega) = \frac{K_+(\omega)}{K_-(\omega)} \frac{\tilde{\Phi}_{LR}}{1 - \tilde{\Phi}_{LL}}.$$

The K_+ and K_- obtained for the massless case are exactly the same as the ones obtained above for the massive one. The extra piece in the above expression for $g(\omega)$ is $-\sinh(\pi\omega/2)/\sinh(\pi t\omega/2)$ here. It results in no other additional poles in the contour because of the zeros in g. Its only effect is to change g_n to $(-1)^n g_n$ (again ignoring the bulk piece). Since the h_n above are not changed, we then find that the series for the massless flow is exactly the same as in ordinary sine-Gordon, except that the signs of every other term are different:

$$\mathcal{E}^{(Im)}(A, M) = \mathcal{E}^{(Im)}(0) - \frac{A^2}{\pi} \sum_{n=0}^{\infty} (-)^n k_n \left(\frac{M}{rQA} \right)^{4n/(1+t)}$$

with precisely the same coefficients k_n as in expansion (7.22).

We conclude that up to the known bulk vacuum energy contributions the massless BA system (7.14)–(7.16) gives the correct analytic continuation of the massive one (7.7)–(7.9) to purely imaginary ξ, providing (7.20) holds. In particular, the low-temperature mass scale M is equal to the high-temperature scale m.

This entire discussion can presumably be put on firmer ground by solving the Thirring model with imaginary mass using the traditional Bethe ansatz approach [45].

8. Conclusions

The examples presented here have mainly been the sine-Gordon and Thirring models, with and without a background field. These calculations can be extended to the truncated (RSOS) cases, the latter situation being completely unitary. See the references [12,31].

It seems that the physics of massless flows is intimately related with the one of *symmetry breaking* and that the conformally-invariant IR fixed points are generally some sort of Goldstone phase. Two of the simplest cases, the flow from tricritical to critical Ising and the flow from dilute to dense polymers have to do respectively with $N = 1$ and $N = 2$ spontaneous SUSY breaking (the latter being possible because of the non-unitarity). Moreover, as already mentioned, the sine-Gordon model with imaginary coupling describes the flow from critical to low temperature $O(n)$ model (with $n = 2 \cos \frac{\pi}{t}$). The Mermin-Wagner theorem preventing spontaneous breaking of continuous symmetry does not apply to the cases n non-integer. As a consequence, $O(n)$ models in two dimensions $-2 < n < 2$ have a low-temperature phase which is massless and has properties reminiscent of a Goldstone phase [51] (for instance they qualitatively agree with what can be deduced from the ϵ-expansion in higher dimensions, extended formally to $D = 2$).

Some important questions remain to be addressed. For instance, can one seriously describe conformal field theories using massless scattering (reconstruct Green functions using form factors)? As explained above, massless scattering is a sort of perturbation of the IR fixed point. Can one, using it (i.e. probably using the conserved quantities) give a meaning to the conformal perturbation theory of an IR fixed point by an irrelevant operator?

Acknowledgments: These lectures were presented by H.S. who thanks the organizers and the students at Trieste for many interesting questions. We have benefited a great deal by interacting with our collaborators K. Intriligator, N. Reshetikhin, S. Skorik and Al.B. Zamolodchikov. P.F. and H.S. were supported by the Packard foundation and DOE grant No. DE-FG03-84ER40618.

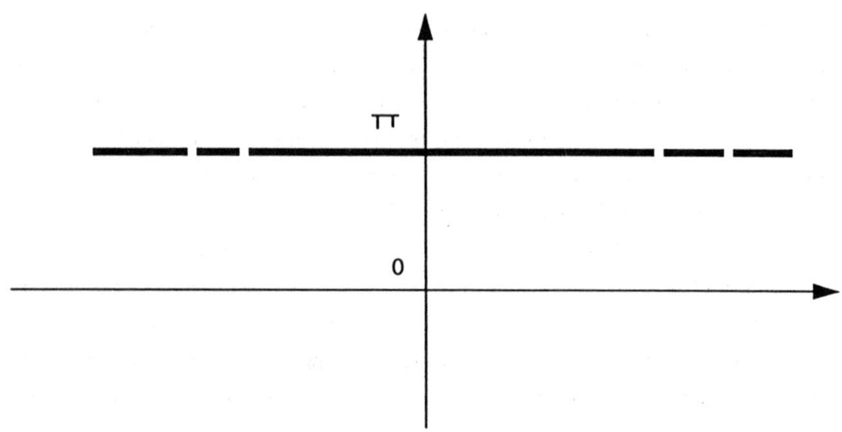

Figure 1: The structure of the ground state in the massive Thirring model. Left and right massless excitations are observed in the limit $\xi \to \pm\infty$. For instance, a pair of left and right solitons is obtained by making two holes for $|\xi| \gg 1$.

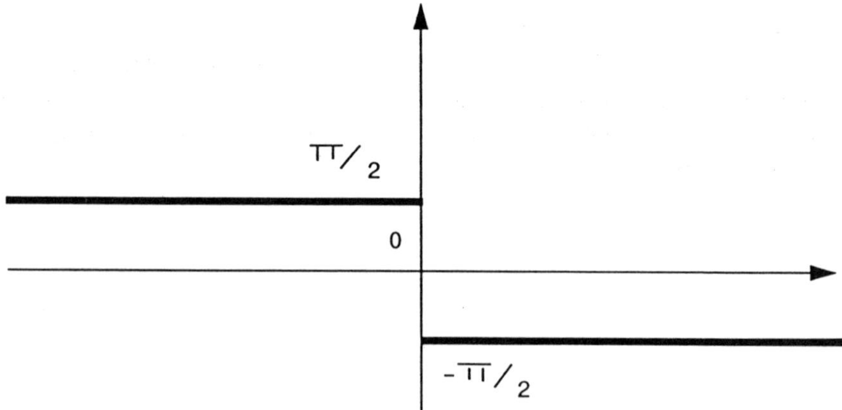

Figure 2: Schematic structure of the ground state for imaginary m_0.

References

[1] N. Andrei, K. Furuya, and J. Lowenstein, Rev. Mod. Phys. 55 (1983) 331; A.M. Tsvelick and P.B. Wiegmann, Adv. Phys. 32 (1983) 453.

[2] I. Affleck, in Les Houches 1988 *Fields, Strings, Critical Phenomena*, ed. by E. Brezin and J. Zinn-Justin, North Holland.

[3] L.D. Faddeev and L.A. Takhtajan, Phys. Lett. 85A (1981) 375.

[4] F. Smirnov, Th. Math. Phys. 60 (1984) 363.

[5] V.E. Korepin, Comm. Math. Phys. 86 (1982) 391.

[6] S. Parke, Nucl. Phys. B177 (1980) 166.

[7] A.B. Zamolodchikov, Al.B. Zamolodchikov, Nucl. Phys. B379 (1992) 602.

[8] A.B. Zamolodchikov and Al.B. Zamolodchikov, Ann. Phys. 120 (1979) 253.

[9] A.B. Zamolodchikov, Adv. Stud. Pure Math. 19 (1989) 1.

[10] G. Mussardo, Phys. Rep. 218 (1992) 215.

[11] Al.B. Zamolodchikov, Nucl. Phys. B358 (1991) 524.

[12] P. Fendley, H. Saleur and Al.B. Zamolodchikov, "Massless Flows I" and "Massless Flows II", hepth #9304050 and 9304051, to appear in Int. J. Mod. Phys. A

[13] P. Fendley, Phys. Rev. Lett. 71 (1993) 2485.

[14] V. Fateev, E. Onofri and Al.B. Zamolodchikov, "The sausage model (integrable deformations of O(3) sigma model)", to appear in Nucl. Phys. B.

[15] P. Fendley and K. Intriligator, "Exact $N=2$ Landau-Ginzburg Flows", hepth #9307166, to appear in Nucl. Phys. B.

[16] N. Reshetikhin, J. Phys. A24 (1991) 3299.

[17] J. Carmelo, P. Horsch, P.A. Bares and A.A. Ovchinnikov, Phys. Rev. B44 (1991) 9967.

[18] F. Essler and V. Korepin, "Scattering Matrix and Excitation Spectrum of the Hubbard Model", ITP-SB-93-40.

[19] B. Nienhuis, Phys. Rev. Lett. 49 (1982) 1062.

[20] J.V. Jose, L.P. Kadanoff, S. Kirkpatrick, D.R. Nelson, Phys. Rev. B16 (1977) 217.

[21] A.B. Zamolodchikov, JETP Lett. 43 (1986) 730.

[22] C. Itzykson and H. Saleur, J. Stat. Phys. 48 (1987) 449.

[23] A. Ludwig and J. Cardy, Nucl. Phys. B285 (1987) 687.

[24] Al.B. Zamolodchikov, "Resonance factorized scattering and roaming trajectories", Ecole Normale preprint ENS-LPS-355

[25] H. Bergknoff and H.B. Thacker, Phys. Rev. D19 (1979) 366.

[26] V.E. Korepin, Th. Math. Phys. 41 (1979) 953.

[27] L. Faddeev, E. Sklyanin and L. Takhtajan, Th. Math. Phys. 40 (1979) 688.

[28] T. Eguchi and S.K. Yang, Phys. Lett. B224 (1989) 373; T. Hollowood and P. Mansfield, Phys. Lett. 226B (1989) 73; M.T. Grisaru, A. Lerda, S. Penati and D. Zanon, Phys.

Lett. B234 (1990) 88; N. Reshetikhin and F. Smirnov, Comm. Math. Phys. 131 (1990) 157.

[29] V. Korepin, Comm. Math. Phys. 76 (1980) 165.

[30] G. Japradize, A. Nersesyan and P. Wiegmann, Nucl. Phys. B230 (1984) 511; P. Wiegmann, Phys. Lett. B152 (1985) 209.

[31] N. Reshetikhin and H. Saleur, "Lattice regularization of massive and massless integrable field theories", preprint USC-93-020, hepth #9309135.

[32] P. Fendley and H. Saleur, Nucl. Phys. B388 (1992) 609.

[33] D. Bernard and A. Leclair, Nucl. Phys. B340 (1990) 721; G. Felder and A. LeClair, Int. J. Mod. Phys A7 (1992) 239.

[34] V. Pasquier and H. Saleur, Nucl. Phys. B330 (1990) 523.

[35] C.N. Yang and C.P. Yang, J.Math. Phys. 10 (1969) 1115.

[36] Al.B.Zamolodchikov, Nucl. Phys. B342 (1991) 695.

[37] H.W. Blote, J.L. Cardy and M.P. Nightingale, Phys. Rev. Lett. 56 (1986) 742; I. Affleck, Phys. Rev. Lett. 56 (1980) 746.

[38] J. Cardy and G. Mussardo, Phys. Lett. 225B (1989) 275.

[39] P. Fendley and K. Intriligator, Nucl. Phys. B372 (1992) 553.

[40] P.P.Kulish and E.R. Nisimov, Th. Math. Phys. 29 (1976) 161.

[41] V. Korepin and L.D. Faddeev, Th. Math. Phys. 25 (1975) 147.

[42] I. Gradshtein and I. Rishnik, *Table of Integrals, Series and Products* (Academic Press, 1980).

[43] T. Klassen and E. Melzer, Nucl. Phys. B350 (1990) 635.

[44] R. Sasaki and I. Yamanaka, Adv. Stud. in Pure Math. 16 (1988) 271.

[45] S. Skorik and H. Saleur, work in progress.

[46] M. Ganin, Izv. Vuzov (Math) 33 (1963) 31.

[47] N.M. Bogoliubov, A. Izergin and V. Korepin, Nucl. Phys. B275 (1986) 687.

[48] N.M. Bogoliubov, A. Izergin and N. Reshetikhin, J.Phys. A 20 (1987) 5361.

[49] F. Woynarovich, H.P. Eckle and T.T. Truong, J.Phys. A22 (1989) 4027.

[50] P. Fendley and K. Intriligator, "Central charges without finite-size effects", to appear in Phys. Lett. B.

[51] H. Saleur, Phys. Rev. B35 (1987) 3657.

Recent Progress in Irrational Conformal Field Theory *

M.B. Halpern [†] [‡]

Department of Physics
University of California
and
Theoretical Physics Group
Physics Division
Lawrence Berkeley Laboratory
1 Cyclotron Road
Berkeley, California 94720
USA

Abstract

In this talk, I will review the foundations of irrational conformal field theory (ICFT), which includes rational conformal field theory as a small subspace. Highlights of the review include the Virasoro master equation, the Ward identities for the correlators of ICFT and solutions of the Ward identities. In particular, I will discuss the solutions for the correlators of the g/h coset constructions and the correlators of the affine-Sugawara nests on $g \supset h_1 \supset \ldots \supset h_n$. Finally, I will discuss the recent global solution for the correlators of all the ICFT's in the master equation.

---

*This work was supported in part by the Director, Office of Energy Research, Office of High Energy and Nuclear Physics, Division of High Energy Physics of the U.S. Department of Energy under Contract DE-AC03-76SF00098 and in part by the National Science Foundation under grant PHY90-21139.

[†]e-mail: MBHALPERN@LBL.GOV, THEORY::HALPERN

[‡]Talk presented at the conference "Strings 1993", Berkeley, May 23-29.

0   Outline of the Talk

1   History of the Affine-Virasoro Constructions

Affine-Virasoro constructions are Virasoro operators constructed with the currents J_a, $a = 1, \ldots, \dim g$ of affine Lie g. All known conformal field theories may be constructed this way.

Here is a brief history of these constructions, which began with two papers[1,2] by Bardakci and myself in 1971. These papers contained the following developments.

a) The first examples of affine Lie algebra, or current algebra on S^1. This was the independent discovery of affine Lie algebra in physics, including the affine central extension some years before it was recognized in mathematics[3]. Following the convention in math, we prefer the neutral names affine Lie algebra or current algebra, reserving "Kac-Moody" for the more general case[4] including hyperbolic algebras.

b) World-sheet fermions (half-integer moded), from which the affine algebras were constructed. Ramond[5] gave the integer-moded case in the same issue of Physical Review.

c) The first affine-Sugawara constructions, on the currents of affine Lie algebra. Sugawara's model[6] was in four dimensions on a different algebra. The affine-Sugawara constructions were later generalized by Knizhnik and Zamolodchikov[7] and Segal[8], and the corresponding world-sheet action was given by Witten[9].

d) The first coset constructions, implicit in Ref. 1 and explicit in Ref. 2, which were later generalized by Goddard, Kent and Olive[10].

What was forgotten for many years was that our first paper also gave another affine-Virasoro construction, the so-called "spin-orbit" model, which was more general than the affine-Sugawara and coset constructions.

2   The General Affine-Virasoro Construction

Motivated by the affine-Sugawara, coset and spin-orbit constructions, Kiritsis and I studied the general affine-Virasoro construction[11,12] in 1989.

The construction begins with the currents of untwisted affine Lie g, [4,1]

$$J_a(z)\, J_b(w) = \frac{G_{ab}}{(z-w)^2} + \frac{if_{ab}{}^c}{z-w}\, J_c(w) + \mathcal{O}(z-w)^0 \tag{2.1}$$

where $a, b = 1, \ldots, \dim g$ and $f_{ab}{}^c$ and G_{ab} are respectively the structure constants and generalized Killing metric of $g = \oplus_I g_I$. To obtain level $x_I = 2k_I/\psi_I^2$ of affine g_I with dual Coxeter number $\tilde{h}_I = Q_I/\psi_I^2$, take

$$G_{ab} = \oplus_I k_I\, \eta_{ab}^I \quad , \quad f_{ac}{}^d f_{bd}{}^c = -\oplus_I Q_I\, \eta_{ab}^I \tag{2.2}$$

where η_{ab}^I and ψ_I are respectively the Killing metric and highest root of g_I.

Given the currents, we consider the general stress tensor

$$T = L^{ab}\, {}^*_*J_a J_b{}^*_* + D^a \partial J_a + d^a J_a \tag{2.3}$$

where the coefficients $L^{ab} = L^{ba}$, D^a and d^a are to be determined. L^{ab} is called the inverse inertia tensor, in analogy with the spinning top. The stress tensor is required to satisfy the Virasoro algebra

$$T(z)\, T(w) = \frac{c/2}{(z-w)^4} + \left(\frac{2}{(z-w)^2} + \frac{\partial_w}{z-w} \right) T(w) + \mathcal{O}(z-w)^0 \tag{2.4}$$

where c is the central charge. I give here only the result at $D^a = d^a = 0$, which is called the Virasoro master equation[11,12]

$$L^{ab} = 2L^{ac}G_{cd}L^{db} - L^{cd}L^{ef}f_{ce}{}^{a}f_{df}{}^{b} - L^{cd}f_{ce}{}^{f}f_{df}{}^{(a}L^{b)e} \qquad (2.5a)$$

$$c = 2G_{ab}L^{ab} \quad . \qquad (2.5b)$$

The master equation has been identified[13] as an Einstein system on the group manifold, with

$$g^{ij} = e_a^i L^{ab} e_b^j \quad , \qquad c = \dim g - 4R \qquad (2.6)$$

where g^{ij} and R are the inverse Einstein metric and Einstein curvature scalar respectively.

Here are some simple facts about the master equation which we will need in this lecture.

a) Affine-Sugawara constructions[1,2,7–9]. The affine-Sugawara construction on g is

$$L_g^{ab} = \oplus_I \frac{\eta_I^{ab}}{2k_I + Q_I} \quad , \qquad c_g = \sum_I \frac{x_I \dim g_I}{x_I + \tilde{h}_I} \qquad (2.7)$$

and similarly for L_h on $h \subset g$.

b) K-conjugation covariance[1,2,10,11]. Solutions of the master equation come in commuting K-conjugate pairs $T = L^{ab} {}^{*}_{*}J_a J_b {}^{*}_{*}$ and $\tilde{T} = \tilde{L}^{ab} {}^{*}_{*}J_a J_b {}^{*}_{*}$, which sum to the affine-Sugawara construction $T_g = L_g^{ab} {}^{*}_{*}J_a J_b {}^{*}_{*}$,

$$L + \tilde{L} = L_g \quad , \qquad T + \tilde{T} = T_g \quad , \qquad c + \tilde{c} = c_g \qquad (2.8a)$$

$$T(z)\,\tilde{T}(w) = \mathcal{O}(z - w)^0 \quad . \qquad (2.8b)$$

K-conjugation is the central feature of affine-Virasoro constructions, and it suggests that the affine-Sugawara construction should be thought of as the tensor product of any pair of K-conjugate conformal field theories. This is the conceptual basis of factorization, discussed in Sections 4-7.

c) Coset constructions[1,2,10]. K-conjugation generates new solutions from old. The simplest examples are the g/h coset constructions

$$\tilde{L} = L_g - L_h = L_{g/h} \quad , \qquad \tilde{T} = T_g - T_h = T_{g/h} \quad , \qquad \tilde{c} = c_g - c_h = c_{g/h} \qquad (2.9)$$

which follow by K-conjugation from L_h on $h \subset g$.

d) Affine-Sugawara nests[14-16]. Repeated K-conjugation on the subgroup sequence $g \supset h_1 \supset \ldots \supset h_n$ gives the affine-Sugawara nests,

$$L_{g/h_1/\ldots/h_n} = L_g - L_{h_1/\ldots/h_n} = L_g + \sum_{j=1}^{n}(-)^j L_{h_j} \qquad (2.10a)$$

$$c_{g/h_1/\ldots/h_n} = c_g - c_{h_1/\ldots/h_n} = c_g + \sum_{j=1}^{n}(-)^j c_{h_j} \qquad . \qquad (2.10b)$$

The nest stress tensors may be rearranged as sums of mutually-commuting Virasoro constructions for g/h and h,

$$T_{g/h_1/\ldots/h_{2n+1}} = T_{g/h_1} + \sum_{i=1}^{n} T_{h_{2i}/h_{2i+1}} \qquad (2.11a)$$

$$T_{g/h_1/\ldots/h_{2n}} = T_{g/h_1} + \sum_{i=1}^{n-1} T_{h_{2i}/h_{2i+1}} + T_{h_{2n}} \qquad (2.11b)$$

so the conformal field theories of the affine-Sugawara nests are expected to be tensor-product field theories (see Section 5).

3 Irrational Conformal Field Theory

Here is an overview of the solution space of the master equation, called affine-Virasoro space. For further details see the review in Ref. 17.

a) Counting[15,18]. The master equation is a set of $\dim g(\dim g + 1)/2$ coupled quadratic equations on the same number of unknowns L^{ab}. This allows us to estimate the number of inequivalent solutions on each manifold. As an example, there are approximately $\frac{1}{4}$ billion conformal field theories on each level of affine $SU(3)$, and exponentially larger numbers on larger manifolds.

b) Exact solutions[15,18-26,16]. Large numbers of new solutions have been found in closed form. On positive integer levels of affine compact g, most of these solutions are unitary with irrational central charge. As an example, the value at level 5 of affine $SU(3)$, [21]

$$c\left((SU(3)_5)_{D(1)}^{\#}\right) = 2\left(1 - \frac{1}{\sqrt{61}}\right) \simeq 1.7439 \qquad (3.1)$$

is the lowest unitary irrational central charge yet observed. See Ref. 16 for the most recent list of exact solutions.

124

c) Systematics[17]. Generically, affine-Virasoro space is organized into level-families of conformal field theories, which are essentially analytic functions of the level. On positive integer level of affine compact g, it is clear from the form of the master equation that the generic level-family has generically irrational central charge.

Moreover, rational central charge is rare in the space of unitary conformal field theories. Indeed, the rational conformal field theories are rare in the space of Lie h-invariant conformal field theories[16], which are themselves quite rare. Many candidates for new rational conformal field theories[23,27,28], beyond the coset constructions, have also been found.

d) Classification[†]. Study of the space by high-level[‡] expansion[21] shows a partial classification by graph theory and generalized graph theories [18,22−26]. In the classification, each graph is a level-family, whose conformal field theories carry the symmetry of the graph. At the present time, seven graph-theory units[17] of generically unitary and irrational conformal field theories have been studied, and it seems likely that many more can be found[26].

Large as they are, the graph theories cover only very small regions of affine-Virasoro space. Enough has been learned however to see that all known exact solutions are special cases of relatively high symmetry, whereas the generic conformal field theory is completely asymmetric[18].

Before going on to the Ward identities, I should mention several other lines of development.

1. Non-chiral CFT's. Adding right-mover copies $\bar{T}$ and $\tilde{\bar{T}}$ of the K-conjugate stress tensors, we may take the usual Hamiltonian $H = L^{(0)} + \bar{L}^{(0)}$ for the L theory. Because the generic CFT has no residual affine symmetry, the physical Hilbert space of the generic theory L is characterized by

$$\tilde{L}^{(m>0)}|L\text{-physical}\rangle = \tilde{\bar{L}}^{(m>0)}|L\text{-physical}\rangle = 0 \quad . \tag{3.2}$$

[†]In the course of this work, a new and apparently fundamental connection between Lie groups and graphs was seen. The interested reader should consult Ref. 16 and especially Ref. 26, which axiomatizes these observations.

[‡]The leading behavior $L^{ab} = (P^{ab}/2k) + \mathcal{O}(k^{-2})$, $\tilde{L}^{ab} = (\tilde{P}^{ab}/2k) + \mathcal{O}(k^{-2})$ is believed[21] to include all unitary conformal field theories on affine compact g, where P and $\tilde{P}$ are projectors which sum to the inverse Killing metric. In the graph theories, the projectors are the adjacency matrices of the graphs.

2. World sheet action[29]. Correspondingly, the world-sheet action of the generic theory L is a spin-two gauge theory, in which the WZW theory is gauged by the K-conjugate theory $\tilde{L}$. An open direction here is the relation with σ-models and the corresponding space-time geometry of irrational conformal field theory. In this connection, see also Ref. 13.

3. Superconformal master equations[30]. The N=1 system has been studied in some detail[23-26], and a simplification[31] of the N=2 system has recently been noted: Eq.(D.5) of Ref. 30 is redundant, so that eq.(D.6) is the complete N=2 system. It is an important open problem to find unitary N=2 solutions with irrational central charge.

Another open question at N=2 is the relation of the master equation to the bosonic constructions of Kazama and Suzuki[27,28].

4. Exact C-functions[32,30,17]. Exact C-functions[33] and associated C-theorems are known for the N=0 and N=1 master equations, but not yet for N=2.

4 The Ward Identities of Irrational CFT

It is clear that the Virasoro master equation is the first step in the study of irrational conformal field theory (ICFT), which contains rational conformal field theory (RCFT) as a small subspace of relatively high symmetry,

$$\text{ICFT} \supset \text{RCFT} \ . \tag{4.1}$$

On the other hand, the correlators of irrational conformal field theory have, until recently, remained elusive. The reason is that most of the computational methods of conformal field theory have been based on null states of extended Virasoro algebras[5,34-36], whereas the generic conformal field theory, being totally asymmetric, is not expected to possess such simple algebras.

Recently, Obers and I have found a set of equations, the affine-Virasoro Ward identities[37], for the correlators of irrational conformal field theory.

The first step is to consider KZ-type null states[7,37], which live in the modules of the universal enveloping algebra of the affine algebra, and which are more general than extended Virasoro null states. As an example, we have

$$0 = L^{(-1)}|R_g\rangle - 2L^{ab}J_a^{(-1)}|R_g\rangle \mathcal{T}_b \tag{4.2}$$

which holds for all affine-Virasoro constructions L^{ab}. Here $|R_g\rangle^\alpha$, $\alpha = 1, \ldots, \dim \mathcal{T}$ is the affine primary state with matrix representation $\mathcal{T}$, and the original KZ null state is recovered from (4.2) by taking $L^{ab} = L_g^{ab}$ and $L^{(-1)} = L_g^{(-1)}$. We also choose a so-called L-basis of representation $\mathcal{T}$, in which the conformal weight matrix is diagonal[37]

$$L^{ab}(\mathcal{T}_a \mathcal{T}_b)_\alpha{}^\beta = \Delta_\alpha(\mathcal{T}) \, \delta_\alpha^\beta \quad , \quad \alpha, \beta = 1, \ldots, \dim \mathcal{T} \quad . \tag{4.3a}$$

$$L^{m \geq 0}|R_g\rangle = \delta_{m,0} \, \Delta_\alpha(\mathcal{T})|R_g\rangle \quad , \quad \tilde{L}^{m \geq 0}|R_g\rangle = \delta_{m,0} \, \tilde{\Delta}_\alpha(\mathcal{T})|R_g\rangle \tag{4.3b}$$

$$\Delta_g(\mathcal{T}) = \Delta_\alpha(\mathcal{T}) + \tilde{\Delta}_\alpha(\mathcal{T}) \quad . \tag{4.3c}$$

In such a basis, the states $|R_g\rangle$ are called the broken affine primary states, because the conformal weights $\Delta_g(\mathcal{T})$ of $|R_g\rangle$ under the affine-Sugawara construction are broken to the conformal weights $\Delta_\alpha(\mathcal{T})$ of the L theory. These states are also examples of Virasoro biprimary states, which are simultaneously Virasoro primary under both of the commuting K-conjugate stress tensors.

To use these null states in correlators, we need something like the Virasoro primary fields of the L theory. Because we have two commuting stress tensors, the natural objects are the Virasoro biprimary fields[38,37]

$$R^\alpha(\bar{z}, z) = e^{(\bar{z}-z)\tilde{L}^{(-1)}} R_g^\alpha(z) \, e^{(z-\bar{z})\tilde{L}^{(-1)}} = e^{(z-\bar{z})L^{(-1)}} R_g^\alpha(\bar{z}) \, e^{(\bar{z}-z)L^{(-1)}} \tag{4.4a}$$

$$R(z, z) = R_g(z) \quad , \quad R(0, 0)|0\rangle = |R_g\rangle \tag{4.4b}$$

which are simultaneously Virasoro primary under T and $\tilde{T}$. In (4.4), R_g^α are the broken affine primary fields, and the independent variable $\bar{z}$ is not necessarily the complex conjugate of z. The averages of these bilocal fields

$$A^\alpha(\bar{z}, z) = \langle R^{\alpha_1}(\mathcal{T}^1, \bar{z}_1, z_1) \ldots R^{\alpha_n}(\mathcal{T}^n, \bar{z}_n, z_n)\rangle \tag{4.5}$$

are called biconformal correlators.

Inserting the null state (4.2) into the biconformal correlator, the Virasoro term gives derivatives with respect to the z's, as usual, while the current term can be evaluated, in terms of the representation matrices, on the affine-Sugawara line $\bar{z} = z$. More generally, we obtain the affine-Virasoro Ward identities[37]

$$\bar{\partial}_{j_1} \ldots \bar{\partial}_{j_q} \partial_{i_1} \ldots \partial_{i_p} A^\alpha(\bar{z}, z)|_{\bar{z}=z} = A_g^\beta(z) W_{j_1 \ldots j_q, i_1 \ldots i_p}(z)_\beta{}^\alpha \tag{4.6}$$

where $W_{j_1\dots j_q, i_1\dots i_p}$ are the affine-Virasoro connections and $A(z,z) = A_g(z)$ is the affine-Sugawara correlator (which satisfies the KZ equation on g). The first-order connections are

$$W_{0,i} = 2L^{ab} \sum_{j\neq i}^{n} \frac{\mathcal{T}_a^i \mathcal{T}_b^j}{z_{ij}} \quad , \quad W_{i,0} = 2\tilde{L}^{ab} \sum_{j\neq i}^{n} \frac{\mathcal{T}_a^i \mathcal{T}_b^j}{z_{ij}} \tag{4.7}$$

which follow from the null state (4.2) and its K-conjugate copy with $L \to \tilde{L}$. The sum of these two connections is the KZ connection W_i^g, with $L = L_g$.

All the connections may be computed by standard dispersive techniques from the formula[37,39]

$$A_g^\beta(z) W_{j_1\dots j_q, i_1\dots i_p}(z)_\beta{}^\alpha =$$

$$\left[\prod_{r=1}^{q} \tilde{L}^{a_r b_r} \oint_{z_{j_r}} \frac{d\omega_r}{2\pi i} \oint_{\omega_r} \frac{d\eta_r}{2\pi i} \frac{1}{\eta_r - \omega_r} \right] \left[\prod_{s=1}^{p} L^{c_s d_s} \oint_{z_{i_s}} \frac{d\omega_{q+s}}{2\pi i} \oint_{\omega_{q+s}} \frac{d\eta_{q+s}}{2\pi i} \frac{1}{\eta_{q+s} - \omega_{q+s}} \right]$$

$$\times \langle J_{a_1}(\eta_1) J_{b_1}(\omega_1) \dots J_{a_q}(\eta_q) J_{b_q}(\omega_q) J_{c_1}(\eta_{q+1}) J_{d_1}(\omega_{q+1}) \dots$$

$$J_{c_p}(\eta_{q+p}) J_{d_p}(\omega_{q+p}) R_g^{\alpha_1}(\mathcal{T}^1, z_1) \dots R_g^{\alpha_n}(\mathcal{T}^n, z_n) \rangle \tag{4.8}$$

since the required averages are in the affine-Sugawara theory on g. The results (4.6) and (4.8) prove the existence of the biconformal correlators (at least as an expansion about the affine-Sugawara line $\bar{z} = z$), but computation of all the connections appears to be a formidable task. So far, we have explicitly evaluated only the first and second-order connections[37] for all theories and all the connections for the coset constructions[37,39] and affine-Sugawara nests[39].

Many general properties of the connections also follow from (4.8), including the high-level form of all the connections[39]

$$W_{j_1\dots j_q, i_1\dots i_p} = W_{j_1\dots j_q, 0} W_{0, i_1\dots i_p} + \mathcal{O}(k^{-2}) \tag{4.9a}$$

$$W_{j_1\dots j_q, 0} = \left(\prod_{r=1}^{q-1} \partial_{j_r} \right) W_{j_q, 0} + \mathcal{O}(k^{-2}) \quad , \quad q \geq 1 \tag{4.9b}$$

$$W_{0, i_1\dots i_p} = \left(\prod_{r=1}^{p-1} \partial_{i_r} \right) W_{0, i_p} + \mathcal{O}(k^{-2}) \quad , \quad p \geq 1 \tag{4.9c}$$

and the crossing symmetry of the connections[39]

$$W_{j_1\dots j_q, i_1\dots i_p}\big|_{k\leftrightarrow l} = W_{j_1\dots j_q, i_1\dots i_p} \tag{4.10}$$

where $k \leftrightarrow l$ includes T's, z's and indices. Moreover, the consistency relations[37]

$$(\partial_i + W_i^g)W_{j_1\ldots j_q,i_1\ldots i_p} = W_{j_1\ldots j_q i,i_1\ldots i_p} + W_{j_1\ldots j_q,i_1\ldots i_p i} \qquad (4.11)$$

show that we need only compute the independent set of connections $W_{0,i_1\ldots i_p}$.

So far we have described only the biconformal correlators. To obtain the conformal correlators, we must factorize the biconformal correlators[§]

$$A^\alpha(\bar{z},z) = (\bar{A}(\bar{z})\,A(z))^\alpha \quad , \quad A_g^\alpha(z) = (\bar{A}(z)\,A(z))^\alpha \qquad (4.12)$$

into the proper correlators $\bar{A}$ and A of the $\tilde{L}$ and L theories respectively. Then the factorized Ward identities[37]

$$(\partial_{j_1}\ldots\partial_{j_q}\bar{A}\,\partial_{i_1}\ldots\partial_{i_p}A)^\alpha = (\bar{A}\,A)^\beta (W_{j_1\ldots j_q,i_1\ldots i_p})_\beta{}^\alpha \qquad (4.13)$$

are an all-order non-linear differential system for the correlators of the K-conjugate pair of conformal field theories.

5 Coset and Nest Correlators

To anchor the new Ward identities, we solved the system first for the rational conformal field theories.

Given the all-order connections for

$$\tilde{L} = L_{g/h} \quad , \quad L = L_h \qquad (5.1)$$

one solves the Ward identities (4.6) or (4.13) to obtain the factorized biconformal correlators[37,39]

$$A^\alpha(\bar{z},z) = A_{g/h}^\beta(\bar{z})A_h(z)_\beta{}^\alpha \qquad (5.2a)$$

$$A_{g/h}^\alpha(\bar{z}) = A_g^\beta(\bar{z})A_h^{-1}(\bar{z})_\beta{}^\alpha \qquad (5.2b)$$

where $A_{g/h}$ are the coset correlators. The two-index symbol A_h is the (invertible) evolution operator of h, which solves the KZ equation for $h \subset g$.

[§]The notation in (4.12) includes a sum $A^\alpha(\bar{z},z) = \sum_\nu(\bar{A}_\nu(\bar{z})\,A_\nu(z))^\alpha$ over the conformal structures $\bar{A}_\nu$, A_ν of the theories and an assignment of Lie algebra indices. See Refs. 37,39 and Section 7.

Using the g and h-invariant tensors of $\mathcal{T}^1 \otimes \cdots \otimes \mathcal{T}^4$, one may change to a conformal-block basis for the four-point coset correlators. Then, the coset blocks[40,37,41]

$$\mathcal{C}(u)_r{}^R = \mathcal{F}_g(u)_r{}^m \mathcal{F}_h^{-1}(u)_m{}^R \tag{5.3}$$

are obtained from (5.2b), where $\mathcal{F}_g$ and $\mathcal{F}_h$ are the KZ blocks of g and h.

Although the affine-Virasoro Ward identities have provided a derivation of these coset blocks from first principles, their form was discussed in 1987 by Douglas[40] who argued that they define consistent non-chiral conformal field theories.

In fact, the coset blocks are the ultimately practical answer for coset correlators. See Ref. 37 for an explicit example on $(SU(n)_{x_1} \times SU(n)_{x_2})/SU(n)_{x_1+x_2}$. More generally, the coset blocks show that coset correlators are always sums of products of generalized hypergeometric functions (i.e., solutions to the KZ equations), a conclusion which seems difficult to obtain in other approaches.

Turning next to the simplest affine-Sugawara nest

$$\tilde{L} = L_{g/h_1/h_2} \quad , \quad L = L_{h_1/h_2} \tag{5.4}$$

one solves the Ward identities to obtain the biconformal nest correlators[39]

$$A^\alpha(\bar{z}, z) = A^\beta_{g/h_1}(\bar{z}) \, A_{h_1/h_2}(z)_\beta{}^\gamma \, A_{h_2}(\bar{z})_\gamma{}^\alpha \tag{5.5}$$

where $A_{h_1/h_2} = A_{h_1} A_{h_2}^{-1}$ is composed of the evolution operators of h_1 and h_2. The $\bar{z}$ dependence of (5.5) suggests that the conformal nest correlators are tensor-product theories

$$A_{g/h_1/h_2}(\bar{z}) = A_{g/h_1}(\bar{z}) \otimes A_{h_2}(\bar{z}) \tag{5.6}$$

and this has been confirmed[39] by analysis of (5.5) with the g, h_1 and h_2-invariant tensors of $\mathcal{T}^1 \otimes \cdots \otimes \mathcal{T}^4$.

Similarly, the biconformal correlators of all the higher nests have been obtained on $g \subset h_1 \subset \ldots \subset h_n$, and the conformal nest correlators $A_{g/h_1/.../h_n}$ are exactly the tensor-product theories indicated in the nest stress tensors (2.11). Further details may be found in Ref. 39.

6 Algebraization of the Ward Identities

We want to solve the Ward identities for all ICFT, but the differential system (4.13) is intimidating. In fact, there is an equivalent algebraic formulation of the system[39], which I will discuss only for four-point correlators.

To begin, we use the $SL(2) \times SL(2)$ covariance of the biconformal correlators to translate our machinery above into invariant form[37]. We obtain the invariant Ward identities

$$\bar{\partial}^q \partial^p Y^\alpha(\bar{u}, u)|_{\bar{u}=u} = Y_g^\beta(u) W_{qp}(u)_\beta{}^\alpha \qquad (6.1a)$$

$$W_{10} = 2\tilde{L}^{ab}\left(\frac{\mathcal{T}_a^1 \mathcal{T}_b^2}{u} + \frac{\mathcal{T}_a^1 \mathcal{T}_b^3}{u-1}\right) \quad , \quad W_{01} = 2L^{ab}\left(\frac{\mathcal{T}_a^1 \mathcal{T}_b^2}{u} + \frac{\mathcal{T}_a^1 \mathcal{T}_b^3}{u-1}\right) \qquad (6.1b)$$

and the invariant factorized Ward identities

$$(\partial^q \bar{Y} \, \partial^p Y)^\alpha = (\bar{Y} \, Y)^\beta (W_{qp})_\beta{}^\alpha \qquad (6.2a)$$

$$Y^\alpha(\bar{u}, u) = (\bar{Y}(\bar{u}) \, Y(u))^\alpha \quad , \quad Y_g^\alpha(u) = (\bar{Y}(u) \, Y(u))^\alpha \qquad (6.2b)$$

where $\bar{u}$ and u are cross ratios among the $\bar{z}$'s and z's respectively. Here, $Y(\bar{u}, u)$ and $Y_g(u)$ are the invariant biconformal correlator and the invariant affine-Sugawara correlator respectively, while W_{qp} are the invariant connections of order $q + p$.

From the results for the n-point connections, the invariant connections are known for all theories thru second order and to all orders for cosets and nests. Similarly, we know the invariant consistency relations

$$(\partial + W^g)W_{qp} = W_{q+1,p} + W_{q,p+1} \qquad (6.3)$$

the crossing relation

$$W_{qp}(1 - u) = (-)^{q+p} P_{23} W_{qp}(u) P_{23} \qquad (6.4a)$$

$$P_{23}\mathcal{T}^2 P_{23} = \mathcal{T}^3 \quad , \quad P_{23}^2 = 1 \qquad (6.4b)$$

and, on simple g, we know the high-level form of the invariant connections

$$W_{qp} = W_{q0}W_{0p} + \Delta_{qp} \quad , \quad \Delta_{qp} = \mathcal{O}(k^{-2}) \qquad (6.5a)$$

$$W_{q0} = \partial^{q-1}W_{10} + \mathcal{O}(k^{-2}) \quad , \quad W_{0p} = \partial^{p-1}W_{01} + \mathcal{O}(k^{-2}) \quad . \qquad (6.5b)$$

The q, p factorized form seen in (6.5a) is true to all orders with $\Delta = 0$ when $\tilde{L} = L_{g/h}$ and $L = L_h$.

Toward solving the Ward identities, we note first that the partially-factorized form of the biconformal correlators[39]

$$Y^\alpha(\bar{u}, u) = \sum_{q,p=0}^{\infty} \frac{(\bar{u} - u_0)^q}{q!} [Y_g^\beta(u_0) W_{qp}(u_0)_\beta{}^\alpha] \frac{(u - u_0)^p}{p!} \tag{6.6}$$

solves the unfactorized Ward identities (6.1a), where u_0 is a regular reference point. The consistency relations (6.3) show that the biconformal correlators are independent of the choice of u_0.

We are left with the problem of factorization, but now we need only factorize the connections $W_{qp}(u_0)$ at the reference point into functions of q times functions of p. The factorized Ward identities (6.2) have been reduced to an algebraic problem.

7 Candidate Correlators for Irrational CFT

In fact, there are many algebraic factorizations of W_{qp}, whose interrelation is not fully understood. I will discuss here only a particular solution, which has good physical properties so far as it has been examined[39].

The invariant connections at the reference point define an eigenvalue problem

$$\sum_p W_{qp}(u_0)_\alpha{}^\beta \bar{\psi}_{p\beta}^{(\nu)}(u_0) = E_\nu(u_0) \bar{\psi}_{q\alpha}^{(\nu)}(u_0) \tag{7.1a}$$

$$\sum_q \psi_{q(\nu)}^\beta(u_0) W_{qp}(u_0)_\beta{}^\alpha = E_\nu(u_0) \psi_{p(\nu)}^\alpha(u_0) \tag{7.1b}$$

where ν, called the conformal structure index, labels the eigenvectors. Then, the spectral resolution

$$W_{qp}(u_0)_\alpha{}^\beta = \sum_{\nu=0}^{\infty} \bar{\psi}_{q\alpha}^{(\nu)}(u_0) E_\nu(u_0) \psi_{p(\nu)}^\beta(u_0) \tag{7.2}$$

gives the desired algebraic factorization of the connections, and we obtain the conformal structures[39]

$$Y^\alpha(\bar{u}, u) = (\bar{Y}(\bar{u}) Y(u))^\alpha = \sum_\nu \bar{Y}_\nu(\bar{u}, u_0) Y_\nu^\alpha(u, u_0) \tag{7.3a}$$

$$\bar{Y}_\nu(u,u_0) = \sqrt{E_\nu(u_0)}\, Y_g^\alpha(u_0)\bar{\psi}_\alpha^{(\nu)}(u,u_0) \quad , \quad \bar{\psi}_\alpha^{(\nu)}(u,u_0) \equiv \sum_{q=0}^\infty \frac{(u-u_0)^q}{q!}\, \bar{\psi}_{q\alpha}^{(\nu)}(u_0)$$

$$(7.3b)$$

$$Y_\nu^\alpha(u,u_0) = \sqrt{E_\nu(u_0)}\, \psi_{(\nu)}^\alpha(u,u_0) \quad , \quad \psi_{(\nu)}^\alpha(u,u_0) \equiv \sum_{p=0}^\infty \frac{(u-u_0)^p}{p!}\, \psi_{p(\nu)}^\alpha(u_0)$$

$$(7.3c)$$

of the $\tilde{L}$ and L theories respectively. Because the eigenvalue problem is infinite dimensional, we find a generically infinite number of conformal structures for each theory - in accord with intuitive notions about ICFT.

This solution verifies the following properties[39].

1. Cosets and nests. The solution reproduces the correct coset and nest correlators above. The mechanism is a *degeneracy* of the conformal structures in which each $\bar{Y}_\nu$ is proportional to the same known correlators

2. Braiding. The solution exhibits a braiding for all ICFT, which follows from the crossing symmetry (6.4) of the connections and the linearity of the eigenvalue problem. Since the coset correlators are correctly included in the solution, this braiding includes and generalizes the braiding of RCFT.

3. Good semi-classical behavior. Because of the factorized high-level form (6.5) of the connections, we find a similar degeneracy among the high-level conformal structures of all ICFT. In this way, we identify the high-level correlators of ICFT,

$$Y_{\tilde{L}}^\alpha(u,u_0) = Y_g^\beta(u_0)(\mathbb{1} + 2\tilde{L}_{ab}\left[T_a^1 T_b^2 \ln\left(\frac{u}{u_0}\right) + T_a^1 T_b^3 \ln\left(\frac{1-u}{1-u_0}\right)\right])_\beta{}^\alpha + \mathcal{O}(k^{-2})$$

$$(7.4)$$

where $\tilde{L}^{ab} = \tilde{P}^{ab}/2k + \mathcal{O}(k^{-2})$ is any affine-Virasoro construction on simple g. The result (7.4) correctly includes the coset and nest correlators, exhibits physical singularities in all channels and shows high-level fusion rules proportional to Clebsch-Gordan coefficients. These correlators should also be analyzed at the level of conformal blocks.

To further analyze the candidate correlators (7.3), it will be necessary to know more about the connections. Of particular interest is the next order in k^{-1}, where the high-level degeneracy of the irrational theories is expected to lift.

8 Conclusions

The Virasoro master equation describes irrational conformal field theory, which includes rational conformal field theory as a small subspace. The affine-Virasoro Ward identities describe the correlators of irrational conformal field theory. The Ward identities are solvable beyond the coset constructions, and a set of candidate correlators have been obtained for all the irrational conformal field theories of the master equation.

Acknowledgement

I am grateful to N. Obers for his generous help in preparing this manuscript.

References

1. K. Bardakçi and M.B. Halpern, *Phys. Rev.* **D3** (1971) 2493.

2. M.B. Halpern, *Phys. Rev.* **D4** (1971) 2398.

3. J. Lepowsky and R.L. Wilson, *Comm. Math. Phys.* **62** (1978) 43.

4. V.G. Kac, *Funct. Anal. App.* **1** (1967) 328; R.V. Moody, *Bull. Am. Math. Soc.* **73** (1967) 217.

5. P. Ramond, *Phys. Rev.* **D3** (1971) 2415.

6. H. Sugawara, *Phys. Rev.* **170** (1968) 1659; C. Sommerfeld, *Phys. Rev.* **176** (1968) 2019.

7. V.G. Knizhnik and A.B. Zamolodchikov, *Nucl. Phys.* **B247** (1984) 83.

8. G. Segal, unpublished.

9. E. Witten, *Comm. Math. Phys.* **92** (1984) 455;

10. P. Goddard, A. Kent and D. Olive, *Phys. Lett.* **B152** (1985) 88.

11. M.B. Halpern and E. Kiritsis, *Mod. Phys. Lett.* **A4** (1989) 1373; Erratum ibid. **A4** (1989) 1797.

12. A.Yu Morozov, A.M. Perelomov, A.A. Rosly, M.A. Shifman and A.V. Turbiner, *Int. J. Mod. Phys.* **A5** (1990) 803.

13. M.B. Halpern and J.P. Yamron, *Nucl. Phys.* **B332** (1990) 411.

14. E. Witten, in *Memorial Volume for V. Knizhnik*, ed. L. Brink et al., World Scientific, 1990.

15. M.B. Halpern, E. Kiritsis, N.A. Obers, M. Porrati and J.P. Yamron, *Int. J. Mod. Phys.* **A5** (1990) 2275.

16. M.B. Halpern, E.B. Kiritsis and N.A. Obers, Proceedings of the RIMS Research Project 1991, "Infinite Analysis", *Int. J. Mod. Phys.* A7, Suppl. **1A** (1992) 339.

17. M.B. Halpern, *"Recent Developments in the Virasoro Master Equation"*, in the proceedings of the Stony Brook conference, *Strings and Symmetries* 1991, World Scientific, 1992.

18. M.B. Halpern and N.A. Obers, *Comm. Math. Phys.* **138** (1991) 63.

19. M.B. Halpern and N.A. Obers, *Int. J. Mod. Phys.* **A6** (1991) 1835.

20. S. Schrans and W. Troost, *Nucl. Phys.* **B345** (1990) 584.

21. M.B. Halpern and N.A. Obers, *Nucl. Phys.* **B345** (1990) 607.

22. M.B. Halpern and N.A. Obers, *Ann. of Phys.* **212** (1991) 28.

23. M.B. Halpern and N.A. Obers, *Int. J. Mod. Phys.* **A7** (1992) 7263.

24. M.B. Halpern and N.A. Obers, *Int. J. Mod. Phys.* **A7** (1992) 3065.

25. M.B. Halpern and N.A. Obers, *J. Math. Phys.* **32** (1991) 3231.

26. M.B. Halpern and N.A. Obers, *J. Math. Phys.* **33** (1992) 3274.

27. Y. Kazama and H. Suzuki, *Mod. Phys. Lett.* **A4** (1989) 235.

28. R. Cohen and D. Gepner, *Mod. Phys. Lett.* **A6** (1991) 2249.

29. M.B. Halpern and J.P. Yamron, *Nucl. Phys.* **B351** (1991) 333.

30. A. Giveon, M.B. Halpern, E.B. Kiritsis and N.A. Obers, *Int. J. Mod. Phys.* **A7** (1992) 947.

31. J.M. Figueroa-O'Farrill, *"Constructing N=2 Superconformal Algebras out of N=1 Affine Lie Algebras"*, Bonn preprint, BONN-HE-93/21.

32. A. Giveon, M.B. Halpern, E.B. Kiritsis and N.A. Obers, *Nucl. Phys.* **B357** (1991) 655.

33. A.B. Zamolodchikov, *JETP Lett.* **43** (1986) 730.

34. A.A. Belavin, A.M. Polyakov and A.B. Zamolodchikov, *Nucl. Phys.* **B241** (1985) 691.

35. A. Neveu and J.H. Schwarz, *Nucl. Phys.* **B31** (1971) 86.

36. A.B. Zamolodchikov, *TMF* **99** (1985) 108; V.A. Fateev and A.B. Zamolodchikov, *Nucl. Phys.* **B280** (1987) 644; V.A. Fateev and S.I. Lykyanov, *Int. J. Mod. Phys.* **A3** (1988) 507.

37. M.B. Halpern and N.A. Obers, *"Ward Identities for Affine-Virasoro Correlators"*, Berkeley/Bonn preprint, UCB-PTH-92/24, BONN-HE-92/21, 1992. To appear in Int. J. Mod. Phys. A

38. M.B. Halpern, *Ann. of Phys.* **194** (1989) 247.

39. M.B. Halpern and N.A. Obers, *"Solving the Ward Identities of Irrational Conformal Field Theory"*, Berkeley/Bonn preprint, UCB-PTH-93/18, BONN-HE-93/17, 1993. To appear in Int. J. Mod. Phys. A

40. M.R. Douglas, *"G/H Conformal Field Theory"*, Caltech preprint, CALT-68-1453, 1987, unpublished.

41. M.Yu. Lashkevich, *"Conformal Blocks of Coset Construction: Zero Ghost Number"*, Landau Inst. preprint, LANDAU-92-TMP-1, 1992.

Affine Lie Algebras in Massive Field Theory
and
Form Factors from Vertex Operators

André LeClair

Newman Laboratory
Cornell University
Ithaca, NY 14853

We present a new application of affine Lie algebras to massive quantum field theory in 2 dimensions, by investigating the $q \to 1$ limit of the q-deformed affine $\widehat{sl(2)}$ symmetry of the sine-Gordon theory, this limit occurring at the free fermion point. Working in radial quantization leads to a quasi-chiral factorization of the space of fields. The conserved charges which generate the affine Lie algebra split into two independent affine algebras on this factorized space, each with level 1 in the anti-periodic sector, and level 0 in the periodic sector. The space of fields in the anti-periodic sector can be organized using level-1 highest weight representations, if one supplements the $\widehat{sl(2)}$ algebra with the usual local integrals of motion. Introducing a particle-field duality leads to a new way of computing form-factors in radial quantization. Using the integrals of motion, a momentum space bosonization involving vertex operators is formulated. Form-factors are computed as vacuum expectation values of vertex operators in momentum space. *Based on talk given at the Berkeley Strings 93 conference, May 1993.*

1. Introduction

The massive integrable quantum field theories in 2 dimensions are characterized as possessing an infinite number of local, commuting integrals of motion, which for instance imply the factorizability of the multi-particle S-matrix[1]. Over the past few years it has been shown that these theories also possess an infinite number of non-abelian symmetries, which generally correspond to q-deformations of affine Lie algebras[2][3][4]. It is now understood for example that the S-matrices are the minimal solutions to the quantum affine symmetry equations.

In order to make further progress toward understanding the full implications of these quantum affine symmetries, the author investigated the following problem[5]. Consider the sine-Gordon (SG) theory, with the action

$$S = \frac{1}{4\pi} \int d^2z \left(\partial_z \phi \partial_{\bar{z}} \phi + 4\lambda \cos(\widehat{\beta}\phi) \right). \tag{1.1}$$

In [2][3] explicit conserved currents were constructed for the 6 generators corresponding to the simple roots of the $q - \widehat{sl(2)}$ affine Lie algebra, where $q = \exp(-2\pi i/\widehat{\beta}^2)$. In this realization the central extension, or level, is zero, and the symmetry is actually a deformed loop algebra. When $\widehat{\beta} = 1$, the SG theory is equivalent to a free massive Dirac fermion[6]. At this value of the SG coupling constant, q becomes 1, and the results of [2] predict the existence of an ordinary undeformed affine $\widehat{sl(2)}$ symmetry in the free Dirac theory. By studying this limit we were able to develop some new structures that we believe can be extended away from $q = 1$. Furthermore, this is not a completely trivial exercise, since there exist fields in the SG description, such as $\exp(i\alpha\phi)$ for $\alpha \notin Z$, which are not simply expressed in terms of the free fermion fields and thus do not have free-field form-factors or correlation functions.

In this talk we will mainly outline the results in [5]. We will first describe the full infinite set of conserved $\widehat{sl(2)}$ charges directly in the free massive Dirac theory, and also the usual infinite number of abelian conserved charges P_n. Radial quantization will then be introduced as the natural way to obtain operators which diagonalize the Lorentz boost operator L. This introduces a fermionic fock space description of the space of fields $\mathcal{H}_F$. Furthermore, $\mathcal{H}_F$ factorizes into $\mathcal{H}_F^L \otimes \mathcal{H}_F^R$, where in the massless limit $\mathcal{H}_F^L$ ($\mathcal{H}_F^R$) is the left (right) 'moving' space of field-states. The conserved charges also factorize in their action on $\mathcal{H}_F^L \otimes \mathcal{H}_F^R$. This leads to two separate algebras $\widehat{sl(2)}_L$ and $\widehat{sl(2)}_R$, which each have level 1 in the anti-periodic sector, and level 0 in the periodic sector. In the same fashion, the integrals of motion $P_n^{L,R}$ satisfy an infinite Heisenberg algebra in the anti-periodic sector, whereas they all continue to commute in the periodic sector. The spectrum of $\mathcal{H}_F$ in the anti-periodic sector can be obtained by supplementing the $\widehat{sl(2)}$ algebra with this Heisenberg algebra, the fields being organized into infinite highest weight modules. By introducing a particle-field duality, we describe a new way to compute form-factors in radial quantization. We then use these algebraic structures to formulate an exact momentum space bosonization. In this operator formulation, non-trivial form-factors of the SG fields $\exp(\pm i\phi/2)$ are computed as expectation values of vertex operators between level 1 highest weight states.

2. Affine $\widehat{sl(2)}$ Symmetry of the Massive Dirac Fermion

The Dirac theory is a massive free field theory of charged fermions. Introducing the Dirac spinors $\Psi_\pm = \begin{pmatrix} \overline{\psi}_\pm \\ \psi_\pm \end{pmatrix}$ of $U(1)$ charge ± 1, the action reads

$$S = -\frac{1}{4\pi} \int dx\,dt \left(\overline{\psi}_- \partial_z \overline{\psi}_+ + \psi_- \partial_{\overline{z}} \psi_+ + i\widehat{m}(\psi_- \overline{\psi}_+ - \overline{\psi}_- \psi_+) \right). \tag{2.1}$$

We have continued to Euclidean space $t \to -it$, and $z, \overline{z}$ are the usual Euclidean light-cone coordinates: $z = (t + ix)/2, \quad \overline{z} = (t - ix)/2$.

Generally, the conserved quantities follow from conserved currents J_μ:

$$\partial_{\overline{z}} J_z + \partial_z J_{\overline{z}} = 0. \tag{2.2}$$

We will often display the two components of conserved currents by writing

$$Q = \int \frac{dz}{2\pi i} J_z - \int \frac{d\overline{z}}{2\pi i} J_{\overline{z}}, \tag{2.3}$$

i.e. without specifying the contour of integration.

Using the equations of motion:

$$\partial_z \overline{\psi}_\pm = i\widehat{m}\psi_\pm, \qquad \partial_{\overline{z}}\psi_\pm = -i\widehat{m}\overline{\psi}_\pm, \tag{2.4}$$

it is easy to find an infinite number of conserved quantities. They are the following:

$$\begin{aligned}
Q_{-n}^\pm &= \frac{(-1)^{n+1}}{2} \left(\int \frac{dz}{2\pi i} (\psi_\pm \partial_z^n \psi_\pm) - \int \frac{d\overline{z}}{2\pi i} (i\widehat{m}\,\overline{\psi}_\pm \partial_z^{n-1} \psi_\pm) \right) \\
Q_n^\pm &= \frac{(-1)^{n+1}}{2} \left(\int \frac{dz}{2\pi i} (-i\widehat{m}\,\psi_\pm \partial_{\overline{z}}^{n-1}\overline{\psi}_\pm) - \int \frac{d\overline{z}}{2\pi i} (\overline{\psi}_\pm \partial_{\overline{z}}^n \overline{\psi}_\pm) \right) \\
\alpha_{-n} &= (-)^n \left(\int \frac{dz}{2\pi i} (\psi_+ \partial_z^n \psi_-) - \int \frac{d\overline{z}}{2\pi i} (i\widehat{m}\,\overline{\psi}_+ \partial_z^{n-1} \psi_-) \right) \\
\alpha_n &= (-)^n \left(\int \frac{dz}{2\pi i} (-i\widehat{m}\,\psi_+ \partial_{\overline{z}}^{n-1}\overline{\psi}_-) - \int \frac{d\overline{z}}{2\pi i} (\overline{\psi}_+ \partial_{\overline{z}}^n \overline{\psi}_-) \right),
\end{aligned} \tag{2.5}$$

where $n \geq 0$, and n is odd for $Q_n^\pm$. (It turns out that $Q_n^\pm = 0$ for n even.) The expressions for $Q_{\pm 1}^\pm$ were originally derived by fermionizing the SG construction in [2] at $\widehat{\beta} = 1$ and noticing that the conservation of the resulting currents was a simple consequence of Dirac equations of motion[7]. Similar conserved charges were constructed for $O(N)$ invariant fermions in [8].

Define

$$P_n \equiv \alpha_n \qquad n \text{ odd}, \qquad T_n \equiv \alpha_n \qquad n \text{ even}. \tag{2.6}$$

Then one can show using the fermionic commutation relations that the charges satisfy the following algebraic relations

$$
\begin{aligned}
&[P_n, P_m] = [P_n, T_m] = [P_n, Q_m^\pm] = [T_n, T_m] = 0 \\
&[T_n, Q_m^\pm] = \pm 2 \, \frac{\widehat{m}^{|n|+|m|}}{\widehat{m}^{|n+m|}} \, Q_{n+m}^\pm \\
&[Q_n^+, Q_m^-] = \frac{\widehat{m}^{|n|+|m|}}{\widehat{m}^{|n+m|}} \, T_{n+m} \\
&[L, \alpha_n] = -n \, \alpha_n, \qquad [L, Q_n^\pm] = -n \, Q_n^\pm,
\end{aligned}
\tag{2.7}
$$

where L is the Lorentz boost operator.

We now interpret this algebraic structure. The P_n's are the usual infinity of commuting integrals of motion with Lorentz spin equal to an odd integer, where $P_z = P_{-1}$, $P_{\bar{z}} = P_1$, and the hamiltonian is $P_1 + P_{-1}$. The additional charges $T_n, Q_n^\pm$ all commute with the P_m's; for $m = \pm 1$ this is just the statement that they are all conserved. The charge T_0 is simply the $U(1)$ charge. The commutation relations of the $T_n, Q_n^\pm$ are the defining relations of the level 0 $\widehat{sl(2)}$ affine Lie algebra. More precisely this is a twisted $\widehat{sl(2)}$ algebra. The twist has a simple explanation: the usual untwisted $\widehat{sl(2)}$ algebra has an $sl(2)$ subalgebra of Lorentz scalars, whereas the Dirac theory has only a $U(1)$ symmetry; the twist breaks $sl(2)$ to $U(1)$. It was shown in [5] how Ward identities for the $\widehat{sl(2)}$ symmetry can fix the correlation functions of the fermion fields.

3. Radial Quantization

Let $\mathcal{F}$ denote the complete space of fields, and let $\mathcal{H}_F$ be spanned by the action of fields on the vacuum:

$$\mathcal{H}_F = \{\Phi_i(0)|0\rangle \equiv |\Phi_i\rangle, \quad \Phi_i \in \mathcal{F}\}. \tag{3.1}$$

The space $\mathcal{H}_F$ diagonalizes the Lorentz boost operator L, but does not diagonalize the momentum operators $P_z, P_{\bar{z}}$. In order to construct the space $\mathcal{H}_F$ explicitly, one can work in radial quantization, since such a construction yields states which diagonalize L. Radial quantization was generally considered in [9], and specifically in 2d free fermion theories in [10][11]. Define the radial coordinates (r, φ) as follows: $z = \frac{r}{2} \exp(i\varphi)$, $\quad \bar{z} = \frac{r}{2} \exp(-i\varphi)$. In radial quantization one treats the r-coordinate as a 'time', and φ as the 'space', and canonical commutation relations are specified at equal φ.

One begins by expanding the fermion fields in a basis of solutions to the Dirac equation of motion in radial coordinates. One finds that one can define both a periodic (p) and an anti-periodic (a) sector in this way. Namely,

$$\Psi_{\pm}^{(a,p)} = \begin{pmatrix} \overline{\psi}_{\pm} \\ \psi_{\pm} \end{pmatrix} = \sum_{\omega} b_{\omega}^{\pm}\, \Psi_{-\omega-1/2}^{(a,p)} + \bar{b}_{\omega}^{\pm}\, \overline{\Psi}_{-\omega-1/2}^{(a,p)}, \tag{3.2}$$

where for the periodic sector $\omega \in Z + 1/2$, and for the anti-periodic sector $\omega \in Z$. The basis spinors have the following explicit expressions:

$$\Psi_{-\omega-1/2}^{(a)} = \Gamma(\tfrac{1}{2} - \omega)\, \widehat{m}^{\omega+1/2} \begin{pmatrix} i\, e^{i(\frac{1}{2}-\omega)\varphi}\, I_{\frac{1}{2}-\omega}(\widehat{m}r) \\ e^{-i(\omega+\frac{1}{2})\varphi}\, I_{-\omega-\frac{1}{2}}(\widehat{m}r) \end{pmatrix}$$

$$\overline{\Psi}_{-\omega-1/2}^{(a)} = \Gamma(\tfrac{1}{2} - \omega)\, \widehat{m}^{\omega+1/2} \begin{pmatrix} e^{i(\frac{1}{2}+\omega)\varphi}\, I_{-\frac{1}{2}-\omega}(\widehat{m}r) \\ -i\, e^{-i(\frac{1}{2}-\omega)\varphi}\, I_{\frac{1}{2}-\omega}(\widehat{m}r) \end{pmatrix}, \tag{3.3}$$

$$\Psi_{-\omega-1/2}^{(p)} = \frac{2\widehat{m}^{\omega+1/2}}{\Gamma(\frac{1}{2}+\omega)} \begin{pmatrix} -i\, e^{i(\frac{1}{2}-\omega)\varphi}\, K_{\omega-\frac{1}{2}}(\widehat{m}r) \\ e^{-i(\omega+\frac{1}{2})\varphi}\, K_{\omega+\frac{1}{2}}(\widehat{m}r) \end{pmatrix} \qquad \omega \geq 1/2$$

$$\overline{\Psi}_{-\omega-1/2}^{(p)} = \frac{2\widehat{m}^{\omega+1/2}}{\Gamma(\frac{1}{2}+\omega)} \begin{pmatrix} e^{i(\frac{1}{2}+\omega)\varphi}\, K_{\omega+\frac{1}{2}}(\widehat{m}r) \\ i\, e^{i(\omega-\frac{1}{2})\varphi}\, K_{\omega-\frac{1}{2}}(\widehat{m}r) \end{pmatrix} \qquad \omega \geq 1/2, \tag{3.4}$$

and $\Psi_{-\omega-1/2}^{(p)}, \overline{\Psi}_{-\omega-1/2}^{(p)}$ for $\omega \leq -1/2$ have the same expression as in the anti-periodic sector. In the quantum theory these modes satisfy:

$$\{b_{\omega}^{+}, b_{\omega'}^{-}\} = \{\overline{b}_{\omega}^{+}, \overline{b}_{\omega'}^{-}\} = \delta_{\omega,-\omega'}, \qquad \{b_{\omega}, \bar{b}_{\omega'}\} = 0. \tag{3.5}$$

In the massless limit $\widehat{m} \to 0$, the operators $b_{\omega}^{\pm}, \bar{b}_{\omega}^{\pm}$ are the familiar fermionic oscillators of conformal field theory. Indeed,

$$\begin{pmatrix} \overline{\psi}_{\pm} \\ \psi_{\pm} \end{pmatrix} \xrightarrow{\widehat{m} \to 0} \sum_{\omega} \begin{pmatrix} \bar{b}_{\omega}^{\pm}\, \bar{z}^{-\omega-1/2} \\ b_{\omega}^{\pm}\, z^{-\omega-1/2} \end{pmatrix}. \tag{3.6}$$

What is remarkable is that since the oscillators are defined in the massive theory, the fermion fock spaces built from these oscillators continue to correspond precisely to the space of fields $\mathcal{H}_F$ in the *massive* theory. Furthermore, $\mathcal{H}_F$ factorizes into quasi-left/right pieces:

$$\mathcal{H}_F = \mathcal{H}_F^L \otimes \mathcal{H}_F^R. \tag{3.7}$$

For more discussion on this point, see [5].

Consider first the periodic sector. We define the physical vacuum as follows:

$$b_\omega^\pm |0\rangle = \overline{b}_\omega^\pm |0\rangle = 0, \qquad \omega \geq 1/2. \tag{3.8}$$

Define the left and right periodic fock spaces:

$$\begin{aligned}
\mathcal{H}_p^L &= \left\{ b_{-\omega_1}^- b_{-\omega_2}^- \cdots b_{-\omega_1'}^+ b_{-\omega_2'}^+ \cdots |0\rangle \right\} \\
\mathcal{H}_p^R &= \left\{ \overline{b}_{-\omega_1}^- \overline{b}_{-\omega_2}^- \cdots \overline{b}_{-\omega_1'}^+ \overline{b}_{-\omega_2'}^+ \cdots |0\rangle \right\},
\end{aligned} \tag{3.9}$$

where $\omega, \omega' \geq 1/2$. One can see directly that in the periodic sector $\mathcal{H}_F = \mathcal{H}_p^L \otimes \mathcal{H}_p^R$. For example, using the explicit expansions (3.2), one finds

$$\partial_z^n \psi_\pm(0)|0\rangle = n!\, b_{-n-\frac{1}{2}}^\pm |0\rangle, \qquad \partial_{\overline{z}}^n \overline{\psi}_\pm(0)|0\rangle = n!\, \overline{b}_{-n-\frac{1}{2}}^\pm |0\rangle. \tag{3.10}$$

Other states in $\mathcal{H}_p^{L,R}$ correspond to composite operators. For example, consider the $U(1)$ current $J_z = \psi_+ \psi_-$, $J_{\overline{z}} = \overline{\psi}_+ \overline{\psi}_-$. One has

$$J_z(0)|0\rangle = b_{-\frac{1}{2}}^+ b_{-\frac{1}{2}}^- |0\rangle, \qquad J_{\overline{z}}(0)|0\rangle = \overline{b}_{-\frac{1}{2}}^+ \overline{b}_{-\frac{1}{2}}^- |0\rangle. \tag{3.11}$$

We now turn to the anti-periodic sector. Due to the existence of the zero modes $b_0^\pm, \overline{b}_0^\pm$, the 'vacuum' in this sector is doubly degenerate for both left and right. Define these vacua as $|\pm \tfrac{1}{2}\rangle_L$ and $|\pm \tfrac{1}{2}\rangle_R$, characterized by

$$\begin{aligned}
b_0^\pm |\mp\tfrac{1}{2}\rangle_L &= |\pm\tfrac{1}{2}\rangle_L, & b_0^\pm |\pm\tfrac{1}{2}\rangle_L &= 0, & b_n^\pm |\pm\tfrac{1}{2}\rangle_L &= 0, & n &\geq 1 \\
\overline{b}_0^\pm |\mp\tfrac{1}{2}\rangle_R &= |\pm\tfrac{1}{2}\rangle_R, & \overline{b}_0^\pm |\pm\tfrac{1}{2}\rangle_R &= 0, & \overline{b}_n^\pm |\pm\tfrac{1}{2}\rangle_R &= 0, & n &\geq 1.
\end{aligned} \tag{3.12}$$

The vacua $\langle \pm\tfrac{1}{2}|$ are defined by the inner products

$$_L\langle \mp\tfrac{1}{2}| \pm\tfrac{1}{2}\rangle_L = {}_R\langle \mp\tfrac{1}{2}| \pm\tfrac{1}{2}\rangle_R = 1. \tag{3.13}$$

142

These vacuum states have $U(1)$ charge $\pm 1/2$. The anti-periodic fock spaces are defined as

$$
\begin{aligned}
\mathcal{H}^L_{a_\pm} &= \left\{ b^-_{-n_1} b^-_{-n_2} \cdots b^+_{-n'_1} b^+_{-n'_2} \cdots | \pm \tfrac{1}{2} \rangle_L \right\} \\
\mathcal{H}^R_{a_\pm} &= \left\{ \bar{b}^-_{-n_1} \bar{b}^-_{-n_2} \cdots \bar{b}^+_{-n'_1} \bar{b}^+_{-n'_2} \cdots | \pm \tfrac{1}{2} \rangle_R \right\},
\end{aligned}
\tag{3.14}
$$

for $n, n' \geq 1$. Based on a study of the massless limit the following identification was proposed in [5]:

$$
e^{\pm i\phi(0)/2} |0\rangle = (| \pm \tfrac{1}{2} \rangle_L \otimes | \mp \tfrac{1}{2} \rangle_R) \equiv | \pm \tfrac{1}{2} \rangle,
\tag{3.15}
$$

where $\phi(z, \bar{z})$ is the local SG field.

Consider now the conserved charges of the last section in radial quantization. Given a conserved current J_z, $J_{\bar{z}}$, the conserved charge is

$$
Q = \frac{1}{4\pi} \int_{-\pi}^{\pi} r \, d\varphi \left(e^{i\varphi} J_z + e^{-i\varphi} J_{\bar{z}} \right).
\tag{3.16}
$$

All of the conserved charges constructed in section 2 can thereby be expressed in terms of the radial modes $b^\pm_\omega$, $\bar{b}^\pm_\omega$. More specifically, define the inner product of two spinors $A = \begin{pmatrix} \bar{a} \\ a \end{pmatrix}$, $B = \begin{pmatrix} \bar{b} \\ b \end{pmatrix}$ as

$$
(A, B) = \frac{1}{4\pi} \int_{-\pi}^{\pi} r d\varphi \left(e^{i\varphi} a b + e^{-i\varphi} \bar{a} \bar{b} \right).
\tag{3.17}
$$

The conserved charges can all be expressed using the above inner product:

$$
\begin{aligned}
Q^\pm_{-n} &= \frac{(-)^{n+1}}{2} (\Psi_\pm, \partial_z^n \Psi_\pm), \qquad Q^\pm_n = \frac{(-)^{n+1}}{2} (\Psi_\pm, \partial_{\bar{z}}^n \Psi_\pm) \\
\alpha_{-n} &= (-)^n (\Psi_+, \partial_z^n \Psi_-), \qquad \alpha_n = (-)^n : (\Psi_+, \partial_{\bar{z}}^n \Psi_-) : \ .
\end{aligned}
\tag{3.18}
$$

One finds that the charges split into left and right pieces:

$$
Q^\pm_n = Q^{\pm,L}_n + Q^{\pm,R}_{-n}, \qquad \alpha_n = \alpha^L_n + \alpha^R_{-n}.
\tag{3.19}
$$

The additional minus sign in the subscript $-n$ of the right piece of the charges in comparison to the left piece is dictated by Lorentz covariance. The explicit expressions are as follows:

$$
\begin{aligned}
\alpha^L_n &= \widehat{m}^{|n|+n} \sum_{\omega \in S_n^{(a,p)}} \frac{\Gamma(\tfrac{1}{2} + \omega - n)}{\Gamma(\tfrac{1}{2} + \omega)} : b^+_{n-\omega} b^-_\omega : \\
Q^{\pm,L}_n &= \widehat{m}^{|n|+n} \sum_{\omega \in S_n^{(a,p)}} \frac{\Gamma(\tfrac{1}{2} + \omega - n)}{\Gamma(\tfrac{1}{2} + \omega)} b^\pm_{n-\omega} b^\pm_\omega,
\end{aligned}
\tag{3.20}
$$

where the sums over ω run over $S_n^{(p)}$ ($S_n^{(a)}$) for the periodic (anti-periodic) sector, and

$$S_n^{(a)} = Z, \quad \forall\, n$$
$$S_n^{(p)} = \{\omega \in Z + 1/2;\ \ 0 > \omega > n \ \text{if}\ n > 0;\ \ n > \omega > 0 \ \text{if}\ n < 0\}. \tag{3.21}$$

Identical expressions apply to $Q_n^{\pm,R}, \alpha_n^R$ with $b_\omega^\pm \to \bar{b}_\omega^\pm$.

Define as before $P_n^{L,R} = \alpha_n^{L,R}$, n odd; $T_n^{L,R} = \alpha_n^{L,R}$, n even. Then one can show that the above operators satisfy the relations

$$[P_n^L, P_m^L] = [T_n^L, T_m^L] = n\, k\, \widehat{m}^{2|n|}\, \delta_{n,-m}$$
$$[P_n^L, T_m^L] = [P_n^L, Q_m^{\pm,L}] = 0$$
$$[T_n^L, Q_m^{\pm,L}] = \pm 2\, \frac{\widehat{m}^{|n|+|m|}}{\widehat{m}^{|n+m|}}\, Q_{n+m}^{\pm,L} \tag{3.22}$$
$$[Q_n^{+,L}, Q_m^{-,L}] = \frac{\widehat{m}^{|n|+|m|}}{\widehat{m}^{|n+m|}}\, T_{n+m}^L + \frac{n}{2}\, k\, \widehat{m}^{2|n|}\, \delta_{n,-m},$$

where in the periodic sector the level k is zero, and in the anti-periodic sector $k = 1$. Identical results apply to the right pieces of the conserved charges, with the same level 1 in the antiperiodic sector. Note that the sum of the left and right operators (3.19) continues to satisfy a level $k = 0$ algebra in either sector, which is required for consistency with section 2.

Due to the non-zero level 1 in the anti-periodic sector, the space of fields in this sector can be organized using infinite highest weight modules. At level 1 there are only 2 highest weight states $|\Lambda_j\rangle$, $j = 0, 1/2$, satisfying

$$Q_n^{\pm,L}\, |\Lambda_j\rangle_L = T_n^L\, |\Lambda_j\rangle_L = 0 \qquad n \geq 1$$
$$T_0^L\, |\Lambda_0\rangle_L = -\tfrac{1}{2}\, |\Lambda_0\rangle_L, \qquad T_0^L\, |\Lambda_{\frac{1}{2}}\rangle_L = \tfrac{1}{2}\, |\Lambda_{\frac{1}{2}}\rangle_L. \tag{3.23}$$

These highest weight states are the 'vacuum' states defined above with the identification:

$$|+\tfrac{1}{2}\rangle_L = |\Lambda_{\frac{1}{2}}\rangle, \qquad |-\tfrac{1}{2}\rangle_L = |\Lambda_0\rangle. \tag{3.24}$$

Let us denote the P_n extension of $\widehat{sl(2)}$ as $\widehat{\widehat{sl(2)}}$, and define the $\widehat{\widehat{sl(2)}}_L$ modules $\widehat{V}_{\Lambda_j}^L$ as follows,

$$\widehat{V}_{\Lambda_i}^L \equiv \left\{ Q_{-n_1}^{\pm,L}\, Q_{-n_2}^{\pm,L} \cdots T_{-n_1'}^L\, T_{-n_2'}^L \cdots P_{-n_1''}^L\, P_{-n_2''}^L \cdots |\Lambda_i\rangle \right\}, \tag{3.25}$$

for $n, n', n'' \geq 1$. Then one can show that

$$\mathcal{H}_a^L = \widehat{V}_{\Lambda_0}^L \oplus \widehat{V}_{\Lambda_{\frac{1}{2}}}^L, \tag{3.26}$$

where $\mathcal{H}_a^L = \mathcal{H}_{a_+}^L \oplus \mathcal{H}_{a_-}^L$. This is proven by comparing the Tr (q^L) characters for the fermionic space $\mathcal{H}_a^L$ with the ones for the (twisted) $\widehat{\widehat{sl(2)}}$ modules.

4. Particle-Field Duality and Form-Factors in Radial Quantization

In conventional canonical approaches to massive quantum field theory, one deals with the space of particles $\mathcal{H}_P$. In 2d, parametrizing the energy and momentum in terms of the usual rapidity, $E = m \cosh\theta$, $P = m \sinh\theta$, the space $\mathcal{H}_P$ is spanned by

$$\mathcal{H}_P = \{|\theta_1, \cdots, \theta_n\rangle_{\epsilon_1 \cdots \epsilon_n}\}, \tag{4.1}$$

where ϵ_i are the quantum numbers of the particles. One can easily define a dual space $\mathcal{H}_P^*$, and a completeness relation

$$1 = \sum_{n=0}^{\infty} \frac{1}{n!} \sum_{\epsilon_i} \int d\theta_1 \cdots d\theta_n \, |\theta_1, \cdots, \theta_n\rangle_{\epsilon_1 \cdots \epsilon_n} {}^{\epsilon_1 \cdots \epsilon_n}\langle \theta_1, \cdots, \theta_n|. \tag{4.2}$$

Form-factors are matrix elements of fields in the space $\mathcal{H}_P$. Consider for example the form-factor

$$ {}^{\epsilon_1 \cdots \epsilon_n}\langle \theta_1 \cdots \theta_n | \Phi(0) | 0 \rangle, \tag{4.3}$$

where $\Phi \in \mathcal{F}$. The completeness relation in $\mathcal{H}_P$ allows one to view $\Phi(0)|0\rangle$ as a element of $\mathcal{H}_P$. Conventionally, one thinks of expressing the fields $\Phi(0)$ in terms of the operators that create the states in $\mathcal{H}_P$, and one computes form-factors in the space $\mathcal{H}_P$.

We now propose an entirely different way to compute the same form-factors. Let us suppose that one can define a non-degenerate inner-product on the space of fields $\mathcal{H}_F$, and thereby construct the dual space $\mathcal{H}_F^*$ and the completeness relation:

$$1 = \sum_i |\Phi_i\rangle\langle \Phi^i|. \tag{4.4}$$

This implies that one can then consider states in $\mathcal{H}_P^*$ as elements of $\mathcal{H}_F^*$:

$$ {}^{\epsilon_1 \cdots \epsilon_n}\langle \theta_1 \cdots \theta_n | \ \in \ \mathcal{H}_F^*. \tag{4.5}$$

Consequently, the form-factors can be computed directly in the space $\mathcal{H}_F$ rather than $\mathcal{H}_P$. Radial quantization provides us with an explicit construction of the spaces $\mathcal{H}_F, \mathcal{H}_F^*$ and the completeness relation, thus by the above reasoning, radial quantization leads to a new way to compute form-factors. The success of the method relies on being able to construct explicitly the map (4.5). We now describe how to determine this map in the model we are considering. Henceforth we will be working exclusively in the antiperiodic sector[1].

---

[1] The periodic sector will be considered in [12]

In the standard temporal quantization, the fermion fields have the following plane wave expansions

$$\Psi_\pm = \begin{pmatrix} \overline{\psi}_\pm \\ \psi_\pm \end{pmatrix} = \pm\sqrt{\widehat{m}} \int_{-\infty}^{\infty} \frac{du}{2\pi i |u|}\, \widehat{b}^\pm(u) \begin{pmatrix} 1/\sqrt{u} \\ -i\sqrt{u} \end{pmatrix} e^{\widehat{m}zu + \widehat{m}\bar{z}/u}, \qquad (4.6)$$

where $u = \exp(\theta)$. We have combined the usual creation-annihilation operators into a single operator $\widehat{b}^\pm(u)$, where $\widehat{b}^\pm(u > 0)$ is a creation operator and $\widehat{b}^\pm(u < 0)$ is an annihilation operator. (See [5] for details.) The states in $\mathcal{H}_P^*$ are constructed as follows:

$$\epsilon_1 \cdots \epsilon_n \langle \theta_1 \cdots \theta_n| = \frac{1}{(2\pi i)^n}\, \langle 0|\widehat{b}^{\epsilon_1}(-u_1) \cdots \widehat{b}^{\epsilon_n}(-u_n), \qquad (4.7)$$

where $u_i = \exp(\theta_i)$.

The key to constructing explicitly the map from $\mathcal{H}_P^*$ to $\mathcal{H}_F^*$ is based on the following fact. By analytically continuing the u integral in (4.6) and appropriately redefining $\widehat{b}^\pm(u)$, one can obtain the expansions in radial quantization. More specifically, let

$$\int_{-\infty}^{\infty} \frac{du}{2\pi i |u|}\, \widehat{b}^\pm(u) \rightarrow \left(\int_{\mathcal{C}_\varphi^L} \frac{du}{2\pi i u}\, b_\pm(u) \; - \; \int_{\mathcal{C}_\varphi^R} \frac{du}{2\pi i u}\, \bar{b}_\pm(u) \right), \qquad (4.8)$$

where $\mathcal{C}_\varphi^L, \mathcal{C}_\varphi^R$ are contours depending on the angular direction φ of the cut displayed in figures 1,2, and

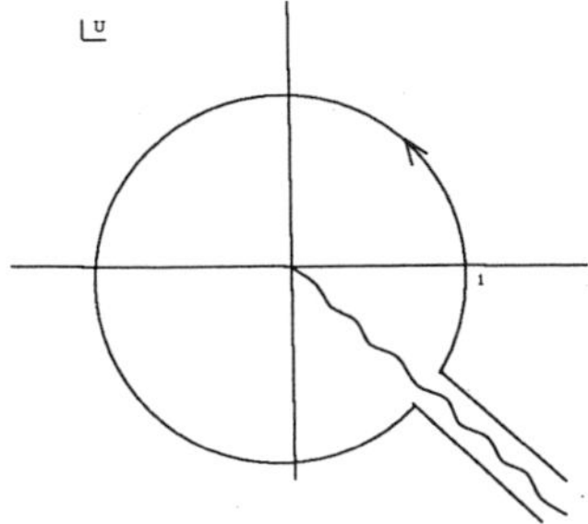

Figure 1. The contour $\mathcal{C}_\varphi^L$. The cut (wavy line) is oriented at an angle φ from the negative y-axis. The circle is at $|u| = 1$.

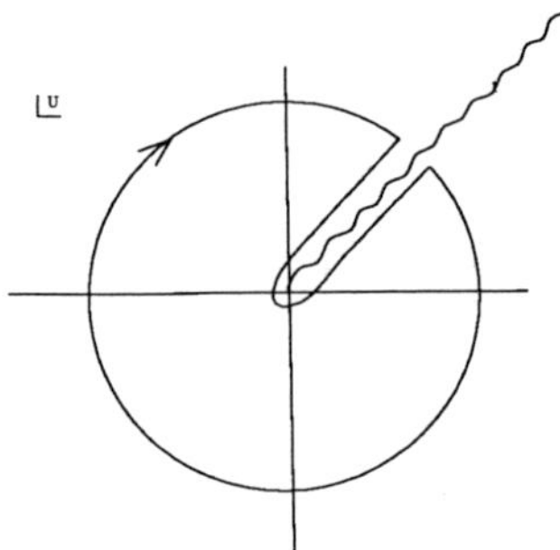

Figure 2. The contour C_φ^R. The cut is oriented at an angle φ from the positive y-axis. The circle is at $|u| = 1$.

$$b^\pm(u) = \pm i \sum_{\omega \in Z} \Gamma(\tfrac{1}{2} - \omega)\, \widehat{m}^\omega\, b_\omega\, u^\omega$$

$$\bar{b}^\pm(u) = \pm \sum_{\omega \in Z} \Gamma(\tfrac{1}{2} - \omega)\, \widehat{m}^\omega\, \bar{b}_\omega\, u^{-\omega}. \tag{4.9}$$

Using integral representations for the Bessel functions, one obtains (3.3). This leads us to propose the following fundamental expression:

$$\langle 0|\widehat{b}^{\epsilon_1}(u_1) \cdots \widehat{b}^{\epsilon_n}(u_n) = \frac{1}{2}\left(\langle +\tfrac{1}{2}| + \langle -\tfrac{1}{2}| \right) \left(b^{\epsilon_1}(u_1) + \bar{b}^{\epsilon_1}(u_1) \right) \cdots \left(b^{\epsilon_n}(u_n) + \bar{b}^{\epsilon_n}(u_n) \right). \tag{4.10}$$

The RHS of the above equation is a state in $\mathcal{H}_F^*$, and this means that the computation of the form-factors can be carried out in $\mathcal{H}_F$.

As an example, consider the form-factors of the fields $\exp(\pm i\phi/2)$ in the SG theory. From the identification (3.15) one has

$$^{\epsilon_1 \cdots \epsilon_n}\langle \theta_1 \cdots \theta_n|\, \exp(\pm i\phi/2)|0\rangle = \frac{1}{(2\pi i)^n}\, {}_L\langle \mp \tfrac{1}{2}|b^{\epsilon_1}(-u_1) \cdots b^{\epsilon_n}(-u_n)| \pm \tfrac{1}{2}\rangle_L$$

$$= \frac{1}{(2\pi i)^n}\, {}_R\langle \pm \tfrac{1}{2}|\bar{b}^{\epsilon_1}(-u_1) \cdots \bar{b}^{\epsilon_n}(-u_n)| \mp \tfrac{1}{2}\rangle_R \tag{4.11}$$

Additional arguments for the above formula were given in [5]. One can compute the RHS explicitly and one finds agreement with the known result[13][14]. We will show this explicitly in the next section. Note that the RHS of the above expression has the structure of a conformal free-fermion correlator, but here we are in momentum space!

5. Vertex Operators and Momentum Space Bosonization

We will now describe a vertex operator construction for the form-factors of the last section. This involves constructing a bosonic representation of the operators $b^{\pm}(u), \bar{b}^{\pm}(u)$. Recall that in the anti-periodic section one has the infinite Heisenberg algebras of the chirally split integrals of motion:

$$\left[\alpha_n^L, \alpha_m^L\right] = n \, \widehat{m}^{2|n|} \, \delta_{n,-m}, \qquad \left[\alpha_n^R, \alpha_m^R\right] = n \, \widehat{m}^{2|n|} \, \delta_{n,-m}. \tag{5.1}$$

Define some momentum space bosonic fields as follows:

$$-i\widehat{\phi}^L(u) = \sum_{n\neq 0} \widehat{m}^{-|n|} \, \alpha_n^L \, \frac{u^n}{n} + \alpha_0^L \log(u) - \widetilde{\alpha}_0^L$$

$$-i\widehat{\phi}^R(u) = \sum_{n\neq 0} \widehat{m}^{-|n|} \, \alpha_n^R \, \frac{u^{-n}}{n} + \alpha_0^R \log(u) + \widetilde{\alpha}_0^R, \tag{5.2}$$

where $\left[\alpha_0^L, \widetilde{\alpha}_0^L\right] = \left[\alpha_0^R, \widetilde{\alpha}_0^R\right] = 1$. The vacua $|\emptyset\rangle_{L,R}$ are defined to satisfy

$$\alpha_n^L |\emptyset\rangle_L = \alpha_n^R |\emptyset\rangle_R = 0, \quad n \geq 0; \qquad \widetilde{\alpha}_0^L |\emptyset\rangle_L, \ \widetilde{\alpha}_0^R |\emptyset\rangle_R \neq 0. \tag{5.3}$$

Note that $|\emptyset\rangle \equiv |\emptyset\rangle_L \otimes |\emptyset\rangle_R$ does not correspond to the physical vacuum since it is not annihilated by P_1 and P_{-1}. One has the following vacuum expectation values

$$_L\langle\emptyset| \, \widehat{\phi}^L(u) \, \widehat{\phi}^L(u') \, |\emptyset\rangle_L = -\log(1/u - 1/u')$$

$$_R\langle\emptyset| \, \widehat{\phi}^R(u) \, \widehat{\phi}^R(u') \, |\emptyset\rangle_R = -\log(u - u') \tag{5.4}$$

$$_L\langle\emptyset| \prod_i e^{i\alpha_i \widehat{\phi}^L(u_i)} \, |\emptyset\rangle_L = \prod_{i<j} (1/u_i - 1/u_j)^{\alpha_i \alpha_j}$$

$$_R\langle\emptyset| \prod_i e^{i\alpha_i \widehat{\phi}^R(u_i)} \, |\emptyset\rangle_R = \prod_{i<j} (u_i - u_j)^{\alpha_i \alpha_j}. \tag{5.5}$$

Using these free momentum space fields one can bosonize all of the ingredients of the last section:

$$b^+(u) = \sqrt{\frac{\pi}{u}} : e^{i\widehat{\phi}^L(u)} :, \qquad b^-(-u) = \sqrt{\frac{\pi}{u}} : e^{-i\widehat{\phi}^L(u)} :$$

$$\overline{b}^+(u) = \sqrt{\pi u} : e^{-i\widehat{\phi}^R(u)} :, \qquad \overline{b}^-(-u) = -\sqrt{\pi u} : e^{i\widehat{\phi}^R(u)} :,$$

$$(5.6)$$

and

$$|\alpha\rangle_L = : \exp(i\alpha\widehat{\phi}^L(\infty)) : |\emptyset\rangle_L, \qquad |\alpha\rangle_R = : \exp(-i\alpha\widehat{\phi}^R(0)) : |\emptyset\rangle_R$$

$$_L\langle\alpha| = \lim_{u\to 0} u^{-\alpha^2} \langle\emptyset| : \exp(i\alpha\widehat{\phi}^L(u)) :, \qquad _R\langle\alpha| = \lim_{u\to\infty} u^{\alpha^2} \langle\emptyset| \exp(-i\alpha\widehat{\phi}^R(u)).$$

$$(5.7)$$

¿From this construction one finds

$$
\begin{aligned}
&_L\langle\mp\tfrac{1}{2}|b^+(u_1)\cdots b^+(u_n)b^-(u_{n+1})\cdots b^-(u_{2n})|\pm\tfrac{1}{2}\rangle_L \\
&= {}_R\langle\pm\tfrac{1}{2}|\overline{b}^+(u_1)\cdots\overline{b}^+(u_n)\overline{b}^-(u_{n+1})\cdots\overline{b}^-(u_{2n})|\mp\tfrac{1}{2}\rangle_R \\
&= \pi^n\sqrt{u_1\cdots u_{2n}}\left(\prod_{i=1}^{n}\left(\frac{u_{i+n}}{u_i}\right)^{\pm 1/2}\right)\left(\prod_{i<j\le n}(u_i - u_j)\right) \\
&\quad \cdot\left(\prod_{n+1\le i<j}(u_i - u_j)\right)\left(\prod_{r=1}^{n}\prod_{s=n+1}^{2n}\frac{1}{u_r + u_s}\right).
\end{aligned}
$$

$$(5.8)$$

These expressions agree with the known form-factors, though they were originally computed using rather different methods[13][14].

6. Conclusions

We have described a new way to compute form-factors based on structures inherent in radial quantization, as vacuum expectation values of vertex operators. The integrals of motion played an essential role in the explicit construction of the vertex operators. We believe that these results can be extended away from the free fermion point, eventually leading to a vertex operator construction of the SG form-factors. These form-factors have been previously computed by Smirnov using bootstrap methods[14]. Recently, Lukyanov has proposed a vertex operator construction for the SG form-factors at all values of the SG coupling[15]. Comparing the latter construction with the results described here one finds some apparent differences. The free boson oscillators in [15] have no apparent connection with the integrals of motion, in contrast with the above results. More importantly, it is proposed in [15] that the form-factors are computed as traces over free-field modules, whereas in our work they are vacuum expection values. It would be interesting to understand more precisely if and how the construction in [15] is related to ours at the free-fermion point.

Acknowledgements

I would like to thank the organizers of the Berkeley Strings 93 conference for the opportunity to present this work. This work is supported by an Alfred P. Sloan Foundation fellowship, and the National Science Foundation in part through the National Young Investigator program.

References

[1] A. B. Zamolodchikov and Al. B. Zamolodchikov, Annals Phys. 120 (1979) 253.

[2] D. Bernard and A. LeClair, Commun. Math. Phys. 142 (1991) 99; Phys. Lett. B247 (1990) 309.

[3] G. Felder and A. LeClair, Int. Journ. Mod. Phys. A7 Suppl. 1A (1992) 239.

[4] F. A. Smirnov, Int. J. Mod. Phys. A7, Suppl. (1992) .

[5] A. LeClair, *Spectrum Generating Affine Lie Algebras in Massive Quantum Field Theory*, CLNS 93/1220, hep-th/9305110

[6] S. Coleman, Phys. Rev. D 11 (1975) 2088.

[7] R. K. Kaul and R. Rajaraman, Int. J. Mod. Phys. A8 (1993) 1815.

[8] E. Abdalla, M. C. B. Abdalla, G. Sotkov, and M. Stanishkov, *Off Critical Current Algebras*, Univ. Sao Paulo preprint, IFUSP-preprint-1027, Jan. 1993.

[9] S. Fubini, A. J. Hanson, and R. Jackiw, Phys. Rev. D7 (1973) 1732.

[10] H. Itoyama and H. B. Thacker, Nucl. Phys. B320 (1989) 541.

[11] P. Griffin, Nucl. Phys. B334, 637.

[12] C. Efthimiou and A. LeClair, *Particle-Field Duality and Form-Factors from Vertex Operators*, in preparation.

[13] B. Schroer and T. T. Truong, Nucl. Phys. B144 (1978) 80 ;
E. C. Marino, B. Schroer, and J. A. Swieca, Nucl. Phys. B200 (1982) 473.

[14] F. A. Smirnov, *Form Factors in Completely Integrable Models of Quantum Field Theory*, in *Advanced Series in Mathematical Physics* 14, World Scientific, 1992.

[15] S. Lukyanov, *Free Field Representation for Massive Integrable Models*, Rutgers preprint RU-93-30, hep-th/9307196.

Super KP Flows in Genus One

Jeffrey M. Rabin
Dept. of Mathematics, U.C.S.D.
La Jolla, CA 92093 USA
jrabin@ucsd.edu

The purpose of this talk is to exhibit a simple, explicit solution to the new super Kadomtsev-Petviashvili (SKP) hierarchy discovered by Mulase [M91] and myself [R91] recently. Unfortunately, it is a "classical" solution, arising from the Krichever construction on a super Riemann surface of genus 1 (supertorus), rather than a "quantum" solution satisfying a string equation and potentially relevant to some matrix model of 2d topological supergravity. However, it provides some understanding of the behavior of the supersymmetric Krichever construction with respect to supermoduli, of interest in the operator formalism for superconformal field theory [AGNSV]. A prime motivation for this work is to investigate the supergeometric structures relevant to the SKP theory and the operator formalism in an explicitly computable example, these structures being super Jacobians, theta functions, tau functions, and Grassmannians. The computations rely on the general SKP theory developed in [R91] and are discussed in full detail in [R93].

To begin, recall the simple solution to the ordinary KP hierarchy arising from the Krichever construction applied to a torus M [Mum]. M is the quotient of the complex plane $\mathbf{C}$ by the lattice generated by $z \to z + 1$ and $z \to z + \tau$. The space of line bundles of fixed degree on M is the Jacobian $\mathrm{Jac}(M)$. It is classical that this is isomorphic to M itself: the line bundle corresponding to a point z_0 of M is the one having the theta function $\Theta(z - z_0|\tau)$ as a holomorphic section. (Theta functions are denoted by a capital letter to avoid later confusion with the supercoordinate θ. The odd theta function with characteristics $[1/2, 1/2]$ is always meant.) The pair of differential operators in x ($d = d/dx$),

$$Q = d^2 - 2\wp(x + a),$$

$$P = Q_+^{3/2} = d^3 - 3\wp(x + a)d - \frac{3}{2}\wp'(x + a),$$

satisfy the "classical string equation" [P, Q] = 0, with $\wp$ being the Weierstrass $\wp$-function, essentially the second derivative of $\log \Theta$. The connection with the torus is that, by the Krichever correspondence, the meromorphic functions on the torus having poles only at $z = 0$ correspond to commuting differential operators. The correspondence makes use of a special section, the Baker function $B(z, x + a)$, of the bundle labeled by the point $x + a$ of $\mathrm{Jac}(M)$. In the language of the operator formalism, the Baker function is a generating function for a set of wave functions for the states of a filled Dirac sea, which are obtained as successive x derivatives of B. The correspondence pairs each meromorphic function F of pole order n with an nth order differential operator D_F constructed so that $FB = D_F B$. In particular, the functions $\wp(z)$ and $-\wp'(z)/2$ correspond to Q and P respectively, and Q and P generate the entire ring of differential operators corresponding to all such meromorphic functions. If the parameter a is given an appropriate linear dependence on the flow parameters t_n, then Q, P satisfy the KP flow equations

$$\frac{\partial X}{\partial t_n} = [Q_+^{n/2}, X].$$

The flows preserve the relation $P = Q_+^{3/2}$, unlike in the "quantum" case of $[P, Q] = 1$ [DK]. Geometrically, they are simply linear motions on $\mathrm{Jac}(M)$. The t_n constitute an extremely simple example of the flat coordinates important in special geometry and topological field theory [LSW]. The tau function corresponding to this solution is simply $\Theta(x + a | \tau)$.

A supersymmetric analogue of this construction begins with the supertorus M obtained as the quotient of $\mathbf{C}^{1,1}$ by the discrete supertranslation group G with generators

$$(z, \theta) \to (z + 1, \theta), \quad (z, \theta) \to (z + \tau + \theta\delta, \theta + \delta).$$

The space of line bundles of fixed degree on M is now a quotient space of M. A point (z_0, θ_0) specifies a bundle having the super theta function $\Theta[z - z_0 - \theta\theta_0 | \tau + (\theta + \theta_0)\delta]$ as a holomorphic section. However, the bundle labeled $(z_0 + \eta\delta, \theta_0)$ for any odd η is equivalent to the one with $\eta = 0$: the ratio of the corresponding theta functions is a meromorphic function on M, namely

$$1 - \eta\delta \frac{d}{dz} \log \Theta(z - z_0 - \theta\theta_0 | \tau).$$

The Jacobian is the quotient of M by this identification $z \equiv z + \eta\delta$. This space, the quotient of a supermanifold by a purely nilpotent identification, is not a supermanifold, nor even a supervariety. However, like any Jacobian, it is an abelian group, the group of line bundles under tensor product. It had to differ from M, because in contrast to an ordinary torus, M is not a group. This is because it is the quotient of the supersymmetry group $\mathbf{C}^{1,1}$ by the discrete subgroup G, which is nonabelian and not a normal subgroup.

The Krichever construction produces a generating set of three supercommuting differential operators Q, P, Σ in an even variable x and an odd variable ξ, corresponding to the meromorphic functions on M,

$$Q \leftrightarrow \wp(z | \tau + \theta\delta),$$

$$P \leftrightarrow -\frac{1}{2}\wp'(z | \tau + \theta\delta),$$

$$\Sigma \leftrightarrow \theta + \frac{\delta}{2\pi i} \frac{d}{dz} \log \Theta(z | \tau).$$

They have the explicit forms,

$$\Sigma = \partial + \frac{\delta}{2\pi i} d,$$

$$Q = d^2 + \omega\partial + u,$$

$$P = Q_+^{3/2} = d^3 + \frac{3}{2}\omega\partial d + \frac{3}{2}ud + \frac{3}{4}\omega'\partial + \frac{3}{4}u',$$

152

where

$$\omega = 2(\alpha\wp' + \frac{\alpha\delta\xi}{2\pi i}\wp'' + \delta\dot\wp),$$

$$u = 2\{-\wp + \frac{\xi\delta}{2\pi i}\wp' - \alpha\delta\dot\wp + \alpha\delta\dot q + \frac{\alpha\delta}{2\pi i}[\wp'\partial_x \log\Theta - (\wp - q)^2]\}.$$

In these equations, all functions have arguments $(x + a|\tau)$, d and ∂ are the differential operators $\partial/\partial x$ and $\partial/\partial\xi$ respectively; a prime on a function denotes its x derivative while a dot denotes its τ derivative; and $q(\tau)$ is a certain theta constant.

The functions and operators correspond in the same fashion as before, with the Baker function $B(z, \theta, \xi, x + a, \alpha)$ being a certain ξ-dependent section of the bundle labeled by the point $(x + a, \alpha)$ of Jac(M). A linear flow on the Jacobian is produced by giving a, α suitable linear dependence on the flow parameters, whereupon the differential operators satisfy the SKP equations

$$\frac{\partial X}{\partial t_n} = [Q_+^{n/2}, X],$$

$$\frac{\partial X}{\partial t_{n-\frac{1}{2}}} = [(\lambda Q^{n/2})_+, X],$$

where $\lambda = S\partial S^{-1}$, S being a suitable wave operator satisfying $Q = Sd^2S^{-1}$. This exhibits the explicit solution promised.

One can also compute the tau function corresponding to this solution. This is the superdeterminant of the change of basis matrix from the filled states $z^{-n}, \theta z^{-n}, n > 0$ of the trivial operator formalism vacuum, to the filled states generated by the Baker function. Using the relation between the Baker and tau functions derived by Dolgikh and Schwarz [DS], I find for the tau function the result [BR]

$$\tau = 1 - \frac{\alpha\delta}{2\pi i}\frac{d^2}{dx^2}\log\Theta(x + a).$$

As a superdeterminant rather than an ordinary determinant, it is a meromorphic rather than holomorphic function of its arguments, and it is also a well-defined function on the Jacobian rather than a section of a bundle as in the ordinary KP theory. The latter feature does not persist [DS] in the generalization to the other SKP hierarchies of Manin–Radul [MR] or Kac–van de Leur [KL].

Work is in progress to generalize the lessons of this example to the algebro-geometric solutions to all three known SKP systems constructed from supermanifolds M of arbitrary genus. This requires a general theory of Jacobians of such supermanifolds and the natural theta functions defined on them. At higher genus the appropriate supermanifolds are *not* super Riemann surfaces, and so some features of the supertorus case will not generalize. In fact, the generic situation is nicer in that Jac(M) is a smooth supermanifold. We need a more explicit solution for the Baker function in terms of the tau function than the Dolgikh-Schwarz formula provides, an expression for both Baker and tau in terms of super theta functions, and a derivation of the Hirota bilinear form of the SKP systems from the Lax/Sato form discussed here. Finally there is the unresolved question of whether *any* nontrivial SKP hierarchy, supplemented by some string equation such as $[P, Q] = 1$, is related to 2d topological supergravity [FS].

References

[AGNSV] L. Alvarez-Gaumé, C. Gomez, P. Nelson, G. Sierra, and C. Vafa, "Fermionic strings in the operator formalism," Nucl. Phys. **B311** (1988) 333–400.

[BR] M.J. Bergvelt and J.M. Rabin, work in progress.

[DK] P. Di Francesco and D. Kutasov, "Integrable models of two-dimensional quantum gravity," pp. 35–51 in *Random Surfaces and Quantum Gravity*, ed. O. Alvarez, E. Marinari, and P. Windey, Plenum Press 1991.

[DS] S.N. Dolgikh and A.S. Schwarz, "Supergrassmannians, super τ-functions, and strings," pp. 231–244 in *Physics and Mathematics of Strings*, ed. L. Brink, D. Friedan, and A.M. Polyakov, World Scientific 1990.

[FS] J.M. Figueroa-O'Farrill and S. Stanciu, "On a new supersymmetric KdV hierarchy in 2-d quantum supergravity," Phys. Lett. **B316** (1993) 282–290.

[KL] V.G. Kac and J.W. van de Leur, "Super boson-fermion correspondence," Ann. Inst. Fourier, Grenoble **37** (1987) 99–137; id., "Super boson-fermion correspondence of type B," pp. 369–406 in *Infinite Dimensional Lie Algebras and Groups*, ed. V.G. Kac, World Scientific 1989.

[LSW] W. Lerche, D.-J. Smit, and N.P. Warner, "Differential equations for periods and flat coordinates in two-dimensional topological matter theories," Nucl. Phys. **B372** (1992) 87–112.

[M91] M. Mulase, "A new super KP system and a characterization of the Jacobians of arbitrary algebraic supercurves," J. Diff. Geom. **34** (1991) 651–680.

[Mum] D. Mumford, "An algebro-geometric construction of commuting operators and of solutions to the Toda lattice equation, Korteweg-de Vries equation, and related non-linear equations," pp. 115–153 in *Symposium on Algebraic Geometry*, ed. M. Nagata, Kinokuniya, Tokyo, 1978.

[MR] Yu. I. Manin and A.O. Radul, "A supersymmetric extension of the Kadomtsev-Petviashvili hierarchy," Commun. Math. Phys. **98** (1985) 65–77.

[R91] J.M. Rabin, "The geometry of the super KP flows," Commun. Math. Phys. **137** (1991) 533–552.

[R93] J.M. Rabin, "Super elliptic curves," preprint hepth-9302105, submitted to J. Geometry and Physics.

CONNECTION ON THE THEORY SPACE*

HIDENORI SONODA

Department of Physics, UCLA
Los Angeles, CA 90024-1547, USA

1. Introduction

The idea of the theory space is essential for qualitative understanding of quantum field theory[1]; only through the notions of relevance, irrelevance, and universality, we understand the meaning of the continuum limit of a field theory. But our use of the theory space has so far been mostly qualitative. It is the purpose of this short article to treat the theory space seriously and introduce a quantitative structure.[2-4]

2. Covariant Quantities in the Theory Space

We consider renormalized field theories in D dimensional euclidean space parametrized by $g^i(i = 1, ..., N)$. The parameters g^i give local coordinates of the theory space. We take the origin $g^i = 0(\forall i)$ to be the UV fixed point. The renormalization group (RG) describes the change of the theory under scale change; under the transformation R_l the renormalization scale is fixed, but the coordinate distance r transforms into re^{-l}. The RG equations of the parameters are given by

$$\frac{d}{dl} g^i = \beta^i(g) \equiv x_i g^i + \mathrm{O}\left(g^2\right),\tag{2.1}$$

where the scale dimension x_i of the parameter g^i is positive. Note that the beta function $\beta^i(g)$ is a vector field on the theory space.

Each point in the theory space is a renormalized field theory which has an infinite number of linearly independent composite fields. Let $\{\Phi_a\}_g$ be a complete basis of composite fields at g. The space of composite fields make a linear space, and therefore the composite fields make an infinite dimensional vector bundle over the theory space. The choice of the local basis is by no means unique, and two bases $\{\Phi_a\}_g$ and $\{\Phi'_a\}_g$ are related by an invertible infinite dimensional matrix $N(g)$ as

$$\Phi'_a = (N(g))_a{}^b \Phi_b.\tag{2.2}$$

Therefore, the correlation function of n composite fields

$$\langle \Phi_{a_1}(r_1)...\Phi_{a_n}(r_n)\rangle_g\tag{2.3}$$

forms a rank-n tensor field on the theory space.

---

* This work was supported in part by the D.O.E. contract DE-AT03-88ER 40384 Mod A006 Task C.

We can write the RG equations of the composite fields as

$$\frac{d}{dl}\,\Phi_a = (\Gamma(g))_a{}^b\,\Phi_b,\tag{2.4}$$

where the matrix

$$(\Gamma(g))_a{}^b = y_a\delta_a^b + O(g)\tag{2.5}$$

gives the scale dimensions y_a and anomalous dimensions of the renormalized fields. The precise meaning of the above RG equations is given by the following RG equations for correlation functions:

$$\frac{d}{dl}\,\langle\Phi_{a_1}(r_1)...\Phi_{a_n}(r_n)\rangle_g \equiv \left(-\sum_{k=1}^{n} r_{k,\mu}\frac{\partial}{\partial r_{k,\mu}} + \beta^i\frac{\partial}{\partial g^i}\right)\langle\Phi_{a_1}(r_1)...\Phi_{a_n}(r_n)\rangle_g$$

$$= \sum_{k=1}^{n}\Gamma_{a_k}{}^b\,\langle\Phi_{a_1}(r_1)...\Phi_b(r_k)...\Phi_{a_n}(r_n)\rangle_g.\tag{2.6}$$

Among the infinite number of composite fields, N composite scalar fields stand out; they are the fields $\mathcal{O}_i$ *conjugate* to the parameters g^i. The notion of conjugacy should be familiar from thermodynamics, where the external magnetic field H has magnetization M as its conjugate field. The expectation value of $\mathcal{O}_i$ gives the derivative of the free energy density $F(g)$ with respect to g^i:

$$\langle\mathcal{O}_i\rangle_g = \frac{\partial}{\partial g^i}\,F(g).\tag{2.7}$$

Since the free energy density satisfies the canonical RG equation

$$\frac{d}{dl}\,F(g) = DF(g)\tag{2.8}$$

(D is the dimensionality of space), we find, from Eqs. (2.1) and (2.8), that

$$\frac{d}{dl}\,\mathcal{O}_i = D\mathcal{O}_i - \frac{\partial\beta^j}{\partial g^i}\,\mathcal{O}_j.\tag{2.9}$$

We will define the conjugate fields more precisely in sect. 3.

3. Variational Formula

We now consider the g-dependence of the correlation functions, which are tensor fields on the theory space. In analogy to thermodynamics and statistical mechanics, the derivative of a correlation function with respect to g^i should be given as a volume integral of the conjugate field $\mathcal{O}_i$:

$$-\frac{\partial}{\partial g^i}\langle\Phi_{a_1}(r_1)...\Phi_{a_n}(r_n)\rangle_g$$

$$= \lim_{\epsilon\to 0}\left[\int_{|r-r_k|\geq\epsilon} d^D r\,\langle(\mathcal{O}_i(r) - \langle\mathcal{O}_i\rangle_g)\,\Phi_{a_1}(r_1)...\Phi_{a_n}(r_n)\rangle_g\right.$$

$$\left.+ \sum_{k=1}^{n}\left(c_{i,a_k}{}^b(g) - \int_{1\geq r\geq\epsilon}\frac{d^D r}{\mathrm{vol}(S^{D-1})}\,C_{i,a_k}{}^b(r;g)\right)\langle\Phi_{a_1}(r_1)...\Phi_b(r_k)...\Phi_{a_n}(r_n)\rangle_g\right],\tag{3.1}$$

where $\mathrm{vol}(S^{D-1})$ is the volume of a unit $D{-}1$ sphere. Two remarks are in order. First, due to the short distance singularity

$$\mathcal{O}_i(r)\Phi_a(0) = \frac{1}{\mathrm{vol}(S^{D-1})}\ C_{i,a}{}^b(r;g)\Phi_b + \mathrm{O}\left(\frac{1}{r^D}\right), \tag{3.2}$$

the integral needs subtractions. In the above operator product expansion (OPE), we keep only composite fields Φ_b whose scale dimension is at most y_a, since we need to subtract only the unintegrable part. Second, we need the finite counterterms $c_{i,a}{}^b(g)$ not only to compensate the arbitrariness of the subtraction scheme, but also to satisfy covariance in the theory space. Under a change of basis, Eq. (2.2), the subtracted integral on the right-hand side of Eq. (3.1) is covariant, but the left-hand side is non-covariant due to the naked derivative. Hence, covariance demands the finite counterterms $c_{i,a}{}^b(g)$ which transform as a connection

$$c_i(g) \to c_i'(g) = N(g)\,(\partial_i + c_i(g))\,N^{-1}(g) \tag{3.3}$$

under the change of basis Eq. (2.2). This is how the connection c_i arises naturally in the theory space. We call Eq. (3.1) as a variational formula.

4. Consistency Conditions

It is difficult to *derive* the variational formula from first principles. Instead, we *demand* the conjugate fields $\mathcal{O}_i$ to satisfy Eq. (2.7) and the variational formula Eq. (3.1). It is an assumption that such fields exist. Nonetheless, we can check the consistency of Eq. (3.1).

First, we demand that Eq. (3.1) be consistent with the RG equations (2.1), (2.4), and (2.9). Then, we obtain

$$C_i(r = 1; g) = \frac{\partial}{\partial g^i}\ \Psi(g) + [c_i(g), \Psi(g)] + \beta^j(g)\Omega_{ji}(g), \tag{4.1}$$

where

$$\Psi(g) \equiv \Gamma(g) + \beta^i(g)c_i(g) \tag{4.2}$$

is a covariant tensor field, and the curvature Ω_{ij} is defined by

$$\Omega_{ij} \equiv \partial_i c_j - \partial_j c_i + [c_i, c_j]. \tag{4.3}$$

Eq. (4.1) gives a geometric expression of the singular part of the OPE coefficients. Second, we demand that Eq. (3.1) satisfy the Maxwell relation:

$$\partial_i \partial_j = \partial_j \partial_i. \tag{4.4}$$

This assures the consistency of using Eq. (3.1) recursively to evaluate higher order derivatives. Eq. (4.4) gives the curvature in terms of a double integral over a finite domain:

$$\Omega_{ij}\ \langle\Phi\rangle_g = \int_{r\leq1} d^D r\ \mathrm{F.P.} \int_{r'\leq1} d^D r'\ \langle\mathcal{O}_j(r)\left(\mathcal{O}_i(r') - \frac{C_i(r')}{\mathrm{vol}(S^{D-1})}\right)\Phi(0) - (i \leftrightarrow j)\rangle_g, \tag{4.5}$$

where F.P. denotes taking the integrable part with respect to r.

5. Conclusions

By studying the variational formula, Eq. (3.1), we have found that the geometrical quantities on the theory space, i.e., the vector field $\beta^i(g)$ (beta functions), the rank-two tensor field $\Psi_a{}^b(g)$ (scale and anomalous dimensions), and the connection c_i (finite counterterms), determine the short-distance singularities of the theory, as is given by Eqs. (4.1) and (4.5). These are analogues of the well-known relation between the renormalization constants and the beta functions and anomalous dimensions in the minimal subtraction scheme.[5] It is important to note that the connection c_i has more than its geometric significance: the connection gives the finite counterterms in the variational formula (3.1).

One promising place where we can apply the variational formula and the connection c_i of sect. 3 is in closed string field theory. To understand the background independence of closed string field theory, we must find the relation between two different formulations based upon two conformal field theories. This requires comparing string states (or vertex operators) in one conformal field theory to those in the other conformal field theory. Hence, we need a connection to parallel transport states from one theory to the other. We have applied the variational formula of sect. 3 to the space of conformal field theories[6,7]; in particular we have studied the properties of the connection c_i and its two modifications. It is not yet clear which precise connection is the right choice for closed string field theory, but a connection, together with the variational formula (3.1), is expected to play an important role.[8]

References

1. K. Wilson and J. Kogut, *Phys. Rep.* **C12** (1974) 75.
2. H. Sonoda, *Nucl. Phys.* **B383** (1992) 173.
3. H. Sonoda, *Nucl. Phys.* **B394** (1993) 302.
4. H. Sonoda, *Current-Current Singularities under External Gauge Fields* (hep-th 9212025).
5. G. 't Hooft, *Nucl. Phys.* **B62** (1973) 444.
6. K. Ranganathan, H. Sonoda, and B. Zwiebach, *Connections on the State-Space over Conformal Field Theories* (hep-th 9304053).
7. B. Zwiebach, contribution to these proceedings.
8. A. Sen and B. Zwiebach, work in progress.

REGULARIZATION AND THE BV FORMALISM *

WALTER TROOST † and ANTOINE VAN PROEYEN ‡

*Instituut voor Theoretische Fysica, K. U. Leuven, Celestijnenlaan 200D,
B-3001 Leuven, Belgium*

ABSTRACT

We summarize the application of the field–antifield formalism to the quantization of gauge field theories. After the gauge fixing, the main issues are the regularization and anomalies. We illustrate this for chiral W_3 gravity. We discuss also gauge theories with a non-local action, induced by a matter system with an anomaly. As an illustration we use induced WZW models, including a discussion of non–local regularization, the importance of the measure and regularization of multiple loops. In these theories antifields become propagating — one of many roles assigned to them in the BV formalism.

1. Introduction

During a dinner a few weeks ago, Mario Tonin reminded us that Cardinal Bellarmino was right in the argument with Galileo. Bellarmino agreed that indeed the picture of the planets going around the earth on circles with epicycles could be simplified by putting the sun in the center. But he insisted that there is not more *truth* in the sun being the center of the universe rather than the earth.

Ten minutes later, Pietro Fré asked what he could expect from the Batalin–Vilkovisky (BV) formalism, if he knows already the BRST framework. We made the comparison with the above situation. One may learn first the BRST formalism for usual Lie algebras. Then one may extend it for the case of structure functions. Learning how to treat open algebras is a further epicycle, and the next one may be the reducible algebras, ... In this way, one does not need the BV formalism. But the latter puts the sun in the center.

We consider the BV formalism for Lagrangian gauge theories. Our main subject will be, how to treat regularization in this framework, but we will first recall the classical theory, in particular how to obtain, from the classical action, a gauge–fixed action including ghosts. This text is complementary to [1]. A longer text is in preparation [2].

First we will indicate the main concepts: antifields, antibrackets, and the extended classical action. This is sufficient for the classical Lagrangian theory. We

---

*presented at the conference by A.V.P.

†Bevoegdverklaard navorser, N.F.W.O. Belgium; e–mail Walter%tf%fys at cc3.kuleuven.ac.be

‡Onderzoeksleider, N.F.W.O. Belgium; e–mail FGBDA19 at cc1.kuleuven.ac.be

will use chiral W_3 gravity [3] as an example, of which a detailed treatment in the present context can be found in [4]. Then we will show how the gauge fixing procedure in this formalism just amounts to a canonical transformation [5]. The fourth section will contain some generalities about the quantum theory, and explain why we use mainly Pauli–Villars (PV) regularization. Anomalies are the central issue of the fifth section. There we will explain how a cohomological theorem leads to drastic simplifications in the calculation of the general form (i.e. including the antifield dependence) of anomalies. Also the cancellation of some anomalies by background charges will be incorporated. In the final part, we will see that there is another role for the antifields, when considering anomaly–induced actions, as bona fide physical fields that correspond to the additional degree of freedom (because the anomalies do not cancel) of the quantum theory. We will illustrate this in Wess–Zumino–Witten (WZW) models. We will treat multiple loops using higher derivative regularization, and will also take into account the issue of locality [6].

Since our main example is W_3 gravity in the chiral gauge, let us recapitulate the (well known) elements that we need (see [3]). We will use only the realization in terms of free bosonic fields. The classical fields are $\phi^i = \{X^\mu, h, B\}$, and their action is

$$S^0 = \int d^2x \left\{ -\tfrac{1}{2}(\partial X^\mu)(\bar{\partial} X^\mu) + \tfrac{1}{2}h(\partial X^\mu)(\partial X^\mu) + \tfrac{1}{3}B d_{\mu\nu\rho}(\partial X^\mu)(\partial X^\nu)(\partial X^\rho) \right\} \tag{1}$$

with h and B Lagrange multipliers, imposing two constraints on the classical level. In order to close the algebra of these constraints, we have to impose

$$d_{\mu(\nu\rho}d_{\sigma)\tau\mu} = \delta_{(\nu\sigma}\delta_{\rho)\tau} \ . \tag{2}$$

There are two gauge symmetries: the remnant of general coordinate and dilatational symmetries, and the spin–3 symmetry. These imply that the path integral

$$Z(J_i) = \int \mathcal{D}\phi^i \ \exp\left[\frac{i}{\hbar}S^0(\phi) + J_i\phi^i\right] \tag{3}$$

(where ϕ denotes X, h and B) is not well defined, some propagators are singular. This problem is characteristic for gauge theories, and solved by gauge fixing.

2. Antifields, antibrackets, and the extended action

The BV formalism for a Lagrangian gauge theory starts by introducing a ghost field for every symmetry, and then an antifield for every field. This leads to table 1. Note that the ghosts are fermionic (statistics opposite to the symmetry), and that the antifields have statistics opposite to that of the fields. The classical fields have ghost number 0, and the ghosts have ghost number 1. The ghost numbers of a field and its antifield always add up to -1. In the space of functionals of fields and antifields, one introduces a Poisson bracket–like structure (called antibrackets), where the antifields take over the role of the canonical momenta. Thus for example,

$$(X^\mu(x), X^*_\nu(y)) = \delta^\mu_\nu \delta(x - y) = -\left(X^*_\nu(x), X^\mu(y)\right) \ . \tag{4}$$

Table 1: Fields and antifields of chiral W_3 gravity

	stat	gh		stat	gh
X^μ	$+$	0	X^*_μ	$-$	-1
B	$+$	0	B^*	$-$	-1
h	$+$	0	h^*	$-$	-1
c	$-$	1	c^*	$+$	-2
u	$-$	1	u^*	$+$	-2

The new feature is that this antibracket is odd, i.e. the Grassmann parity of (A, B) is opposite to that of the product AB. Equipped with this structure, the BV formalism is analogous to the canonical formalism of classical mechanics in many respects: brackets, canonical transformations,.... Note however that we will stay within the Lagrangian framework, thus sticking to the covariant formulation. For the Hamiltonian approach, we refer to the literature.

The extended classical action is a function of fields and antifields: $S(\Phi^A, \Phi^*_A)$. The part of the extended action which does not contain antifields is the classical action:

$$S(\Phi^A, \Phi^*_A = 0) = S^0(\phi^i) \ . \tag{5}$$

In our example it is Eq. (1). This classical action depends only on the classical fields ϕ^i, which in our example (with μ running over D values) means that it depends on $D+2$ fields. The problem of singular propagators for the quantum theory can be restated as

$$rank \ \left(\partial_i \partial_j S^0\right) = D < D+2 \ . \tag{6}$$

To remedy this problem, we add a term

$$S^1 = \phi^*_i R^i{}_a c^a \ , \tag{7}$$

where $R^i{}_a$ is the variation of ϕ for the symmetry transformation a. In W_3 this is §

$$
\begin{aligned}
S^1 = \ & X^*_\mu \left[(\partial X^\mu)c + d_{\mu\nu\rho}(\partial X^\nu)(\partial X^\rho)u\right] \\
& + h^* \left[\nabla^{(-1)}c + (\partial X^\mu)(\partial X^\mu)(u\partial B - B\partial u)\right] + B^* \left[\nabla^{(-2)}u - 2B\partial c + c\partial B\right]
\end{aligned}
\tag{8}
$$

with $\nabla^{(j)} = \bar{\partial} - h\partial - j(\partial h)$. Note that for each symmetry this gives an extra contribution 2 to the rank of $\partial_\alpha \partial_\beta S$. (To be more precise, we look at this matrix at the surface where all fields of non–negative ghost number vanish, and those of ghost number 0 satisfy the equations of motion from S^0). In our example this contributes 4, and the total rank is $D+4$, while introducing the 2 ghosts and all the antifields, raised the dimension of this matrix to $2(D+4)$. The rank is thus half the dimension, which turns out to be the maximum possible, due to the master equation, Eq. (9). If the rank would still be lower, this would signal that the transformations were dependent. This would be remedied by introducing a term $c^*_a Z^a{}_{a_1} c^{a_1}$ where Z stands

---

§We omit $\int d^2x$, here and in the sequel, for actions and anomaly expressions.

for the dependence relations between the symmetries and c^{a_1} would be new ghosts (so–called ghosts for ghosts, having ghost number 2).

The statement that S^0 is invariant under the transformations defined by $R^i{}_a$ is neatly expressed with the help of antibrackets: $(S^0, S^1) = 0$. This is a special consequence of imposing the more general relation

$$(S, S) = 0 \tag{9}$$

which contains, apart form the invariance of the action, also the commutators, non–closure relations, (modified) Jacobi identities and any other relevant data of the symmetry algebra. This relation is called the (classical) master equation, and is a cornerstone of the formalism. The extended action that satisfies this relation will in general have, apart from S^0 and S^1, other pieces. Also necessary are the following terms (written symbolically):

$$S^2 = (-)^b \tfrac{1}{2} c_a^* T^a{}_{bc} c^c c^b + (-)^{i+a} \tfrac{1}{4} \phi_i^* \phi_j^* E^{ji}{}_{ab} c^b c^a \ , \tag{10}$$

where T are structure functions and E appear in the non–closure relations. For W_3 for example, the master equation is solved by

$$\begin{aligned}
S^2 \ = \ & c^* \left[(\partial c)c + (1 - \alpha)(\partial X^\mu)(\partial X^\mu)(\partial u)u \right] + u^* \left[2(\partial c)u - c(\partial u) \right] \\
& -2\alpha h^* (-2B\partial B^* - 3B^* \partial B + \nabla^2 h^*)(\partial u)u - 2(\alpha + 1) X_\mu^* h^* (\partial u)u(\partial X^\mu) \ ,
\end{aligned} \tag{11}$$

where α is a free parameter. In general one can prove that a solution (of ghost number 0) of the master equation always exists (possibly containing terms with three or more antifields). This was proven in [7, 8] under some regularity conditions. In [2] it will be proven that there is always a *local* solution (local in space-time). The essential ingredients and steps have been given already in [4]. Also there, the regularity conditions for this theorem have been weakened: previous versions were not applicable to, e.g., W_2 (chiral gravity) and chiral W_3. The theorem says moreover that the solution of the master equation is unique up to a canonical transformation . In [4] it is shown explicitly that indeed the parameter α in Eq. (11) can be removed by a canonical transformation.

For W_3 one finds that $S = S^0 + S^1 + S^2$ is the full solution of the master equation. It satisfies the three requirements: master equation, properness (rank of the matrix of second derivatives is on–shell equal to the number of fields) and classical limit (for vanishing antifields), which were discussed also in section 3.1 of [1].

In this section we used a basis where the fields have $gh(\Phi) \geq 0$, while the antifields satisfy $gh(\Phi^*) < 0$. Such a basis always exists, and will be called the '*classical basis*'. We define the antifield number (afn), as zero for the fields in this basis, and equal to minus the ghost number for the antifields. It is often convenient, as for example in the construction of the extended action above, to work perturbatively in antifield number.

3. Gauge fixing

The extended action has been constructed such that the rank of $\partial_\alpha \partial_\beta S$ is half its dimension, i.e. equal to the number of fields. Now imagine making a canonical transformation between fields and antifields such that in the new basis (to be called the *gauge-fixed* basis) the submatrix $\partial_A \partial_B S$ (derivatives w.r.t. fields only) is invertible (on shell). The path integral with the new fields as integration variables no longer suffers from the gauge problem above: the propagators of the new 'fields' have no zero-modes. This canonical transformation is the gist of the gauge fixing procedure in the BV framework.

In our example the canonical transformation needed is extremely simple: we just have to interchange the names field and antifield for h and B (like changing a momentum into a coordinate and vice versa in the canonical formalism). Then $b \equiv h^*$ and $v \equiv B^*$ are fields and $b^* = -h$, $v^* = -B$ are antifields. The fields Φ^A of the gauge-fixed basis are then $\{X^\mu, v = B^*, b = h^*, c, u\}$. For the extended action this implies

$$S = -\tfrac{1}{2}(\partial X^\mu)(\bar{\partial} X^\mu) + b\nabla^{(-1)}c + v\nabla^{(-2)}u + \text{ antifield dependent terms} \qquad (12)$$

and in the antifield independent terms one recognizes the gauge fixed action. This has no gauge symmetries left.

It may often be preferable (for example, to keep explicit Lorentz covariance) to introduce extra fields to fix the gauge ('non-minimal' solutions of the master equation), and to perform the canonical transformation with the help of a generating function called the gauge fermion $\Psi(\Phi)$:

$$\Phi' = \Phi \; ; \qquad \Phi_A^* = \Phi_A'^* + \frac{\partial}{\partial \Phi^A}\Psi \; . \qquad (13)$$

For the examples we consider, this is a detour, which leads back to the same gauge fixed action after eliminating auxiliary fields.

4. Quantum theory

4.1. Measure and counterterms

The quantum theory is not determined by the classical action alone. First, there is the question what is the measure of the path integral, or equivalently one needs a regularization. Secondly, what is the quantum action? The (extended) action S may have terms of order $\hbar$. For now, we will use the following expansion (later we will see that half-integer powers of $\hbar$ may also be needed):

$$W = S + \hbar M_1 + \hbar^2 M_2 + \dots \, ,$$

where M_1, ... are integrals of local functionals. The extra terms may have infinite coefficients related to the regularization, or may be finite. The 'counterterms' M_i are restricted by extra requirements, e.g. rigid symmetries, renormalization conditions, or experimental numbers. Symmetry usually fixes the structure of these terms, leaving only some undetermined constants.

The two questions above are connected. A change in the regularization can be compensated by a change of the M_i. E.g. the regularization scheme which will be introduced below depends on a mass matrix, and there is a formula [9] giving the change of M_i which is connected to a change of the mass matrix.

4.2. Regularization procedure

The quantum theory starts from the path integral

$$Z(J, \Phi^*) = \int \mathcal{D}\Phi \; \exp\left[\frac{i}{\hbar} S(\Phi, \Phi^*) + J\Phi\right] \; .$$

For a variety of reasons it is advisable not to put the antifields equal to zero (or, as is usually done, to put $\Phi^* = \partial\Psi/\partial\Phi$, which is the same as putting them to zero after a gauge–fixing canonical transformation). We recall that this antifield dependence was already used, in the form of dependence on sources for the BRST transformations, in [10], to study the renormalization properties of ordinary non–abelian gauge theories (where S is linear in Φ^*). For anomalous theories this dependence will describe additional degrees of freedom.

For the regularized theory, the path integral is considered to be the limit when some cutoff M is removed from a well–defined path integral, $Z_R = \lim_{M\to\infty} Z_M$. One might read this M as $1/(n-4)$ in dimensional regularisation, but it is hard to see how to make sense of Z_M as a path integral based on an action for n not an integer. Another proposal would be a point–splitting or lattice regularization, but as we try to work in the context of local field theories, we prefer not to use these. We will use a two-step procedure, first modifying propagators with higher derivative terms, and then applying the method of Pauli and Villars (PV). For the present formalism, both steps have the advantage that they can be incorporated in a Lagrangian framework. The higher derivatives serve to regularize all the diagrams apart from the one–loop contributions, while PV should regularize the remaining one–loop divergences. When we only consider the quantum corrections of order $\hbar$ (one–loop diagrams), then PV alone suffices.

4.3. PV action

To determine the PV action, we introduce for every field Φ^A, another field χ^A as PV–partner, with statistics such that a minus sign is introduced in each loop. For a more detailed description, see section 5.1 of [1]. Each PV-field comes with its own antifield. We will collectively denote the original fields and antifields as z^α, the PV partners as w^α: $z^\alpha = \{\Phi^A, \Phi_A^*\}$; $w^\alpha = \{\chi^A, \chi_A^*\}$. Then suppose that we have found an action with higher derivatives which regularizes the higher loops. If S_Λ is this extended action, (when considering only one loop, one may replace S_Λ with S), then we propose for the full action

$$S_M = S_\Lambda(z) + w^\alpha \left(\frac{\overrightarrow{\partial}}{\partial z^\alpha} S_\Lambda \frac{\overleftarrow{\partial}}{\partial z^\beta} \right) w^\beta - \tfrac{1}{2} M^2 \chi^A T_{AB} \chi^B \; . \tag{14}$$

Here, T_{AB} is an matrix that is arbitrary but invertible for the propagating fields. Note that in the mass term only the PV fields, and not the PV antifields appear. On the other hand, T may depend on ordinary fields and antifields. The general prescription given here is certainly not the most general one, but it satisfies some nice properties: to one–loop order it commutes with canonical transformations, satisfies the master equation automatically, and corresponds to the addition (classically) of a cohomologically trivial system when $M \to \infty$.

5. Anomalies

5.1. General theory

Since we keep the antifield dependence, all the usual generating functionals will generically depend on Φ^* also. The one–particle irreducible functional, defined as usual, is then a functional both of some classical fields Φ_{cl} and on Φ^*. The Ward identities, as derived by Zinn–Justin [10] for the usual non–abelian gauge theories, find a very natural expression in this framework. The general formal derivation is simply a matter of filling in the definitions and doing a partial integration:

$$(\Gamma, \Gamma) = \Delta \cdot \Gamma , \tag{15}$$

where Γ is the effective action, and the antibracket is now with respect to $\{\Phi_{cl}, \Phi^*\}$. The right hand side symbolizes an operator insertion which computes the eventual anomalies, its path integral expression is

$$(\Delta \cdot \Gamma) = \frac{-2i\hbar}{Z} \int \mathcal{D}\Phi \; \mathcal{A}(\Phi, \Phi^*) \; \exp \frac{i}{\hbar}(W + J\Phi) .$$

Using formal manipulations on the path integral, the anomaly $\mathcal{A}(\Phi, \Phi^*)$ takes the form

$$A = \Delta W + \frac{i}{2\hbar}(W, W) , \tag{16}$$

where

$$\Delta = \frac{\partial}{\partial \Phi^A} \frac{\partial}{\partial \Phi_A^*} . \tag{17}$$

However, for a local action ΔS gives an expression proportional to $\delta(0)$, and is thus ill–defined, indicating where regularization necessarily enters. Continuing anyway formally for a moment, we expand $\mathcal{A} = 0$ in $\hbar$, obtaining the equations

$$(S, S) = 0 ; \qquad \Delta S + i(S, M_1) = 0 ; \; \ldots .$$

Now we consider this in a regularized version, using the PV framework and including only single loops. For the regularized path integral we take

$$Z_R(J, \Phi^*) = \exp -\frac{i}{\hbar} W_c(J, \Phi^*) = \lim_{M \to \infty} \int \mathcal{D}\Phi \mathcal{D}\chi \; \exp \left[\frac{i}{\hbar} W_M(z, w) + J\Phi \right] \Big|_{\chi^*=0} , \tag{18}$$

and Γ introduced above is of course the Legendre transform of W_c. The measure we define so that ΔW_M vanishes (up to terms quadratic in PV (anti)fields). Indeed, the contributions of the ordinary fields and the PV fields in Eq. (17) cancel (again see section 5.1 of [1] for more details). The anomaly arises from the fact that $(S_M, S_M) \neq 0$ generically, due to the non–invariance of the PV mass term. The integral over the

PV fields gives for the terms quadratic in PV fields a propagator, which is of order $\hbar$, and we obtain

$$\mathcal{A} = Tr \left[J \exp(\mathcal{R}/M^2) \right] + i(S, M_1) + \mathcal{O}(\hbar) \ . \tag{19}$$

where J is like the jacobian matrix of the BRST transformations and $\mathcal{R}$ is a regulator constructed from the action. The full expressions are [11]

$$J^A{}_B = \frac{\overrightarrow{\partial}}{\partial \Phi_A^*} S \frac{\overleftarrow{\partial}}{\partial \Phi^B} + \frac{1}{2}(T^{-1})^{AC}\,(T_{CB},S)\,(-)^B \ ; \qquad \mathcal{R}^A{}_B = (T^{-1})^{AC}\left(\frac{\overrightarrow{\partial}}{\partial \Phi^C} S \frac{\overleftarrow{\partial}}{\partial \Phi^B} \right) \ .$$

Comparing Eq. (19) with Eq. (16), we have effectively obtained a one–loop regularized definition of ΔS. Its value depends on the matrix T. Different choices of T give values of ΔS which differ by (S, G) for some local function G. In practice, Gilkey's formulas for the heat kernel [12, 2] are very useful.

Both for the formal expression and for the regularized one, one can prove that

$$(S, \mathcal{A}) = 0 \ . \tag{20}$$

This is the expression in BV language of the Wess–Zumino consistency conditions. Thus we automatically obtain the 'consistent' anomalies. This was to be expected, since these conditions are satisfied for finite values of M also.

5.2. Simplifications using a theorem on antibracket cohomology

The calculations of the one–loop anomalies for chiral W_3 has been greatly simplified [4] by using a theorem on the cohomology of the nilpotent operator $SF \equiv (F, S)$. Whereas in [8] this was discussed in general terms, in [4] it has been shown to be applicable also within the set of local functions.

The theorem (see [4], theorem 3.1) states that the cohomology of S in the set of local functions of fields and antifields is equivalent to the 'weak' cohomology of an operator D^0 which acts in the set of local functions of fields (or antifields) with non–negative ghost number. Thus it is most useful in the classical basis, where D^0 acts on fields only: $D^0 F^0 = (F^0, S)|_{\Phi^*=0}$. 'Weak' cohomology means that field equations of S^0 (i.e. the part of the action which depends on fields of ghost number 0) may be used freely.

The theorem implies that it is sufficient to calculate the part of the possible anomaly which depends only on the fields of the classical basis, call it $\mathcal{A}^0$. Given $\mathcal{A}^0$, the WZ consistency condition Eq. (20) implies that $\mathcal{A}$ is then fixed up to terms (S, M), with M the integral of a local function. Such terms are removable anyway by including M as a local counterterm.

This way, the same results are obtained as in [13]. Removing all dependence on the antifields of the classical basis (such as the antighosts $b = h^*$ and $v = B^*$) one finds

$$\begin{aligned}
\mathcal{A}^0 \ &\approx \ -\frac{1}{24\pi} \left(c\delta_\nu^\mu + 2u d_{\mu\nu\rho}(\partial X^\rho) \right) \partial^3 \left(b^* \delta_{\mu\nu} + 2 d_{\mu\nu\rho} v^* \partial X^\rho \right) \\
&\quad + \frac{100}{24\pi} c\,\partial^3 v^* + \frac{\kappa}{2\pi}(u\partial v^* - v^* \partial u)(\partial^3 X^\mu)(\partial X^\mu) \ ,
\end{aligned}$$

where $\approx$ means 'up to field equations'(the complete result is in [4]). We call attention to the fact that the anomaly depends on b^* and v^*, which are antifields in the gauge–fixed basis. In other treatments one has to include background fields in the gauge fixing to see the anomalies (for example, in the gauge $h = 0$ the anomaly would not show up). In the BV formalism, this task is taken over by the antifields, without such extra ingredients.

5.3. Background charges

It is well known that the anomalies of the chiral W_3 model of Eq. (1) can be cancelled by introducing background charges. In our framework these correspond to terms with $\sqrt{\hbar}$ in the quantum action, which takes the form

$$W = S + \sqrt{\hbar}\, M_{1/2} + \hbar\, M_1 + \dots . \tag{21}$$

Then the master equation $\mathcal{A} = 0$ expands as

$$(S, S) = (S, M_{1/2}) = 0 \; ; \qquad (S, M_1) = i\Delta S - \frac{1}{2}(M_{1/2}, M_{1/2}) \; . \tag{22}$$

In view of the cohomology theorem, $M_{1/2}$ is fixed when the terms independent of the antifields (of the classical basis) are given. These are

$$M_{1/2}{}^0 = a_\mu h(\partial^2 X^\mu) + e_{\mu\nu} B(\partial X^\mu)(\partial X^\nu) \; ,$$

where a_μ and $e_{\mu\nu}$ are constants (background charges). The theorem says that $(S, M_{1/2}) = 0$ has a solution for $M_{1/2}$ if and only if $D^0 M_{1/2} \approx 0$. This gives some conditions on the background charges [14]. Taking into account Eq. (2), there is a unique solution for $e_{\mu\nu}$ for each value of D (the range μ), and a_μ is still arbitrary. The latter is also fixed by requiring that Eq. (22) has a solution for M_1, so that we have an anomaly–free action of the form Eq. (21). The simplification brought about by the theorem is that the analysis could be performed in the restricted space of fields of non–negative ghost number.

We have used PV regularization for one loop, to be combined with higher derivatives for more loops as in the next section. The reason is that these regularizations are possible at the level of the Lagrangian, thus leaving the BV–setup intact. If the regularization does not break symmetries, then there are no anomalies. In the usual cases the higher derivatives can be taken to be covariant, so that anomalies only come from the PV mass term, and are genuine one–loop anomalies. For W_3 it is not known how to define fully covariant derivatives (with a finite number of terms) and this clarifies the existence of anomalies beyond 1 loop in these theories.

6. Induced theories

Now we will consider theories where the anomalies are not cancelled, i.e. $(\Gamma, \Gamma) \neq 0$, implying that $\Gamma(\Phi_{cl}, \Phi^*)$ describes new quantum degrees of freedom in addition to the classical ones. They manifest themselves as propagating antifields. For conformal matter coupled to 2–d gravity, there is generically an anomaly proportional to $\mathcal{A} = c\,\partial^3 h$. The h–dependence of effective action obeying Eq. (15) is proportional to the non–local expression:

$$\Gamma \sim \partial^2 h \frac{1}{\partial\bar\partial - \partial h \partial} \partial^2 h \ .$$

It was noticed by Polyakov that this is local in f, where $h = \frac{\bar\partial f}{\partial f}$. This fact can be understood [15] by first expressing the original action in f and performing the regularization using as a mass term $M^2 X^\mu X^\mu(\partial f)$. This mass term is invariant under $\delta X = \epsilon\partial X, \delta f = \epsilon\partial f$, there is no anomaly and zero induced action for h. However, we really wanted a regularisation that is local *in* h, not just in f, which can be done using the mass term $M^2 X^\mu X^\mu$. This is just a change of regularization, and, as mentioned before, this amounts to changing the theory with a local counterterm, local in f that is. Actually, there is another integral left, over a parameter interpolating between the regularizations, but this is trivial here. Therefore, the final induced action is local in f.

A similar reasoning can be given for the WZW model, to which we now turn. This will again illustrate some aspects of the quantization procedure, like the importance of the definition of the measure, non–local regularization including also higher loops[6]. This model was also discussed, from a different point of view, at this conference by P. van Nieuwenhuizen [16]. The WZW action arises as an induced action from

$$S = \psi^t(\bar\partial - A)\psi - \psi^* c\psi + A^* \bar D(A)c - c^* cc \ , \tag{23}$$

with $\bar D(A)c \equiv \bar\partial c - [A, c]$, where the fields A and c are in the adjoint representation of a Lie–algebra, and ψ is in an arbitrary representation. The gauge fixing of this model can be done by considering A^* as a field (antighost), and A is then an antifield. Integrating out the fermions, the anomaly implies that the resulting action is a non–local action for the 'antifield' A:

$$\Gamma[A] = \frac{k}{2\pi x} \left\{ \frac{1}{2} A \frac{\partial}{\bar\partial} A - \frac{1}{3} \frac{1}{\bar\partial} A \left[A, \frac{\partial}{\bar\partial} A\right] + \ldots \right\} \tag{24}$$

Again it is possible to compare different regularizations [15] to arrive at the conclusion that this action is (almost) local when written in terms of a group element g. Almost, since this time the integral over the interpolating parameter remains (as the third space coordinate below):

$$\Gamma[A = \bar\partial g g^{-1}] = k\Gamma^0[A] = -k \ S^+(g) \ , \tag{25}$$

with

$$S^+(g) = \frac{1}{4\pi x} \left(\int d^2 x \ \{\partial g^{-1}\bar\partial g\} + \frac{1}{3} \int d^3 x \ \epsilon^{\alpha\beta\gamma} \{g_{,\alpha}g^{-1}g_{,\beta}g^{-1}g_{,\gamma}g^{-1}\} \right) \ . \tag{26}$$

To investigate the antifield degree of freedom, one takes the action Eq. (24) as a starting point and quantizes. The generating functional of interest is then

$$Z[u] = e^{-W[u]} = \int \mathcal{D}A \, e^{-\Gamma[A] + \frac{1}{2\pi x} \{u\,A\}} \ . \tag{27}$$

It can be argued in different ways ([17, 18], see [6] for a review) that it enjoys the remarkable renormalization property

$$W[u] = Z_W W^0[Z_u u] \,, \tag{28}$$

where the Legendre transform of $\Gamma^0[A]$ defines $W^0[u_0]$:

$$W^0[u] = \min_{\{A\}} \left(\Gamma^0[A] - \frac{1}{2\pi x}\{u\,A\} \right) \,. \tag{29}$$

Different formal arguments lead to $Z_W = k + 2\tilde{h}$, but the values obtained for Z_u vary. We will discuss now a regularized argument, which makes use of the antifields in the regularization. To start [18], consider the non-chiral induced action depending on the gauge fields A and $\bar{A}$:

$$\Gamma^v[A, \bar{A}] = \Gamma[A] + \overline{\Gamma}[\bar{A}] - \frac{k}{2\pi x}\{A\,\bar{A}\} \,. \tag{30}$$

It can be viewed as arising from two chirally conjugated matter systems like Eq. (23). If one separately regulates both chiral parts (a separate mass term for each chiral fermion) then one obtains of course just the first two terms in Eq. (30). Regulating in a vector invariant way (a vector invariant mass term for the fermions), the result differs from the previous one by a finite local counterterm, which is the third term in Eq. (30). This term can be computed explicitly, example 4.1 in [9] does it in 4 dimensions. To quantize this action, one has to take the vector invariance into account, with corresponding ghost c, and furthermore the antifields A^*, $\bar{A}^*$ and c^*. The gauge fixing is done by considering $\bar{A}$ as an antifield, and $\bar{A}^*$ as a field. We thus consider the path integral

$$Z_{nc}(\bar{A}) \equiv \int \mathcal{D}A\,\mathcal{D}\bar{A}^*\,\mathcal{D}c\ e^{\Gamma^v(A, \bar{A}) + \bar{A}^* D(\bar{A})c} \,. \tag{31}$$

For our present purpose it is sufficient to take this path integral at $A^* = c^* = 0$, and not to introduce sources for the fields A, $\bar{A}^*$ and c. The vector gauge invariance allows us to fix $\bar{A}$ to an arbitrary value, and if there is no anomaly then Z_{nc} will not depend on it. Thus by investigating the vector anomaly we obtain the dependence on $\bar{A}$ (an antifield from the present point of view). For a one–loop calculation, we introduce PV fields with the mass terms [6]

$$M^2 \underline{A}\frac{1}{\bar{\partial}^2}\underline{A} + M\underline{\bar{A}^* c} \,,$$

where the underscore denotes PV–partners. These mass terms are not invariant, but the anomalies cancel. In [6] also more loops were considered. In the spirit of the method of higher derivative regularization, the added term was taken to be the vector–covariant

$$\frac{1}{\Lambda^2} \left(\partial A - \bar{\partial}\bar{A} + [A, \bar{A}] \right)^2 \,. \tag{32}$$

Note the $\bar{A}$–dependence, which from the present point of view is really an antifield–dependence. This term regularizes all but the one–loop diagrams. Then one could add the PV sector as in Eq. (14), but it is shown in [6] that a simpler action regularizes all the diagrams. It turns out that also for multiple loops there is no anomaly, and thus Eq. (31) is independent of $\bar{A}$. One can therefore normalize $Z_{nc}(\bar{A}) = 1$. We can now compare Eq. (31) with Eq. (27). Identifying $\bar{A}$ with the source u,

$$W[u] = (k + 2\tilde{h})W^0 \left[\frac{1}{k}u\right] . \tag{33}$$

The above amounts to a specific method to fix what we mean by the functional integral measure $\mathcal{D}A$. The result for the field renormalization factor is valid for this specific regularization, but not necessarily for others. To develop some understanding for other approaches, one may relate this measure to an integration measure over group variables by the usual formula:

$$\mathcal{D}A = \mathcal{D}g \, e^{2\tilde{h}S^+[g]} . \tag{34}$$

Checking the consequences for this $\mathcal{D}g$, one finds that it is not Haar invariant:

$$\left[\mathcal{D}(hg)(hg)^{-1}\right] \neq \left[\mathcal{D}g \, g^{-1}\right] . \tag{35}$$

It differs from the Haar invariant measure only by a counterterm that is local in g however. If one uses that Haar invariant measure, then for the new path integrals one finds

$$W_{\text{Haar}}(u) = (k + 2\tilde{h})W^0 \left(\frac{1}{k + 2\tilde{h}}u\right) ,$$

differing from Eq. (33) in the field renormalization factor. This shows once more the importance of the definition of the measure.

7. Conclusions

We have seen that the BV formalism unifies many aspects of the quantization in one formalism. It is not just that it offers the possibility to treat all sorts of gauge theories, but also it phrases Ward identities, WZ consistency conditions, induced actions, ... in a very natural language. Another indication that the sun is in the center is the multiple role played by the antifields, viz. first to generate the (Faddeev–Popov) antighosts, then also as sources for BRST transformations, as background fields in general gauges, and finally to describe propagating quantum degrees of freedom appearing in induced actions arising from antifield–dependent effective actions.

8. Acknowledgments

We thank the organizers of the conference for providing a stimulating atmosphere. Also discussions with Joaquim Gomìs, Alexander Sevrin, Ruud Siebelink and Stefan Vandoren are gratefully acknowledged.

9. References

1. A. Van Proeyen, in *Strings & Symmetries 1991*, eds. N. Berkovits et al. (World Sc. Publ. Co., Singapore, 1992), p. 388.

2. W. Troost and A. Van Proeyen, *An introduction to Batalin–Vilkovisky Lagrangian quantization*, (Leuven Notes in Math. Theor. Phys.), in preparation.

3. Hull, *Phys. Lett.* **B240** (1990) 110.

4. S. Vandoren, A. Van Proeyen, preprint KUL-TF-93/23, hep-th/9306147.

5. W. Siegel, *Int. J. Mod. Phys.* **A3** (1988) 2707.

6. A. Sevrin, R. Siebelink and W. Troost, preprint LBL-33777, UCB-PTH-93/08, KUL-TF-93/25, hep-th/9306135.

7. I.A. Batalin and G.A. Vilkovisky, *J. Math. Phys.* **26** (1985) 172.

8. J. Fisch, M. Henneaux, J. Stasheff and C. Teitelboim, *Comm. Math. Phys.* **120** (1989) 379;
   J. Fisch and M. Henneaux, *Comm. Math. Phys.* **128** (1990) 627;
   M. Henneaux and C. Teitelboim, *Quantization of gauge systems*, (Princeton University Press, Princeton, 1992).

9. W. Troost, P. van Nieuwenhuizen and A. Van Proeyen, *Nucl. Phys.* **B333** (1990) 727.

10. J. Zinn–Justin, in *Trends in Elementary Particle Theory*, Lecture notes in Physics **37**, eds. H. Rollnik and K. Dietz, (Springer, Berlin, 1975); *Nucl. Phys.* **B246** (1984) 246.

11. M. Hatsuda, W. Troost, P. van Nieuwenhuizen and A. Van Proeyen, *Nucl. Phys.* **B335** (1990) 166.

12. P.B. Gilkey, *J. Diff. Geo.* **10** (1975) 601; *Invariance Theory, the Heat Equation and the Atiyah–Singer Index Theorem* (Publish or Perish, Inc., Math. Lect. series no. 11, 1984).

13. K. Schoutens, A. Sevrin and P. van Nieuwenhuizen, preprint ITP–SB–91–13, to appear in the proceedings of the January 1991 *Miami workshop on Quantum Field Theory, Statistical Mechanics, Quantum Groups and Topology*;
    C.M. Hull, *Int. J. Mod. Phys.* **A8** (1993) 2419;
    C.N. Pope, L.J. Romans and K.S. Stelle, *Phys. Lett.* **B268** (1991) 167.

14. L.J. Romans, *Nucl. Phys.* **B352** (1990) 829.

15. F. De Jonghe, R. Siebelink and W. Troost, *Phys. Lett.* **B288** (1992) 47.

16. P. van Nieuwenhuizen, contribution to this conference;
    B. de Wit, M.T. Grisaru and P. van Nieuwenhuizen, preprint THU-93/15, ITP-SB 93-353, BRX-TH-349, hep-th/9307027

17. A. M. Polyakov, in *Fields, Strings and Critical Phenomena*, eds. E. Brézin and J. Zinn-Justin, (North-Holland, Amsterdam, 1990)

Al.B. Zamolodchikov, preprint ITEP 84-89 (1989).

K. Schoutens, A. Sevrin and P. van Nieuwenhuizen, in *Strings & Symmetries 1991*, eds. N. Berkovits et al. (World Sc. Publ. Co., Singapore, 1992), p. 558.

18. V.G. Knizhnik, A.M. Polyakov and A.B. Zamolodchikov, *Mod. Phys. Lett.* **A3** (1988) 819.

3. QCD

QCD AND STRINGS IN 2D[1]

Itzhak Bars[2]
Department of Physics and Astronomy
University of Southern California
Los Angeles, CA 90089-0484, USA

Abstract

In two dimensions large N QCD with quarks, defined on the plane, is equivalent to a modified string theory with quarks at the ends and taken in the zero fold sector. The equivalence that was established in 1975 was expressed in the form of an interacting string action that reproduces the spectrum and the 1/N interactions of 2D QCD. This action may be a starting point for an analytic continuation to a four dimensional string version of QCD. After reviewing the old work I discuss relations to recent developments in the pure QCD-string equivalence on more complicated background geometries.

1 Introduction

During the past year, starting with the work of Gross [1], there has been new progress in establishing a definite relationship between large-N QCD and strings in 2-dimensions [2, 3, 4, 5, 6]. Pure QCD (without quarks) is trivial in 2 dimensions if it is defined on the plane. However, when defined on more complicated geometries, such as the cylinder, torus, pretzel, etc. there remains non-contractable Wilson lines (color glue strings) that survive as degrees of freedom. They have shown, to all orders in the 1/N expansion, that the path integral for large N QCD can be reorganized as a sum over

---

[1]Based on lecture delivered at the Strings '93 Conference, Berkeley, CA, May 1993.
[2]Research supported in part by the DOE Grant No. DE-FG03-84ER-40168.

176

surfaces analogous to the Polyakov path integral over all Riemann surfaces in string theory (all string loops). The parameter $1/N$ plays the role of the string coupling constant. This shows that there must be a string action from which this string path integral may be derived. However, so far it has not been possible to identify this QCD-string action.

On the other hand, back in 1975 we established other definite quantitative relationships between large-N QCD and strings [7, 8, 9, 10]. In that case quarks were included, but QCD was defined on the plane. Therefore, the new results are complementary to the old ones. Furthermore, in contrast to the recent progress, the old correspondance established a many-body formulation of an *interacting string action* [9, 10] that reproduced all the features of QCD on the plane, including the large-N propagators, and the $1/N$ interaction vertices. In explicit computations to first order in $1/N$ this string action gave the same results as QCD for the spectrum, the 3-meson form factors and the electroweak form factors. For agreement with QCD the string had to be limited to one topological sector, namely to the no-fold sector. This sector was self contained and consistent within the string theory. The interacting Hamiltonian that was derived from the string action implemented the restriction to the no fold sector consistently [10]. The recent developments have also emphasized the no fold sector.

The string theory version allowed additional topological sectors in which the string folded on itself. The folded string was analysed both classically and quantum mechanically [7]. The semi-classical spectrum of the folded string which was derived in [7] was later reproduced by the semi-classical spectrum of the Liouville mode discovered by Polyakov [11]. Thus, there is a close correspondance between the degrees of freedom provided by the folds on the one hand and the degrees of freedom of the Liouville field on the other hand. One may therefore associate the folds with the dynamics that relate to 2D gravity. The fold degrees of freedom were obviously absent in 2D QCD on the plane. On the other hand, from another point of view the folds could also be interpreted as points carrying color charge (i.e. dynamical degree of freedom, such as gluons or other adjoint matter), while the strings in between could be interpreted as the glue on a Wilson line [9] [3] . These were also obviously absent in 2D QCD on the plane. But in fact, this interpretation of folds

---

[3]This point of view, first suggested in Ref.[9], seems to have found real applications in four dimensions. Our model was later adopted by several phenomenological groups that applied the 2D string dynamics to the fragmentation of hadrons and hadronization

is realized in a hybrid lattice formulation of 4D QCD later developed by Bardeen and Pierson [14]. In trying to make a connection to the more recent developments that involve more complicated geometries or more complicated versions of QCD, there seems to be room for the folds that fit either kind of interpretation.

In this lecture I will first briefly review the old QCD-string correspondance on the plane, which includes quarks. In particular I will emphasize the action which may be analytically continued to four dimensions. Other lecturers in this conference will review the very interesting recent progress for pure QCD on more complicated fixed (non-dynamical) geometries, but without quarks or an action. In the second part of my lecture I will discuss the folds, and in the last part of my lecture I will speculate on possible relations between the new results and the old ones in a way that considers the folds.

2 Interacting action of strings and quarks and equivalence to 2D QCD on the plane

In 1975 'tHooft [15] derived an integral equation that determines the spectrum of large-N QCD. Soon afterwards Bardeen, Bars, Hanson and Peccei [7], working in 2D string theory, showed that the Nambu string modified *with dynamical points added to the ends*, and taken in the *zero fold sector* has the same spectrum as large-N QCD [4]. The zero fold sector is a self consistent sector of the string+points theory. This sector has a massless "pion" (not a tachyon) in the zero quark mass limit [15, 16]. Bars and Hanson [8] modified the action at the end points by adding quark degrees of freedom that carry spin and color/flavor charges and showed how the string+quark system can

---

of quark and gluon jets in various high energy reactions in four dimensions. It is my understanding that the best phenomenological fit to hadronization or fragmentation is a model that utilizes our 2D string theory (with folds representing gluons) for the transverse dynamics of quarks and gluons. The model is known to phenomenologists as the "Lund Model" [12]. The 2D string theory with no folds has also been useful in the discussion of other aspects of hadrons in four dimensions [13].

[4]It is important to emphasize that the 'tHooft equation for bosonic quarks (i.e. scalars in the fundamental representation) is different than the one for fermions, and it does not match any known string theory.

interact with electroweak gauge bosons. Furthermore, Bars [9, 10] formulated an interacting string action in a many body formalism, by adding a term to the action which fused the end point quarks on different strings, or created a pair of quarks by splitting a string. The fusing/splitting occurs when the end points on different strings touch each other, and involves the spins of the quarks in a non-trivial manner. Making all terms gauge invariant with respect to the electroweak interactions gave a further modification of the total action. This final form of the action is

$$S_{total} = \sum_{mesons} S_{meson} + \sum_{I,J} S_{fuse/split}(I, J) + \sum_{I,J} S_{electroweak}(I, J) \qquad (1)$$

The action for a single meson is

$$S_{meson} = \int d\tau\, L_0(x_1(\tau)) + \int d\tau\, L_0(x_2(\tau)) + \gamma_M \int d\tau \int_0^\pi d\sigma \sqrt{-g} \qquad (2)$$

where the string $x^\mu(\tau, \sigma)$ has end points $x_1^\mu(\tau) = x^\mu(\tau, \sigma = 0)$ and $x_2^\mu(\tau) = x^\mu(\tau, \sigma = \pi)$. The last term is the Nambu action, while the first two terms are the world line actions for the quarks at the end points

$$L_0(x_I(\tau)) = \overline{\psi}_I(\tau) i \overleftrightarrow{\partial_\tau} \gamma_\mu \psi_I(\tau) \frac{x_{\tau I}^\mu(\tau)}{2\sqrt{-x_{\tau I}^2}} - \sqrt{-x_{\tau I}^2(\tau)}\, \overline{\psi}_I(\tau)\, m\, \psi_I(\tau) \qquad (3)$$

where $x_{\tau I}^\mu(\tau) = \partial x_I^\mu(\tau)/\partial \tau$ is the velocity vector of the end point labelled by I, and $\psi_{\alpha I}^{ia}(\tau)$ is the Dirac wavefunction for the quark attached to the same end point, with its spin "α", color "i" [5] and flavor "a" labels. Furthermore, $m_{ab} = m_a \delta_{ab}$ is a mass matrix in flavor space. The 'tHooft spectrum for mesons, in the leading term of the $1/N$ expansion of QCD, is identically reproduced by the string meson action S_{meson} in the zero fold sector (which is a self consistent sector even after interactions are included).

The index I runs over all end points on all the strings that represent the different mesons in the many body formalism. The interactions for joining and/or splitting of any two end points is

---

[5] The color index is used only as a counting device in the 2D string theory. It is assumed that there are only color singlet states. Factors of 1/N eventually appear because of the color counting.

$$\sum_{I,J} S_{join/split}(I,J) = -(f/4)\sum_{I\neq J} \int d\tau \, d\tau' \, \delta\left(x_I^\mu(\tau) - x_J^\mu(\tau')\right) \sqrt{-x_{\tau I}^2(\tau)} \sqrt{-x_{\tau J}^2(\tau')}$$
$$\times \left\{\overline{\psi}_I(\tau) \left(i\overleftarrow{\partial}_\tau \frac{\gamma \cdot x_{\tau I}}{x_{\tau I}^2} + m\right) \psi_J(\tau') + \overline{\psi}_J(\tau') \left(i\frac{\gamma \cdot x_{\tau I}}{x_{\tau I}^2}\overrightarrow{\partial}_\tau + m\right) \psi_I(\tau)\right\}$$

$$(4)$$

where the delta function insures that the points are at the same location, while the spins of the quarks (I,J) are also fused. The factors $\sqrt{-x_{\tau I}^2(\tau)}$ insure τ and τ' reparametrization invariance. The operator sandwiched between the (I,J) quark wavefunctions is the Dirac operator in which the derivative is taken along the tangent to the worldline (or velocity of the end point). The equation of motion that follows from 3 also involves the same operator. Therefore, if the string coupling for mesons γ_M in 2 vanishes, the quarks can perform only free motion and do not interact. Thus, after accounting for the quark equations of motion, it is evident that the interaction above can take place only in the presence of strings. Indeed, the interaction Hamiltonian derived from 4 in [10] turns out to be proportional to γ_M. Furthermore, the overall coefficient in 4 is fixed to $f = \pi/2$ by crossing symmetry. This interaction turns out to reproduce exactly the $1/N$ corrections in 2D QCD. This was shown by comparing the detailed QCD results [17] and string results [10] for the 3-meson vertex, which is a form factor describing the decay of one meson to two mesons.

Electroweak interactions of the quarks can be introduced by replacing ordinary derivatives by gauge covariant derivatives in the actions above, as follows

$$\frac{x_{\tau I}^\mu(\tau)}{x_{\tau I}^2} i\partial_\tau \longrightarrow \frac{x_{\tau I}^\mu(\tau)}{x_{\tau I}^2} i\partial_\tau - eA^\mu(x_I(\tau)) \tag{5}$$

where the gauge field A couples to the flavor indices. This produces the electroweak interactions

$$\sum_{I,J} S_{electroweak} = \sum_I \int d\tau \sqrt{-x_{\tau I}^2(\tau)} \, \overline{\psi}_I(\tau) \, e\gamma \cdot A(x_I(\tau)) \, \psi_I(\tau)$$
$$+ (f/4)\sum_{I\neq J} \int d\tau \int d\tau' \, \delta\left(x_I^\mu(\tau) - x_J^\mu(\tau')\right) \sqrt{-x_{\tau I}^2(\tau)} \sqrt{-x_{\tau J}^2(\tau')}$$
$$\times e\{\overline{\psi}_I(\tau) \, A(x_I(\tau)) \, \psi_J(\tau') + \overline{\psi}_J(\tau') \, A(x_I(\tau)) \, \psi_I(\tau)\}$$

$$(6)$$

where the first term comes from covariantizing 3 and the second term comes from covariantizing 4. A smooth parametrization in τ or τ' require that each worldline be timelike. Then, the first term in this expression describes the interaction of the electroweak field with a timelike quark line, while the second term corresponds to the creation or annihilation of a pair, where each member of the pair has a timelike propagation. Therefore these two terms must be crossing symmetric to each other. After computing some diagrams in momentum space, it was found that crossing symmetry requires $f = \pi/2$; so gauge invariance plus crossing symmetry fixes the overall coefficient of the interaction term in 4. Computations in both QCD [17, 18] and string theory [9, 10] have shown that the electroweak form factors of mesons agree in detail in both theories.

Thus the string action 1, which is a modification of the Nambu action, taken in the zero fold sector, agrees exactly with QCD on the plane, including the $1/N$ corrections. The zero fold condition is implemented in the derivation of the Hamiltonian [10]. The spectrum is given by 2 and the $1/N$ interaction vertices come from (4,6). Furthermore, the perturbative expansion of the total Hamiltonian has diagrams that are in one to one correspondance with the quark loops in QCD. Since the propagator (i.e. spectrum), the vertex, and the set of diagrams are the same in both theories, one expects agreement to all orders of $1/N$ for all quark loops, without introducing gluon loops; however, this latter point has not been checked in detail since the computations are not available in either QCD or string theory.

Therefore, according to the old results, we have an action formulation for the correspondance between string theory with quarks at the ends *on the plane,* and QCD with quarks *on the plane.* Note that the string action is written in any number of dimensions in flat spacetime, although it was worked out in detail in only two dimensions. The question naturally arises whether this string action may be valid for the correspondance between strings and QCD in higher dimensions? Of course, the whole point of this kind of exercise in two dimensions is based on the hope that a QCD-string formulation that is valid in two dimensions might be "analytically continued" to four dimensions and be used as a good approximation to understand hadronic physics. We know that the form of string theory given above is incomplete for all applications in four dimensions since closed strings that represent glueballs and their interactions are not yet included. However, one can still attempt a comparison of 4D QCD to the above string action, by restricting oneself to only planar graphs of QCD which do not include glue balls. With this in mind, I believe it is a good exercise for someone to attempt a comparison in four dimensions of the string theory defined above to *planar* QCD (but with $1/N$ corrections due to quark loops, which make holes on the sheet, included in the comparison).

If such a limited comparison in 4D is successful to begin with, then the following additional suggested modifications of the string action may reproduce the gluonic aspects of QCD: (1) Closed strings representing glueballs should be included along with the meson action 2, (2) The interaction 4 allows open strings to make transitions to closed string, (3) An interaction that allows the string to interact at interior points should be included. Such a many body type action formulation, to be used at low orders of $1/N$, could be used for phenomenological applications to low energy hadronic physics since it would account for confinement and Regge behaviour, etc.. A more ambitious step would be to find a field theoretic string action instead of a many body type string action, but that formalism would be harder to work with for computations of standard processes.

3   Folds in 2D string theory and longitudinal modes

In [7] folded 2D strings were discussed using several approaches. It was shown that the classical solutions for folded strings emerged in different gauges and the gauge invariant motions were displayed graphically. The quantum theory of folded open massless strings was solved semi-classically in a formalism of action-angle variables (closed strings could be done just easily, see below). The normal modes were identified and shown to be equivalent to those of an additional longitudinal string degree of freedom, and the Lorentz invariance of the system was proven in the second paper in [7]. The *semi-classical* mass operator was expressed as $\alpha'(M^2 - m_0^2) = \sum_{n=1}^{\infty} \alpha_{-n}\alpha_n$ with $m_0^2 = const..$ Notice that the $n = 0$ term is excluded [7]. It has the spectrum $\alpha'(M^2 - m_0^2) = \sum_n n l_n$ with $l_n = 0, 1, 2, \cdots$. The classical solutions showed that each normal mode corresponds to an open string that has $n - 1$ folds, wraps around by folding upon itself in equal lengths, and oscillates. Later, with Polyakov's discovery of the Liouville mode, one could associate semi-classically the Liouville degrees of freedom with those of folded strings.

The classical solutions for folded strings are most easily given in the conformal gauge (for the same physics in other gauges see [7]) . The classical equations of motions and constraints for a free string in 2D are

$$\partial_+\partial_- x^\mu = 0, \qquad (\partial_+ x^\mu)^2 = 0 = (\partial_- x^\mu)^2. \tag{7}$$

The solutions are $x^\mu(\tau,\sigma) = x_L^\mu(\sigma^+) + x_R^\mu(\sigma^-)$, where the left and right movers depend on $\sigma^\pm = \frac{1}{2}(\tau \pm \sigma)$ respectively, and they are constrained by $\partial x_L^0 = \pm\partial x_L^1$, $\partial x_R^0 = \pm\partial x_R^1$. The $\pm$ can generally switch signs discontinuously in various regions of $\sigma^\pm$, but the x^μ's have to be continuous to obtain a continuous string. For a physical description, the time coordinate $x^0(\tau,\sigma) = x_L^0(\sigma^+) + x_R^0(\sigma^-)$ must be monotonically increasing with τ , while both $x^\mu(\tau,\sigma)$ must be periodic in σ (this periodicity is obvious for closed strings, but is also required for open strings to satisfy boundary conditions). Using the remaining gauge invariance of conformal transformations one can fix the gauge as $x_L^0 = \frac{\alpha'E}{2\pi}(\tau+\sigma)/2$, $x_R^0 = \frac{\alpha'E}{2\pi}(\tau-\sigma)/2$, so that $x^0 = \frac{\alpha'E}{\pi}\tau$ satisfies the periodicity and monotonicity conditions. The other string coordinate may be written as

$$x^1(\tau,\sigma) = q + \frac{\alpha'E}{2\pi}\left(f(\tau+\sigma)+g(\tau-\sigma)\right), \qquad f'(\tau+\sigma) = \pm 1, \quad g'(\tau-\sigma) = \pm 1. \tag{8}$$

Thus, the functions f,g see-saw with slopes ± 1 in the range $[-\pi,\pi]$ and are periodic with a period of 2π . For the closed string the two functions are independent, but for the open string $f = g$. The simplest example of such a function is $f = |\tau + \sigma|_{per}$, i.e. the absolute value taken in the range $[-\pi,\pi]$ and then periodically repeated (see [7, 8] for pictures) . For the open string one gets the solution

$$x^1(\tau,\sigma) = q + \frac{\alpha'E}{2\pi}\left(|\tau + \sigma|_{per} + |\tau - \sigma|_{per}\right) \tag{9}$$

This function describes an open string stretched out without any folds, the two ends oscillate against each other while the intermediate points move in such a way as to allow the whole string to twist around when the two end points pass each other (remember that $\sigma = [0,\pi]$ for open strings). The parameter E is the energy of the system. For a closed string with $\sigma = [-\pi,\pi]$, taking for illustration $f = g$, one obtains a closed string looped around once, each point moving such that the loop twists while the two end points (folds) oscillate against each other. At the end of one period $\tau \to \tau + 2\pi$ the string comes back to its original shape. There are an infinite number of such periodic see-saw functions which may be distinguished by the locations of the see-saw points in the range $[-\pi, \pi]$. The general folded string may fold and unfold during different time intervals in one period $\tau = [0, 2\pi]$. However, the normal modes correspond to strings that remain completely folded in equal lengths and oscillate in that form [7]. Although we have referred to such strings as "folded", it is hard to distinguish between strings that fold or strings that wrap in the present formalism, since the string lies in one space dimension. The physical content of these two cases is quite different from the point of view of oriented color flux, and this point will be addressed below.

To quantize the theory one may identify physical degrees of freedom by fixing the gauge completely. This was done in [7] in a gauge that differs from the conformal gauge above. Starting with the generally covariant formalism, and using the τ, σ reparametrization invariance, one may choose arbitrarily two functions. Thus, one may choose (i) $x^0(\tau, \sigma) = \frac{\alpha' E}{2\pi}\tau$ and (ii) $\partial_\sigma x^1(\tau, \sigma) =$independent of σ in between Z folds at $\sigma = \sigma_i$, $i = 1, 2, \cdots, Z$ (i.e. $x^1(\tau, \sigma)$ is linear in σ in between the folds) . This leaves only the end points $(x^1(\tau, 0) \equiv x_0(\tau),\ x^1(\tau, \pi) \equiv x_{Z+1}(\tau))$ and the folds at $x^1(\tau, \sigma_i) \equiv x_i(\tau)$ and their momenta p_i as canonical degrees of freedom (note that at a fold or end point $\partial_\sigma x^\mu = 0$ and therefore these points must move with the speed of light to satisfy the constraint $(\partial_\tau x^\mu)^2 = (\partial_\sigma x^\mu)^2$). It was shown that, in the Z-fold sector, the open string Hamiltonian is given by

$$H = \sum_{i=0}^{Z+1} |p_i| + \frac{1}{\alpha'} \sum_{i=0}^{Z} |x_i - x_{i+1}| . \tag{10}$$

The classical solutions of this Hamiltonian coincide with those displayed above in the conformal gauge [7]. For a closed string with $2Z$ folds, one gets the same form, with a small modification that replaces the lower limit by $i = 1$, the upper limit on both sums by $2Z$ instead of $Z + 1$ and Z respectively, and also requires $x_{2Z+1} = x_1$ (redefining $x^1(\tau, 0) \equiv x_1(\tau)$) .

One may also discuss the theory in the lightcone gauge, $x^+(\tau, \sigma) = \frac{\alpha' P^+}{\pi}\tau$, provided one slows down the motion of the folds or end-points by making them massive, and then taking the mass to zero at the end of the calculation (this is necessary because of the singular lightcone description of massless particles that move with the speed of light). The canonical lightcone degrees of freedom of the folds or end-points are (x_i^-, p_i^+). The lightcone gauge open

184

string Hamiltonian takes the form

$$P^- = \sum_{i=0}^{Z+1} \frac{m_i^2}{2p_i^+} + \frac{1}{\alpha'} \sum_{i=0}^{Z} |x_i^- - x_{i+1}^-| \; . \tag{11}$$

Again, the classical motions coincide with those of the conformal gauge *in the limit* of zero masses $m_i \to 0$ [7]. The advantage of the lightcone formalism is its obvious Lorentz covariance, but it can be argued that in a certain sense the timelike gauge is Lorentz covariant as well [7]. The closed string Hamiltonian is obtained with the same modifications mentioned in the previous paragraph is

$$P^- = \sum_{i=1}^{2Z} \frac{m_i^2}{2p_i^+} + \frac{1}{\alpha'} \sum_{i=1}^{2Z} |x_i^- - x_{i+1}^-| \; . \tag{12}$$

with $x_{2Z+1}^- = x_1^-$.

The semi-classical quantization was carried out by transforming to action-angle variables (ϑ_i, J_i) , and showing that the mass operator takes the form $\alpha' M^2 = \sum_i J_i$. It was shown that the i'th mode describes a string that is wrapped around i times, and that it covers a basic phase space i times, thus its Bohr-Sommerfeld quantization was given as $J_i = i(l_i + const.)$ where $l_i = 0, 1, 2, \cdots$. This is the same spectrum that would be obtained from string oscillators, by identifying $J_i = \alpha_{-i} \alpha_i$. For a closed string the same arguments lead to the *semi-classical* mass spectrum

$$\alpha'(M^2 - m_0^2) = \sum_{n=1}^{\infty} \alpha_{-n} \alpha_n + \sum_{n=1}^{\infty} \tilde{\alpha}_{-n} \tilde{\alpha}_n \; . \tag{13}$$

The exact quantum spectrum of 10 or 11 is not known for any value of Z . But it is known that, for the zero fold sector $Z = 0$, 11 coincides with the 'tHooft equation, and that in the limit of $m_i^2 = 0$ there is an exact zero mass eigenstate $(M^2 = 0)$ that may be interpreted as a pion constructed from massless quarks in the chiral symmetry limit. This interpretation is more appropriate for the string theory version of the previous section (which gives the same equation) since it includes fermionic degrees of freedom, thus introducing the concept of chiral symmetry in the context of string theory. Also, numerical studies of the spectrum of the 'tHooft equation with massive quarks shows that its spectrum becomes linear after the first few excited states, thus coming into agreement with the semi-classical massless string spectrum described by oscillators that give linearly rising Regge trajectories.

One may also carry out a covariant quantization of the D=2 open or closed massless string by applying the Virasoro constraints on the Fock space constructed from *two* covariant oscillators α_n^μ, $\mu = 0, 1$. This was done a long time ago [20], and here we give the main results. By keeping $c = d = 2$ (rather than $c = 26$) [6], and taking an intercept $\alpha_0 = \frac{d-2}{24} = 0$ (that is $L_0 = 0$), it is seen that the only states that satisfy the Virasoro constraints, $L_n|\psi> = 0$, $n \geq 0$, have positive or zero norm. The positive norm states are in one-to-one correspondance with the states described by 13, and have the same spectrum and same degeneracy as the one longitudinal oscillator of 13, provided $m_0^2 = 0$. The vacuum state $|p^\mu >$ is a massless "tachyon" that may be interpreted as the "pion" of the zero-mass 'tHooft equation. Thus, in an interacting theory (i.e. the $1/N$ corrections) this "tachyon" or "pion" must decouple (suggestive of Adler zeroes for pions) since massless particles cannot exist in two dimensions due to their infrared behaviour. The absence of the "tachyon" in the interacting theory has also been deduced from a very different point of view in the recent work of Gross [1, 3].

4 Correspondence to recent developments

In the previous sections I described strings propagating in flat spacetime and exhibited an action that reproduces 2D QCD interacting with quarks in flat spacetime. How does this relate to the recent studies of QCD that are conducted without quarks and in curved spacetime with non-trivial topology? There are at least two important observations that have been rediscovered from a different point of view in the recent developments : (i) The string-QCD correspondance seems to require strings without folds, (ii) The interacting string, including the $1/N$ correction, has no massless "tachyon" or "pion" (i.e. even if it is there in the free spectrum, it must decouple in the interacting theory, or the theory must undergo a phase transition).

Is there a role that the folds may play to make some further connections? Note that the mathematical formalism of "folds" that we have outlined may describe different physical situations. A fold may occur by (1) the string folding back on itself in the opposite orientation, thus making a worldsheet that folds back on itself, or by (2) wrapping around, which makes a world-

---

[6]The spectrum is dramatically different if c=26. The positive norm states become then zero norm states.

sheet with non-trivial topology. For a folded string lying in one dimension at a constant time one cannot tell the difference between the two cases. Therefore, the folded string formalism that we have discussed may apply to either possibility, however they will have different physical interpretations from the point of view of color flux tubes. The first possibility would correspond to having a color charge at the point of the fold, so that at constant time the color flux goes in and out along the same line in opposite orientations (one color in, and another color or the same color out). The second possibility would correspond to the same flux line (same color) continuing to wrap around until it finds a color charge or its own starting point. The second possibility can produce a big color flux passing through the same point, while the first possibility cannot do that. The QCD-string correspondance tested so far excludes the first kind of fold, while allowing the second kind to occur only with non-trivial geometrical backgrounds. This is understandable and expected since the theories considered do not have any color charges in the adjoint representation of the color group (no dynamical gluons).

Thus, to test for either kind of folds in the QCD-string correspondance we can consider enlarging our present 2D theories in two possible directions: (1) Consider 2D QCD coupled to adjoint matter in either flat or non-trivial geometries. One example is to consider a sector of 4D QCD in the hybrid formulation given in [14]; another example is 2D QCD coupled to adjoint fermions, with or without quarks in the fundamental representation. This will introduce color charges that can absorb and emit color flux, thus allowing a fold to occur. (2) Consider strings propagating in non-trivial topologies (like the recent work on QCD) and allow the string to wrap around the topology thus producing "folds" at the edges of the geometry. For example a cylinder which is squashed to lie on the table, that is representing the worldsheet of a 2D string, has the string wrapped around it at any fixed time, and the string is forced to fold at the edges of the squashed cylinder. The cylinder may be multiple sheeted, which would lead to the interpretation that the string is wrapped around many times, thus giving strings with many folds. Let me say a bit more about tentative results for either case.

For the enlarged model of QCD coupled to adjoint matter let us consider bosonic versus fermionic matter. The string-like 'tHooft equation is easily derived in a Hamiltonian approach in the lightcone gauge $A_+ = 0$. Such equations were derived in [14] for the case of bosons, and in [21] for the case of fermions. For all cases, the Hamiltonian contains kinetic terms for the matter degrees of freedom and an interaction term that consists of a linear potential between the total color densities. In the large N limit this Hamiltonian produces coupled 'tHooft-like equations for multi-particle states. Thus the two particle state mixes with the three particle states, etc. The coupling occurs because one can draw leading planar graphs that include the propagation of the matter fields in addition to the glue [7]. One may split these equations into two parts: the first part may be considered "zeroth order" in the sense that it involves only the wavefunctions with a fixed number of matter particles, and the second part is the coupling. The zeroth order may be associated with strings binding the particles, and making worldsheets during their propagation. Then the coupling may be viewed as string-string interactions that are *not* suppressed by factors of $1/N$ [8]. The zeroth order term may be compared to a string theory before interactions. For bosonic adjoint matter the details of the 'tHooft-like equation do not match any known string theory. However, for fermions the 'tHooft-like equation for Z fermions is identically reproduced by the folded string Hamiltonian given in 12! This is a new observation that relates strings and large N gauge theory. However, one must emphasize that the wavefunctions for the Z fermions must be completely antisymmetric, so that only the antisymmetric solutions of these equations can be admitted. This additional input that comes from the QCD formulation is missing in the string formulation of 12. So, there is a bit more to understand.

One may also study strings in curved spacetime in 2D and compare to QCD. One possibility is to consider higher Riemann surfaces (e.g. cylinder) that are squashed to lie on the plane since we are in 2D. A Wilson line that is wrapped around such a cylinder must fold at the end of it. If this cylinder changes its radius as a function of time, then the folds appear to oscillate just like the fold solutions of the string theory given in (8,9). A cylinder that changes as a function of time is possible in QCD provided it is coupled to dynamical Gravity in 2D. Therefore, one possibility for including the folds in the string-QCD correspondance is to consider the larger theory of QCD plus gravity in 2D and compare it to a string theory on the same curved manifold. So far 2D QCD has been analyzed on static 2D manifolds. As

---

[7]For quarks in the fundamental representation alone there are only 2-body bound states, and thus no complicated coupling to many body bound states in the large N limit

[8]However, for reasons that are not well understood this coupling seems to be numerically small [21]. I thank M. Douglas for making me aware of this paper.

another example, it should be fun to couple QCD with or without quarks to the 2D SL(2,R)/R black hole, and study the relations to the corresponding string theory. I point out that the 2D black hole string theory has classical solutions that describe folded strings moving in the vicinity of the black hole.

It is evident that there is much to do even in two dimensions, but I would urge those interested in this subject to begin to seriously explore the possible correspondance between 4D QCD and the string action given in 1, as outlined at the end of section 2.

References

[1] D. Gross, Nucl. Phys. **B400** (1993) 161.

[2] J. Minahan, Phys. Rev. **D47** (1993) 3430.

[3] D. Gross and W. Taylor, Nucl. Phys. **B400** (1993) 181; **B403** (1993) 395.

[4] J. Minahan and Polychronachos, **hepth** 9303153, 9309044.

[5] M. Douglas, **hepth** 9303159.

[6] M. Douglas and V.A.Kazakov, **hepth** 9305047.

[7] W.A.Bardeen, I.Bars, A.Hanson and R.Peccei, Phys. Rev. **D13** (1976) 2364.
For the Poincare invariance of the quantum theory see also Phys. Rev. **D14** (1976) 2193.

[8] I. Bars and A. Hanson, Phys. Rev. **D13** (1976) 1744.

[9] I. Bars, Phys. Rev. Lett. **36** (1976) 1521.

[10] I. Bars, Nucl. Phys. **B111** (1976) 413.

[11] A. Polyakov, Phys. Lett. **103B** (1981) 207.

[12] B. Andersson et. al., Zeit. Phys. **C1** (1979) 105; Zeit. Phys. **C3** (1980) 223; J. Math. Phys. **26** (1985) 112; Nucl. Phys. **B355** (1991) 82; J. Phys.**G17** (1991) 1507.

[13] R. Pisarski and J.D. Stack, Nucl. Phys. **B286** (1987) 657.

[14] W. A. Bardeen and R. Pierson, Phys. Rev. **D14** (1976) 547.
W.A. Bardeen, R. Pierson and E. Rabinovici, Phys. Rev. **D21** (1980) 1037.

[15] G. 'tHooft, Nucl. Phys. **B72** (1974) 461; **B75** (1974) 461.

[16] M.Durgut and N. Pak, Phys. Lett. **117B** (1982) 453.

[17] C. Callan, N. Coote and D. Gross, Phys. Rev. **D13** (1976) 3451.

[18] M. Einhorn, Phys. Rev. **D13** (1976) 3451.

[19] C. Thorn, Nucl. Phys. **B248** (1974) 551.

[20] I. Bars, unpublished.

[21] S. Dalley and I. Klebanov, Phys. Rev. **D47** (1993) 2517.
D. Kutasov, **hepth** 9306013.

NEW QCD RESULTS FROM STRING THEORY*

ZVI BERN

Department of Physics, UCLA, Los Angeles, CA 90024

LANCE DIXON

Stanford Linear Accelerator Center, Stanford University, Stanford, CA 94309

and

DAVID A. KOSOWER

Service de Physique Théorique de Saclay, Centre d'Etudes de Saclay
F-91191 Gif-sur-Yvette cedex, France

ABSTRACT

We discuss new results in QCD obtained with string-based methods. These methods were originally derived from superstring theory and are significantly more efficient than conventional Feynman rules. This technology was a key ingredient in the first calculation of the one-loop five-gluon amplitude. We also present a conjecture for a particular one-loop helicity amplitude with an arbitrary number of external gluons.

1. Introduction.

Calculations beyond the leading order in quantum chromodynamics are important to refining our understanding of known physics in present-day and future collider experiments. The discovery of new physics relies to a large extent on the subtraction of known physics from the data. In particular, QCD loop corrections are important but are in general quite formidable to calculate. Intermediate expressions can be many thousands of times larger in size than final expressions. This explosion of terms has been a major obstacle in performing computations required by experiment. Here we discuss new techniques based on string theory which bypass much of the algebra associated with one-loop Feynman diagram computations in gauge theories [1–4]. With the new string-based techniques the one-loop five-gluon amplitudes have been computed yielding a compact form [3]. These amplitudes have not been obtained with traditional techniques.

Recent years have seen substantial progress in improving the situation in tree-level calculations [5-8]. Tree-level matrix elements have been essential for checks of QCD processes and for estimates of QCD backgrounds to new physics searches. Here we discuss the one-loop corrections to the amplitudes. The next-to-leading order corrections are important because leading tree-level calculations miss essential physics. Two problems are that the tree-level results exhibit a strong renormalization scale dependence (which does not make physical sense) and that the cone

---

* Talk presented by Z.B.

angle dependence is incorrect. The next-to-leading order corrections to a large extent remedy this situation.

The one-loop method discussed here was originally derived from string theory. Although based on string theory it has been summarized in terms of simple rules which require no knowledge of string theory [1]; the structure of these rules can also be understood from conventional field theory through particular gauge choices and organizations of the amplitude [9].

Using methods developed through string theory we have obtained results which have previously not been obtained through conventional means. Besides the five-gluon amplitudes [3], a first calculation of a four-point one-loop gravity amplitude has been performed [10]. We have used the explicit five-gluon results to jump-start a conjecture for a particular one-loop helicity amplitude but with arbitrary numbers of external legs [38].

Using direct string methods we have also calculated the one-loop four-point helicity amplitudes with two external quarks [11], which agree with the calculation of ref. [12].

Is string theory 'required' for field theory calculations? To develop and extend the methods string theory has been crucial. To actually evaluate amplitudes there is no need to turn to string theory. The main role of string theory is to provide a principle for discovering compact representations for field theory amplitudes. In particular, given the string-based rules for the one-loop n-gluon amplitudes and the understanding of these rules in field theory [9,13], there does not appear to be a clear way to extend the rules to multi-loops, or to gravity, without referring back to string theory to at least some extent. It is, however, possible to formulate a conventional field theory framework for obtaining much of the efficiency of the string-based method by working backwards from the string-based rules. These field theory ideas can then be applied more generally to gauge theory amplitude calculations which involve non-abelian vertices [9]. For example, some of the ideas obtained from string theory were used to aid the calculation of one-loop five-point amplitudes with external fermions and gluons [14]. Another example is the application of these ideas [15] to weak interaction processes such as $Z \to 3\gamma$ [16].

2. Difficulty of Loop Computations

An underlying cause of the complexity of QCD calculations is that the non-abelian vertices are relatively complicated. Since the vertices each contain six terms, one encounters a rapidly growing number of terms as one sews together vertices with propagators to form Feynman diagrams. Furthermore, the integrals associated with larger numbers of legs become increasingly complicated.

As a simple example consider the pentagon diagram one would encounter in a brute force five-point computation. A naive count of the number of terms gives about 6^5 terms. (This count is reduced by the use of on-shell conditions but increased since each internal momentum turns into a sum of external momenta.) Each term is associated with an integral which evaluates to an expression on the order of

192

a page in length. This means that one is faced with about 10^4 pages of algebra for this single diagram. As bad as this situation might seem, it is actually much worse because of the structure of the results. After evaluating the integrals and summing over diagrams one obtains expressions of the form

$$\frac{N_1}{D_1} + \frac{N_2}{D_2} + \cdots \tag{1}$$

where the N_i and D_i are the numerators and denominators one encounters when performing the integrals. In general the denominators contain spurious singularities which cancel only after putting large numbers of terms on a common denominator; this unfortunately causes an explosion of terms in the numerators.

The basic observation for being able to improve on conventional computations is that Feynman diagram computations always involve large cancellations amongst the various terms. Anyone who has done a Feynman diagram computation has undoubtedly asked themselves why vasts amounts of algebra are required when answers tend to be quite small. A nice example of this is the one-loop four-gluon helicity amplitude

$$A_{4;1}(1^-, 2^+, 3^+, 4^+) = -\left(1 + \frac{n_s}{N} - \frac{n_f}{N}\right) \frac{i}{48\pi^2} \frac{[2\,4]^2 u}{[1\,2] \langle 2\,3 \rangle \langle 3\,4 \rangle [4\,1]} \tag{2}$$

where n_s and n_f are the number of massless complex scalars and Weyl fermions in the fundamental representation of an $SU(N)$ theory. The plus and minus signs associated with each leg denote the helicity and $u = 2k_1 \cdot k_3$ is a Mandelstam variable. The various brackets are spinor inner products defined by [7]

$$\langle j\,l \rangle = \langle k_j{}^- | k_l{}^+ \rangle, \qquad\qquad [j\,l] = \langle k_j{}^+ | k_l{}^- \rangle$$

where the $|k_l{}^\pm\rangle$ are Weyl spinors. The spinor inner products are a convenient way to represent helicity amplitudes. (This amplitude has been decomposed using the one-loop Chan-Paton color decomposition [17].) Although this expression fits on a line, a brute force computation performed in the conventional way would start with expressions containing about 10^4 terms. Clearly there is considerable room for improving Feynman diagram computations at one loop. The claim is that string theory coupled with other ideas such as spinor helicity methods provides a way for doing this.

3. Why String Theory is Helpful.

It does not take much string theory for one to realize that string theory can be helpful in gauge theory computations. String theory satisfies a number of properties which indicates its usefulness in gauge theory computations.

1) String amplitudes are compact. At each order of perturbation theory there is only a single diagram; this provides for a compact organization, which might

lead one to suspect (at least at a hand-waving level) that string theory might be useful for gauge theory calculations.

2) In string theory one can switch between fermions and bosons in the loop by changing the world-sheet boundary conditions. This means that whether bosons or fermions circulate in the loop, the same basic string equations describe the two situations. This leads one to suspect that it should be possible to use information from a calculation with fermions in the loop to aid calclations with bosons in the loop. This may be contrasted to the usual Feynman rules one finds in textbooks, where the fermion and gluon Feynman rules do not resemble each other.

3) The $N = 4$ superstring amplitude integrands are simple [18]. In particular for the four-point amplitude the kinematic expression factors out of the string integrand. In the field theory limit this implies that the various contributions to a gluon amplitude A satisfy

$$A^{\text{gluon}} + 4A^{\text{fermion}} + 3A^{\text{scalar}} = \text{simple} \qquad (3)$$

where the particle labels refer to the particles circulating in the loop. (Note that the massless spectrum of $N = 4$ super-Yang-Mills is one gluon, four Weyl fermions and three complex scalars.) This relation then imples that not only are the integrands of the various contributions related via the previous point, but that they in fact satisfy a simple linear relation. These relations can be used to prevent duplication of effort when calculating the various contributions to a process.

4) In string theory (closed string) $\sim$ (open string)^2. Since closed strings contain gravity and open strings contain Yang-Mills, one might expect that (gravity) $\sim$ (Yang-Mills)^2. This can be made precise and be used to turn string theory into an extremely efficient computational tool for gravity [10].

5) In string theory the loop momentum is implicitly integrated out. This is useful because helicity techniques [7] are most efficient after loop momentum is integrated out.

The strategy is to build a string theory with a spectrum relevant for the field theory of interest, and then to take the infinite string tension limit in order to extract the organizational efficiency of the string in the field theory limit. Once the essential reorganization has been summarized in terms of a set of field theory rules or ideas there is no need to turn to string theory for every new calculation.

4. String Theory and Extraction of Field Theory Limit

The infinite tension limit of a string theory is a field theory [19,18]. In order to use string theory as a computational tool, control of the massless matter content of the string theory is required, because colored massless matter particles can run around the loops. It is possible to build consistent heterotic string theories [20] whose infinite-tension limit is a non-abelian gauge theory where one of the

194

factors is an $SU(N)$ with no matter fields [21]. The technology needed for such a construction is the one used to construct four-dimensional string models [22]; the formulation of Kawai, Lewellen and Tye is particularly simple, although any of the other formulations can be used depending on one's taste. In the original derivation of the string-based rules [1], it was essential to use a consistent string in order to prevent extraneous problems from entering. Without full consistency there would be no guarantee that the final results obtained would be correct. A heterotic string was used in the original derivation of the string-based rules because bosonic strings always contain unwanted massless scalars and tachyons, while four-dimensional type II [23,24] and type I [25] superstrings do not have a rich enough variety of fully consistent models.

Bosonic string constructions are generally much simpler than super or heterotic string constructions so for pedagogical purposes that is what will be discussed here. Heterotic string constructions are needed to provide a consistent derivation of the rules [1], but the procedure for extracting the field theory limit is similar. The open bosonic string discussed here is identical to the one used by Metsaev and Tseytlin [26] to obtain the Yang-Mills β-function from string theory. This string is given by a naive truncation of an oriented open bosonic string to four-dimensions. In this way all massless colored scalars arising from the dimensional compactification are simply thrown away. This string is inconsistent as a fundamental string theory because of the naive truncation of the spectrum. Another technicality is that the string does contain a tachyon, which might be worrisome; however, one can handle this with the prescription that exponentially large terms due to the tachyon should be dropped in the same way that exponentially small terms from the higher mass states are dropped. These potential inconsistencies of the bosonic string are irrelevant in the field-theory limit, as can be verified explicitly using an independent calculation with the fully-consistent heterotic-string formalism. What is important here is the basic structure that emerges from string theory without facing the full technicalities of heterotic string constructions.

In general, an amplitude in string theory is evaluated by performing the Polyakov surface integral [27]

$$A_n \sim \int DX \exp\left[\frac{1}{\alpha'} \int d^2\nu \, \partial_\alpha X^\mu \partial_\alpha X_\mu\right] V_1 V_2 \cdots V_n \tag{4}$$

where the $V_i \sim \varepsilon_i \cdot \partial X e^{ik \cdot X}$ are the vertex operators for external gluons with polarizations ε. At one-loop this path integral is performed on a world-sheet annulus. Since the world-sheet bosons are free, Wick's theorem can be used to evaluate the string n-gluon amplitude in terms of the two-point correlation on the annulus

$$\langle X_\mu(\nu_1) X^\nu(\nu_2) \rangle = \delta_\mu{}^\nu G_B(\nu_{12}) = -\delta_\mu{}^\nu \left[\log|2\sinh(\nu_{12})| - \frac{(\nu_{12})^2}{\tau} - 4q\sinh^2(\nu_{12})\right]$$
$$+ \mathcal{O}(q^2)$$

$$\tag{5}$$

where $\tau = -\log(q)/2$ is the real modular parameter of the annulus, ν_i represents the location of the vertex operator on the annulus and $\nu_{ij} = \nu_i - \nu_j$. (These parameters are π/i times the conventional one in refs. [23,28].) As discussed in ref. [1], in the field theory limit these parameters are proportional to sums of Schwinger proper time parameters. A repeated application of Wick's theorem to evaluate the Polyakov integral yields the string partial amplitude associated with the color trace $\mathrm{Tr}(T^{a_1} T^{a_2} \cdots T^{a_n})$

$$
\begin{aligned}
A_{n;1} =& i \frac{(4\pi)^{\epsilon/2}}{16\pi^2} (\sqrt{2})^n (\alpha')^{n/2-2} \int_0^\infty d\tau \int \prod_{i=1}^{n-1} d\nu_i \theta(\nu_i - \nu_{i+1})\, \tau^{-2+\epsilon/2} Z \\
& \times \prod_{i<j}^n \exp\Big\{ \alpha' k_i \cdot k_j G_B(\nu_{ij}) + \sqrt{\alpha'}(k_i \cdot \varepsilon_j - k_j \cdot \varepsilon_i) \dot{G}_B(\nu_{ij}) \qquad (6) \\
& \qquad\qquad\qquad - \varepsilon_i \cdot \varepsilon_j \ddot{G}_B(\nu_{ij}) \Big\}\Big|_{\mathrm{multi-linear}}
\end{aligned}
$$

where

$$
\dot{G}_B(\nu) = \frac{1}{2} \frac{\partial}{\partial \nu} G_B(\nu) , \qquad\qquad \ddot{G}_B(\nu) = \frac{1}{4} \frac{\partial^2}{\partial \nu^2} G_B(\nu) \qquad (7)
$$

and ν_n is fixed at τ. The 'multi-linear' signifies that after expanding the exponential only terms which are linear in all n polarizations vectors are to be kept. The string oscillator contributions to the partition function are

$$
Z = q^{-1} \prod_{n=1}^{\infty} (1 - q^n)^{-2(1-\delta_R\epsilon/2)} . \qquad (8)
$$

Full consistency of the string demands that the dimension $D = 26$ [29], but for the purposes of obtaining field theory amplitudes $D = 4 - \epsilon$ where ϵ is the dimensional regularization parameter necessary to handle infrared divergences; the regularization parameter δ_R, included in the string partition function, determines the precise form of the regularization [1]. In order to obtain a sensible field theory limit, the leading q^{-1} has been maintained by hand independent of the number of dimensions. (A fully consistent heterotic string such as the one used in ref. [1] does not require any adjustments, such as this one.) The field theory limit of the amplitude (6) yields the pure Yang-Mills contributions to the amplitude including Faddeev-Popov ghosts. The conventions have been adjusted so that in the field theory limit the number of π's and 2's which need to be shuffled around are minimized and so that these normalizations agree with the ones used in the heterotic string analysis of refs. [1].

Partial amplitudes associated with two color traces are a bit different since the string vertex operators are located on both boundaries of the annulus; examples can be found in chapter 8 of ref. [28].

In order to take the infinite string tension limit $\alpha' \to 0$ of the string amplitude (6), it is convenient to first integrate by parts on the string world-sheet in order to remove all $\ddot{G}_B$ from the kinematic factor [1]. (The analysis of the field theory limit can also be performed without the integration-by-parts step so it should not be taken as an essential ingredient to the string-based method.) In open string theory there are potential boundary terms, but these can be removed by an appropriate analytic continuation in external momenta since all the boundary terms contain a factor of $|\nu_i - \nu_j|^{-n-\alpha' k_i \cdot k_j}|_{\nu_i \to \nu_j} = 0$. (One technicality is that the periodicity on the annulus under $\nu \to \nu + \tau$ must be used to remove some of the surface terms.) In appendix B of ref. [30] it was proven that all $\ddot{G}_B$'s can always be eliminated from the kinematic function, by appropriate integration by parts.

In the field theory limit, the contributions to an integrated-by-parts one-loop amplitude can be classified in terms of tree and loop parts. The tree parts are obtained by first extracting the massless poles in the S-matrix before taking the field theory limit of the loop. Examples of these kinematic poles are found in the regions where $\nu_i \to \nu_j$ and are of the form

$$\int d\nu_i \frac{1}{\nu_{ij}^{1+\alpha' k_i \cdot k_j}} \longrightarrow -\frac{1}{\alpha' k_i \cdot k_j} \qquad (\alpha' \to 0). \qquad (9)$$

In general, the kinematic poles extracted in this way correspond to the poles of a scalar ϕ^3 diagram.

After kinematic poles have been extracted, the field theory limit of the loop is needed. This is obtained by taking $\tau, |\nu| \to \infty$ which corresponds to squeezing the annulus down to a field theory loop. The values of the Green functions in this limit are

$$\exp(G_B(\nu)) \to \exp\left(\frac{\nu^2}{\tau} - |\nu|\right) \times \text{constant}$$
$$\dot{G}_B(\nu) \to \frac{\nu}{\tau} - \text{sign}(\nu)(\tfrac{1}{2} + e^{-2|\nu|} - qe^{2|\nu|}). \qquad (10)$$

The exponentiated bosonic Green function was not expanded beyond $\mathcal{O}(q^0)$; after carrying out the integration by parts procedure the higher order terms do not contribute since they carry too many explicit powers of α'. For $\dot{G}_B$, terms through $\mathcal{O}(q)$ should be kept due to the presence of the overall q^{-1} in the string amplitude (6).

In the field theory limit two types of loop contributions are obtained depending on whether a power of q is extracted from the string partition function or from the Green functions. For the former contribution one simply keeps the leading order contributions from the bosonic Green functions. This type of contribution is described by the bosonic zero-mode [28] or loop momentum integral of the string [9]. A product of $\dot{G}_B$'s contains exponentially growing and decaying terms as well as terms which are constant. In general, when terms proportional to $q = e^{-2\tau}$ are extracted from a product of $\dot{G}_B$ in order to cancel the overall q^{-1}, a factor of the form

$$\exp\left[\left(|x_k - x_l| - \sum |x_i - x_j|\right)\tau\right] \tag{11}$$

is obtained where $x_i \equiv \nu_i/\tau$. In order to avoid exponential suppression or growth as $\alpha' \to 0$ the sum must add up to cancel within the exponential exactly. This will happen only if each x_i which appears with a positive sign also appears with a negative sign after expressing the absolute values in terms of the x_is directly. The correct prescription for dealing with exponentially growing terms due to the tachyon is to simply drop them in the same way that exponentially decaying terms are dropped. (The exponential growth is an artifact of the Schwinger proper time representation of tachyonic propagators.)

The result of collecting those terms where the exponential terms completely cancel is that only those which form a *cycle* of $\dot{G}_B$'s, defined to be a product of $\dot{G}_B$'s with indices arranged in the form

$$\dot{G}_B(\nu_{i_1 i_2})\dot{G}_B(\nu_{i_2 i_3})\cdots\dot{G}_B(\nu_{i_m i_1})\,, \tag{12}$$

will not vanish. Furthermore, the cyclic ordering of the indices must follow the same ordering of the corresponding legs in the partial amplitude.

The analysis of the field theory limit of a superstring is quite similar. A superstring is essential in order to be able to include space-time fermions into the string formalism and provides a more consistent framework than bosonic strings. A detailed discussion of the field theory limit of a heterotic string has been given in ref. [1].

By organizing the contributions obtained in the field theory limit a set of string-based rules for calculating gluon amplitudes can be obtained [1,2]. The string-based rules work by specifying a set of substitution rules on the string kinematic expression

$$\mathcal{K} = \int \prod_{i=1}^{n} dx_i \prod_{i<j}^{n} \exp\left(k_i \cdot k_j G_B^{i,j} + (k_i \cdot \varepsilon_j - k_j \cdot \varepsilon_i)\dot{G}_B^{i,j} - \varepsilon_i \cdot \varepsilon_j \ddot{G}_B^{i,j}\right)\Bigg|_{\text{multi-linear}} \tag{13}$$

to obtain the contributions to various diagrams which represent the various corners of moduli space. This 'master formula' contains all information for all diagrams and particle types which can circulate in the loop. Although this is the kinematic formula for a bosonic string, one can use world-sheet supersymmetry in a superstring to relate the contributions of the world-sheet fermions to those of the world-sheet bosons [2]; in this way the kinematic expression associated with the bosons can be used to summarize all contributions. Alternatively one can specify rules which act directly on the superstring kinematic expression, which was the original form of the rules presented in ref. [1]. An example of a substitution rule is

$$\dot{G}_B^{i,j} \to \frac{1}{2} + \sum_{k=i+1}^{j} a_k \tag{14}$$

where the a_k are ordinary Feynman parameters. There are a variety of other substitution rules which depend on the particular corner of string moduli space under consideration and are described more fully in refs. [1,4].

Because the contribution of any type of particle is contained in the master formula, relationships between fermion and boson contributions become apparent within the integrands of each diagram. This can be used to obtain even further simplifications; once the fermion loop contribution to the n-gluon amplitude has been computed, calculating the gluon loop contribution is relatively simple [3].

In this way one obtains a set of Feynman parameter polynomials which are far more compact than one would obtain by traditional Feynman diagram methods. The Feynman parameters must then be integrated in order to obtain the amplitudes.

5. Field Theory Interpretation

Since a conventional field theory interpretation of string-based rules has been developed [9], one can use string-motivated ideas directly in a conventional field theory setting. The following strategy incorporates the ideas that were extracted from the mapping between field theory and string theory and greatly improves the calculational efficiency over traditional Feynman diagram computations. First background field Feynman gauge [31] should be used in calculations where a nonabelian vertex appears in the loop. This gauge is useful for constructing the one-particle-irreducible diagrams, since the vertices are particularly well suited for loop calculations. These one-particle irreducible diagrams describe a gauge-invariant effective action. For sewing trees onto the one-particle irreducible loop diagrams the Gervais-Neveu gauge [32] is a particularly efficient gauge because of the simplicity of the three- and four-point vertices. (Although it might seem strange that two different gauge choices are used for the loop and tree parts of the Feynman diagrams, in the background field method this has been justified by Abbott, Grisaru and Schaeffer [31].) In general, one should use color ordered [5,30] vertices in order to minimize the number of diagrams which must be explicitly computed. For internal fermions it is best to use the second order formalism described in ref. [9] because then the gauge boson and fermion contributions are quite similar. In particular for $N = 4$ supersymmetry there are large cancellations between the various contributions to the loops [18]. In this way a good fraction of the work does not have to be duplicated for each type of particle circulating in the loop in agreement with the string theory expectations.

Spinor helicity methods [7] are also important to help minimize the amount of required algebra. Since spinor helicity methods do not handle off-shell loop momentum efficiently it should be integrated out early in the calculation to obtain a representation in terms of Feynman parameters. In order to minimize the number of terms which appear, spinor helicity should be applied on a term-by-term basis in the numerator as one integrates out the loop momentum. An alternative approach which implicitly and systematically integrates out the loop momentum is the electric circuit analogy discussed in refs. [33].

In this way one can attain many of the simplifications of a more direct string approach. Gravity does, however, provide a concrete example where a direct string-based computation is significantly more efficient than a computation based on the above field theory ideas [10].

6. Explicit Calculations

Using the string-based methods discussed above we have performed a computation of the one-loop five gluon amplitudes [3]. Additional ingredients which enter into this calculation are a simple integration method for the pentagon parameter integrals [34] and improvements in the spinor helicity method.

The finite helicity one-loop amplitudes associated with the color trace $\text{Tr}(T^{a_1} \cdots T^{a_5})$ are

$$
A_{5;1}(1^+, 2^+, 3^+, 4^+, 5^+) = \left(1 + \frac{n_s}{N} - \frac{n_f}{N}\right)
$$
$$
\times \frac{i}{48\pi^2} \frac{\langle 1\,2\rangle\,[1\,2]\,\langle 2\,3\rangle\,[2\,3] + \langle 4\,5\rangle\,[4\,5]\,\langle 5\,1\rangle\,[5\,1] + \langle 2\,3\rangle\,\langle 4\,5\rangle\,[2\,5]\,[3\,4]}{\langle 1\,2\rangle\,\langle 2\,3\rangle\,\langle 3\,4\rangle\,\langle 4\,5\rangle\,\langle 5\,1\rangle}
$$
$$(15)$$

$$
A_{5;1}(1^-, 2^+, 3^+, 4^+, 5^+) = \left(1 + \frac{n_s}{N} - \frac{n_f}{N}\right)
$$
$$
\times \frac{i}{48\pi^2} \frac{1}{\langle 3\,4\rangle^2} \left[-\frac{[2\,5]^3}{[1\,2]\,[5\,1]} + \frac{\langle 1\,4\rangle^3\,[4\,5]\,\langle 3\,5\rangle}{\langle 1\,2\rangle\,\langle 2\,3\rangle\,\langle 4\,5\rangle^2} - \frac{\langle 1\,3\rangle^3\,[3\,2]\,\langle 4\,2\rangle}{\langle 1\,5\rangle\,\langle 5\,4\rangle\,\langle 3\,2\rangle^2} \right].
$$
$$(16)$$

The infrared-divergent amplitudes (which are the ones which interfere with the tree diagrams to produce the next-to-leading order corrections to the cross-section) are given in ref. [3] and are more complicated. These amplitudes have not been obtained with traditional techniques used, for example, by Ellis and Sexton [35].

We have also calculated four-point matrix elements with two external quarks by a direct string approach [11]. Five-point matrix elements have also been obtained, using field theory. An example of one of the amplitudes is

$$
A_{5;1}(1_{\bar{q}}^-, 2_q^+, 3^+, 4^+, 5^+) = \frac{i}{16\pi^2} \left[-\frac{1}{2}\left(1 + \frac{1}{N^2}\right) \frac{\langle 1\,2\rangle\,(\langle 1\,2\rangle\,[2\,3]\,\langle 3\,1\rangle + \langle 1\,4\rangle\,[4\,5]\,\langle 5\,1\rangle)}{\langle 1\,2\rangle\,\langle 2\,3\rangle\,\langle 3\,4\rangle\,\langle 4\,5\rangle\,\langle 5\,1\rangle} \right.
$$
$$
\left. + \frac{1}{3}\left(1 - \frac{n_f}{N} + \frac{n_s}{N}\right)\left(\frac{\langle 1\,3\rangle\,[3\,4]\,\langle 4\,1\rangle^2}{\langle 1\,2\rangle\,\langle 3\,4\rangle^2\,\langle 4\,5\rangle\,\langle 5\,1\rangle} + \frac{\langle 1\,4\rangle\,\langle 2\,4\rangle\,[4\,5]\,\langle 5\,1\rangle}{\langle 1\,2\rangle\,\langle 2\,3\rangle\,\langle 3\,4\rangle\,\langle 4\,5\rangle^2} + \frac{[2\,3]\,[2\,5]}{[1\,2]\,\langle 3\,4\rangle\,\langle 4\,5\rangle}\right) \right].
$$
$$(17)$$

The remaining five-point amplitudes with external fermions are more complicated and will be presented elsewhere [14]. The calculational ideas motivated by the string organization were useful in obtaining these results.

200

7. String-Based Methods for Gravity

Another application of the string-based technique is to gravity [10]. Roughly speaking the structure of string theory implies that

$$(\text{Closed String}) \sim (\text{Open String})^2 . \tag{18}$$

Since closed strings contain gravity and open strings contain gauge theory one might expect that

$$(\text{Gravity}) \sim (\text{Yang-Mills})^2 . \tag{19}$$

This relationship can be made precise and turned into an extremely efficient computational tool for perturbative gravity amplitudes. At tree-level Berends, Giele and Kuijf [36] have made use of this relationship to obtain tree-level gravity amplitudes from known Yang-Mills amplitudes. At one-loop this relationship can also be made precise [10]; in particular, the calculation of the one-loop four-graviton amplitude with one minus and three plus helicities is rather easy by making use of string-based rules modified for the case of gravity coupled to massless matter. The result of such a calculation is given by

$$A(1^-,2^+,3^+,4^+) = \frac{i\kappa^4}{(4\pi)^2}\frac{1}{5760}(N_b - N_f)\frac{s^2 t^2}{u^2}(u^2 - st)\left(\frac{[24]^2}{[12]\langle 23\rangle\langle 34\rangle[41]}\right)^2 \tag{20}$$

where κ is the gravitational coupling, N_b is the number of physical bosonic states and N_f is the number of fermionic states in the particular theory of gravity under consideration. (Here s, t, u are Mandelstam variables.) The fact that any massless state gives an identical contribution up to a sign is in agreement with the supersymmetry identities [37].

This type of calculation would be exceedingly difficult with conventional techniques, given the complexity of the gravity three- and four-point field theory vertices. This may be compared to the string-based technique where the calculation of the above helicity amplitude is reduced to an elementary exercise. It is amusing that the string-based gravity calculation is only slightly more difficult than the gluon calculation. It is intriguing that in terms of conventional field theory the required reorganization to obtain this simplicity is fairly difficult to guess without some input from string theory.

8. A Conjecture for an Arbitrary Number of External Legs.

It is possible to construct a conjecture for the n-point all plus gluon helicity amplitude [38], using properties of the amplitude as two legs become collinear, and using the explicitly calculated five-point amplitude $A_{5;1}(1^+,2^+,3^+,4^+,5^+)$ to jump-start the conjecture. This conjecture is analogous to the one for maximal helicity violating tree-amplitudes formulated by Parke and Taylor [39] and proven by Berends and Giele [6].

The conjecture for the gluon contribution to the loop is given by

$$A_{n;1}(1^+, 2^+, \ldots, n^+) \;=\; \frac{i}{96\pi^2} \, \frac{E_n + O_n}{\langle 1\,2\rangle \langle 2\,3\rangle \cdots \langle n\,1\rangle} \;, \tag{21}$$

where the parity-odd terms are given by

$$O_n \;=\; 4i \sum_{1 \le i_1 < i_2 < i_3 < i_4 \le n-1} \varepsilon_{\mu\nu\rho\sigma} k_{i_1}^\mu k_{i_2}^\nu k_{i_3}^\rho k_{i_4}^\sigma. \tag{22}$$

To describe the even terms define

$$\begin{aligned}
t_i^{(p)} &= (k_i + k_{i+1} + \cdots + k_{i+p-1})^2 \;, \\
t_i^{(2)} &= (k_i + k_{i+1})^2 \;=\; s_{i,i+1} \;, \\
t_i^{(1)} &= 0 \;,
\end{aligned} \tag{23}$$

and with all indices mod n:

$$\begin{aligned}
X_{p,p}^{(n)} &= \sum_{i=1}^{n} t_i^{(p)} t_{i+1}^{(p)} \;, \\
X_{p-1,p+1}^{(n)} &= \sum_{i=1}^{n} t_{i+1}^{(p-1)} t_i^{(p+1)} \;.
\end{aligned} \tag{24}$$

Then the ansätze for odd and even n are

$$\begin{aligned}
E_{2m+1} &= \sum_{p=2}^{m} \left(X_{p,p}^{(2m+1)} - X_{p-1,p+1}^{(2m+1)} \right) \;, \\
E_{2m} &= \sum_{p=2}^{m-1} \left(X_{p,p}^{(2m)} - X_{p-1,p+1}^{(2m)} \right) + \frac{1}{2} \left(X_{m,m}^{(2m)} - X_{m-1,m+1}^{(2m)} \right) \;.
\end{aligned} \tag{25}$$

Note that $X_{1,3}^{(n)} = 0$, since $t_i^{(1)} = 0$. The expression (21) is cyclicly symmetric, and in the limit that two legs become collinear, is proportional to the corresponding $(n-1)$-point amplitude. This conjecture will be discussed more fully elsewhere [38]. (A very recent paper by Mahlon gives the result for the corresponding all-n expression in QED [40].)

9. Summary and Conclusions

In this talk, we reviewed the string-based method for evaluating one-loop n-gluon amplitudes [1]. The method was originally derived by taking the field theory limit of an appropriately constructed [21] four-dimensional string theory [22]. Using the string-based method, together with a simple integration method [34] and improvements in the spinor helicity method, a first calculation of all five-gluon helicity amplitudes has been performed [3]. The five-point amplitude can be used to jump-start a conjecture for the all plus helicity amplitude with an arbitrary number of external legs [38]. Using a direct string approach we have also calculated four-point helicity amplitudes with two external quarks and gluons [11] which agree with ref. [12]. The first calculation of a one-loop four-point graviton amplitude has also been performed in a direct string theory approach [10].

The string based rules have been reinterpreted in terms of field theory in refs. [9,13]. The field theory ideas obtained in this way can then be applied more widely to a variety of problems to aid in computations. This was used as an aid in the calculation of five-point one-loop amplitudes with external gluons and quarks [14]. Another example is the application of these ideas to certain weak interaction processes to significantly improve their calculational efficiency [15]. Some progress has also been made on the extension of string-based methods to multiloops [41].

In summary, string theory has been useful for calculations of gauge theory amplitudes required by experiments.

The work of ZB was supported in part by the US Department of Energy Grant DE-FG03-91ER40662 and in part by the Alfred P. Sloan Foundation Grant BR-3222. The work of LD was supported by the Department of Energy under contract DE-AC03-76SF00515. The work of DAK was supported by the *Direction des Sciences de la Matière* of the *Commissariat à l'Energie Atomique* of France.

References

[1] Z. Bern and D.A. Kosower, Phys. Rev. Lett. 66:1669 (1991); Nucl. Phys. B379:451 (1992);
Z. Bern and D.A. Kosower, in *Proceedings of the PASCOS-91 Symposium*, eds. P. Nath and S. Reucroft.

[2] Z. Bern, Phys. Lett. 296B:85 (1992).

[3] Z. Bern, L. Dixon and D.A. Kosower, Phys. Rev. Lett. 70:2677 (1993).

[4] Z. Bern, UCLA/93/TEP/5, hep-ph/9304249, proceedings of TASI 1992.

[5] F.A. Berends and W.T. Giele, Nucl. Phys. B294:700 (1987);
M. Mangano and S.J. Parke, Nucl. Phys. B299:673 (1988);
M. Mangano, S. Parke and Z. Xu, Nucl. Phys. B298:653 (1988);
M. Mangano, Nucl. Phys. B309:461 (1988).

[6] F.A. Berends and W.T. Giele, Nucl. Phys. B306:759 (1988);
D.A. Kosower, Nucl. Phys. B335:23 (1990).

[7] F. A. Berends, R. Kleiss, P. De Causmaecker, R. Gastmans and T. T. Wu, Phys. Lett. 103B:124 (1981);
P. De Causmaeker, R. Gastmans, W. Troost and T. T. Wu, Nucl. Phys. B206:53 (1982);
R. Kleiss and W. J. Stirling, Nucl. Phys. B262:235 (1985);
J. F. Gunion and Z. Kunszt, Phys. Lett. 161B:333 (1985);

R. Gastmans and T.T. Wu, *The Ubiquitous Photon: Helicity Method for QED and QCD* (Clarendon Press) (1990) ;
Z. Xu, D.-H. Zhang and L. Chang, Nucl. Phys. B291:392 (1987).

[8] M. Mangano and S.J. Parke, Phys. Rep. 200:301 (1991).

[9] Z. Bern and D.C. Dunbar, Nucl. Phys. B379:562 (1992).

[10] Z. Bern, D.C. Dunbar and T. Shimada, Phys. Lett. B312:277 (1993).

[11] Z. Bern, L. Dixon and D. Kosower, unpublished.

[12] Z. Kunszt, A. Signer and Z. Trocsanyi, preprint ETH-TH/93-11, hep-ph/9305239, to appear in Nucl. Phys. B.

[13] M.J. Strassler, Nucl. Phys. B385:145 (1992);
M.G. Schmidt and C. Schubert, preprint HD-THEP-93-24, hep-th/9309055.

[14] Z. Bern, L. Dixon and D.A. Kosower, in preparation.

[15] Z. Bern and A.G. Morgan, preprint UCLA/93/TEP/36, DTP/93/80.

[16] M. Baillargeon and F. Boudjema, Phys. Lett. B272:158 (1991);
X.Y. Pham, Phys. Lett. B272:373 (1991);
F.-X. Dong, X.-D. Jiang and X.-J. Zhou, Beijing preprint BIHEP-TH-92-32;
E.W.N. Glover and A.G. Morgan, DPT/93/4.

[17] J.E. Paton and H.M. Chan, Nucl. Phys. B10:516 (1969).

[18] M.B. Green, J.H. Schwarz and L. Brink, Nucl. Phys. B198:472 (1982).

[19] J. Scherk, Nucl. Phys. B31 (1971) 222;
A. Neveu and J. Scherk, Nucl. Phys. B36 (1972) 155;
J. Minahan, Nucl. Phys. B298:36 (1988).

[20] D.J. Gross, J.A. Harvey, E. Martinec and R. Rohm, Nucl. Phys. B256:253 (1985); Nucl. Phys. B267:75 (1986).

[21] Z. Bern and D.A. Kosower, Phys. Rev. D38:1888 (1988);
Z. Bern and D. A. Kosower, in proceedings, *Perspectives in String Theory*, eds. P. Di Vecchia and J. L. Petersen, Copenhagen 1987.

[22] L. Dixon, J. Harvey, C. Vafa and E. Witten, Nucl. Phys. B261:678 (1985), B274:285 (1986);
K.S. Narain, Phys. Lett. 169B:41 (1986);
W. Lerche, D. Lust and A.N. Schellekens, Nucl. Phys. B287:477 (1987);
H. Kawai, D.C. Lewellen and S.-H.H. Tye, Nucl. Phys. B288:1 (1987);
K.S. Narain, M.H. Sarmadi and C. Vafa, Nucl. Phys. B288:551 (1987);
I. Antoniadis, C.P. Bachas and C. Kounnas, Nucl. Phys. B289:87 (1987).

[23] J.H. Schwarz, Phys. Reports 89:223 (1982).

[24] L. Dixon, V. Kaplunovsky and C. Vafa, Nucl. Phys. B294:43 (1987);
H. Kawai, D.C. Lewellen and S.-H. H. Tye, Phys. Lett. 191B:63 (1987).

[25] Z. Bern and D.C. Dunbar, Phys. Rev. Lett. 64:827 (1990); Phys. Lett. 242B:175 (1990).

[26] R.R. Metsaev and A.A. Tseytlin, Nucl. Phys. B298:109 (1988).

[27] A.M. Polyakov, Phys. Lett. 103B:207 (1981); 103B:211 (1981).

[28] M.B. Green, J.H. Schwarz, and E. Witten, *Superstring Theory* (Cambridge University Press) (1987).

[29] C. Lovelace, Phys. Lett. 34B:500 (1971).

[30] Z. Bern and D.A. Kosower, Nucl. Phys. B362:389 (1991).

[31] G. 't Hooft, *in* Acta Universitatis Wratislavensis no. 38, 12th Winter School of Theoretical Physics in Karpacz; *Functional and Probabilistic Methods in Quantum Field Theory*, Vol. 1 (1975);
B.S. DeWitt, *in* Quantum gravity II, eds. C. Isham, R. Penrose and D. Sciama (Oxford, 1981);
L.F. Abbott, Nucl. Phys. B185:189 (1981);
L.F. Abbott, M.T. Grisaru and R.K. Schaefer, Nucl. Phys. B229:372 (1983).

[32] J.L. Gervais and A. Neveu, Nucl. Phys. B46:381 (1972).

[33] J.D. Bjorken, Stanford Ph.D. thesis (1958);
J.D. Bjorken and S.D. Drell, *Relativistic Quantum Fields* (McGraw-Hill, 1965);
J. Mathews, Phys. Rev. 113:381 (1959);
S. Coleman and R. Norton, Nuovo Cimento 38:438 (1965);
C.S. Lam and J.P. Lebrun, Nuovo Cimento 59A:397 (1969;
C.S. Lam, Nucl. Phys. B397:143 (1993).

[34] Z. Bern, L. Dixon and D.A. Kosower, Phys. Lett. B302:299,1993; SLAC–PUB–5947, Nucl. Phys. B, to appear;
R.K. Ellis, W. Giele and E. Yehudai, in progress.

[35] R.K. Ellis and J.C. Sexton, Nucl. Phys. B269:445 (1986).

[36] F.A. Berends, W.T. Giele and H. Kuijf, Phys. Lett. 211B:91 (1988).

[37] M.T. Grisaru, H.N. Pendleton and P. van Nieuwenhuizen, Phys. Rev. D15:996 (1977);
M.T. Grisaru and H.N. Pendleton, Nucl. Phys. B124:81 (1977);
S.J. Parke and T. Taylor, Phys. Lett. B157:81 (1985);
Z. Kunszt, Nucl. Phys. B271:333 (1986);
M.L. Mangano and S.J. Parke, Phys. Rep. 200:301 (1991).

[38] Z. Bern, G. Chalmers, L. Dixon, D. Kosower, in preparation.

[39] S.J. Parke and T.R. Taylor, Phys. Rev. Lett. 56:2459,1986.

[40] G.D. Mahlon, preprint Fermilab-Pub-93/327-T, hep-ph/9311213.

[41] K. Roland, Phys. Lett. B289:148 (1992); preprint SISSA/ISAS 131-93-EP;
G. Cristofano, R. Marotta and K. Roland, Nucl. Phys. B392:345 (1993).

Some Comments on QCD String

Michael R. Douglas

Dept. of Physics and Astronomy

Rutgers University

We try to draw lessons for higher dimensions from the string representations recently derived for large N Yang-Mills theory by Gross and Taylor, Kostov, and others, and call attention to three characteristics that should be expected of a string theory precisely equivalent to a higher dimensional gauge theory: continuous world-sheets; strong coupling at short distances; and negative weights.

Other speakers at this conference have reviewed the long history of attempts to precisely reformulate large N $SU(N)$ gauge theory as a string theory, and the recent line of attack, initiated by D. Gross,[1] who has proposed that since we expect QCD string, if it exists, to exist in dimensions $2 \leq D \leq 4$, and since Yang-Mills theory in $D = 2$ is exactly solvable, it behooves us to understand that case as fully as possible, by reformulating it as a string theory and attempting to identify features which generalize to arbitrary D.

My own recent work, with V. Kazakov,[2] has concentrated on understanding the validity of the string representation of [3] in $D = 2$ for the case which seems to us most representative of higher dimensions, namely where two-dimensional (Euclidean) space-time is a sphere. We consider this case more representative than the other Riemann surfaces studied in [3] because it is the only case in which the free energy displays the $O(N^2)$ behavior we expect in higher dimensions, and the only case in which a saddle point dominates the functional integral, recovering the usual reason for expecting the large N limit to be simple.

The basic result is very simple. One has the formula of Migdal and Rusakov[5] for the partition function on a sphere with area A as a sum over representations, which labelling representations by the distinct integers $n_i = \alpha_i + \rho_i$ (the highest weight plus half the sum of positive roots in a diagonal Cartan basis) becomes

$$\lim_{N \to \infty} Z_G(A) = \exp[N^2 F(A)] = \sum_{n_i \neq n_j} \prod_{i>j} \left(\frac{n_i - n_j}{i - j} \right)^{2-2G} \exp -\frac{g^2 A}{2N} \sum_{i=1}^{N} n_i^2. \qquad (1)$$

Since the n_i are distinct they are $O(N)$ and the sum $(1/N) \sum n_i^2 \sim N^2$. For the case of $G = 0$ only we can regard the prefactor as an 'entropic' term of $O(\exp N^2)$ which competes with the exponential to make the total summand highly peaked for a single 'master representation' with particular values n_i. Minimizing the resulting effective action is the same problem as for an integral over an $N \times N$ hermitian matrix, with a single difference: in both cases we can describe the saddle point distribution by a continuous density $\rho(n/N)$, but in the present case the discreteness of the variables n_i implies a maximum bound on the density,

$$\rho(n/N) \leq 1. \qquad (2)$$

Disregarding this constraint, the saddle point is the standard one for a Gaussian integral:

$$\rho(x) = \frac{g^2 A}{2\pi} \sqrt{\frac{4}{g^2 A} - x^2}. \qquad (3)$$

However, for $g^2 A \geq g^2 A_{crit} = \pi^2$, this saddle point violates the constraint. It is not hard to enforce the constraint by hand and find another saddle point, with density expressed in terms of elliptic functions.[2]

The string representation of [3] is a valid series expansion for this second, strong coupling phase. The weak coupling answer (first found this way in [6]) is very much simpler and not immediately suggestive of any string representation.

Much more can be said about this problem,* but I will just mention the main lesson I drew from this work (which I admit may be too pessimistic): it is that the evidence is against any precise correspondance between the present string constructions (in any D) and weak coupling, continuum Yang-Mills theory.

On the other hand, one feels that these string constructions do capture something real about gauge theory, and as such deserve interpretation, in particular hypotheses about which features survive the continuum limit (both world-sheet and space-time) and are responsible for the drastic differences between QCD string and the known critical and non-critical strings. In fact, after trying to make sense of the idea of QCD string, the known critical and non-critical strings start to all look so similar that I will lump them all together in the following under the rubric 'fundamental string.' (What I mean by this will become more clear below.) My own thinking has largely been in the context of [3,4,9] and in particular about features which are apparent in any D in the formalism of [4] but I suspect that other derivations of string representations would also suggest these hypotheses, especially the first one. I also believe that what I have to say is already known to some workers in the field, but since it has not been written down in one place, it seems of value to collect it here. Let me also mention that many interesting statements have been made about QCD string which I will not repeat here; see [10] for a review of some of these.

On the most basic level are statements which are not specific to Yang-Mills theory but would also be true for 'Weingarten theory' (a lattice theory with link variables allowed to take arbitrary complex hermitian values and a quadratic term in the action to make this integral well defined).[11] A striking feature of the string representations (and of the strong coupling expansion at finite N) is that the partition function is expressed as a sum over *continuous* embeddings of surfaces in space-time. Now continuity is certainly part of the definition of the word 'embedding' and one might think it only natural for a string theory. But, as is well known,[12] it is not true for the fundamental string, here taken to be any string with embedding described by world-sheet variables $X^\mu(z)$ and an action whose short-distance behavior is controlled by the term $(1/\alpha') \int d^2z \, (\partial X)^2$.

What do we mean by this? A continuous embedding $X^\mu(z)$ by definition satisfies the following: for any $\epsilon > 0$, there exists a δ such that $|z - z'| < \delta$ implies $|X(z) - X(z')| < \epsilon$. If we are integrating over embeddings, we can take this statement over to the expectation

---

* And most of my lecture at Berkeley was devoted to this, but that material will be published in [7]. See as well [8]. The remainder of this article is based on my introductory comments at Berkeley, expanded and presented in talks at ENS, CERN and Princeton in July 1993.

values $\langle |X(z) - X(z')| \rangle$ or $\langle (X(z) - X(z'))^2 \rangle$ if we have a uniform bound $\delta \geq \delta_0 > 0$ for all embeddings.

For the fundamental string, we can crudely estimate

$$\langle (X(z) - X(0))^2 \rangle \sim z^2 \langle \partial X(z)\, \partial X(0) \rangle \sim \alpha' \tag{4}$$

which does not go to zero as $z \to 0$. This crude argument can be improved in many ways (it is a well-known fact of quantum field theory that at and above the critical dimension 2 continuous configurations do not dominate the functional integral) and in the context of strings for quantum gravity this fact has been offered as a sense in which string theory has a 'minimum length' $\sqrt{\alpha'}$ below which conventional ideas of space-time are not appropriate.[12]

We see that we should not get such a result by taking a limit of a sum over continuous embeddings, as we expect for QCD string. To make this precise we would need to establish the uniform bound referred to above, and to do this we would need a natural choice of local coordinate z on the world-sheet. This brings us to a related point: it seems that the induced metric $g_{ab} = \partial_a X^\mu \partial_b X_\mu$ plays a fundamental role in QCD string. This is clearest in the string representation of [4] (or any string modelled on the lattice strong coupling expansion). We can think of the lattice as embedded in D-dimensional flat space and literally induce the metric and local coordinates onto each plaquette of a surface. Such a string manifestly satisfies the bound $|X(z) - X(z')| \leq d(z, z')$ where d is the minimum world-sheet distance between two points. More generally, we should expect the coordinate z to be compatible with the induced metric in the very minimal sense that we can locally write $g_{ab} ds^a ds^b = e^\phi dz d\bar{z}$ with continuous ϕ, in which case we have a bound of the sort described.

How can we replace the very natural world-sheet action $(1/\alpha') \int d^2z \, (\partial X)^2$? It is not hard to come up with candidates which satisfy continuity as defined above. A very simple one is just $\int \sqrt{h}(\Delta X)^2$ where Δ is the two-dimensional Laplacian for the metric h. The metric h_{ab} may be related to g_{ab}, but to make the point at hand let's make it independent and non-singular around a point $z = 0$, so we can use a coordinate in which $h_{ab} ds^a ds^b \sim dz d\bar{z}$, and the world-sheet theory is still free. (Perhaps this can be thought of as the zeroth order in an expansion around $h_{ab} = \langle g_{ab} \rangle$.) Then the short distance behavior we expect is

$$\langle X(z)X(0) \rangle \sim z^2 \log z \tag{5}$$

which is compatible with continuity.

Another well-known difference between fundamental string theory and field theory (and therefore with QCD string) is that in the former we cannot easily define correlation functions local in space-time. The simplest argument for this is that a vertex operator must be dimension 2 so that its world-sheet integral will be covariant, and that the anomalous dimension of a vertex operator $O(z, \bar{z}) :\, e^{ikX(z,\bar{z})}:$ has a contribution $\alpha' k^2$ which is fixed by this constraint. We find it significant that the above considerations of continuity drastically change this constraint as well. If the Green's function $\langle X(z)X(0)\rangle$ is non-singular at short distance, the operator $e^{ikX(z,\bar{z})}$ does not need renormalization and therefore has no anomalous dimension. Thus we can define integrated world-sheet correlation functions for any external momenta, and Fourier transform them in the usual way to get local correlation functions in space-time.

Although I appealed to the strong coupling expansion, I believe that the property of continuity has a more basic origin, and is already strongly suggested by the simple observation that the gauge-invariant observables are Wilson loops, which must be continuously embedded in space-time.

The second hypothesis I would propose is much more closely tied to the strong coupling expansion, and furthermore assumes that the QCD string does reproduce the correct short-distance physics (e.g. asymptotic freedom in $D = 4$). It is that the QCD string, as a two-dimensional quantum field theory on the world-sheet, is strongly coupled at short distances. This would certainly explain why nobody has yet found the appropriate string theory! More generally, weak bare coupling in field theory, implies strong coupling on the world-sheet for the string theory. The observation behind this is the following: we know that short distance QCD physics is well described by perturbation theory, and that the link variables fluctuate but only very near the identity (up to pure gauge fluctuations). To get such a result out of the strong coupling expansion, we need to first reproduce the zeroth order behavior, in other words the rough form $\exp{-\mathrm{tr}U/g^2}$ of the action in this limit. Now if the strong coupling expansion converged this would certainly work, but we would clearly need a growing number of terms in the expansion as $g \to 0$, and no one term in the expansion would dominate. But, reinterpreting the expansion as a sum over world-sheets, this is just the situation we would interpret as strong coupling. In a weakly coupled theory a semiclassical treatment of the path integral would be valid, meaning that a single configuration (or a finite-dimensional space of configurations) would dominate, and we would expand in fluctuations around this. This is not the situation here.

Another way to see this is to observe the close similarity between the derivation of the string (or strong coupling) expansions, which start from a character expansion of the Boltzmann weights, and the duality transform in abelian gauge theory. In abelian gauge theory, once we do the link integrals, we have precisely reformulated a weakly coupled theory of link variables as a strongly coupled theory of the dual 'representation' variables. In non-abelian gauge theory, we have additional information to keep at this point, which in the string representation is what defines the connectivity of the world-sheet, but the strong-weak coupling relation is still present.

The two-dimensional case might at first be thought to contradict this, if we accept the popular hypothesis that a continuum world-sheet theory exists for which the functional integral can be reduced to a sum over classical solutions, reproducing the explicit sum of (1). However this in itself is not incompatible with strong coupling (as evidenced by a topological field theory with partition function independent of $\hbar$) and we might ask whether the $g^2 A \to 0$ limit is such that no single classical solution dominates. A good example where this is true is the $O(N^0)$ free energy for the torus space-time, with asymptotic behavior[1,17]

$$F_{G=1} = -2\log \eta(ig^2 A) \sim \log g^2 A + \pi/6g^2 A. \tag{6}$$

Even the sphere is an example, though the large N phase transition means that the limit of the string answer need not reproduce the 'correct' weak coupling asymptotics (it behaves as $F \sim 1/(g^2 A)^2$, and is real, surprisingly enough.)

The reader may have observed that my second hypothesis tends to undercut the example (the free, higher derivative action) I gave to illustrate the first hypothesis. I would agree and consider this a failing of the example, not the hypothesis.

Finally, I want to recall the results of the work on sums over discretized surfaces with positive weights, especially the work of [14]. There are very general arguments that in $D \geq 2$ such strings will always have a 'branched polymer' critical behavior. Here is a clear difference between a string derived from the Weingarten model and a QCD string, which cannot have branched polymer behavior. Now the easiest way to escape the arguments of [14] (as they point out) is simply to give negative weights to some surfaces, and the string representations of [3,4] have negative weights in profusion. In the $D = 2$ theory of [3] the solution to the branched polymer problem is not this, however. Rather, it is that they explicitly restrict their sum over surfaces to embeddings of the world-sheet in the target space without folds.

In what sense can this restriction generalize to arbitrary D ? If it does, could it solve the branched polymer problem in $D > 2$? If one simply thinks of excluding configurations which fold from the statistical sum, since the condition can be stated locally on the world-sheet, general considerations suggest that in the continuum limit this constraint would renormalize to a local operator expressible in terms of the embedding variables X^μ. Much attention has been devoted to the search for new theories with embedding (and no other) degrees of freedom, especially in the condensed matter (membrane) context.[16]

A possible clue is given by the formalism of [4]. There the 'no-fold' constraint is not explicit but emerges after cancellations between surfaces of positive and negative weight. This formalism applies in $D > 2$, and there it is not known how to state the result as a simple cancellation: rather, it appears that some surfaces must enter the sum with negative weights.

Another indication of the role of negative weights follows from an important feature of [3,4] not yet mentioned: the world-sheet cosmological constant can be identified with the bare gauge theory coupling constant. This is one of the most significant differences between this expansion and the strong coupling expansion derived from the Wilson action. For readers more familiar with the $D = 2$ case, the starting point for the $D > 2$ case is to write each plaquette Boltzmann weight as a sum over terms each with an interpretation as a cover by some number of string world-sheets with specified topology. The parameter $g^2 a^2$ for a plaquette of area a^2 in two dimensions is replaced by $g_0^2 a^{4-D}$ where g_0 is the bare coupling and a the lattice spacing. In $D = 2$ the reproducing property of the heat kernel action implies that this description is equally valid on all scales including the total system, and the partition function will be a sum of terms corresponding to an n-fold cover to the total space with weight $\exp -g^2 n A$ (times polynomial corrections). In $D > 2$ this is not true but string world-sheets are still formed by sewing the covers of each plaquette, and a world-sheet covering plaquettes of total area na^2 (with multiplicity, i.e. a locally m-fold cover counts ma^2) has weight $\exp -g_0^2 a^{4-D} n$ times polynomial corrections. If we assume that the exponential dependence is the important dependence, clearly we want to call n the bare world-sheet area and $g_0^2 a^{4-D}$ the 'bare cosmological constant'. The simplest estimate for the partition function at fixed bare world-sheet area would then be to count each cover once, giving the asymptotic behavior $\exp cn$ for some positive $c > 0$. The constant is determined by the choice of lattice and local considerations, for example whether we exclude folds. (The exponential growth would have been present even in counting random walks, and in $D > 2$ any reasonable class of surfaces will include 'fat random walks' where a cutoff-scale loop traces a random walk.)

If we assume that a string representation of this type converges to correct weak coupling answers, we can work backward from our weak coupling expectations (in particular those given by the RG) to determine the dependence of the string partition function on bare world-sheet area. The relation is just Laplace transform, and a simple example to make the point is to imagine that there is a scaling term in the $D = 4$ (with total volume L^4) free energy such as $F \sim (L/a)^4 \exp -\beta/g_0^2$. This could be the Laplace transform of a term in the fixed area partition function such as $Z(A) \sim \cos \sqrt{A}$. (The subexponential dependence on A is more significant than the non-positive definite nature of this result; although both features are necessary to get this $F(g_0^2)$, there could also be additive non-universal contributions to F which make it positive.)

The essential point is that with positive weight for each configuration, the usual exponential growth in the number of configurations as a function of the bare area will lead to a critical point at finite bare coupling. As in [14], this implies that the string tension remains at the cutoff scale, preventing a continuum string interpretation (and generically leading to branched polymer behavior). It is clearly not the QCD critical point at zero bare coupling. A theory with the critical behavior appropriate to QCD must have negative Boltzmann weights to produce a subexponential fixed area partition function. The no-fold constraint in higher dimensions does not suffice.

To summarize, there are major differences between QCD string (if it exists) and all known consistent string theories. The most striking differences follow if we try to reproduce field-theoretic short distance behavior. One way out is to look for a string which only reproduces long-distance physics, and whose short-distance physics is different from field theory. If one thinks of the large N phase transition on the two-sphere as a non-analyticity of an observable as a function of length scale, it might be taken as evidence that such a string exists, reproducing the analytic continuation of observables (such as the Wilson loop expectation value as a function of enclosed area) from their long distance behavior. This is to be contrasted to the more traditional interpretation of the transition as associated with a critical coupling and therefore posing a barrier to any contact with the weak coupling continuum limit. It would be important to find a tractable higher dimensional calculation to distinguish between these scenarios. A good example would be the finite temperature free energy; the hope would be that the higher dimensional analog of the two dimensional large N transition was the deconfinement transition, and that a string representation could reproduce the high temperature continuation of the confining phase [18] even at weak coupling.

On the other hand it might really be that a string theory exists with very different short-distance properties than the fundamental string. An example of such a theory may be the 'rigid' string with an extrinsic curvature term in the action, as advocated in [13], and we argued that theories of this form are more likely than fundamental string theory to reproduce certain qualitative features of field theory, such as the existence of Green's functions of local operators. To our present understanding, this string suffers a fatal flaw: it is non-unitary. Related to this, the classical theory is unstable. We have no new answer to these problems, but we wonder if their ultimate resolution might be as subtle as the problem of showing that non-abelian gauge theory was renormalizable and unitary once was.

The two-dimensional examples allow us to make precise statements but do not capture all the physics, even qualitatively. Perhaps it would be fruitful to consider a larger class of models, for example gauge theory defined on lattices more general than discretizations of D-dimensional space, in hopes of finding tractable models illustrating the points discussed above.

I am indebted to many people for helping me learn about this subject, and I would like to express my special thanks to I. Kostov for explaining his work, and to V. Kazakov and M. Staudacher for their unflagging interest. I would also like to express my thanks for the hospitality of the Institute for Theoretical Physics, Santa Barbara, and of the Ecole Normale Supérieure, Paris.

References

[1] D. J. Gross, Nucl. Phys. B400 (1993) 161.

[2] M. R. Douglas and V. A. Kazakov, to appear in Phys. Lett. B. hep-th/9305047.

[3] D. J. Gross and W. Taylor, Nucl. Phys. B400 (1993) 181 and Nucl. Phys. B403 (1993) 395.

[4] I. K. Kostov, SACLAY-SPHT-93-050, June 1993. hep-th/9306110.

[5] A.A.Migdal, Sov. Phys. JETP 42 (1975) 413, 743.
B.Rusakov, Mod.Phys.Lett. A5 (1990) 693.

[6] B.Rusakov, Phys.Lett. B303 (1993) 95.

[7] V. A. Kazakov, M. R. Douglas, to appear in the proceedings of the May 1993 Cargèse workshop on Strings, Conformal Models and Topological Field Theories.

[8] J. A. Minahan and A. P. Polychronakos, CERN-TH-7016-93, Sept. 1993. hep-th/9309119.
M. Caselle, A. D'Adda, L. Magnea, S. Panzeri, DFTT-50-93, Sept. 1993. hep-th/9309107.
J.-M. Daul and V. A. Kazakov, LPTENS-93-37, Oct. 1993. hep-th/9310165.
D. V. Boulatov, NBI-HE-93-57, Oct. 1993. hep-th/9310041.

[9] V. A. Kazakov and I. K. Kostov, Nucl. Phys. B176 (1980) 199.

[10] J. Polchinski, UT Austin preprint UTTG-16-92, hep-th/9210045, and references there.

[11] D. Weingarten, Phys. Lett. 90B (1980) 280.

[12] I. Klebanov and L. Susskind, Nucl. Phys. B309 (1988) 175;
K. Konishi, G. Paffuti, and P. Provero, Phys. Lett. B234 (1990) 276.

[13] A. M. Polyakov, Nucl.Phys. 268B (1986) 406;
A. M. Polyakov, "Gauge Fields and Strings," Harwood (1987).

[14] B. Durhuus, J. Fröhlich and T. Jonsson, Nucl. Phys. B240[FS12] (1984) 453;
J. Ambjorn and B. Durhuus, Phys. Lett. 188B (1987) 253;
J. Ambjorn, in Random Surfaces and Quantum Gravity, Plenum 1991, pp. 327-336.

[15] J. Ambjorn, A. Irback, J. Jurkiewicz, B. Petersson Nucl. Phys. B393 (1993) 571-600.

[16] "Statistical Mechanics of Membranes and Surfaces," vol. 5, eds. D. Nelson, T. Piran and S. Weinberg, World Scientific (1989);
F. David, B. Duplantier, S. Leibler, L. Peliti, Phys. Rep. 184 pp. 221-282.

[17] M. R. Douglas, RU-93-57, Oct. 1993. hep-th/9311130.

[18] J. Polchinski, Phys. Rev. Lett. 68 (1992) 1267.

TWO-DIMENSIONAL QCD AND STRINGS*

DAVID J. GROSS
Joseph Henry Laboratories, Princeton University
Princeton, New Jersey 08544, USA

and

WASHINGTON TAYLOR
Center for Theoretical Physics, Laboratory for Nuclear Science and Department of Physics
Massachusetts Institute of Technology; Cambridge, Massachusetts 02139, USA

ABSTRACT

A review is given of recent research on two-dimensional gauge theories, with particular emphasis on the equivalence between these theories and certain string theories with a two-dimensional target space. Some related open problems are discussed.

1. Introduction

There has been a recent renewal of interest in two-dimensional gauge theories. In two dimensions any pure gauge theory is locally trivial and has no propagating modes. However, by either considering the theory on a compact 2-manifold or introducing external Wilson loop sources, a nontrivial character of the theory emerges which is almost topological in nature. These two-dimensional theories have provided an interesting simplified model with which to study certain properties of gauge theories in any dimension.

One aspect of gauge theories which has long been of interest is the connection between a gauge theory in d dimensions and a string theory with d-dimensional target space. Speculations about how such a connection might be made have motivated a wide range of research in both QCD and strings. In this talk I will describe some recent work in which this connection is made rigorously for two-dimensional gauge theories. The string theories which will be discussed are of a very special type, being described by covering maps from a string world-sheet onto the two-dimensional target space with a finite number of singularities and no folds.

I will begin by giving a brief review of some of the previous research on two-dimensional gauge theories which is relevant to the work at hand. I will then give a simplified description of the proof that these gauge theories are equivalent to string theories. Following this, I will discuss a variety of recent related work, and then close with a brief description of several outstanding problems in this area.

---

*This work was supported in part by the divisions of Applied Mathematics of the U.S. Department of Energy under contracts DE-FG02-88ER25065 and DE-FG02-88ER25066 and in part by the National Science Foundation under grant PHY90-21984.

2. History

Almost 20 years ago[1], 't Hooft pointed out that in certain circumstances it is reasonable and interesting to expand the partition function and correlation functions of a $U(N)$ gauge theory in powers of N. The guiding principle behind this expansion is that when one represents the gluon propagator by double lines corresponding to the adjoint representation of the gauge group, each Feynman diagram has associated with it a natural genus g; it turns out that the power of N associated with a Feynman diagram of genus g is precisely N^{2-2g}. As a particularly simple example, 't Hooft considered the case of 2-dimensional QCD. He included fermionic matter fields in the fundamental representation of the gauge group (quarks), and was able to calculate the masses of mesons in the theory to leading order in N (the planar approximation).

In the following years, further progress was made in understanding the large N meson spectrum, and in connecting the two-dimensional QCD theory to a stringy theory. Bars and Hanson[2] showed that in the large N limit, the result of 't Hooft for the meson spectrum can also be derived by assuming that the quarks interact by a linear potential; this condition is equivalent to taking the Nambu action for a theory of strings connecting the quarks, and neglecting folds and singular points in the string. A major tool for the study of gauge theories was then developed, namely the Makeenko-Migdal loop equations. The loop equations for 2-dimensional gauge theories were first explicitly written down by Kazakov and Kostov[3], who used these relations to compute the VEV's of Wilson loops on the plane in a large N expansion. Similar results were achieved by Bralic using a nonabelian version of Stokes' theorem[4]. The connection between the Wilson loop VEV's and physical observables of the two-dimensional gauge theory was made explicit by Strominger, who showed that the Green's functions for quark bilinears could be explicitly written in terms of an integral over Wilson loops[5].

In a different thread of research, progress was made in understanding the pure gauge theories in 2 dimensions by exact solution. It was first shown by Migdal that by using the heat kernel action for the gauge theory, one arrives at a theory on the lattice which is invariant under triangulations, and which is equivalent to the usual gauge theory as the triangulation becomes arbitrarily fine[6]. Using this approach, it was shown by Rusakov, Witten, and others that the partition function of the Euclidian gauge theory on a compact Riemann surface could be exactly expressed in terms of the group theory of the gauge group[7]. Explicitly, they showed that the partition function on a manifold $\mathcal{M}$ of genus G and area A is given by

$$Z(G, \lambda A, N) = \int [\mathcal{D}A^\mu] e^{-\frac{1}{4\tilde{g}^2} \int_{\mathcal{M}} d^2 x \sqrt{g} \; TrF^{\mu\nu}F_{\mu\nu}} = \sum_R (\dim R)^{2-2G} e^{-\frac{\lambda A}{2N} C_2(R)}, \tag{1}$$

where the sum is taken over all irreducible representations of the gauge group, with $\dim R$ and $C_2(R)$ being the dimension and quadratic Casimir of the representation R. (λ is related to the gauge coupling $\tilde{g}$ by $\lambda = \tilde{g}^2 N$.)

3. String Theory

We will now explain how the exact expression Eq. (1) for the gauge theory partition function can be rewritten as a string theory partition function to all orders in $1/N$. The results in this section were originally described in the papers [8-10]. Recently, Kostov has described a similar equivalence for a lattice version of the theory[11].

3.1. Statement of Main Result

In the string theory which we describe here, the partition function is given by a sum over topologically distinct maps from a two-dimensional string world sheet $\mathcal{N}$ of any genus g to the target space $\mathcal{M}$. Essentially, the string partition function is written

$$Z = \int_{\Sigma(\mathcal{M})} d\nu \, W(\nu),\tag{2}$$

where $\Sigma(\mathcal{M})$ is a set of covering maps $\nu : \mathcal{N} \to \mathcal{M}$, which are allowed to have singularities at a finite set of points in $\mathcal{M}$. The specific types of singular points allowed in the maps in $\Sigma(\mathcal{M})$ depend upon the choice of gauge group and the genus G. The weight $W(\nu)$ associated with a certain string map ν is given by

$$W(\nu) = \pm \frac{N^{2-2g}}{|S_\nu|} e^{-\frac{n\lambda A}{2}},\tag{3}$$

where g is the genus of $\mathcal{N}$, n is the degree of the map ν, and $|S_\nu|$ is the symmetry factor of the map ν (the number of diffeomorphisms π of $\mathcal{N}$ which satisfy $\nu\pi = \nu$). In general, the manifold $\mathcal{N}$ need not be connected; the genus of a disconnected manifold is defined so that the Euler characteristic is additive for disjoint unions of connected manifolds. Apart from the sign, which depends upon the types of singularity points in the map ν, this is a very natural weight for a string theory. The power of N is just the usual power of the string coupling $1/N$, the exponent of $n\lambda A/2$ is just the usual Nambu action, and the symmetry factor is the usual one associated with Feynman diagrams in any field theory. The unusual feature about this string theory is that the sum over string maps is restricted to the set of maps $\Sigma(\mathcal{M})$. In particular, the strings are not allowed to have folds other than at the finite number of singular points. Of course, the gauge theory we are studying has no propagating degrees of freedom; allowing folds into the string theory would clearly violate this characteristic, so the absence of folds is in some sense not surprising.

3.2. Outline of Proof – Simple Case

We will now give a brief description of the essential features in a proof of Eq. (2); we will also describe in more detail the set of allowed maps $\Sigma(\mathcal{M})$. We will assume for the remainder of this section that the gauge group is $SU(N)$. For now, we will also restrict attention to the torus $G = 1$, where the contribution from the dimension terms vanishes in Eq. (1). We will return later to the corrections due to these dimension terms.

In order to write an asymptotic expansion of Eq. (1) in powers of $1/N$, we would like to proceed by computing the quadratic Casimir of each representation R

of $SU(N)$ as a function of N, and thus writing an asymptotic expansion separately for each term in the sum over representations. Because as N varies, the set of representations of $SU(N)$ itself changes, we must find a way of implementing this procedure which is well-defined. The theory of Young tableaux gives us such an approach. We can associate each representation R with a certain Young tableau, which we also denote by R. Given a Young tableau R with n boxes, the quadratic Casimir of the representation is given by

$$C_2(R) = nN + \frac{n(n-1)\chi_R(T_n)}{\chi_R(1)} - \frac{n^2}{N}, \tag{4}$$

where $\chi_R(T_n)$ and $\chi_R(1) = d_R$ are characters of the symmetric group S_n in the representation associated with the Young tableau R. We denote by T_n the conjugacy class of elements in the symmetric group which have one cycle of length 2 and $n-2$ cycles of length 1.

We would now like to insert Eq. (4) into Eq. (1) to form the asymptotic expansion of the partition function. However, we must be careful about which representations we include in the sum. Naively, one might expect that in the asymptotic $1/N$ expansion, it would suffice to include all representations corresponding to Young tableaux with a finite number of boxes. In fact, however, it is necessary to include all representations whose quadratic Casimir is of the form $C_2(T) = nN + \mathcal{O}(1)$. In addition to the Young tableaux with a finite number of boxes, we must include another set of representations corresponding to the conjugates of these representations, and "composite" representations arising from tensor products of these two types of representations. To simplify the presentation, we will temporarily assume that the sum in Eq. (1) can be replaced by a sum over representations with a finite number of boxes. We denote this simplified partition function by Z_Y. We will return to the correct sum over composite representations shortly; the simplifying assumption of only including Young tableaux with finite boxes corresponds to only considering a single "chiral" sector of that complete theory.

Replacing the sum over representations by a sum over all Young tableaux in each set Y_n of tableaux with a finite number n of boxes, we have

$$Z_Y(1, \lambda A, N) = \sum_n \sum_{R \in Y_n} \exp\left[-\frac{\lambda A}{2N}\left(nN + \frac{n(n-1)\chi_R(T_n)}{d_R} - \frac{n^2}{N}\right)\right]. \tag{5}$$

Expanding the exponential and using some elementary identities from the theory of characters of the symmetric group, we have

$$Z_Y(1, \lambda A, N) = \sum_{n,i,t,h} e^{-\frac{n\lambda A}{2}} N^{2-2g} \frac{(\lambda A)^{i+t+h}}{i!\, t!\, h!} \frac{(-1)^i n^h (n^2-n)^t}{2^{t+h}}$$
$$\cdot \sum_{p_1,\ldots,p_i \in T_n} \sum_{s,t \in S_n} \left[\frac{1}{n!}\delta(p_1 \cdots p_i sts^{-1}t^{-1})\right], \tag{6}$$

where $2 - 2g = -2(t+h) - i$.

We would like to now interpret this expression in terms of a sum over covering maps of the torus. This interpretation follows from a simple theorem which holds for any genus G. Define $\Sigma(G, n, i)$ to be the set of all topologically distinct covering maps onto a genus G target space of degree n and with i elementary branch point singularities at a fixed set of points q_j. Then

$$\sum_{\nu \in \Sigma(G,n,i)} 1/|S_\nu| = \sum_{p_1,\ldots,p_i \in T_n} \sum_{s_1,t_1,\ldots,s_G,t_G \in S_n} \left[\frac{1}{n!} \delta(p_1 \cdots p_i \prod_{j=1}^{G} s_j t_j s_j^{-1} t_j^{-1}) \right]. \tag{7}$$

This theorem can be proven as follows. We can cut the surface along the usual

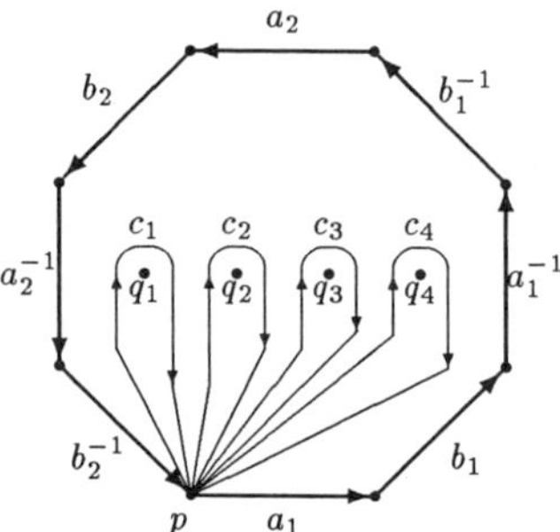

Figure 1: Surface with Genus $G = 2$, $i = 4$ Branch Points

homotopy generators a_j, b_j to form a $4G$-gon. A covering space with branch points at q_j can be described by choosing a labeling $1, 2, \ldots, n$ of the sheets of the covering space at a point p, and determining the permutation on this set of labels which is realized by moving around the homotopy generators and a set of loops c_j which encircle the branch points. Denote the permutations associated with the loops a_j, b_j and c_j by s_j, t_j and p_j respectively. The statement that the branch points are elementary is equivalent to the condition that p_j is an element of the conjugacy class T_n for all j. The single condition on the homotopy group $\pi_1(\mathcal{M} \setminus \{q_1, \ldots, q_i\})$ is that

$$c_1 \cdots c_i a_1 b_1 a_1^{-1} b_1^{-1} a_2 b_2 a_2^{-1} b_2^{-1} \cdots a_G b_G a_G^{-1} b_G^{-1} = 1. \tag{8}$$

The permutation associated with the cycle on the left hand side of this equation must therefore be the the identity permutation. We can now view the sum over all p_j, s_j, t_j in Eq. (7) as a sum over all distinct labeled coverings with branch points at q_j, where the δ function enforces the condition that the association of permutations to homotopy generators be a homeomorphism into the symmetric group. By noting that the number of distinct labelings for a particular topological type of covering map is precisely $n!/S_\nu$, we see that the combinatorial factor in the sum over coverings works out correctly, and we have proven Eq. (7).

We are now in a position to rewrite the partition function Z_Y in the form of Eq. (2). We choose an orientation on $\mathcal{M}$, and define a set of covers $\Sigma_+(\mathcal{M})$ to be the set of orientation-preserving covering maps from an oriented Riemann surface $\mathcal{N}$ to $\mathcal{M}$, which have a finite number of branch point singularities, and in addition a finite number of singularities corresponding to handles and tubes in $\mathcal{N}$ which are contracted to points in $\mathcal{M}$. We define a measure $d\nu$ on the space $\Sigma_+(\mathcal{M})$ by using the positions of the singular points as parameters in each connected component, and giving the position each of singularity a measure proportional to λdA. With respect to this measure, each topological map with i branch points carries a factor of $(\lambda A)^i/i!$, where the denominator arises from the indistinguishability of the branch points. Similarly, each map with h contracted handles carries a measure factor of $(n\lambda A)^h/h!$, and each map with t contracted tubes carries a measure factor of $(n(n-1)\lambda A)^t/(2^t t!)$. Each contracted handle carries a symmetry factor of $1/2$.

Combining these weights, we have the result that the asymptotic expansion for the partition function on the torus restricted to Young tableaux with a finite number of boxes can be written

$$Z_Y(1, \lambda A, N) = \int_{\Sigma_+(\mathcal{M})} d\nu\, W(\nu), \tag{9}$$

where the weight $W(\nu)$ is given by

$$W(\nu) = (-1)^i \frac{N^{2-2g}}{|S_\nu|} e^{-\frac{n\lambda A}{2}}. \tag{10}$$

3.3. Coupled Theory

As we remarked above, to correctly derive the asymptotic expansion of the complete theory it is necessary to consider all Young tableaux with a quadratic Casimir of leading order N. The set of Young tableaux which satisfy this condition are precisely those tableaux containing a finite number of columns with $N - k$ boxes where k is finite, and a finite number of columns with a finite number of boxes. We will call the representations associated with these tableaux composite representations. We write composite representations as $\bar{S}R$ where S, R are Young tableaux with a finite number of boxes. The representation $\bar{S}R$ is constructed by taking the Young tableau with the maximum number of boxes in the tensor product of the representation R with the conjugate of the representation S. Although the quadratic Casimir for any representation which is not of the composite type scales as N^2 and therefore gives a contribution to the partition function proportional to e^{-N}, one might question whether it is really acceptable to neglect such representations when constructing the asymptotic $1/N$ expansion. Recent work by Douglas and Kazakov[12] which will be discussed in more detail later demonstrates that in fact these representations can indeed be neglected except on spheres of area $\lambda A < \pi^2$.

Given Young tableaux S, R with $\tilde{n}, n$ boxes respectively, we can compute the quadratic Casimir of the composite representation $\bar{S}R$; we find that $C_2(\bar{S}R) = C_2(R) + C_2(S) + 2n\tilde{n}/N$. We can write the complete asymptotic expansion for the partition function for any genus G as a sum over composite representations,

$$Z(G, \lambda A, N) = \sum_{n,\tilde{n}} \sum_{R,S \in Y_n, Y_{\tilde{n}}} (\dim \bar{S}R)^{2-2G} e^{-\frac{\lambda A}{2N}[C_2(\bar{S}R)]}. \tag{11}$$

Except for the term $2n\tilde{n}/N$, this partition function for genus $G = 1$ factorizes into two components, each equal to the partition function Z_Y. We can reproduce this factorization geometrically by including in the set of allowed maps $\Sigma(\mathcal{M})$ maps from an oriented Riemann surface $\mathcal{N}$ which are either orientation-preserving or orientation-reversing maps, relative to a fixed orientation on $\mathcal{M}$. We refer to these two types of maps as two "chiral sectors" of the string theory. To see how the coupling term proportional to $n\tilde{n}$ can be incorporated, one simply expands the exponential of the quadratic Casimir in the complete partition function as in Eq. (6). Just as the terms containing h and t were interpreted as arising from handles and tubes in the string world-sheet contracted to points in the target space, we can interpret the contribution from the coupling term as arising from infinitesimal tubes which connect sheets of opposite orientation. The number of ways in which one of these tubes can connect n sheets of one orientation with $\tilde{n}$ sheets of the opposite orientation is clearly $n\tilde{n}$, so the counting is correct. From the sign on the coupling term, we see that each orientation-reversing tube carries a factor of -1.

Thus, in the complete theory containing a sum over all composite representations, we can define the set of covering maps $\Sigma(\mathcal{M})$ to be the set of all maps from a (possibly disconnected) oriented world-sheet onto $\mathcal{M}$ which is locally a covering map at all but a finite number of singular points; the allowed types of singular points consist of (i) elementary branch points, (h) contracted handles, (t) contracted tubes between sheets of identical relative orientation, and $(\tilde{t})$ contracted orientation-reversing tubes. The partition function is then given by an equation of the form of Eq. (2), where the weight $W(\nu)$ is given by $W(\nu) = (-1)^{(i+\tilde{t})} N^{2-2g} \exp(-\frac{n\lambda A}{2})/|S_\nu|$. This completes the description of the asymptotic $1/N$ expansion of the $SU(N)$ gauge theory on the torus as a string theory.

3.4. Example

As a simple example of the equivalence between the gauge theory partition function and the sum over covering maps, let us consider the leading order terms in the partition function of a single chiral sector on the torus. To order N^0, the partition function Z_Y is given by

$$Z_Y(1, \lambda A, N) = \sum_n x^n(\pi_n + \mathcal{O}(1/N)) = \prod_i \frac{1}{(1 - x^i)} + \mathcal{O}(1/N), \qquad (12)$$

where $x = \exp(-\frac{\lambda A}{2})$, and π_n is the number of ways of partitioning an integer n into a sum of integers (the number of distinct Young tableaux with n boxes). In the complete theory with both chiral sectors, the leading order term is simply given by squaring this expression.

As in any field theory, by taking the logarithm of the partition function, we get a free energy which is given by a sum over connected diagrams. In this case,

$$W_Y(1, \lambda A, N) = \ln Z_Y(1, \lambda A, N) = \sum_n \omega_n x^n + \mathcal{O}(1/N), \qquad (13)$$

where $\omega_n = \sum_{k|n} k/n$. We expect from the string description of the partition function, that the quantity ω_n should be precisely the sum of $1/|S_\nu|$ over all unbranched connected n-fold covers of the torus. This equality can easily be seen to hold, using the fact that all such covers have a symmetry group of order n. Thus, we have a verification of the string interpretation in this simple case.

3.5. Higher Genus

We will now briefly describe the correction to the simple string theory description which arises when $G \neq 1$ from the insertion of the $2 - 2G$th power of the dimension $\dim R$ for each representation. In terms of the string geometry of the theory, the only change which occurs from these extra terms can be described by including in the set $\Sigma(\mathcal{M})$ maps containing additional singularities at a set of $|2 - 2G|$ fixed points in $\mathcal{M}$. When $G = 0$ we refer to these points as Ω-points; when $G > 1$, we call the points Ω^{-1}-points. The unusual feature of these points is that they are not allowed to move on the manifold $\mathcal{M}$, and do not carry factors of the area A as do the other singularities. In addition, the types of singularities allowed at Ω-points and Ω^{-1}-points are different.

In a single chiral sector of the theory, the singularity structure at an Ω-point is extremely simple. The map ν can contain a singularity which gives rise to an arbitrary permutation on the sheets of the covering space when one follows a closed loop around the singularity. In the coupled theory, at an Ω-point, in each sector an arbitrary permutation of the sheets of the covering space is allowed. However, in addition, the sets of sheets which are connected by the cycles of these permutations can be connected in pairs by orientation-reversing tubes.

The set of singularities allowed at a Ω^{-1}-point is very similar to that of an Ω-point, but slightly more complicated. Essentially, at an Ω^{-1}-point, there may be an arbitrary number of nontrivial singularity points of the type from an Ω-point, each carrying a factor of -1. This description holds both in the complete theory and in a single chiral sector.

In sum, then, by defining the set $\Sigma(\mathcal{M})$ of covering spaces to include covers with 2 Ω-point type singularities when $G = 0$, and to include $2G - 2$ Ω^{-1}-point type singularities when $G > 1$, we can write the gauge theory partition function for arbitrary genus as Eq. (2); in this general case, $W(\nu)$ is given by

$$W(\nu) = (-1)^{(i+\tilde{i}+\sum_j x_j)} \frac{N^{2-2g}}{|S_\nu|} e^{-\frac{n\lambda A}{2}}, \tag{14}$$

where x_j is the number of distinct Ω-point type singularities at the jth Ω^{-1}-point. Thus, we have defined a string theory representation of the partition function for the $SU(N)$ gauge theory on an arbitrary genus Riemann surface.

4. Further Results

4.1. Other Gauge Groups

Up to this point, we have restricted attention to the gauge group $SU(N)$. However, for other gauge groups it is also possible to construct a string interpretation of the gauge theory partition function. In general, for a given gauge group, it is necessary to first determine the set of Young tableaux which correspond to representations whose quadratic Casimir is of order N. By then expressing the quadratic

Casimir and dimension of these representations in terms of characters of the symmetric group, one can ascertain the types of singularity structures which are allowed in the set of string maps $\Sigma(\mathcal{M})$ associated with that particular theory. Generally, the subleading terms in the quadratic Casimir correspond to "mobile" singularity types, which carry factors of the area, and the subleading terms in the dimension correspond to the "static" types of singularities which appear in the Ω-points of the theory.

The simplest example of this general analysis is for the gauge group $U(N)$. For $U(N)$, the expression for the dimension is the same as for $SU(N)$, so the Ω-points are identical in this theory. The quadratic Casimir, however, differs from that for $SU(N)$ in that the final term $-n^2/N$ associated with vanishing $U(1)$ charge is absent. Thus, in the $U(N)$ theory the mobile tube and handle singularities do not occur; on the torus, the only allowed types of singularities are branch points.

A similar analysis of the string theories for gauge groups $SO(N)$ and $Sp(N)$ has been carried out by Naculich, Riggs, and Schnitzer, and by Ramgoolam[13]. They found that for these gauge groups, the string world sheet is not necessarily orientable, and that there are additional singularities corresponding to infinitesimal cross-caps in the string maps. The insertion of a cross-cap singularity essentially corresponds to cutting out an infinitesimal disk and replacing it with a projective plane minus a point; moving along a loop which surrounds this point gives a reversal of orientation. The singularity structure at Ω-points for these gauge groups is remarkably similar to that for Ω-points in the $SU(N)$ theory; however, cross-cap singularities also appear at these Ω-points.

4.2. Wilson Loops

Just as the partition function of any 2D gauge theory on an arbitrary Riemann surface can be written as a weighted sum over closed string maps, it is possible to show that the VEV of any Wilson loop γ can be expressed as a weighted sum over *open* string maps, where the boundary of the string world-sheet is taken to γ by the covering map. There are some technical complications with the calculation for Wilson loops with self-intersections; one finds that each of the disconnected regions into which the Wilson loop cuts the manifold $\mathcal{M}$ must be associated with some fixed number of Ω-point singularities, even for a Wilson loop on the torus. The details of the calculation of Wilson loop VEV's in terms of open strings are described in [10].

4.3. Phase Transition

An important related development is the recent demonstration by Douglas and Kazakov[12] of a phase transition in the partition function on the sphere at the point $\lambda A = \pi^2$ (the trivial small area phase was previously observed by Rusakov[14]). They used techniques familiar from matrix models to study which representations contribute to the partition function in the large N regime. In the phase with $\lambda A > \pi^2$, they showed that the set of representations which contribute to the partition function is precisely the set of composite representations. Below the critical value of the area, nonperturbative effects of other representations simplify the partition function and render invalid the restriction to composite representations. This phase transition is analogous to the phase transition which occurs for the Wilson action at large N[15].

The significance of these results for the understanding of QCD through a string interpretation is not yet clear. The existence of the phase transition clearly represents an obstacle to expanding the string interpretation to small coupling in certain situations; however, the fact that this phase transition does not occur for higher genus surfaces or for finite values of N gives hope that it is not a fundamental obstacle to progress in this direction.

4.4. Equivalence of QCD_2 to Other Theories

Other recent work has shown that due to its simple group-theoretic structure, QCD_2 can be related to many other theories of current interest. In its Hamiltonian formulation, pure QCD_2 on a cylinder is essentially equivalent to quantum mechanics on the manifold of the gauge group. The theory was recently studied from this perspective by Douglas[16], and by Minahan and Polychronakos[17], and shown to be equivalent to a theory of free fermions on the circle. Minahan and Polychronakos showed that by writing the string formulation of the theory in Hamiltonian form, one arrives at the bosonization of the fermion theory. Using the collective field formalism, they related this theory to a $c = 1$ matrix model. It has also been shown by several authors that the introduction of a Wilson loop source in QCD_2 on the cylinder gives a theory of interacting fermions with a Sutherland-type interaction[17,18]. In a related work, Caselle *et al.* related QCD_2 on the cylinder to a Kazakov-Migdal model with periodic boundary conditions[19]. Their approach was to explicitly write the heat kernel on the cylinder in terms of the invariant angles of group elements. An interesting result of their analysis is a formula for the grand canonical partition function of QCD_2 on the cylinder, containing the correct expressions for the partition function for finite values of N.

4.5. Finite N Results

Most of the work described so far connecting gauge theories to string theory has been done in the context of an asymptotic $1/N$ expansion. For the gauge theories in two dimensions, however, the explicit form of the exact solution as a sum over representations makes it possible to consider making such a correspondence for a fixed finite value of N. For a finite value of N, the sum over representations in the partition function must be restricted to the set of irreducible representations of the group. In the case of $SU(N)$, this corresponds to summing over only Young tableaux with less than N rows. By extending the group-theoretic analysis described here to the finite N case, one finds that the effect of this restriction in the summation can be fairly easily described in the string picture. The essential result is that one restricts to only orientation-preserving string maps, and for any genus one must introduce a new type of static singularity point. Only one such point must be introduced for any genus; at this "projection" point, a singularity corresponding to an arbitrary permutation of the covering sheets can occur. The weight of a singularity associated

224

with a permutation σ is given by

$$P^{(N)}(\sigma) = \sum_{R \in Y_n^{(N)}} \frac{d_R}{n!} N^{n-K_\sigma} \chi_R(\sigma), \tag{15}$$

where $Y_n^{(N)}$ is the set of Young tableaux with n boxes and less than N rows, and K_σ is the number of cycles in the permutation σ. The details of the construction of the string theory in the finite N $SU(N)$ gauge theory are described in [20].

5. Open Questions

5.1. Meson Spectrum

It should be possible to reproduce the planar approximation to the meson spectrum derived by 't Hooft from the string point of view. The work of Bars and Hanson[2], and of Strominger[3] represent a partial result in this direction. However, using a purely string-theoretic argument based on the work presented here, one would like to rederive this result, and calculate the lower order corrections to this meson spectrum. This problem presents an interesting and nontrivial challenge.

5.2. Action Formulation

The primary reason that we are interested in studying gauge theories in two dimensions is for the insight which they give into the structure of gauge theories in 4 dimensions. Unfortunately, the formulation presented here of the string theory associated to gauge theories in two dimensions is highly dependent upon the unique characteristics of maps from 2-manifolds to other 2-manifolds. In order to extend this work to higher dimensions, one natural approach is to attempt to describe the theory by an action description where there is no restriction on the string maps allowed. In such a formulation, the fact that maps with folds do not contribute to the partition function would hopefully arise naturally from an integration over fermion zero modes in the theory or some other familiar mechanism. We have several clues to the form that such an action formulation of the string theory might take. Clearly, the action will contain in some way the Nambu action. The signs associated with branch points indicate the possible existence of a fermionic structure in the theory. The existence of infinitesimal tubes and handles presents an obstruction to a holomorphic characterization of the string maps, since no holomorphic map takes an entire handle of a Riemann surface into a point. The invariance of the theory under the group of area-preserving diffeomorphisms indicates that the string action should also have this invariance. Finally, the static singularities associated with Ω-points may give some hint as to some global structure in the theory we are looking for; the similarity between the Ω-points for different gauge groups is particularly striking in this regard. Unfortunately, as yet no one has successfully described the string theory presented here with an action formalism of the desired type, or even made clear progress towards such a formalism.

References

1. G. 't Hooft, *Nucl. Phys.* **B72**, 461 (1974); *Nucl. Phys.* **B75**, 461 (1974).

2. I. Bars and A. Hanson, *Phys. Rev.* **D13**, 1744 (1976); *Nucl. Phys.* **B111**, 413 (1976).

3. V. Kazakov and I. Kostov, *Nucl. Phys.* **B176**, 199 (1980); V. Kazakov, *Nucl. Phys.* **B179**, 283 (1981).

4. N. Bralic, *Phys. Rev.* **D22**, 3090 (1980).

5. A. Strominger, *Phys. Lett.* **101B**, 271 (1981).

6. A. Migdal, *Zh. Eksp. Teor. Fiz.* **69**, 810 (1975) (Sov. Phys. JETP. **42** 413)

7. B. Rusakov, *Mod. Phys. Lett.* **A5**, 693 (1990); E. Witten, *Comm. Math. Phys.* **141**,153 (1991); D. Fine, *Comm. Math. Phys.* **134**, 273-292 (1990); *Comm. Math. Phys.* **140**, 321-338 (1991); M. Blau, G. Thompson, *Int. J. Mod. Phys.* **A7**, 3781-3806 (1992).

8. D. Gross, *Nucl. Phys.* **B400**, 161 (1993).

9. J. Minahan *Phys. Rev* **D47**, 3430 (1993).

10. D. Gross, W. Taylor, *Nucl. Phys.* **B400**, 181 (1993); *Nucl. Phys.* **B403**, 395 (1993).

11. I. K. Kostov, *Continuum QCD_2 in terms of discretized random surfaces with local weights*, preprint SPhT/93-050, HEPTH/9306110, June 1993.

12. M. R. Douglas, V. A. Kazakov *Large N Phase Transition in Continuum QCD_2*, preprint LPTENS-93/20, RU-93-17, HEPTH/9305047, May 1993.

13. S. G. Naculich, H. A. Riggs, H. J. Schnitzer *Mod. Phys. Lett.* **A8**, 2223 (1993); S. Ramgoolam, *Comment on Two-dimensional $O(N)$ and $Sp(N)$ Yang-Mills Theories as String Theories*, preprint YCTP-P16-93, HEPTH/9307085, July 1993.

14. B. Rusakov, *Phys. Lett.* **B303**, 95 (1993).

15. D. Gross and E. Witten, *Phys. Rev.* **D21**, 446(1980).

16. M. R. Douglas, *Conformal Field Theory Techniques for Large N Group Theory*, preprint RU-93-13, NSF-ITP-93-39, HEPTH/9303159, March 1993.

17. J. A. Minahan, A. P. Polychronakos *Equivalence of Two Dimensional QCD and the $c = 1$ matrix model* preprint CERN-TH-6843/93, UVA-HET-93-02, HEPTH/9303153, March 1993; *Interacting Fermion Systems from Two Dimensional QCD* preprint CERN-TH-6994/93, UVA-HET-93-11, HEPTH/9309044, September 1993;

18. A. Gorsky, N. Nekrasov, *Hamiltonian Systems of Calogero Type and Two Dimensional Yang-Mills Theory*, preprint UUITP-6/93, HEPTH/9304047, May 1993.

19. M. Caselle, A. D' Adda, L. Magnea, S. Panzeri, *Two Dimensional QCD is a One Dimensional Kazakov-Migdal Model*, preprint DFTT 15/93, HEPTH/9304015, April 1993.

20. J. Baez, W. Taylor, *Strings and Two Dimensional QCD for finite N*, preprint in preparation.

Some Aspects of Two Dimensional QCD[*]

JOSEPH A. MINAHAN[†]

Department of Physics,
University of Southern California, Los Angeles, CA 90089-0484 USA

and

ALEXIOS P. POLYCHRONAKOS[‡]

Theory Division, CERN
CH-1211, Geneva 23, Switzerland

ABSTRACT

This talk is comprised of two basic parts. In the first part we consider two dimensional QCD with the spatial dimension compactified to a circle. We show that the states in the theory consist of interacting strings that wind around the circle and using rules developed by Gross and Taylor derive the complete Hamiltonian for this theory in the large N limit. We then express this Hamiltonian in terms of a continuous field. For a $U(N)$ gauge group, we recover the collective field Hamiltonian found by Das and Jevicki for a $d = 1$ matrix model, except the spatial coordinate is on a circle. In the second part we consider the gauge theory on the sphere. We then show how the phase transition of Douglas and Kazakov occurs from Witten's point of view that the partition function is a sum over classical saddle points. We also generalize the work of Douglas and Kazakov to other string solutions found for a $U(N)$ gauge group.

[*] Talk presented by J. Minahan at Strings '93, Berkeley, CA
[†] minahan@physics.usc.edu
[‡] poly@dxcern.cern.ch

Over the past year there has been a flurry of activity in two dimensional QCD (QCD$_2$). One direction of this work concerns the string picture of QCD$_2$[1-3]. Another interesting development is the realization that continuum QCD on the sphere has a third order phase transition in the large N limit[4]. In this talk we wish to discuss some issues concerning both of these points. More complete explanations can be found in [5,6]. We also refer you to [7,8,9] where results similar to ours were found.

In [3], Gross and Taylor developed a set of rules for computing the QCD$_2$ string theory partition function. In the first part of this talk we will show that these rules lead to a Das-Jevicki-Sakita collective coordinate Hamiltonian for a unitary matrix model in the large N limit. To begin, consider $SU(N)$ QCD$_2$ living on a torus with area βL, where β is the inverse temperature. The partition function is given by[10]

$$Z = \sum_{\text{reps}} \exp(-\beta L g^2 C_{2R}/N), \tag{1}$$

where the sum is over all representations of $SU(N)$, $g/\sqrt{N}$ is the QCD coupling, and C_{2R} is the quadratic Casimir of the representation. A given representation R is associated with a Young tableau described by m rows, with n_i boxes in row i, which satisfy $n_i \geq n_j$ if $i < j$. C_{2R} is then given by

$$C_{2R} = \frac{N}{2}(n + \frac{\tilde{n}}{N} - \frac{n^2}{N^2}) \qquad \text{where} \qquad n = \sum_{i=1}^{m} n_i, \qquad \tilde{n} = \sum_{i=1}^{m} n_i(n_i - 2i + 1). \tag{2}$$

¿From (1) it is clear that every representation of $SU(N)$ corresponds to a physical state of the theory, with energy $(g^2 L/N)C_{2R}$. A state in representation R is created and destroyed by a Wilson loop in representation R that wraps once around the spatial dimension. To see this, we can consider a cylindrical surface with Euclidean length β and Wilson loops with representations R and R' inserted at the two ends of the cylinder. This partition function is given by

$$Z = \int d\Omega d\Omega' \chi_R(\Omega)\chi_{R'}(\Omega'^{-1}) \sum_{R''} \chi_{R''}(\Omega)\chi_{R''}(\Omega'^{-1})e^{-g^2\beta L C_{2R''}} = \delta_{RR'}e^{-g^2\beta L C_{2R}}, \tag{3}$$

where Ω and Ω' are the $SU(N)$ elements around the circles at the ends of the cylinder and $\chi_R(\Omega)$ and $\chi_{R'}(\Omega'^{-1})$ are the corresponding characters. Clearly, the partition function in (3) represents the propagation of one state into itself over a euclidean time β.

The string theory partition function of QCD_2 describes maps of world-sheets into target spaces. The maps can multiply cover the surface, and such maps can contain branch cuts or small tubes that connect the different sheets of the world-sheet. For a given representation R, the expression $\exp[-(g^2A/2)(n+\tilde{n}/N - n^2/N^2)]$ can be expanded in powers of $1/N$. The leading term is $\exp(-g^2An/2)$, hence this representation describes an n-covered map, with the leading term coming from the integration of the Nambu-Goto action over the world-sheet. The expansion of $\exp[-(g^2A/2)(\tilde{n}/N)]$ is the contribution of the branch cuts connecting the sheets and the expansion of $\exp[-(g^2A/2)(-n^2/N^2)]$ gives the contributions of the tubes and small handles. There are also complex conjugate representations, $\bar{R}$, to consider. The sheets corresponding to these representations have the opposite chirality. Tensor products of R and $\bar{R}'$ have no branch points connecting sheets of opposite chirality but can have tubes that connect such sheets but with an overall minus sign[3].

Now consider the string picture for the cylinder with Wilson loops inserted at the ends. A chiral representation R, with n boxes in its tableau is a linear combination of string states that wrap around the compact dimension a total of n times, all in the same direction. Hence, there could be n strings that wrap once, or one string that wraps around n times. The total number of such states is $P(n)$, the number of partitions of n. The branch points on the world-sheet correspond to interactions where two strings join to form one string or *vice versa*. The tubes correspond to interactions where two strings join at a point and break apart again, or a multiwound string which bumps into itself. Since there are no branch points joining sheets of opposite chirality, two strings of opposite winding will not join to form a single string, nor will a string break into strings with opposite winding. However, two states with opposite winding can have a pointlike interaction, but the sign is opposite to that of two strings with the same winding.

Since the interactions for two strings with opposite winding are almost trivial, we can essentially separate the two sectors. Consider a state with n windings in one direction. Following the work of Gross and Taylor[3], a string state can be described by an element of the permutation group for n elements, S_n. At the spatial point $x = 0$, a label can be assigned to each of the n strands of string. Tracing the strands form $x = 0$ to $x = L$, the strands are mapped to themselves or to different strands, with the mapping described an element s, of S_n. For any t in S_n, the state tst^{-1} corresponds to a relabeling of the strands, hence this state is equivalent to the state described by s. Therefore, the inequivalent states are given by the conjugacy classes of S_n.

We can define an inner product

$$\langle s'|s\rangle = \sum_{t\in S_n} \delta_{s',tst^{-1}} = \frac{n!}{C_s} \qquad \text{if } s \simeq s'$$
$$= 0 \qquad \text{otherwise.} \tag{4}$$

where s and s' are elements of S_n. The symbol $\simeq$ means that the elements are equivalent up to a conjugacy and C_s is the number of elements in the conjugacy class. Each conjugacy class is described by a partition of n, with each element of the class a cycle within the elements of the partition. If $s = s'$, then the elements of S_n which commute with s are those elements which are cycles of s, multiplied by those elements which exchange cycles with equal number of elements. If the partition is given by $\prod_{l=1}^{n}(l)^{nl}$, where $\sum_{l=1}^{n} ln_l = n$, then the order of the subgroup that commutes with s is $\prod_{l=1}^{n}(l)^{nl}n_l!$. Hence this particular state can be written as

$$\prod_{l=1}^{n}(a_l^\dagger)^{n_l}|0\rangle, \tag{5}$$

where $a_l^\dagger$ is the creation operator for a string with winding l and $|0\rangle$ is the vacuum state. The a_l satisfy the commutation relations $[a_l, a_m^\dagger] = |l|\delta_{l,m}$, thus the inner products of these states will reproduce the result in (4).

To leading order in $1/N$, the energy of such a state is given by $g^2 L(n_l + n_r)/2$, where n_l is the number of left windings and n_r is the number of right windings. Hence, the leading order Hamiltonian is given by

$$H_0 = \frac{g^2 L}{2} \sum_{n\neq 0} a_n^\dagger a_n. \tag{6}$$

To compute the interaction term, consider a branch point where two strings join or break apart. As far as the permutations of the strands are concerned, this corresponds to inserting an element of S_n which has one cycle of order 2 and $n-2$ cycles of order 1. One should sum over all possible branch points, which corresponds to summing over the entire conjugacy class of these elements. Therefore, the matrix element describing this interaction is given by

$$\sum_{p\in S_{n2}} \langle s'|p|s\rangle = \sum_{\substack{t\in S_n \\ p\in S_{n2}}} \delta_{s'p,tst^{-1}}, \tag{7}$$

where S_{n2} are the elements of S_n in the conjugacy class with one 2-cycle and the rest 1-cycles. If s is comprised of two cycles of order n_1 and n_2 and s' is comprised of one cycle of order

$n_1 + n_2$, then there is a unique p such that $s'p = s$. Consider the set of elements in S_n which are given by $t = rq$, where $qsq^{-1} = s$ and $r^{-1}s'r = s'$. The elements q form a subgroup of order $n_1 n_2$, while the elements r form a subgroup of order $n_1 + n_2$. Moreover, the conjugates of p, $r^{-1}pr$ form $n_1 + n_2$ distinct elements. Hence the sum in (7) is given by $(n_1 + n_2)n_1 n_2$. s and s' could also have additional cycles, but these are basically spectators as far as p is concerned, so the matrix elements for these states can be determined using (4). Since each branch point comes with a factor $g^2/2N$, and since the branch point can occur anywhere along the circle of length L, then in terms of the creation and annihilation operators, the operator that leads to the matrix element in (7) is

$$\frac{g^2 L}{2N} a^\dagger_{n_1+n_2} a_{n_1} a_{n_2}. \tag{8}$$

Including the windings in both sectors, one finds that the general Hamiltonian describing this class of interactions is given by

$$H_b = \frac{g^2 L}{2N} \left(\sum_{n,n'>0} + \sum_{n,n'<0} \right) (a^\dagger_{n+n'} a_n a_{n'} + \text{c.c.}). \tag{9}$$

Finally, it is easy to see that the interaction term that describes the handles and tubes on the world-sheet is given by $H_t = \frac{g^2 L}{2N^2} \left[\sum_{n>0} (a^\dagger_n a_n - a^\dagger_{-n} a_{-n}) \right]^2$. The operator inside the square brackets counts the net winding number of the state. Actually, we can get rid of this last term by enlarging the gauge group to $U(N)$, which we will do for the rest of this talk.

The creation and annihilation operators for the winding states in (6) and (9) are conjuage to operators which create and destroy momentum states. Hence, define a new length $\tilde{L} = 4\pi/(g^2 L)$, and a momentum variable, $k = 2\pi n/\tilde{L}$, where n is the erstwhile winding number. Letting $a_k = a_n$, the commutation relation becomes $[a_k, a^\dagger_{k'}] = \frac{\tilde{L}}{2\pi}|k|\delta_{k,k'}$. We then define a field $\varphi(x)$ and its canonical conjugate field $\Pi(x)$, where $[\varphi(x), \Pi(y)] = i\delta(x - y)$. We can then write a_k as

$$a_k = \frac{1}{2} \int dx\, e^{-ikx} [\varphi(x) \pm \frac{1}{\pi}\partial\Pi(x)], \qquad k \gtrless 0 \tag{10}$$

which one can easily show satisfies the commutation relations.

We now use these expressions in the full Hamiltonian given in (6) and (9). First substituting k for n and then using the expression for a_k in (10), one finds that the complete

Hamiltonian is given by

$$H = \frac{1}{2\pi} \int dx (\pi^2 \varphi^2 + (\partial \Pi)^2) + \frac{\tilde{L}}{4\pi N} \int dx (\pi^2 \varphi^3 + 3\partial \Pi \varphi \partial \Pi) + \Delta H$$
$$- \frac{1}{4\pi \tilde{L} N} \left\{ \int dx \pi^2 \varphi(x) \left[\int dy \varphi(y) \cot \frac{\pi}{\tilde{L}}(x-y) \right]^2 \right.$$
$$+ \int dx \left[\int dy \partial \Pi(y) \cot \frac{\pi}{\tilde{L}}(x-y) \right] \varphi(x) \left[\int dz \partial \Pi(z) \cot \frac{\pi}{\tilde{L}}(x-z) \right]$$
$$\left. + 2 \int dx \partial \Pi(x) \int dy \varphi(y) \cot \frac{\pi}{\tilde{L}}(x-y) \int dz \partial \Pi(z) \cot \frac{\pi}{\tilde{L}}(x-z) \right\}, \tag{11}$$

where ΔH is a singular term.

The Hamiltonian we are left with is still non-local, but this can be dealt with since QCD_2 contains no zero winding excitations and in fact, no such terms appear in (6) or (9). Therefore, $\varphi(x)$ and $\partial \Pi(x)$ can not contain zero modes. Thus, one must impose the constraints $\int dx \varphi(x) = \int dx \partial \Pi(x) = 0$.

Since the integrals in the constraint are finite, the non-local pieces can be expressed in terms of local terms by using a somewhat modified trick of collective coordinate field theories[11]. Defining $\tilde{f}(x)$ as

$$\tilde{f}(x) = \frac{\pi}{\tilde{L}} \int dx \cot \frac{\pi}{\tilde{L}}(x-y) f(y), \tag{12}$$

one can derive the relation

$$\int dx f g h = \int dx [f \widetilde{gh} + \tilde{f} g \tilde{h} + \tilde{f} \tilde{g} h] + \frac{\pi^2}{\tilde{L}^2} \int dx f(x) \int dy g(y) \int dz h(z). \tag{13}$$

Using (13) and the zero-mode constraint, we are now able to rewrite the Hamiltonian as

$$H = \frac{1}{2\pi} \int dx \left\{ \pi^2 \varphi^2 + (\partial \Pi)^2 + \frac{\tilde{L}}{N} \left[\frac{\pi^2}{3} \varphi^3 + \partial \Pi \varphi \partial \Pi \right] \right\} + \Delta H. \tag{14}$$

Shifting φ to $\varphi + N/\tilde{L}$, the Hamiltonian becomes

$$H = \frac{4}{g^2 LN} \int dx \left\{ \tfrac{1}{2} \partial \Pi \varphi \partial \Pi + \frac{\pi^2}{6} \varphi^3 - \left(\frac{g^2 LN}{4} \right)^2 \varphi \right\} + \Delta H, \tag{15}$$

up to a constant. Moreover, the new constraint becomes

$$\int dx \varphi(x) = \tilde{L} N/\tilde{L} = N. \tag{16}$$

Except for a missing potential term and the fact that the fields live on a circle as opposed to in a box, the Hamiltonian in (15) and the constraint in (16) are precisely those found by Das and Jevicki for the collective coordinate field of the $c = 1$ matrix model[12]. ΔH is the quantum correction to the free energy. ¿From (15) we see that the bare string coupling constant is $4/(g^2 LN)$, thus strong coupling QCD leads to a weak coupling string theory.

In the remaining time, we would like to discuss some more recent work concerning QCD_2 on the sphere. One question to consider is how the result of Douglas and Kazakov[4] relates to some recent work of Witten's on QCD_2. The sphere partition function can be derived from the partition function on the cylinder. Basically the partition function is the amplitude of N fermions all at a point $x = 0$ and time $t = 0$ that propagate back to the same point at time $t = T$. Some elementary arguments give the result

$$Z_{\text{sphere}} = C \sum_{p_i} \prod_{i>j} (p_i - p_j)^2 \exp(-\tfrac{1}{2} g^2 LT \sum_i p_i^2), \tag{17}$$

where the p_i are quantized momentum modes on the circle. At first it would appear that Z_{sphere} is simply the partition function for a $d = 0$ matrix model in a quadratic potential, so one might not expect much of interest to occur. As was shown by Douglas and Kazakov, this is *not* correct. The point is that since the variables p_i that appear in (17) are discrete, the density of eigenvalues are bounded in the classical solutions. When this bound is reached, a phase transition occurs.

There should be a classical field configuration, termed master field, which dominates the path-integral in each phase. In fact, as was shown by Witten[13] using the localization theorem of Duistermaat and Heckman (DH), the full QCD_2 path integral can be written as a suitable sum over classical saddle points. In what follows we will give a very simple demonstration of Witten's result using the fermion picture and identify the classical configuration which dominates, that is, the master field. The key observation is that the exact propagator of a free particle is proportional to the exponential of the action corresponding to the classical (straight) path connecting the initial and final points. Since QCD_2 on the torus is equivalent to N free fermions, its partition function will also be given by an appropriate classical path of the free fermions. Since the particles live on a circle there are several possible classical paths for each, differing by their winding around the circle with fixed initial and final positions. Also, since the particles are identical fermions, the final positions can be permuted, weighted by a factor $(-)^C$, where C is the number of times the paths of the particles cross. The total partition function will then be the (weighted) sum of the actions of all these classical configurations. Since to each path corresponds a (diagonal) matrix $W(t)$, and to that (up to gauge transformations) a classical field configuration satisfying the field equations of motion, this is the sum over saddle points of the action of Witten. For the sphere the same picture holds, with the difference that all paths start and end at the point $x = 0$, and that each path is further weighted by an extra factor, due to the division by the Vandermonde determinants. This is, again, the sum over saddle points of Witten, and the extra weighting factors are the determinants which appear in the DH theorem. These paths are characterized by their winding numbers $\{n_i\}$ (up to permutation) and thus this is a sum of the form

$$Z_{\text{sphere}} = \sum_{n_i} w(n_i) \exp\left(-\frac{2\pi^2}{g^2 LT} \sum_i n_i^2\right) \tag{18}$$

where $w(n_i)$ are the (as yet undetermined) weighting factors.

The easiest way to obtain the full expression in (18) is to Poisson resum (17). Using the formula

$$\sum_n f(n) = \sum_n \tilde{f}(2\pi n) \tag{19}$$

where $f(x)$ is any function and $\tilde{f}$ is its Fourier transform, we obtain

$$Z_{\text{sphere}} = C \sum_{n_i} F_2(2\pi n_i), \tag{20}$$

where

$$\begin{aligned}
F_2(x_i) &= \int \prod_i dp_i\, e^{-i\sum_i x_i p_i} \Delta^2(p_i) \exp(-\tfrac{1}{2}g^2 LT \sum_i p_i^2) \\
&= C \exp(\frac{1}{2g^2 LT} \sum_i x_i^2) \int \prod_i dy_i \prod_{i<j}(y_{ij}^2 - x_{ij}^2) \exp\left(-\frac{1}{2g^2 LT} \sum_i y_i^2\right)
\end{aligned} \tag{21}$$

and where $x_{ij} = x_i - x_j$ and $y_{ij} = y_i - y_j$. Substituting (21) in (20) we recover expression (18), with the weights $w(n_i)$ given by

$$w(n_i) = C \int \prod_i dy_i \prod_{i<j}(y_{ij}^2 - n_{ij}^2) \exp\left(-\frac{1}{2g^2 LT} \sum_i y_i^2\right). \tag{22}$$

The above determines the expression of the sphere partition function in terms of classical saddle point configurations.

In the large N limit one particular classical configuration in (18) should dominate. For the moment ignore the fact that the n_i are discrete and replace the sum in (18) with an integral. Then we can bring the expression for Z_{sphere} into the form

$$Z_{\text{sphere}} = C \int \prod_i ds_i dt_i \Delta(s_i) \Delta(t_i) \exp\left(-\tfrac{1}{2}N\sum_i (s_i^2 + t_i^2)\right), \tag{23}$$

where we rescaled g^2 to g^2/N, in order to have a nontrivial large N limit, and changed variables to $s = (y+2\pi n)/\sqrt{2g^2 LT}$, $t = (y-2\pi n)/\sqrt{2g^2 LT}$. We see that in (23) all explicit dependence on the coupling constant and area have disappeared. The area enters this picture indirectly, through the discreteness of n_i. (In fact, if it were not for this discreteness the above integral would vanish.) Due to this, the plane (s,t) is discrete in the $s - t$ direction and consists of parallel diagonal lines with distance $D = 2\pi/\sqrt{g^2 LT}$. If this spacing is such that the distribution of s and t contributing to the above integral in the large N limit is entirely within the lines $n = 1$ and $n = -1$, then only the sector $n = 0$ contributes. This signals a phase transition. Unlike the quantum case, there is no exclusion principle for the windings n_i and thus no maximum density to be saturated. The phase transition in the classical expansion is more like a Bose condensation to the ground state $n = 0$. To estimate the critical area, substitute the Vandermondes in (23) with their absolute value. This has no effect for the $n_i = 0$ term, while it overestimates the contribution of the other sectors. (23) then becomes the product of two independent integrals involving a gaussian factor and one power of the Vandermonde. The large N saddle point of these integrals is found in a standard way and the distribution of s_i and t_i is, in fact, again a Wigner semicircle with radius $R = \sqrt{2}$. This on the (s,t) plane creates a "square" distribution with side $2R$ in each direction. When this square lies entirely within the lines $n = \pm 1$, the sectors $n \neq 0$ do not contribute and we are in the Boson condensate phase. For an estimate of the critical area, assume that the transition occurs when the corners of the square start touching the lines $n = \pm 1$, that is, when the half-diagonal becomes D. Putting $2R\sqrt{2}/2 = D$ we obtain $g^2 A_{\text{crit}} = \pi^2$. We see that the classical configuration in the weak-coupling (small area) phase is $n_i = 0$. This corresponds to the master field $A = 0$, up to gauge transformations.

Another consideration of QCD_2 on the sphere is a subtlety that occurs if the gauge group is $U(N)$ instead of $SU(N)$. There is one difficulty with the string picture for the $U(N)$ gauge group, namely, the sum is not asymptotic with the exact answer in the limit $1/N \to 0$. This is because there exist states with finite energy, but which will never appear in the perturbative sum in (17). These are the states that correspond to N fermions with momenta shifted by a constant finite amount. Such states have finite energy in the large N limit, but do not show up in a perturbative sum over surfaces since the corresponding

Young tableau has at least N boxes. Moreover, these states are local minima, in the sense that in order to find a state with lower energy, it is necessary to shift the total momentum by a large amount.

We can address this problem by including other sectors in the sum. For finite N, this will eventually lead to overcounting, but the answer will be asymptotic. To this end, let us define Z_m

$$Z_m = \sum_{\text{Reps}} d_R^2 \exp\left(-\frac{Ag^2}{2}\sum_i (n_i + m)\right) \exp\left(-\frac{Ag^2}{2N}\sum_i (n_i - 2i + 1 + m)(n_i + m)\right), \quad (24)$$

where we have basically added m boxes to each row. After some manipulation, Z_m reduces to

$$Z_m = \sum_{\text{Reps}} d_R^2 e^{-\frac{Ag^2}{2}(1+2m/N)n_R} e^{-\frac{Ag^2}{2N}\tilde{n}_R} e^{\frac{-Ag^2 m^2}{2}}, \quad (25)$$

Hence we see that Z_m has the same form as Z_0, except that the area term that appears in front of n has been shifted and there is an extra factor of m^2 in the energy. But the $\tilde{n}$ term is the same, meaning that the Gross-Taylor rules are exactly the same as in the Z_0 case, except that the area that will be used for the Nambu-Goto term has been renormalized, with a different shift for the chiral and antichiral sectors. The sum

$$\mathcal{Z} = \sum_m Z_m \quad (26)$$

is now an asymptotic sum for the QCD_2 string.

We can interpret the sum in (26) as a sum over different classical string solutions, since around each sector m there is a sum over $1/N$, the string coupling. These different sectors are basically the conjugates of the $U(1)$ instanton sectors. We can Poisson resum (26), giving the expression

$$\mathcal{Z} = \sqrt{\frac{\pi}{Ag^2}} \sum_m \sum_{\text{Reps}} d_R^2 e^{-(\frac{Ag^2}{2}+\frac{2\pi im}{N})n_r} e^{-\frac{Ag^2}{2N}\tilde{n}_R} e^{\frac{Ag^2 n_R^2}{2N^2}} e^{-\frac{m^2}{2g^2 A}}. \quad (27)$$

The m^2 term now has a factor of $1/g^2$ in front of it, which is the contribution one would expect from $U(1)$ instantons. Furthermore, the area that now appears in the Nambu-Goto action is now complex and there is also an additional n^2 term, which appears in the $SU(N)$ case and is attributed to contributions from tubes and handles on the world-sheet.

We can then generalize the work of Douglas and Kazakov to these other solutions. Hence, consider the solutions to the equations of motion for the path integral in (17),

$$\frac{A p_i}{2N} = \sum_{j \neq i} \frac{1}{p_i - p_j}, \tag{28}$$

where we have absorbed the QCD coupling into the area and rescaled it by a factor of N. In the large N limit we can define the new variables $x = i/N$, and $h(x) = p_i/N$, giving the new equation,

$$Ah/2 = \mathrm{P} \int d\lambda \frac{u(\lambda)}{h - \lambda}, \tag{29}$$

where $u(\lambda)$ is the density of eigenvalues, $u(\lambda) = dx/d\lambda$. Due to the discreteness of the p_i the density $u(\lambda)$ is bounded, satisfying $u(\lambda) \leq 1$.

DK treated this problem by dividing up the possible real values of λ into three regions, where $u(\lambda) = 0$, $0 < u(\lambda) < 1$ and $u(\lambda) = 1$. They chose the ansatz that the first region occurs for $|\lambda| > a$, the second for $b < |\lambda| < a$, and the third for $|\lambda| < b$. We will modify this somewhat and choose the three regions to be given by $\lambda > a$ or $\lambda < d$, $b < \lambda \leq a$ or $d \leq \lambda < c$, and $c \leq \lambda \leq b$. If we define $\widetilde{u}(\lambda)$ to be the density of eigenvalues minus the contribution from the region where $u(\lambda) = 1$, then (29) can be rewritten as

$$Ah/2 + \log \frac{h - b}{h + b} = \mathrm{P} \int d\lambda \frac{\widetilde{u}(\lambda)}{h - \lambda}, \tag{30}$$

The problem has been reduced to a two cut eigenvalue problem. To solve this, define

$$f(h) = \int d\lambda \frac{\widetilde{u}(\lambda)}{h - \lambda}. \tag{31}$$

$f(h)$ then must satisfy

$$f(h) = \frac{1}{2\pi i} \sqrt{(h - a)(h - d)(h - b)(h - c)} \oint ds \frac{As/2 - \log \frac{b-s}{b+s}}{(h - s)\sqrt{(a - s)(s - d)(b - s)(s - c)}}, \tag{32}$$

where the contour surrounds the cuts from the square roots, but not the cut from the log nor the singularity at h. Pulling the contour in (32) back leads to the equation

$$f(h) = h\frac{A}{2} + \log \frac{h - b}{h - c} - \int_c^b ds \frac{\sqrt{(a - h)(b - h)(c - h)(d - h)}}{(h - s)\sqrt{(a - s)(b - s)(s - c)(s - d)}}. \tag{33}$$

From (31) and properties of elliptic integrals one finds that the density of eigenvalues is given by

$$u(\lambda) = \frac{2}{\pi(\lambda - c)(\lambda - d)} \frac{\sqrt{(a - \lambda)(\lambda - b)(\lambda - c)(\lambda - d)}}{\sqrt{(a - c)(b - d)}}$$
$$\times \left((c - d)\Pi(\frac{b - c}{b - d}\frac{\lambda - d}{\lambda - c}, q) + (\lambda - c)K(q) \right), \tag{34}$$

where K and Π are the complete elliptic integrals of the first and third kind.

Let us now expand the righthand side (33) in powers of $1/h$ and compare it to the expansion in (31). This gives the five following equations

$$A = \frac{4K(q)}{\rho}, \tag{35}$$

$$\frac{c - d}{\rho}\Pi(\alpha, q) = \frac{a + b + c - d}{2\rho}K(q), \tag{36}$$

$$1 = \frac{(a - b - c + d)^2}{4\rho}K(q) + \rho E(q), \tag{37}$$

$$Q = \frac{W}{4} + \frac{K(q)}{16}XYZ, \tag{38}$$

$$F'(A, Q) - 1/24 = \frac{K(q)}{12\rho}(6WXYZ + Y^2Z^2 + X^2Y^2 + X^2Z^2)$$
$$+ \frac{1}{48}(3W^2 + X^2 + Y^2 + Z^2). \tag{39}$$

where $\qquad \rho = \sqrt{(a - c)(b - d)}, \qquad q = \sqrt{\frac{(a - d)(b - c)}{(a - c)(b - d)}}, \qquad \alpha = \sqrt{\frac{b - c}{b - d}},$

$$W = a + b + c + d \qquad\qquad X = a - b - c + d$$
$$Y = a + b - c - d \qquad\qquad Z = a - b + c - d.$$

$E(q)$ is the complete elliptic integral of the second kind, Q is the rescaled $U(1)$ charge of the classical solution and $F'(A, Q)$ is the specific heat.

Given the area A and the charge Q, in principle, one should be able to solve for a, b, c and d using the four equations (35), (36), (37) and (38). However, we should not expect solutions for all possible values of A and Q. For instance, from DK, we know that there are no solutions for $Q = 0$ and $A < \pi^2$. This of course is the weak coupling regime. But for nonzero values of Q, the minimum value of A should be higher. In fact, in the infinite area limit, the upper value of Q is bounded. Naively, this should happen for the sector when the fermion with the smallest or largest momentum is zero. One can show that the maximum charge when the area is near its critical value is $Q_{max} = \frac{32\sqrt{2}}{27\pi^4}(A - A_c)^{3/2}$. Deep in the strong coupling region, the maximum allowed charge is $Q = 1$, which corresponds to the fermion state where all momenta between 0 and N are occupied.

Acknowledgements: The research of J.A.M. was supported in part by D.O.E. grant DE-FG03-84ER-40168. J.A.M. thanks the organizers for the invitation to speak and acknowledges very helpful discussions with I. Bars, S. Chaudhuri, M. Douglas, R. Myers, V. Periwal and W. Taylor.

REFERENCES

1. D. Gross, *Nucl. Phys.* **B400** (1993) 161.

2. J. Minahan, *Phys. Rev. D* **47** (1993) 3430.

3. D. Gross and W. Taylor, *Nucl. Phys.* **B400** (1993) 181; *Nucl. Phys.* **B403** (1993) 395.

4. M. Douglas and V. Kazakov, hep-th/9305047.

5. J. Minahan and A. Polychronakos, *Phys. Lett.* **B312** (1993) 155.

6. J. Minahan and A. Polychronakos, hep-th/9309119

7. M. Douglas, RU-93-13 hepth/9303159.

8. M. Caselle, A. D'Adda, L. Magnea, and S. Panzeri, hep-th/9304015, 1993.

9. M. Caselle, A. D'Adda, L. Magnea, and S. Panzeri, hep-th/9309107, 1993.

10. A. Migdal, *Zh. Eksp. Teor. Fiz.* **69** (1975) 810; B. Rusakov, *Mod. Phys. Lett. A* **5** (1990) 693.

11. A. Jevicki and B. Sakita, *Nucl. Phys.* **165** (1980) 510; B. Sakita, *Quantum Theory of Many-Variable Systems and Fields*, World Scientific, 1985

12. S. Das and A. Jevicki, *Mod. Phys. Lett. A* **5** (1990) 1639.

13. E. Witten, *J. Phys.* **G9** (1992) 303.

14. J. Duistermaat and G. Heckman, *Invent. Math* **69** (1982) 259.

4. String Theory: Compactification, Phenomenology, . . .

Spacetime Topology Change:
The Physics of Calabi-Yau Moduli Space*

Paul S. Aspinwall,[†] Brian R. Greene[‡] and David R. Morrison[§]

We review recent work which has significantly sharpened our geometric understanding and interpretation of the moduli space of certain $N=2$ superconformal field theories. This has resolved some important issues in mirror symmetry and has also established that string theory admits physically smooth processes which can result in a change in topology of the spatial universe.

* Lecture delivered by B.R.G. at Strings '93, Berkeley.

† School of Natural Sciences, Institute for Advanced Study, Princeton, NJ 08540.

‡ School of Natural Sciences, Institute for Advanced Study, Princeton, NJ 08540. On leave from: F.R. Newman Laboratory of Nuclear Studies, Cornell University, Ithaca, NY 14853.

§ School of Mathematics, Institute for Advanced Study, Princeton, NJ 08540. On leave from: Department of Mathematics, Duke University, Box 90320, Durham, NC 27708.

1. Introduction

The essential lesson of general relativity is that the geometrical structure of spacetime is governed by dynamical variables. That is, the metric changes in time according to the Einstein equations. In the usual formulations of general relativity, the spacetime metric is defined on a space of fixed topological type – the "size" and "shape" of the space can smoothly change, but the underlying topology does not. A natural question to ask is whether this formulation is too restrictive; might the topology of space itself be a dynamical variable and hence possibly change in time? This issue has long been speculated upon. Heuristically, one suspects that topology might be able to change by means of the violent curvature fluctuations which would be expected in any quantum theory of gravity. Just as the fluctuations of the magnetic field in a box of size L are on the order of $(\hbar c)^{1/2}/L^2$, those of the curvature of the gravitational field are on the order of $(\frac{\hbar G}{c^3})^{1/2}/L^3$. Thus, on extremely small scales, say $L \sim L_{\text{Planck}}$, huge curvature fluctuations are unsuppressed. One can imagine that such curvature fluctuations could "tear" the fabric of space resulting in a change of topology. The expected discontinuities in physical observables accompanying the discontinuous operation of a change in topology would be hidden, one hopes, behind the smoothing effects of quantum uncertainty. Of course, without a true theory of quantum gravity, one cannot make quantitative sense of such hypothesized processes.

With the advent of string theory, we are led to ask whether any new quantitative light is shed on the issue of topology change. Two works over the last year [1,2] have carried out studies, from somewhat different points of view, which definitively establish that there are *physically smooth* processes in string theory which result in a change in the topology of spacetime. Furthermore, as phenomena in string theory, these processes are not at all exotic. Rather, they correspond to the most basic kind of operation arising in conformal field theory: deformation by a truly marginal operator. From a spacetime point of view, this corresponds to a slow variation in the VEV of a scalar field which has an exactly flat potential.[1] It is crucial to emphasize that these physically smooth topology changing processes occur even at the level of *classical* string theory. It is not, as had been suspected from point particle intuition, that quantum effects give rise to topology change,

---

[1] To avoid confusion, we remark that the present study focuses on static vacuum solutions to string theory. One expects that configurations involving the generic slow variation of such scalar fields are solutions as well.

but, rather, it is the extended structure of the string which bears responsibility for this effect.

We can immediately summarize here the essential content of [1] and [2]. From the viewpoint of classical general relativity or the classical nonlinear sigma model, we know that there are constraints on the metric tensor which appears in the action. Namely, since the metric is used to measure lengths, areas, volumes, etc., it must satisfy a set of positivity conditions. For instance, if we have a nonlinear sigma model on a Kähler target space M with metric (in complex coordinates) $g_{\mu\bar\nu}$, we can write the Kähler form of the metric as $J = ig_{\mu\bar\nu}dX^\mu \wedge dX^{\bar\nu}$ (a real closed 2-form). The latter must satisfy

$$\int_{M_r} J^r > 0 \tag{1.1}$$

where M_r is an r (complex) dimensional submanifold of M and J^r represents the r-fold wedge product of J with itself. The set of real closed 2-forms which satisfy (1.1) is a subset of $H^2(X, \mathbb{R})$ known as the *Kähler* cone and is schematically depicted in figure 1a. Such Kähler forms manifestly span a cone because if J satisfies (1.1) then so does sJ for any positive real s. The burden of [1] and [2] is that, in string theory, (1.1) can be relaxed and still result in perfectly well behaved physics. In fact, the Kähler form of a target Calabi-Yau space is one of the moduli fields of the associated conformal field theory. Investigation of the conformal field theory moduli space reveals that the corresponding geometrical description *necessarily* involves configurations in which the (supposed) Kähler form lies outside of the Kähler cone of the particular Calabi-Yau being studied. In fact, *any and all* choices of an element of $H^2(X, \mathbb{R})$ give rise to well-defined conformal field theories. In [1,2] it was shown that some of these configurations can be interpreted as nonlinear sigma models on Calabi-Yau manifolds of topological type distinct from the original. With respect to this Calabi-Yau of new topology, the Kähler modulus satisfies (1.1) and hence may be thought of as residing in a new Kähler cone which shares a common wall with the original. Furthermore, there is no physical obstruction to continuously deforming the underlying conformal field theory so that its geometrical description passes from one Kähler cone to another and hence results in a change in topology of the target space – i.e. of space itself.

These results were established in [2] by means of mirror symmetry. Hence, in the next section we shall briefly review the phenomenon of mirror manifolds. In section III we will give a discussion of moduli spaces of both conformal theories and Calabi-Yau manifolds in order to fill in a bit more detail required for the discussion of topology change. We will see

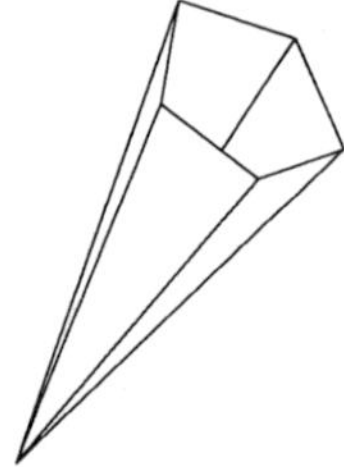

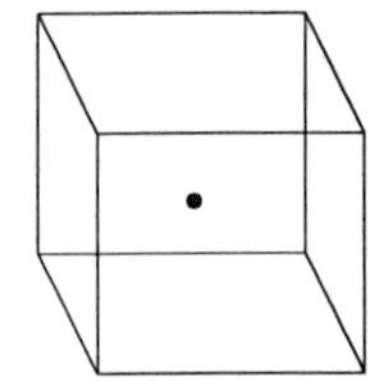

Figure 1a. Kähler cone.   Figure 1b. Domain of w_l's.

that this discussion raises an interesting puzzle whose resolution, discussed in section IV, directly leads to the necessity of physically smooth topology changing processes. In section V we shall verify the abstract discussion of the preceding sections in an explicit example which provides a highly sensitive confirming test of the picture we present. Finally, in section VI we shall give our conclusions.

2. Mirror Manifolds

Nonlinear sigma models on Calabi-Yau target spaces, at their infrared fixed point, provide the geometric interpretation for a class of conformal field theories. As is well known, these conformal field theories have $N = (2, 2)$ world sheet supersymmetry. One can turn this identification around and inquire as to whether every $N = (2, 2)$ superconformal field theory with central charge, say, equal to 9 is interpretable as a nonlinear sigma model with some Calabi-Yau as target. The answer to this question is not known; however, it is known that one conformal field theory can sometimes be interpretable in terms of nonlinear sigma models on *two* very different Calabi-Yau target spaces [3]. The possible existence of this phenomenon was raised in [4] and in [5] based on the fact that there is an unnatural asymmetry between the identification of an abstract conformal field theory with a Calabi-Yau nonlinear sigma model which is resolved when a second Calabi-Yau interpretation exists. Namely, the truly marginal operators in the abstract conformal field theory can be labeled with the $U(1)_L \times U(1)_R$ quantum numbers of the lowest components of the supermultiplet to which they belong. These eigenvalues divide the space of truly marginal operators into those with charges $(1, 1)$ and $(-1, 1)$ (and their complex conjugates). Now, a Calabi-Yau sigma model also has two types of truly marginal operators: the complex structure deformations and the Kähler deformations. Mathematically, these two types are vastly different objects; nonetheless, since they correspond to the truly marginal operators in the associated conformal field theory, the abstract formulation only distinguishes them by the sign of a $U(1)$ charge. It is surprising that an important mathematical distinction finds such a trivial conformal field theory manifestation. It was suggested in [4] and [5] that a natural resolution of this asymmetry would be the existence of a second Calabi-Yau manifold giving rise to the same conformal field theory but with the identification of geometrical deformations and conformal field theory marginal operators reversed (relative to the $U(1)$ charges) with respect to the first Calabi-Yau. One consequence of the existence of such a second Calabi-Yau interpretation is that the Hodge numbers of the first, say M, and those of the second, say $\widetilde{M}$, are related via

$$h_M^{p,q} = h_{\widetilde{M}}^{3-p,q}. \tag{2.1}$$

Although an interesting speculation, there was no evidence for the existence of this phenomenon until the simultaneous works of [6] and [3]. The authors of [6] performed a computer survey of a large number of Calabi-Yau manifolds realized as hypersurfaces in weighted projective four space. They found that almost every Calabi-Yau in the resulting list had a counterpart with Hodge numbers related as above. This falls short of establishing that these pairs of Calabi-Yau manifolds correspond to the same conformal field theory but it is at least consistent with this possibility. In [3], on the other hand, a constructive proof was given for the existence of certain pairs of Calabi-Yau manifolds M and $\widetilde{M}$ whose Hodge numbers are related by (2.1) *and which give rise to isomorphic conformal field theories*. Such pairs of Calabi-Yau spaces were named "mirror manifolds" in [3] because the relation between the Hodge numbers corresponds to a reflection in a diagonal plane of the corresponding Hodge diamonds. The construction of [3] applies to any conformal field theory built up from the $N=2$ minimal models and these include Fermat hypersurfaces in weighted projective space. At the present time, this construction supplies the only known examples of mirror manifolds.[2]

Before proceeding, there are two points which, although not directly relevant to our present study, are worth emphasizing here. First, mirror manifolds are not the first nor the only examples of distinct geometrical spaces which give rise to the same conformal

---

[2] There have been other conjectured constructions of mirror manifolds in both the physics [7] and mathematics [8,9] literatures, but as yet no one has been able to establish that these constructions yield pairs of Calabi-Yau manifolds corresponding to the same conformal field theory.

field theory. For example, string theory on a circle of radius R and on a distinct circle of radius α'/R give rise to isomorphic physics [10]. The same is true for certain pairs of toroidal orbifolds [11]. The most general term describing the phenomenon in quantum geometry of distinct spaces giving rise to identical physical models is *string equivalent spaces*. That is, two distinct background spaces X and Y on which string propagation is physically isomorphic are called *string equivalent*. From this definition, it is clear that M and $\widetilde{M}$ are a mirror pair if they are string equivalent *and* if they are Calabi-Yau manifolds whose Hodge diamonds are mirror reflected and hence related by (2.1). In keeping with this definition, for instance, circles of radii R and α'/R are string equivalent but are not a mirror pair. Second, it has sometimes been asserted that the phenomenon of mirror manifolds amounts to nothing more than the fact that there is a trivial automorphism of $N = (2,2)$ conformal field theory obtained by changing the sign of one of the $U(1)$ charges. There is a misleading imprecision here. It is true that there is a trivial automorphism of these conformal theories arising from such a change in sign. However, this trivial conformal field theory operation has an *equally trivial geometrical interpretation*. Namely, to specify a supersymmetric nonlinear sigma model we need to supply not only a Calabi-Yau manifold but also a vector bundle on it (to which the world sheet fermions couple), meeting certain conditions. The simplest solution to these conditions and the solution implicitly chosen in most studies is that of the tangent bundle to the Calabi-Yau manifold. There is, however another equally valid and physically equivalent choice: the *cotangent* bundle to the same Calabi-Yau space. These two equivalent choices differ, from the conformal field theory viewpoint, by a change in sign of one of the $U(1)$ charges in the theory. Thus, as promised, a trivial conformal field theory operation has a trivial geometric interpretation. *This is not mirror symmetry.* Rather, mirror symmetry is a phenomenon in which the space changes, not simply the bundle. It is true that this isomorphism is proved [3] by making use of the fact that the two relevant conformal theories differ by the trivial automorphism associated with the $U(1)$ charge. However, the existence of such a trivial automorphism does not (at our present level of understanding) by any means establish the existence of mirror manifolds — in fact, as just mentioned, there is a far more trivial geometric interpretation which immediately presents itself.

There are a number of interesting and important implications of mirror symmetry which we will not have time to discuss here. However, one particular result will be useful in our later discussion, so we briefly record it now.

Since a mirror pair M and $\widetilde{M}$ correspond to the same conformal field theory, every correlation function in the latter has two geometric interpretations: one on M and one on $\widetilde{M}$. Typically, these geometric realizations will be quite different; however, since they mathematically represent one and the same correlation function they must be identically equal. This fact gives rise to some highly nontrivial identities between particular geometrical formulas on M and others on $\widetilde{M}$. One such identity, originally shown in [3] and later employed to remarkable ends in [12], arises from the study of three point functions amongst fields $\{O_i\}$ associated with, say, the (chiral, chiral) primary fields in the conformal theory. On $\widetilde{M}$ such fields are associated with harmonic $(0,1)$ forms taking values in the tangent bundle, $B^\alpha_{(i)}$, and it has been shown that [13]

$$\langle O_i O_j O_k \rangle = \int_{\widetilde{M}} \Omega \wedge B^\alpha_{(i)} \wedge B^\beta_{(j)} \wedge B^\gamma_{(k)} \Omega_{\alpha\beta\gamma}. \tag{2.2}$$

On M, the O_i are associated with harmonic $(0,1)$ forms taking values in the cotangent bundle which are isomorphic to harmonic $(1,1)$ forms $A_{(i)}$ and it has been shown that [14,15,12,16]

$$\langle O_i O_j O_k \rangle = \int_M A_{(i)} \wedge A_{(j)} \wedge A_{(k)} + \\ \sum_{m,\{u\}} e^{\int_{\mathbb{P}^1} u_m^* K} \left(\int_{\mathbb{P}^1} u^* A_{(i)} \int_{\mathbb{P}^1} u^* A_{(j)} \int_{\mathbb{P}^1} u^* A_{(k)} \right), \tag{2.3}$$

where $\{u\}$ is the set of holomorphic maps to rational curves on M, $u : \mathbb{P}^1 \to C$ (with C such a holomorphic curve), π_m is an m-fold cover $\mathbb{P}^1 \to \mathbb{P}^1$ and $u_m = u \circ \pi_m$.

As each of these mathematical expressions on the right hand side is equal to the same correlation function in a single conformal field theory, they must be equal to each other. Hence we have

$$\int_{\widetilde{M}} \Omega \wedge B^\alpha_{(i)} \wedge B^\beta_{(j)} \wedge B^\gamma_{(k)} \Omega_{\alpha\beta\gamma} = \tag{2.4}$$

$$\int_M A_{(i)} \wedge A_{(j)} \wedge A_{(k)} + \sum_{m,\{u\}} e^{\int_{\mathbb{P}^1} u_m^* K} \left(\int_{\mathbb{P}^1} u^* A_{(i)} \int_{\mathbb{P}^1} u^* A_{(j)} \int_{\mathbb{P}^1} u^* A_{(k)} \right).$$

In fact, although for ease of discussion we have focused on a single conformal field theory, if we deform that theory to any point in its moduli space, there exist corresponding choices for the Kähler class on M and the complex structure of $\widetilde{M}$ such that this equality continues to hold. Notice that the leading term in (2.2) is the topological intersection form on M and that this term is the only one which contributes to the correlation function in (2.2) if the integral $\int_C K$ goes to infinity, for every rational curve C on M. This occurs if the Kähler form K approaches a "large radius limit" — a concept which will be made precise in the sequel.

3. Moduli Spaces

Quite generally, as mentioned, the conformal field theories we study here come in continuously connected families related via deformations by truly marginal operators. For the $N{=}2$ theories, more specifically, these truly marginal operators come in two varieties which are distinguished by their $U(1) \times U(1)$ charges with the latter being a subalgebra of the $N{=}2$ superconformal algebra. In particular, the two types of marginal operators have charges $(1,1)$ and $(-1,1)$ respectively. (Actually, the marginal operators are chargeless — they lie in supermultiplets whose lowest component has the given charges.) Invoking standard usage, we refer to the space of all conformal theories related by such truly marginal deformations as the *conformal theory moduli space*.

When an $N{=}2$ conformal theory arises from a nonlinear sigma model with a Calabi-Yau target space, the marginal operators just referred to have geometrical counterparts. The two types of marginal operators correspond to the two types of deformations of the Calabi-Yau space which preserve the Calabi-Yau condition (of Ricci flatness). These are deformations of the complex structure and deformations of the Kähler structure. Concretely, one can think of the latter as Ricci flat deformations of the metric of the form $\delta g_{\mu\bar{\nu}}$ and the while the former are of the form $\delta g_{\mu\nu}$.

As our analysis will involve a close study of these moduli spaces, let us now describe each in a bit more detail.

3.1. Kähler Moduli Space

Given a Kähler metric $g_{\mu\bar{\nu}}$ we can construct the Kähler form $J = i g_{\mu\bar{\nu}} dX^\mu \wedge dX^{\bar{\nu}}$. As discussed earlier, the set of allowed J's forms a cone known as the Kähler cone of M. One additional important fact is that string theory instructs us to work not just with J but also with $B = B_{\mu\bar{\nu}}$ the antisymmetric tensor field. The latter, which is a closed two-form, combines with J in the form $B + iJ$ to yield the highest component of a complex chiral multiplet we shall call K. K can therefore be thought of as a *complexified* Kähler form. The precise way in which B enters the conformal field theory is such that if B is replaced by $B + Q$ with $Q \in H^2(M, \mathbb{Z})$, then the resulting physical model does not change. Thus, a convenient way to parametrize the space of allowed and physically distinct K's is to introduce

$$w_l = e^{2\pi i (B_l + i J_l)} \tag{3.1}$$

where we have expressed

$$B + iJ = \sum_l (B_l + iJ_l)e^l \tag{3.2}$$

with the e^l forming an integral basis for $H^2(M, \mathbb{Z})$. The w_l have the invariance of the antisymmetric tensor field under integral shifts built in; the constraint that J lie in the Kähler cone bounds the norm of the w_l. Thus, the Kähler cone and space of allowed and distinct w_l are schematically shown in figures 1a and 1b. Notice that any choice of complexified Kähler form in the interior of figure 1b is physically admissible. Choices of K which correspond to points on the walls in figure 1b (or 1a) correspond to metrics on M which fail to meet (1.1) and hence are degenerate in some manner.

3.2. Complex Structure Moduli Space

All of the Calabi-Yau spaces we shall concern ourselves with here are given by the vanishing locus of homogeneous polynomial constraints in some projective space (or possibly a weighted projective space and products thereof). For ease of discussion, and in preparation for an explicit example we will examine shortly, let's assume we are dealing with a Calabi-Yau manifold given by the vanishing locus of a homogeneous polynomial P of degree d in weighted projective four space $\mathbb{P}^4_{\{k_1,\ldots,k_5\}}$. The Calabi-Yau condition translates into the requirement that $d = \sum_i k_i$. Let's call the homogeneous weighted projective space coordinates $(z_1, \ldots, z_5)$ and write down the most general form for P:

$$P = \sum a_{i_1 i_2 \ldots i_5} z_1^{i_1} \ldots z_5^{i_5} \tag{3.3}$$

where $\sum_j k_j i_j = d$. Different choices for the constants $a_{i_1 i_2 \ldots i_5}$ correspond to different choices for the complex structure of the underlying Calabi-Yau manifold. There are two important points worthy of emphasis in this regard. First, not all choices of the $a_{i_1 i_2 \ldots i_5}$ give rise to distinct complex structures. For instance, distinct choices of the $a_{i_1 i_2 \ldots i_5}$ which can be related by a rescaling of the z_j of the form $z_j \to \lambda_j z_j$ with $\lambda_j \in \mathbb{C}^*$ manifestly correspond to the same complex structure (as they differ only by a trivial coordinate transformation). The most general situation would require that we consider $a_{i_1 i_2 \ldots i_5}$'s related by general linear transformations on the z_j's. Second, not all choices of $a_{i_1 i_2 \ldots i_5}$ give rise to smooth Calabi-Yau manifolds. Specifically, if the $a_{i_1 i_2 \ldots i_5}$ are such that P and $\frac{\partial P}{\partial z_j}$ have a common zero (for all j), then the space given by the vanishing locus of P is not smooth. The set of all choices of the coefficients $a_{i_1 i_2 \ldots i_5}$ which correspond to such singular spaces comprise the *discriminant locus* of the family of Calabi-Yau spaces associated with P. The precise equation of the discriminant locus is generally quite complicated; however, the only fact we need is that it forms a complex codimension one subspace of the complex structure moduli space. From the viewpoint of conformal field theory, the nonlinear sigma model associated to points on the discriminant locus appears to be ill defined. For example, the chiral ring becomes infinite dimensional. It is an interesting and important question to thoroughly understand whether there might be some way of making sense of such theories. For the present purposes, though, all we need to know is that at worst the space of badly behaved physical models is complex codimension one in the complex structure moduli space. We illustrate the form of the complex structure moduli space in figure 2.

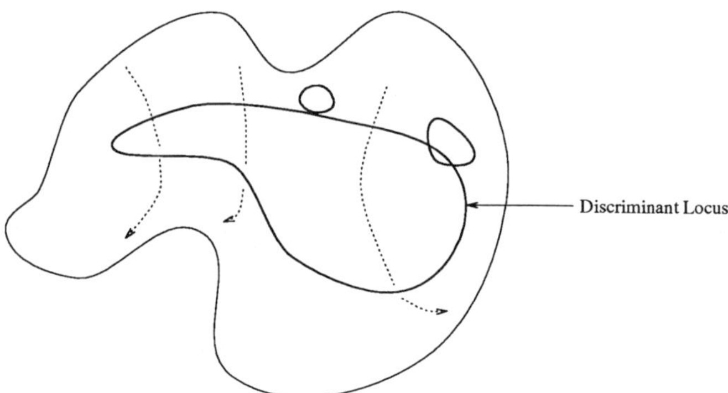

Figure 2. The moduli space of complex structures.

3.3. Implications of Mirror Manifolds

Locally the moduli space of Calabi-Yau deformations is a product space of the complex and Kähler deformations (in fact, up to subtleties which will not be relevant here, we can think of the moduli space as a global product). Thus, we expect

$$\mathcal{M}_{\text{CFT}} \equiv \mathcal{M}_{\text{complex structure}} \times \mathcal{M}_{\text{Kähler structure}} \tag{3.4}$$

with $\mathcal{M}_{(\ldots)}$ denoting the moduli space of $(\ldots)$. Pictorially, we can paraphrase this by saying that the conformal field theory moduli space is expected to be the product of figure 1b and figure 2.

This, in fact, was the picture which had emerged from much work over the last few years and was generally accepted. The advent of mirror symmetry, however, raised a serious puzzle related to this description (as first observed in [17]). Let M and $\widetilde{M}$ be a mirror pair of Calabi-Yau spaces. As we discussed before, such a pair correspond to isomorphic conformal theories with the explicit isomorphism being a change in sign of, say, the right moving $U(1)$ charge. From our description of the moduli space, it then follows that the moduli space of Kähler structures on M should be isomorphic to the moduli space of complex structures on $\widetilde{M}$ and vice versa. That is, both M and $\widetilde{M}$ correspond to the same family of conformal theories and hence yield the same moduli space on the left hand side of (3.4). Therefore, the right hand side of (3.4) must also be the same for both M and $\widetilde{M}$. The explicit isomorphism of mirror symmetry shows this to be true with the two factors on the right hand side of (3.4) being interchanged for M relative to $\widetilde{M}$.

The isomorphism of the Kähler moduli space of one Calabi-Yau and the complex structure of its mirror is a statement which appears to be in direct conflict with the form of figure 1b and that of figure 2. Namely, the former is a bounded domain while the latter is a quasi-projective variety. More concretely, the subspace of theories which appear possibly to be badly behaved are the boundary points in figure 1b (where the metric on the associated Calabi-Yau fails to meet (1.1)) and the points on the discriminant locus in figure 2. The former are real codimension 1 while the latter are real codimension 2. Therefore, how can these two spaces be isomorphic as implied by mirror symmetry?

4. Topology Change

As the puzzle raised in the last section was phrased in terms of those points in the moduli space which have the potential to correspond to badly behaved theories, it proves worthwhile to study the nature of such points in more detail. We will first do this from the point of view of the Kähler moduli space of M.

Consider a path in the Kähler moduli space which begins deep in the interior and moves towards and finally reaches a boundary wall as illustrated in figure 3. More specifically, we follow a path in which the area of a $\mathbb{P}^1$ (a rational curve) on M is continuously shrunk down to zero, attaining the latter value on the wall itself. The question we ask ourselves is: does this choice for the Kähler form on M yield an ill defined conformal theory and furthermore, what would happen if we try to extend our path beyond the wall where it appears that the area of the rational curve would become negative? (We note the

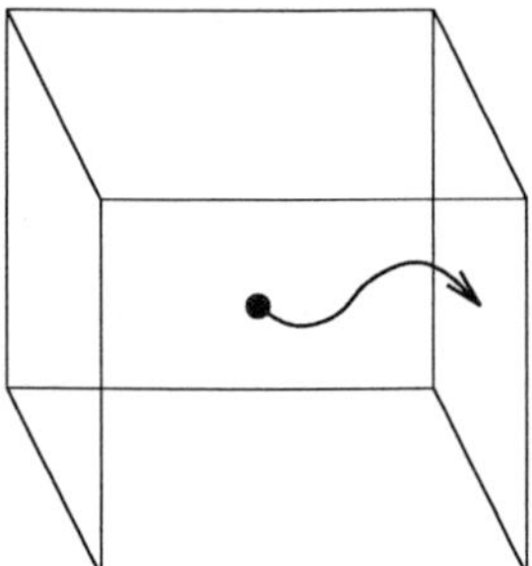

Figure 3. A path to the wall.

linguistically awkward phrase "area of a curve" arises since we are dealing with complex curves which therefore are real dimension two.)

As a prelude to answering this physical question, we note that precisely this operation is well known and thoroughly studied from the viewpoint of mathematics. Namely, in algebraic geometry there is an operation called a *flop* in which the area of a rational curve is shrunk down to zero (*blown down*) and then expanded back to positive volume (*blown up*) in a "transverse" direction. Typically (although not always) this operation results in a change of the topology of the space in which the curve is embedded. Thus, when we say that the blown up curve has positive volume we mean positive with respect to the Kähler metric on the new ambient space. That is, the flop operation involves first following a path like that in figure 3 which blows the curve down, and then continuing through the wall (as in figure 4) by blowing the curve up to positive volume on a new Calabi-Yau space. The latter space, M' also has a Kähler cone whose complexification in the exponentiated w_l coordinates is another bounded domain. Thus, the operation of the flop corresponds to a path in moduli space beginning in the Kähler cone of M, passing through one of its walls and landing in the *adjoining* Kähler cone of M'. Although M and M' can be topologically distinct, their Hodge numbers are the same; they differ in more subtle topological invariants such as the intersection form governing the classical homology ring. Mathematically, they are said to be topologically distinct but in the same birational equivalence class.

The mathematical formulation of what it means to pass to a wall in the Kähler moduli space has led us to a more detailed framework for studying the corresponding description in conformal field theory. We see that from the mathematical point of view, distinct

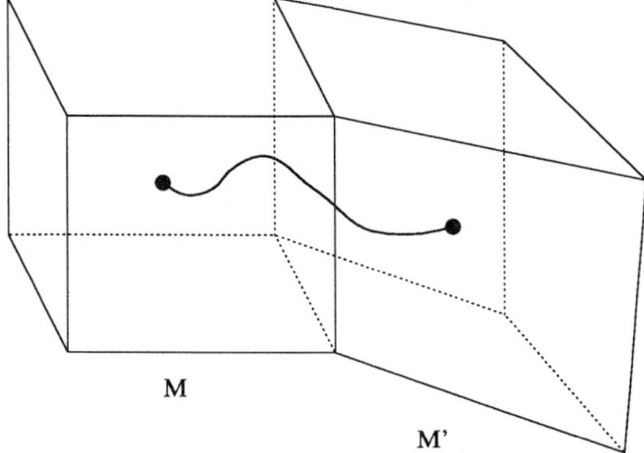

Figure 4. A topology-changing path.

Kähler moduli spaces naturally adjoin along common walls. We can rephrase our initial motivating question of two paragraphs ago as: does the operation of flopping a rational curve (and thereby changing the topology of the Calabi-Yau under study) have a physical manifestation? That is, does a path such as that in figure 4 correspond to a family of well behaved conformal theories?

This is a hard question to answer directly because our main tool for analyzing nonlinear sigma models is perturbation theory. The expansion parameters of such perturbative studies are of the form $\sqrt{\alpha'}/R$ where R refers to the set of Kähler moduli on the target manifold. Now, when we approach or reach a wall in the Kähler moduli space, at least one such moduli field R is going to zero (namely the one which sets the size of the blown down rational curve). Hence, sigma model perturbation theory breaks down and we are hard pressed to answer directly whether the associated conformal theory makes nonperturbative sense.

This situation — one in which we require a nonperturbative understanding of observables on M — is tailor made for an analysis based upon mirror symmetry. Perturbation theory breaks down on M because of the degenerate (or nearly degenerate) choice of its Kähler structure. Note that all of our discussion could be carried through for any convenient (smooth) choice of its complex structure. Via mirror symmetry, this implies that the relevant analysis for answering the question raised two paragraphs ago should be carried out on $\widetilde{M}$ for a particular form of the *complex structure* (namely, that which is mirror

254

to the degenerate Kähler structure on M) but for any convenient choice of the Kähler structure. The latter, though, determines the applicability of sigma model perturbation theory on $\widetilde{M}$. Thus, we can choose this Kähler structure to be arbitrarily "large" (that is, distant from any walls in the Kähler cone) and hence arrange things so that we can completely trust perturbative reasoning. In other words, by using mirror symmetry we have rephrased the difficult and necessarily nonperturbative question of whether conformal field theory continues to make sense for degenerating Kähler structures in terms of a purely perturbative question on the mirror manifold.

This latter perturbative question is one which is easy to answer and, in fact, we have already done so in our discussion of the complex structure moduli space. For large values of the Kähler structure (again, this simply means that we are far from the walls of the Kähler cone) the only choices of the complex structure which yield (possibly) badly behaved conformal theories are those which lie on the discriminant locus. As noted earlier, the discriminant locus is complex codimension one in the moduli space (real codimension two). Thus, the complex structure moduli space is, in particular, path connected. Any two points can be joined by a path which only passes through well behaved theories; in fact, the generic path in the complex structure moduli space has the latter property. This is the answer to our question. By mirror symmetry, this conclusion must hold for a generic path in Kähler moduli space and hence it would seem that a topology changing path such as that of figure 4 (by a suitable small jiggle at worst) is a physically well behaved process. Even though the metric degenerates, the physics of string theory continues to make sense. We are already familiar from the foundational work on orbifolds [18] that degenerate metrics can lead to sensible string physics. Now we see that physically sensible degenerations of other types (associated to flops) can alter the topology of the universe. In fact, the operation being described — deformation by a truly marginal operator — is amongst the most basic and common physical processes in conformal field theory.

To summarize the picture of moduli space which has emerged from this discussion we refer to figure 5. The conformal field theory moduli space is geometrically interpretable in terms of the product of a complex structure moduli space and an enlarged Kähler moduli space $\mathcal{M}_{\text{enlarged Kähler}}$. The latter contains numerous complexified Kähler cones of birationally equivalent yet topologically distinct Calabi-Yau manifolds adjoined along common walls.[3] There are two such geometric interpretations, via mirror symmetry, with

---

[3] The union of such regions constitutes what we call the "partially enlarged" Kähler moduli space. The enlarged Kähler moduli space includes additional regions as we shall mention shortly.

the roles of complex structure and Kähler structure being interchanged. This is also indicated in figure 5.

We should stress that from an abstract point of view this is a compelling picture. Although we do not have time or space to discuss it here, the augmentation of the Kähler moduli space in the manner presented (and, more precisely, as we will generalize shortly) gives it a mathematical structure which is *identical* to that of the complex structure moduli space of its mirror. In the important case of Calabi-Yau's which are toric hypersurfaces, both of these moduli spaces are realized as identical compact *toric varieties*. Hence, the picture presented resolves the previous troubling asymmetry between the structure of these two spaces which are predicted to be isomorphic by mirror symmetry.

Although compelling, we have not proven that the picture we are presenting is correct. We have found a natural mathematical structure in algebraic geometry which *if* realized by the physics of conformal field theory resolves some thorny issues in mirror symmetry. We have not established, as yet, whether conformal field theory makes use of this compelling mathematical structure. If conformal field theory does avail itself of this structure, though, there is a very precise and concrete conclusion we can draw: every point in the (partially) enlarged Kähler moduli space of M must correspond under mirror symmetry to some point in the complex structure moduli space of $\widetilde{M}$. This implies, of course, that any and all observables calculated in the theories associated to these corresponding points must be identically equal. Let's concentrate on the three point functions we introduced earlier in (2.4). As we discussed, if we choose a point in the Kähler moduli space for which the instanton corrections are suppressed, the correlation function approaches the topological intersection form on the Calabi-Yau manifold. For ease of calculation, we shall study the correlation functions of (2.3) in this limit. This analysis will be similar to that presented in [19] although in this case in the (partially) enlarged Kähler moduli space, there is not a single unique "large radius" point of the sort we are looking for. Rather, every cell in the (partially) enlarged moduli space supplies us with one such point. Since these cells are the complexified Kähler cones of topologically distinct spaces, the intersection forms associated with these large radius points are different. If the moduli space picture we are presenting in figure 5 is correct, then there must be points in the complex structure moduli space of the mirror whose correlation functions exactly reproduce each and every one of these intersection forms. This is a precise and concrete statement whose veracity would provide a strong verification of the picture presented in figure 5. In the next section we carry out this verification in a particular example.

256

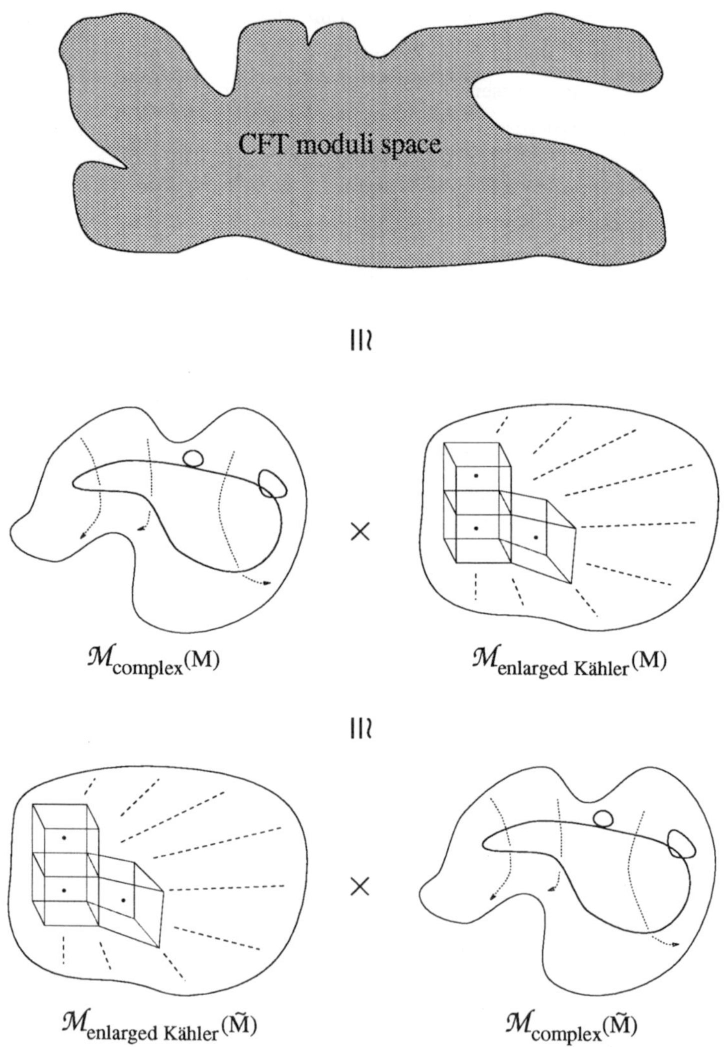

Figure 5. The conformal field theory moduli space.

5. An Example

In this section we briefly carry out the abstract program discussed in the last few sections in a specific example. We will see that the delicate predictions just discussed can be explicitly verified.

We focus on the Calabi-Yau manifold M given by the vanishing locus of a degree 18 homogeneous polynomial in the weighted projective space $\mathbb{P}^4_{\{6,6,3,2,1\}}$ and its mirror $\widetilde{M}$. For the former we can take the polynomial constraint to be

$$z_0^3 + z_1^3 + z_2^6 + z_3^9 + z_4^{18} + a_0 z_0 z_1 z_2 z_3 z_4 = 0 \tag{5.1}$$

where the z_i are the homogeneous weighted space coordinates and a_0 is a large and positive constant (whose value, in fact, is inconsequential to the calculations which follow). The mirror to this family of Calabi-Yau spaces is constructed via the method of [3] by taking an orbifold of M by the maximal scaling symmetry group $\mathbb{Z}_3 \times \mathbb{Z}_3 \times \mathbb{Z}_3$.

A study of the Kähler structure of M reveals that there are five cells in its (partially) enlarged Kähler moduli space, each corresponding to a sigma model on a smooth topologically distinct Calabi-Yau manifold. In each of these cells there is a large radius point for which instanton corrections are suppressed and hence the correlation functions of (2.3) are just the intersection numbers of the respective Calabi-Yau's. We have calculated these for each of the five birationally equivalent yet topologically distinct Calabi-Yau spaces and we record the results in table 1. To avoid having to deal with issues associated with normalizing fields in the subsequent discussion, in table 1 we have chosen to list our results in terms of ratios of correlation functions for which such normalizations are irrelevant. (The D_i and H are divisors on M, corresponding to elements in $H^1(M, T^*)$ by Poincaré duality.)

Resolution	Δ_1	Δ_2	Δ_3	Δ_4	Δ_5
$\dfrac{(D_1^3)(D_4^3)}{(D_1^2 D_4)(D_1 D_4^2)}$	-7	$0/0$	$0/0$	∞	9
$\dfrac{(D_2^2 D_4)(D_3^2 D_4)}{(D_2 D_3 D_4)(D_2 D_3 D_4)}$	2	4	0	$0/0$	$0/0$
$\dfrac{(D_2 D_3 D_4)(H D_2^2)}{(D_2^2 D_4)(H D_2 D_3)}$	1	1	1	0	$0/0$
$\dfrac{(D_2 D_3 D_4)(H D_1^2)}{(D_1^2 D_4)(H D_2 D_3)}$	2	1	∞	$0/0$	0

Table 1: Ratios of intersection numbers

Following the discussion of the last section, our goal now is to find five limit points in the complex structure moduli space of $\widetilde{M}$ such that appropriate ratios of correlation functions yield the same results as in table 1. To do so, we note that the most general complex structure on $\widetilde{M}$ can be written

$$W = z_0^3 + z_1^3 + z_2^6 + z_3^9 + z_4^{18}$$
$$+ a_0 z_0 z_1 z_2 z_3 z_4 + a_1 z_2^3 z_4^9 + a_2 z_3^6 z_4^6 + a_3 z_3^3 z_4^{12} + a_4 z_2^3 z_3^3 z_4^3 = 0. \tag{5.2}$$

We will describe these limit points by parametrizing the complex structure as $a_i = s^{r_i}$ for real parameters s and r_i and we send s to infinity. The limit points are therefore distinguished by the *rates* at which the a_i approach infinity. Our task, therefore, is to find appropriate values for the r_i (if they exist) such that we obtain mirrors to the five large radius Calabi-Yau spaces of the last paragraph. The technique we use to do this is to describe both the complex structure moduli space of $\widetilde{M}$ and the enlarged Kähler moduli space of M in terms of toric geometry. This description, at a fundamental level, makes it manifest that these two moduli spaces are isomorphic. We do not have time to present such analysis here — rather, we refer the reader to [2]. For the present purpose we note that a direct outcome of this analysis is a prediction for five choices of the vector $(r_0, \ldots, r_4)$ which should yield the desired mirrors. As we have discussed, a sensitive test of these predictions is to calculate the mirror of the ratios of correlation functions in table 1 (using (2.2) and the method of [20]) for each of these complex structure limits and see if we get the same answers. We have done this and we show the results in table 2. Note that in the limit s goes to infinity we get precisely the same results. (The φ_i are elements of $H^1(\widetilde{M}, T)$.)

Resolution	Δ_1	Δ_2	Δ_3	Δ_4	Δ_5
Direction	$(\frac{11}{9}, 1, \frac{4}{3}, \frac{5}{3}, \frac{5}{3})$	$(\frac{7}{6}, \frac{1}{2}, 1, 1, \frac{5}{2})$	$(\frac{3}{2}, \frac{1}{2}, 2, 3, \frac{3}{2})$	$(\frac{13}{9}, 2, \frac{5}{3}, \frac{4}{3}, \frac{4}{3})$	$(\frac{11}{6}, \frac{7}{2}, 1, 1, \frac{3}{2})$
$\dfrac{\langle \varphi_1^3 \rangle \langle \varphi_4^3 \rangle}{\langle \varphi_1^2 \varphi_4 \rangle \langle \varphi_1 \varphi_4^2 \rangle}$	$-7 - 181s^{-1} + \ldots$			$-\frac{2}{5}s^2 - \frac{129}{250}s + \ldots$	$9 + 289s^{-1} + \ldots$
$\dfrac{\langle \varphi_2^2 \varphi_4 \rangle \langle \varphi_3^2 \varphi_4 \rangle}{\langle \varphi_2 \varphi_3 \varphi_4 \rangle \langle \varphi_2 \varphi_3 \varphi_4 \rangle}$	$2 - 5s^{-1} + \ldots$	$4 - 22s^{-1} + \ldots$	$0 + 2s^{-1} + \ldots$		
$\dfrac{\langle \varphi_2 \varphi_3 \varphi_4 \rangle \langle \varphi_0 \varphi_2^2 \rangle}{\langle \varphi_2^2 \varphi_4 \rangle \langle \varphi_0 \varphi_2 \varphi_3 \rangle}$	$1 + \frac{1}{2}s^{-1} + \ldots$	$1 + \frac{3}{2}s^{-2} + \ldots$	$1 + 4s^{-1} + \ldots$	$0 - 2s^{-1} + \ldots$	
$\dfrac{\langle \varphi_2 \varphi_3 \varphi_4 \rangle \langle \varphi_0 \varphi_1^2 \rangle}{\langle \varphi_1^2 \varphi_4 \rangle \langle \varphi_0 \varphi_2 \varphi_3 \rangle}$	$2 + 27s^{-1} + \ldots$	$1 - \frac{1}{2}s^{-1} + \ldots$	$-2s - 33 + \ldots$		$0 + 4s^{-2} + \ldots$

Table 2: Asymptotic ratios of 3-point functions

This, in conjunction with the abstract and general isomorphism we find between the complex structure moduli space of a Calabi-Yau and the *enlarged* Kähler moduli space of its mirror (using toric geometry), provides us with strong evidence that our understanding of Calabi-Yau conformal field theory moduli space is correct. In particular, as our earlier discussion has emphasized, this implies that the basic operation of deformation by a truly marginal operator (from a spacetime point of view, this corresponds to a slow variation in the vacuum expectation value of a scalar field with an exactly flat potential) can result in a change in the topology of the Calabi-Yau target space. This discontinuous mathematical change, however, is perfectly smooth from the point of view of physics. In fact, using mirror symmetry, such an evolution can be reinterpreted as a smooth, topology preserving, change in the "shape" (complex structure) of the mirror space.

There are two important points we need to mention. First, for ease of discussion we have focused on the case in which the only deformations are those associated with the Kähler structure of M and, correspondingly, only the complex structure of $\widetilde{M}$. This may have given the incorrect impression that the topology changing transitions under study can always be reinterpreted in a topology preserving manner in the mirror description. The generic situation, however, is one in which the complex structure *and* the Kähler structure of M and $\widetilde{M}$ both change. Again, from a spacetime point of view this simply corresponds to a slow variation of the expectation values of a set of scalar fields with flat potentials. Under such circumstances, topology change can occur in both the original and the mirror description. Our reasoning will ensure that such changes are physically smooth. Clearly there is *no* interpretation — the original or the mirror — which can avoid the topology changing character of the processes.

Second, we have used the terms "enlarged" and "partially enlarged" in our discussion of the Kähler moduli space. We now briefly indicate the distinction. The central result of the present work (and that of Witten [1]) is that the proper geometric interpretation of conformal field theory moduli space requires that we augment the previously held notion of a single complexified Kähler cone associated with a single topological type of Calabi-Yau space. In the previous sections we have focused on *part* of the requisite augmentation: we need to include the complexified Kähler cones of Calabi-Yau spaces related to the original by flops of rational curves (of course, it is arbitrary as to which Calabi-Yau we call the original). These Kähler cones adjoin each other along common walls. The space so created is the *partially* enlarged Kähler moduli space. It turns out, though, that conformal field theory moduli space requires that even more regions be added. Equivalently, the partially enlarged Kähler moduli space is only a subregion of the moduli space which is mirror to the complex structure moduli space of the mirror Calabi-Yau manifold. The extra regions which need to be added arise directly from the toric geometric description and were first identified in the two dimensional supersymmetric gauge theory approach of [1]. These regions correspond to the moduli spaces of conformal theories on orbifolds of the original smooth Calabi-Yau, Landau-Ginzburg orbifolds, gauged Landau-Ginzburg theories and hybrids of the above. The union of all of these regions (which also join along common walls) constitutes the *enlarged* Kähler moduli space. For instance, in the example studied in section V we found that there were five regions in the partially enlarged Kähler moduli space. The enlarged Kähler moduli space, as it turns out, has 100 regions. One of these is a Landau-Ginzburg orbifold region, 27 of these are sigma models on Calabi-Yau orbifolds, and 67 of these are hybrid theories consisting of Landau-Ginzburg models fibered over various compact spaces. It is worthwhile emphasizing that in contrast with previously held notions, orbifold theories are not simply boundary points in the moduli space of smooth Calabi-Yau sigma models but, rather, they have their own regions in the enlarged Kähler moduli space and hence are more on equal footing with the smooth examples.

6. Conclusions

In this talk we have focused on the proper and complete geometric interpretation of points in the moduli spaces of a class of $N{=}2$ superconformal field theories. We have uncovered a surprisingly rich structure. Previously it was believed that any such moduli space was interpretable in terms of the complex structure and Kähler structure of an associated Calabi-Yau manifold. We now see that this is but a small fragment of the full story. The Kähler moduli space must be augmented to the *enlarged* Kähler moduli space of which the former is one of many cells adjoined along various common walls. The new cells correspond to nonlinear sigma models on smooth topologically distinct Calabi-Yau's (related by flops of rational curves) as well Calabi-Yau orbifolds, Landau-Ginzburg theories and hybrid models.

There are a number of implications of this augmented picture. First, we have shown that deformations by truly marginal operators can take us in a physically smooth manner from any region to any other. In particular, this means that the topology of the target space (the universe in a theory of strings) can change with no more exotic physical impact than mere geometric expansion. Second, the enlargement of the moduli space harmoniously clears up some troubling puzzles in mirror symmetry. More precisely, whereas it proved difficult to understand how the complexified Kähler moduli space of a Calabi-Yau could be isomorphic to the complex structure moduli space of its mirror, there is a manifest isomorphism when the enlarged Kähler moduli space is used. Third, we have seen how we are led to a shift in perspective regarding orbifolds. Rather than being boundary points in moduli space — and hence less than generic — orbifolds occupy their own regions just like the smooth Calabi-Yau manifolds. Fourth and finally, we have mentioned that the enlarged Kähler moduli space generally contains numerous regions whose most natural interpretation is not in terms of nonlinear sigma model field theories. The geometric properties of such models are presently under study.

Acknowledgements

The work of P.S.A. was supported by DOE grant DE-FG02-90ER40542, the work of B.R.G. was supported by the Ambrose Monell Foundation, by a National Young Investigator award and by the Alfred P. Sloan foundation, and the work of D.R.M. was supported by NSF grant DMS-9304580 and by an American Mathematical Society Centennial Fellowship.

References

[1] E. Witten, Nucl. Phys. **B403** (1993) 159.

[2] P.S. Aspinwall, B.R. Greene and D.R. Morrison, Phys. Lett. **303B** (1993) 249; "Calabi-Yau moduli space, mirror manifolds and spacetime topology change in string theory", IASSNS-HEP-93/38, CLNS-93/1236.

[3] B.R. Greene and M.R. Plesser, Nucl. Phys. **B338** (1990) 15.

[4] L. Dixon, in *Superstrings, Unified Theories and Cosmology 1987* (G. Furlan et al., eds.), World Scientific, 1988, p. 67.

[5] W. Lerche, C. Vafa and N. Warner, Nucl. Phys. **B324** (1989) 427.

[6] P. Candelas, M. Lynker and R. Schimmrigk, Nucl. Phys. **B341** (1990) 383.

[7] P. Berglund and T. Hübsch, in *Essays on Mirror Manifolds*, (S.-T. Yau, editor), International Press, 1992, p. 388.

[8] V. Batyrev, "Dual Polyhedra and Mirror Symmetry for Calabi-Yau Hypersurfaces in Toric Varieties", Essen preprint, November 18, 1992.

[9] L. Borisov, "Towards the Mirror Symmetry for Calabi-Yau Complete intersections in Gorenstein Toric Fano Varieties", Univ. of Michigan preprint, October, 1993.

[10] K. Kikkawa, M. Yamasaki, Phys. Lett. **149B** (1984) 357; N. Sakai, I. Senda, Prog. Theor. Phys. **75** (1986) 692.

[11] R. Dijkgraaf, E. Verlinde and H. Verlinde, Commun. Math. Phys. **115** (1988) 649; A. Shapere and F. Wilczek, Nucl. Phys. **B230** (1989) 669.

[12] P. Candelas, X.C. de la Ossa, P.S. Green, and L. Parkes, Phys. Lett. **258B** (1991) 118; Nucl. Phys. **B359** (1991) 21.

[13] A. Strominger and E. Witten, Commun. Math. Phys. **101** (1985) 341.

[14] A. Strominger, Phys. Rev. Lett. **55** (1985) 2547.

[15] M. Dine, N. Seiberg, X.-G. Wen, and E. Witten, Nucl. Phys. **B278** (1987) 769; Nucl. Phys. **B289** (1987) 319.

[16] P.S. Aspinwall and D.R. Morrison, Commun. Math. Phys. **151** (1993) 245.

[17] P.S. Aspinwall and C.A. Lütken, Nucl. Phys. **B355** (1991) 482.

[18] L. Dixon, J. Harvey, C. Vafa and E. Witten, Nucl. Phys. **B261** (1985) 678; Nucl. Phys. **B274** (1986) 285.

[19] P.S. Aspinwall, C.A. Lütken, and G.G. Ross, Phys. Lett. **241B** (1990) 373.

[20] P. Candelas, Nucl. Phys. **B298** (1988) 458.

A New Approach to the Green-Schwarz Superstring

Nathan Berkovits
Maths Dept., King's College, Strand, London, WC2R 2LS, United Kingdom

Abstract

By replacing two of the bosonic scalar superfields of the N=2 string with fermionic scalar superfields (which shifts $d_{critical}$ from (2,2) to (9,1)), a quadratic action for the ten-dimensional Green-Schwarz superstring is obtained. Using the usual N=2 super-Virasoro ghosts, one can construct a BRST operator, picture-changing operators, and covariant vertex operators for the Green-Schwarz superstring. Superstring scattering amplitudes with an arbitrary number of loops and external massless states are then calculated by evaluating correlation functions of these vertex operators on N=2 super-Riemann surfaces, and integrating over the N=2 super-moduli.

These multiloop superstring amplitudes have been proven to be SO(9,1) super-Poincaré invariant (by constructing the super-Poincaré generators and writing the amplitudes in manifest SO(9,1) notation), unitary (by showing agreement with amplitudes obtained using the light-cone gauge Green-Schwarz formalism), and finite (by explicitly checking for divergences in the amplitudes when the Riemann surface degenerates). There is no multiloop ambiguity in these Green-Schwarz scattering amplitudes since spacetime-supersymmetry is manifest (there is no sum over spin structures), and therefore the moduli space can be compactified.

There are two main reasons for studying the Green-Schwarz superstring. The first reason is to obtain a more efficient method for calculating superstring amplitudes than is possible using the NSR formalism. Because of manifest spacetime-supersymmetry in the Green-Schwarz formalism, there is no need to perform GSO projections or sum over spin structures. As will be described later in this talk, this simplifies calculations involving external fermions (there are no square-root cuts and fermionic vertex operators do not require ghosts), removes the multiloop ambiguity (the moduli space can be compactified since there is no need for a cutoff associated with summing over spin structures), and allows a direct proof of finiteness (the amplitudes can be explicitly checked to be free of divergences).[1-4] Although these amplitude calculations are perturbative in the string coupling constant, they may be useful for studying possible quantum corrections to general relativity, or for finding new symmetries in superstring theory.

The second reason for studying the Green-Schwarz superstring is to get a better understanding of super-Yang-Mills and supergravity. Since two-dimensional non-linear sigma models provide a natural framework for studying the massless fields of the string, one would expect that by constructing the appropriate sigma model for the Green-Schwarz superstring, one could learn something about off-shell super-Yang-Mills and supergravity. For the case of the four-dimensional Green-Schwarz superstring, this expectation has been confirmed (four-dimensional super-Yang-Mills and supergravity fields are scalar potentials and Kahler vectors of N=2 non-linear sigma models)[5,6], while for the ten-dimensional case, work is still in progress. Hopefully, a better understanding of these massless supersymmetric field theories will be useful in unraveling how superstring theory produces a consistent quantum theory of gravity.

Until recently, the only method available for calculating Green-Schwarz superstring amplitudes was the light-cone gauge method in which all world-sheet symmetries, including conformal invariance, are non-manifest.[7-9] Amplitudes are calculated in this method by evaluating correlation functions of light-cone vertex operators and interaction-point operators on a two-dimensional surface, and integrating over the positions of the vertex-operator punctures and the moduli of the surface. These interaction-point operators are required for Lorentz invariance and can be understood as the light-cone analog of picture-changing operators, which come from integrating out the world-sheet gravitini.[1] However unlike picture-changing operators, their locations on the surface are completely fixed, and in fact are extremely complicated functions of the puncture positions, the surface moduli, and the P^+ momenta of the external states. Because of this complication, the light-cone Green-Schwarz method has not yet produced manifestly Lorentz-invariant expressions for any amplitude with more than one loop or more than four external states. An additional problem of the light-cone Green-Schwarz method is that in order to remove non-physical divergences when interaction-points coincide, one needs to introduce contact-terms whose precise form has not yet been determined.[10]

Recently, a new method[2] has been developed for calculating Green-Schwarz superstring amplitudes which starts from the manifestly N=(2,0) worldsheet supersymmetric action:

$$S = \int d^2z d^2\kappa [X^a \partial_{\bar{z}} X^{\bar{a}} + W^- \partial_{\bar{z}} \Theta^+ - W^+ \partial_{\bar{z}} \Theta^-], \tag{1}$$

subject to the chirality constraints, $D_+ X^{\bar{a}} = D_+ \Theta^- = D_- X^a = D_- \Theta^+ = 0$ ($a, \bar{a}$ range from 1 to A and $D_\pm = \partial_{\kappa^\pm} + \kappa^\mp \partial_z$), the N=2 superconformal constraint, $D_+ X^a D_- X^{\bar{a}} + D_- W^- D_+ \Theta^+ + D_+ W^+ D_- \Theta^- = 0$, and the global constraint, $D_+ \Theta^+ D_- \Theta^- - \frac{1}{2}(\Theta^+ \partial_z \Theta^- + \Theta^- \partial_z \Theta^+) = \partial_z X^+$ for some real superfield X^+. This action is manifestly invariant under an $SU(A) \times U(1)$ subset of the super-Poincaré group, which includes the $2A + 2$ spacetime-supersymmetry transformations, $\delta X^a = \epsilon^a \Theta^+$, $\delta X^{\bar{a}} = \epsilon^{\bar{a}} \Theta^-$, $\delta \Theta^+ = \epsilon^+$, $\delta \Theta^- = \epsilon^-$, $\delta W^- = \epsilon^a X^{\bar{a}}$, $\delta W^+ = \epsilon^{\bar{a}} X^a$. Although only the heterotic superstring (ignoring lattice degrees of freedom) will be discussed in this talk, the new method easily generalizes to non-heterotic Green-Schwarz superstrings. However up to now, it is possible only for the heterotic case to obtain this quadratic action by partially gauge-fixing a manifestly Lorentz-covariant action.[1,11]

The action of equation 1 is just the usual N=2 string action,[12,13] except that the two "longtitudinal" pairs of bosonic scalar superfields, $(X^0, X^{\bar{0}})$ and $(X^d, X^{\bar{d}})$, have been exchanged for two pairs of fermionic scalar superfields, (Θ^+, W^-) and (Θ^-, W^+). Note that although the $W^\pm$ superfields are not chiral, only $D_+ W^+$ and $D_- W^-$ contribute to the action. Since the central charge contribution of the longtitudinal superfields is thereby flipped from $+6$ to -6, the critical N=2 string now contains four pairs of transverse superfields, $X^a = x^a + \kappa^+ \Gamma^a$ and $X^{\bar{a}} = x^{\bar{a}} + \kappa^- \Gamma^{\bar{a}}$ for $a, \bar{a} = 1$ to 4, which describe the usual light-cone Green-Schwarz content of eight scalar bosons and eight spin-$\frac{1}{2}$ fermions (because of spectral flow, these eight spin-$\frac{1}{2}$ fermions could alternatively be treated as four spin-0 and four spin-1 fermions).

The longtitudinal degrees of freedom of the Green-Schwarz superstring are described by the four bosonic and four fermionic components of the superfields, $\Theta^\pm = \theta^\pm + \kappa^\pm \lambda^\pm$ and $D_+ W^\pm = w^\pm + \kappa^\mp \varepsilon^\pm$, which are subject to the global constraint, $\lambda^+ \lambda^- - \frac{1}{2}(\theta^+ \partial_z \theta^- + \theta^- \partial_z \theta^+) = \partial_z x^+$ for some real field x^+. Since this global constraint on $\lambda^\pm$ commutes with the N=2 stress-energy tensor, it does not affect the conformal anomaly calculation. However in order to construct vertex operators and calculate scattering amplitudes, it is necessary to solve the constraint in the following way:

$$\lambda^+ = (\partial_z x^+ + \frac{1}{2}(\theta^+ \partial_z \theta^- + \theta^- \partial_z \theta^+))e^{h^+} + e^{-h^-}, \quad \lambda^- = e^{-h^+}, \tag{2}$$

$$w^+ = e^{h^+}[\partial_z(h^+ + h^-) + x^-(\partial_z x^+ + \frac{1}{2}(\theta^+ \partial_z \theta^- + \theta^- \partial_z \theta^+)] + x^- e^{-h^-}, \quad w^- = x^- e^{-h^+},$$

where x^+ and x^- are the usual longtitudinal bosonic scalars, and $h^\pm$ are chiral bosons of screening charge -1 with the operator-product $\partial_z h^+ \partial_z h^- = z^{-2}$ (note that the relationship between $w^\pm$, $\lambda^\pm$ and $x^\pm$ closely resembles the twistor condition of Penrose,[14,15] $w = x\lambda$). Because this field redefinition preserves the operator-product relations of $\lambda^\pm$ and $w^\pm$, the free-field action of equation 1 is still a free-field action when $\lambda^\pm$ and $w^\pm$ are replaced by $x^\pm$ and $h^\pm$. As will be shown later, equation 2 has the effect of transforming a global constraint on the fields into a constraint on the U(1) moduli of the surface.[2]

Using the full SO(9,1) vector, x^μ, one can now construct N=2 superconformally invariant Lorentz generators out of the free fields.[3] This is done by combining the free fields $[x^\mu, \Gamma^a, \Gamma^{\bar{a}}, \theta^\pm, \varepsilon^\pm, h^\pm]$ into a pair of bosonic worldsheet-scalar spacetime-vector superfields, $X^\mu_\pm$ for $\mu = 0$ to 9, and a pair of fermionic worldsheet-scalar spacetime-spinor superfields, $\Theta^\alpha_\pm$ for $\alpha = 1$ to 16. These covariantly transforming superfields satisfy the chirality conditions, $D_+ X^\mu_- = D_+ \Theta^\alpha_- = D_- X^\mu_+ = D_- \Theta^\alpha_+ = 0$, and the relation, $(X^\mu_+ - X^\mu_-)\gamma^{\alpha\beta}_\mu = \Theta^\alpha_+ \Theta^\beta_-$. They are not free fields on the worldsheet and are similar in philosophy to the NSR spin field,[16] which can only be expressed in terms of free fields by breaking the manifest Lorentz covariance (note however that unlike the NSR spin field, they are constructed entirely out of matter fields). In terms of these covariantly transforming superfields, the Lorentz generators are $M^{\mu\nu} = \int dz d^2 \kappa X^\mu_+ X^\nu_-$, the supersymmetry generators are $S_\alpha = \int dz d^2 \kappa X^\mu_+ \gamma_{\mu\alpha\beta} \Theta^\beta_-$, and the massless vertex operators are $\eta_\mu \int dz d^2 \kappa \Theta^\alpha_+ \gamma^\mu_{\alpha\beta} \Theta^\beta_- e^{ik_\nu X^\nu_+}$ for the vector boson, and $u_\alpha \int dz d^2 \kappa \Theta^\alpha_+ e^{ik_\nu X^\nu_-}$ for the spinor fermion.[3]

The next step is to introduce N=2 super-Virasoro ghosts and a BRST charge, to bosonize the super-reparametization ghosts, and to construct picture-changing operators as in the NSR formalism.[16] Scattering amplitudes are then calculated by evaluating correlation functions of the vertex operators on N=2 super-Riemann surfaces of genus g, and integrating over the N=2 super-moduli of the surface. Integration over the fermionic moduli brings down $2(2g-2)$ picture-changing operators whose locations are arbitrary on the surface. The region of integration for the U(1) moduli, m_j for $j = 1$ to g, is the Jacobian variety $C^g/(Z^g + \tau Z^g)$, where $e^{2\pi i m_j}$ measures the change in phase when Γ^a or e^{h^+} goes around the jth B-cycle on the surface.

The only correlation function that is not straightforward to evaluate is that of the $h^\pm$ fields. Since $e^{h^+ - h^-}$ has negative conformal weight, it is not possible on a general surface to define the holomorphic correlation function $< \exp(\sum_k (c_k h^-(z_k) + d_k h^+(z_k))) >$ without allowing unphysical poles (this situation also arises with the fields, ϕ, that come from bosonizing the bosonic super-reparameterization ghosts, but in that case, the residues of the unphysical poles are BRST trivial[17]). However on a surface with the special values of the U(1) moduli, $m_j = \sum_k c_k \int^{z_k} \omega_j$ where ω_j is the jth canonical holomorphic one-form, the unphysical poles are not present. Therefore, it is necessary for BRST invariance to define the correlation function $< \exp(\sum_k (c_k h^-(z_k) + d_k h^+(z_k))) >$ to be proportional to $\prod_{j=1}^g \delta(m_j - \sum_k c_k \int^{z_k} \omega_j)$, where the proportionality factor is

266

completely fixed by its conformal properties. In this way, the global constraint on the $\lambda^\pm$ fields has transformed into a restriction on the U(1) moduli of the surface.[2]

After performing the functional integral over the free fields, one is left with an integrand which depends on the $2(2g-2)$ arbitrary points where the picture-changing operators have been inserted, and which must be integrated over the usual $(6g-6+2N)$ Teichmuller parameters and puncture locations. Although the integrand is not manifestly Lorentz-invariant, it is possible to use knowledge of the Lorentz invariance (recall that Lorentz generators have been constructed which commute with the BRST operator and transform the vertex operators covariantly) to rewrite the integrand in a manifestly SO(9,1) invariant form.[3]

To prove that these Lorentz-invariant expressions are unitary, one must show agreement with amplitudes obtained using the manifestly unitary light-cone Green-Schwarz formalism.[7-9] This has been done by choosing the moduli (and corresponding Beltrami differentials) for the surface to be the light-cone interaction points, twists, and internal P^+ momenta, and choosing the locations of the picture-changing operators to be precisely the interaction points. With this choice for the surface moduli, it is straightforward to show that the path integrals over the longtitudinal matter fields $[x^\pm, h^\pm, \theta^\pm, \varepsilon^\pm]$ precisely cancel the path integrals over the ghost fields $[c, b, u, v, \beta^\pm, \gamma^\pm]$, since each bosonic/fermionic longtitudinal matter field couples in the same way as a corresponding fermionic/bosonic ghost field.[4] The remaining path integrals over the transverse matter fields give precisely the light-cone gauge prescription for calculating amplitudes, with the transverse part of the picture-changing operators becoming the light-cone Green-Schwarz interaction-point operators.[1]

Finally, it has been shown[4] that these superstring amplitudes (for external massless boson-boson states in the Type II superstring) are finite by explicitly checking for divergences when the Riemann surface degenerates either into one surface with a pinched handle, or into two surfaces connected by a thin tube (all other possibly divergent regions in moduli space are related to these by modular transformations). Note that a similar analysis has not yet been done in the NSR formalism since before summing over spin structures, the NSR amplitudes are not divergence-free.[18,19] When the surface degeneracy corresponds to a pinched handle of radius R, the amplitude behaves like $(log R)^5 R^{-1} dR$ as $R \to 0$, and is therefore divergence-free. When the surface degenerates into two surfaces connected by a tube of radius R, the amplitude factorizes into $A_1 A_2 R^{k^2-1} dR$ as $R \to 0$, where k^μ is the momentum flowing through the tube. If there are vertex operators on both surfaces, this divergence corresponds to the physical massless pole that is present in all field theories containing massless particles. In the case when all vertex operators are located on one of the surfaces, A_1 vanishes since there are no zero modes for the $\theta^\pm$ path integrals (these are two of the manifest spacetime supersymmetries), and the amplitude is divergence-free.

It should be noted that under a change in the locations of the picture-changing operators, the integrand of the scattering amplitude changes by a total derivative in the Teichmuller parameters which can lead to surface term contributions if the moduli space has a boundary. For example, in the NSR formalism the need to sum over spin structures before obtaining a finite amplitude means that a cutoff in the moduli space has to be introduced, which causes a multiloop ambiguity in the amplitude.[20] However such an ambiguity does not occur in the above Green-Schwarz amplitudes since the integrand is well-behaved when the surface degenerates, and therefore there is no need to introduce a cutoff and the moduli space contains no boundary.

References

(1) Berkovits,N., Nucl.Phys.B379 (1992), p.96., hep-th bulletin board 9201004.

(2) Berkovits,N., Nucl.Phys.B395 (1993), p.77, hep-th 9208035.

(3) Berkovits,N., Phys.Lett.B300 (1993), p.53, hep-th 9211025.

(4) Berkovits,N., Finiteness and Unitarity of Lorentz-Covariant Green-Schwarz Superstring Amplitudes, King's College preprint KCL-TH-93-6, March 1993, submitted to Nucl.Phys.B, hep-th 9303122.

(5) Delduc,F. and Sokatchev,E., Class.Quant.Grav.9 (1992), p.361.

(6) Berkovits,N., Phys.Lett.B304 (1993), p.249, hep-th 9303025.

(7) Green,M.B. and Schwarz,J.H., Nucl.Phys.B243 (1984), p.475.

(8) Mandelstam,S., Prog.Theor.Phys.Suppl.86 (1986), p.163.

(9) Restuccia,A. and Taylor,J.G., Phys.Rep.174 (1989), p.283.

(10) Greensite,J. and Klinkhamer,F.R., Nucl.Phys.B291 (1987), p.557.

(11) Tonin,M., Phys.Lett.B266 (1991), p.312.

(12) Ademollo,M., Brink,L., D'Adda,A., D'Auria,R., Napolitano,E., Sciuto,S., Del Giudice,E., DiVecchia,P., Ferrara,S., Gliozzi, F., Musto,R., Pettorini,R., and Schwarz,J., Nucl.Phys.B111 (1976), p.77.

(13) Ooguri,H. and Vafa,C., Nucl.Phys.B361 (1991), p.469.

(14) Penrose,R. and MacCallum,M.A.H., Phys.Rep.6C (1972), p.241.

(15) Sorokin,D.P., Tkach,V.I., Volkov,D.V., and Zheltukhin,A.A., Phys.Lett.B216 (1989), p.302.

(16) Friedan,D., Martinec,E., and Shenker,S., Nucl.Phys.B271 (1986), p.93.

(17) Verlinde,E. and Verlinde,H., Phys.Lett.B192 (1987), p.95.

(18) Christofano,G., Musto,R., Nicodemi,F., and Pettorino,R., Phys.Lett.B217 (1989), p.59.

(19) Mandelstam,S., Phys.Lett.B277 (1992), p.82.

(20) Atick,J. and Sen,A., Nucl.Phys.B296 (1988), p.157.

Topics In String Phenomenology[*]

MICHAEL DINE

Santa Cruz Institute for Particle Physics
University of California, Santa Cruz, CA 95064

Abstract

We consider two questions in string "phenomenology." First, are there any generic string predictions? Second, are there any general lessons which string theory suggests for thinking about low energy models, particularly in the framework of supersymmetry? Among the topics we consider are the squark and slepton spectrum, flavor symmetries, discrete symmetries including CP, and Peccei-Quinn symmetries. We also note that in some cases, discrete symmetries can be used to constrain the form of supersymmetry breaking.

---

[*] Invited Talk Presented at Conference STRINGS 93, University of California, Berkeley 1993
[†] Work supported in part by the U.S. Department of Energy.

1. Introduction

The title of this talk is rather presumptious. In my view, we are far from possessing any string phenomenology; we still have very little idea how the observable world might emerge from string theory. This is not to minimize the fact that many compactifications of string theory have been discovered with desirable features: standard model gauge groups, three generations of quarks and leptons, low energy supersymmetry, intricate patterns of discrete symmetry and more. Yet in detail, it is probably safe to say that all of these models have serious flaws. Moreover, we don't presently have a clue as to how or why one of these models might be picked out over another or why the cosmological constant should vanish after supersymmetry breaking.

As a result, my goals today will be far more modest. I won't review the many interesting efforts to develop a detailed string phenomenology. Rather I will

1. Look for features of string compactifications which might be generic, leading to qualitative or quantitative predictions

2. Seek insights from string theory into more conventional model building.

Virtually all of what I have to say will be in the framework of models with low energy supersymmetry. In the first category, I will consider the minimal supersymmetric standard model. This model has become a paradigm for a low energy supersymmetric world. Yet it rests on a set of strong assumptions about the underlying microscopic theory. Generically, these assumptions are not true in string theory. Recently, however, Kaplunovsky and Louis have noted that there is one scenario for string dynamics in which they *are* true.[1] In this scenario, one obtains a two (or three) parameter description of the low energy world, and predicts a significant degree of degeneracy among squarks and sleptons (necessary to suppress rare processes). We will also describe alternative, string-inspired scenarios for obtaining squark degeneracy based on flavor symmetries.[2,3,4]

Next I will turn to a variety of questions in the second category, mostly involving symmetries. Recently there has been renewed discussion of the plausibility of global symmetries, both continuous and discrete. In string theory, it has been known for some time that there are no (unbroken) continuous global symmetries.[5] String theory exhibits intricate patterns of discrete symmetries, and it is often conjectured that these symmetries are also gauge symmetries. We will see, however, that this is not always the case. String perturbation theory often exhibits anomalous, global discrete symmetries. Such symmetries could be of great phenomenological importance; they could, for example, lead to $m_u = 0$. (Other scenarios, with stringy features, which lead naturally to $m_u = 0$ have been considered in ref. 6.) These symmetries are closely related to certain gauged, non-anomalous discrete symmetries. These gauge symmetries are spontaneously broken; the model-independent axion transforms under them non-linearly. As a result, they would seem to be rather uninteresting. In fact, however, we will see that using these symmetries, one can hope to restrict the form of any non-perturbative superpotential which might be generated in string theory. Indeed, we will see that under these circumstances, one can argue (subject to some plausible assumptions) that "gluino condensation" is the largest supersymmetry-violating effect at weak coupling. These considerations may bear on the possibility of a "duality symmetry" involving the dila-

270

ton, discussed by Joanne Cohn and John Schwarz at this meeting,[7] as well as on the general question of "stringy non-perturbative effects."[8]

Finally, we will comment on CP and the strong CP problem. We will see, first, that CP is a discrete gauge symmetry in string theory (it can be thought of as a combination of a general coordinate transformation and a non-Abelian gauge transformation in the higher dimensional space). As a result, there can be no non-perturbative CP-violating parameters. CP-violation is necessarily spontaneous, and any CP-violation is in principle, calculable. It is probably premature to develop a detailed theory of the origin of the KM angle in string theory. But it is of interest to consider how the strong CP problem might be resolved in string theory. We have already remarked that string theory may be a framework in which to understand $m_u = 0$. We will comment on various aspects of axions in string theory, particularly in light of recent criticisms of the axion idea.[9] We will also see that, quite generally in the framework of supersymmetric models (not just strings), it is difficult to implement scenarios of the Nelson-Barr type, in which θ is arranged to be zero at tree level.[10,11] The problem is that generically radiative corrections to θ are unacceptably large, unless the soft breakings satisfy certain striking constraints.[2]

2. Supersymmetry Breaking and the Problem of the Dilaton

It is appropriate to begin by reviewing the suggestions which have been made for how supersymmetry breaking might arise in string theory. These revolve large around the dynamics of "gaugino condensation" in some hidden sector gauge group.[12] The basic idea is very simple. Suppose that below the Planck scale the model contains a gauge group, $\mathcal{G}$, under which none of the matter fields transform. The scale of the group is

$$\Lambda \sim e^{-\frac{8\pi^2}{g^2 b_o}}.$$

$$(2.1)$$

At this scale one expects that gluinos condense, i.e.

$$< \lambda\lambda > \propto \Lambda^3.$$

$$(2.2)$$

In the low energy theory, the dilaton superfield, S, couples universally to the gauge fields. In particular, the auxiliary component of S, F_S, couples to $\lambda\lambda$, so one finds a potential

$$V \sim |< \lambda\lambda >|^2.$$

$$(2.3)$$

However, in string theory, the coupling, g, is itself determined by the expectation value of the dilaton, $g \sim e^{-D}$. Thus V is a potential for the dilaton. This potential can be understood as arising from a non-perturbatively generated superpotential of the form

$$W \sim e^{-S/b_o}$$

$$(2.4)$$

where $S = e^{-D} + ia$, a being the axion field.

Unfortunately, however, while the appearance of this superpotential constitutes dynamical supersymmetry breaking, at least at weak coupling, this potential has no minimum except at zero coupling, i.e. infinite value of the dilaton vev.[13] One can object that perhaps this problem is an artifact of our assumptions. Indeed, except in special cases (to be discussed below) there is no argument that high energy, non-perturbative string effects cannot be *larger* than any effects which can be seen in the low energy theory. For example, integrating out massive string modes might lead to a superpotential with a different dependence on the dilaton than that of eqn. (2.4). A little thought, however, makes clear that the problem of the runaway dilaton is generic; the potential will always go to zero in weak coupling. One can hope that there is a minimum at strong coupling. Such a prospect, however, is disturbing, not merely because we have at present no idea how to treat strongly coupled string theories, but also because it is precisely the features of string theory at weak coupling which make it so attractive.

There has been at least one interesting proposal to remedy this problem.[14] Suppose one has several hidden sector groups, each producing a gluino condensate. Then, for some choices of phase of the condensate, the non-perturbative superpotential has the form

$$W = \alpha e^{-aS} - \beta e^{-bS}. \tag{2.5}$$

Now, because of the negative sign, one can have a local minimum in the potential. If $a \approx b$, then it is possible that this minimum lies at large S, i.e. weak coupling. Note, however, that there is no parameter at our disposal which can be made arbitrarily small. Instead, one tries to find compactifications which lead to numerically large values of S, values which, from a field-theoretic perspective, one expects to correspond to weak coupling.* In practice, examples have been constructed with rather small effective coupling. However, even though supersymmetry turns out to be broken in some of these minima, the cosmological constant is typically non-zero even in the leading approximation.

---

* One can debate whether, in the absence of a small parameter which can formally be taken to zero, there can be a sense in which weak coupling is valid. To fully justify this procedure one must be able to argue that even though the effects of the high energy theory are not under control, one can still integrate them out, obtaining an effective action of a known form, and that the largest supersymmetry breaking effects are those which are visible in the low energy theory. In the case of anomalous discrete symmetries, discussed below, this may not be necessary.

3. The Minimal Supersymmetric Standard Model

The MSSM, as noted in the introduction, has become the paradigm for a supersymmetric model of nature. In this section, I would like to review the features of this model. We will see that the basic assumptions of the model are very strong, but that if supersymmetry breaking in string theory takes one very particular form, string theory can yield precisely such a theory.

The particle content of theMSSM is that of a supersymmetrized ordinary standard model with two Higgs doublets (i.e. one takes the gauge group to be $SU(3) \times SU(2) \times U(1)$, introduces a chiral superfield for every quark and lepton, and introduces two $SU(2)$ doublet fields, H_1 and H_2). Of course, in string models one typically obtains additional fields, such as gauge singlet chiral fields and additional gauge interactions, but such a spectrum might emerge from a string model.

To specify the model, it is also necessary to make some statements about soft breakings. There are a variety of experimental constraints on these parameters. Obviously, the masses of the superpartners must be large enough that these particles not have been seen. Presumably, these masses should not be arbitrarily large, since in that case it will be necessary to fine tune parameters in order to obtain electroweak breaking. Operationally, the most severe constraint of this type arises from corrections to the H_1 mass containing a top squark; these are proportional to $3m_{\tilde{t}}^2 g_t^2/(16\pi^2)$ times a logarithmic factor. Here g_t is the top quark Yukawa coupling, and $m_{\tilde{t}}^2$ is the top squark mass. One also requires significant degeneracy of squark masses to avoid flavor changing neutral currents (FCNC's). Indeed, from the real part of the $K - \bar{K}$ mass difference, one obtains a constraint (ref. 15)

$$\frac{\delta m^2}{m_{susy}^2} < 10^{-2} - 10^{-3} \tag{3.1}$$

where δm^2 represents a typical splitting and m_{susy} a typical susy-breaking mass.

Without specifying further details of the soft breakings, it should be noted that this model has already scored some striking successes.[16] Perhaps most dramatic is the successful unification of gauge (and some Yukawa) couplings which results in this picture.[17] Other good features include the presence of suitable dark matter candidates.

Of course, the model as it stands possesses a large number of free parameters, and a number of theoretical approaches have been adopted to narrowing this parameter space. The most common assumption is that all of the squarks and sleptons are degenerate at the unification scale, and that the soft-breaking cubic couplings are simply proportional to the superpotential. One then evolves to low energies using the renormalization group, trading one parameter for M_Z. This leaves four parameters (excluding m_t). Further constraints which are often imposed include: $m_b = m_\tau$ at the high scale; presence of a suitable dark matter candidate, and absence of fine tuning. Such programs lead to suggestive values of the soft-breaking parameters.[16]

These are strong assumptions, and they are not easy to justify within the most popular framework for supersymmetry model building: supergravity theories with supersymmetry broken in a hidden sector. In such models one has a set of "hidden sector fields," Φ and "visible sector fields, " Q^I(quarks, leptons, etc.; here we are adopting the notation of ref. 1). The general theory of this type is specified by three functions: the Kahler potential, $K(\phi, \phi^*)$, the superpotential, $W(\phi)$, and a function $f(\phi)$ which describes the coupling of matter to gauge fields. In this approach, one assumes that some of the hidden sector auxiliary fields obtain vacuum expectation values, $F_\Phi \sim m_{3/2} M_p$. Taylor expanding the Kahler potential about the origin of the visible sector fields,

$$K = K(\Phi^*, \Phi) + Z_{\bar{I}J}(\Phi^*, \Phi)Q^{*I}Q^J + (\frac{1}{2}H_{IJ}Q^I Q^J + cc.) + ... \qquad (3.2)$$

The term $Z_{\bar{I}J}$ is the origin of the squark masses. Squark degeneracy means

$$Z_{\bar{I}J} \propto \delta_{\bar{I}J}. \qquad (3.3)$$

But there is no reason, in general, for this condition to hold; there is certainly no symmetry of the low energy theory which enforces it. There is no reason one would expect this to hold generically in string theory; indeed, Ibanez and Lust[18] have shown in particular orbifold examples that such a relation does not hold.

While a number of suggestions have been made over the years to understand degeneracy, Kaplunovsky and Louis have recently pointed out a possible stringy solution.[1] These authors tried to study the question of soft breakings in the context of string theory, without making detailed assumptions about the origin of supersymmetry breaking. They assumed only

1. The potential has a stable minimum at reasonably weak coupling.

2. $V \approx 0$ at the minimum

3. SUSY breaking is primarily due to expectation values either for the moduli or the dilaton, $< F_M >$ or $< F_S >$.

These assumptions are modest in the sense that they are probably the minimal assumptions required for any plausible phenomenology; whether or not these conditions are actually achieved in string theory is another matter. Without further assumptions, one can make only modest statements (see V. Kaplunovsky's talk at this meeting). In particular, there is no explanation for squark degeneracy. However, these authors noted that if one assumes $< F_S >\gg< F_M >$, then, because of the universality of the dilaton couplings, one *does* obtain a significant degree of squark and slepton degeneracy, and one in fact obtains a highly predictive scenario. Examining the tree level lagrangian, one finds that the gaugino masses, m_g, the squark and slepton masses, m_ϕ^2, and the A parameter (the coefficient of the cubic soft breaking terms in the potential) are given by

$$m_g = \frac{\sqrt{3}}{2}m_{3/2} \qquad m_\phi^2 = m_{3/2}^2 \qquad A = -\sqrt{3}m_{3/2}. \qquad (3.4)$$

This is precisely the structure of the soft-breaking parameters in the minimal supersymmetric standard model! Moreover, instead of the four parameters listed above, there is now only one (or two, depending one what one assumes about the origin of the so-called μ parameter).[1] One expects that these relations will be corrected by one loop effects of

274

order α_{GUT}/π. Whether or not this is good enough to explain the suppression of FCNC's will depend on the precise value of the one loop coefficients. Naively, however, one does not expect these corrections to be small enough, since at the large scale α_s/π, for example, is likely to be larger than 10^{-2}, so something more is likely to be required, particularly to understand the smallness of the imaginary part of $K\bar{K}$. Still, this looks tantalizingly close.

We do not know, of course, whether string theory dynamics satisfy this condition. Kaplunovsky and Louis, in fact, argue that they do not for known susy breaking schemes. However, we know very little about susy breaking in string theory; no known mechanism even satisfies the modest set of assumptions listed above. Given that the dilaton breaking scenario looks close to what one wants, and that only one assumption is required, it is interesting to explore its consequences. The required renormalization group analysis has been performed in ref. 19. Assuming MSSM particle content, one finds that it is difficult to implement this scenario given current experimental constraints; significant fine tunings (at the part in 10^{-2} level) are required. Non-minimal models are presumably not so highly constrained (and of course not so predictive); it will be of interest to explore their phenomenology.

4. Flavor Symmetries: An Alternative Solution to the Degeneracy Problem

An alternative approach to the problem of degeneracy, about which string theory also offers some suggestive clues, is to assume that there is some underlying flavor symmetry.[2,6] One possibility is that there exists a non-Abelian, gauged horizontal symmetry. The obvious problem with this proposal is that whatever flavor symmetry there is must be badly broken in order to explain the ordinary quark mass matrix. The simplest model which one can use to examine this question possesses an extra gauge symmetry $SU(2)_H$. Under this symmetry we assume that the quarks transform as doublets and singlets: $Q^a, \bar{u}^a, \bar{d}^a$ and singlets, $Q_s, u_s, \bar{d}_s$ and similarly for the leptons. The Higgs fields are assumed to be $SU(2)_H$ singlets.

In addition, to break the horizontal symmetry, we assume that we have some $SU(2)_H$ doublets which are standard model singlets, $\Phi^a_{(i)}$. From a stringy perspective, the presence of such particles is quite plausible. We might imagine that the $SU(2)_H$ symmetry corresponds to an enhanced gauge symmetry at some particular point in the moduli space. The fields $\Phi^a_{(i)}$ then represent moduli. Giving them expectation values corresponds to moving away from the special point. If this identification is correct, these fields have no potential, and can easily obtain vev's of order M_p. For what follows, we will assume

$$\frac{< \Phi^a_{(i)} >}{M_p} \sim 10^{-1}. \tag{4.1}$$

Such an assumption is not unnatural. Vev's of this order might arise in the presence of Fayet-Iliopoulos terms generated at one loop, for example.

To proceed, we need to adopt a set of rules about the sizes of various couplings. We will enforce 't Hooft's notion of naturalness:[20] couplings can be small only if the theory becomes more symmetric in that limit. Ultimately, we would like to explain such small parameters through additional symmetries (e.g. discrete symmetries of the type to be discussed shortly).

Let us consider, then, the allowed couplings in the lagrangian. The superpotential just below M_p contains dimension-four terms:

$$W_q = \lambda_1 \epsilon_{ab} Q_a \bar{d}_b H_1 + \lambda_2 \epsilon_{ab} Q_a \bar{u}_b H_2 + \lambda_3 Q_s \bar{d}_s H_1 + \lambda_4 Q_s \bar{u}_s H_2. \tag{4.2}$$

These give rise to $SU(2)_H$ symmetric terms in the mass matrix. Clearly we need to assume that λ_1 and λ_2 are small (this might be arranged by means of a discrete symmetry). $SU(2)_H$-violating terms arise at the level of dimension five and dimension six operators:

$$\frac{1}{M_p}(\lambda_5^i \epsilon_{ab} \Phi_a^i Q_b \bar{d}_s H_1 + \lambda_6^i \epsilon_{ab} \Phi_a^i Q_s \bar{d}_b H_1) + \frac{1}{M_p^2}(\lambda_7^{ij} \epsilon_{ab} \epsilon_{cd} \Phi_a^i \Phi_c^j Q_b \bar{d}_d H_1 + ...). \tag{4.3}$$

Note that the charmed-quark mass must arise from these operators, and is thus of order $(\Phi/M_p)^2$, so Φ/M_p can't be much smaller than 0.1.

The breaking of the squark degeneracy can also be understood in terms of the effective action at scales slightly below M_p. This lagrangian contains dimension-four, soft-breaking terms which give $SU(2)_H$-symmetric contributions to the squark mass matrices:

$$V_{soft} = m_1^2 |Q_a|^2 + m_2^2 |Q_s|^2 + m_3^2 |\bar{u}_a|^2 + m_4^2 |\bar{u}_s|^2 + ...$$

$$+ A_1 \lambda_1 Q \bar{d} H_1 + A_2 \lambda_2 Q \bar{u} H_1 + ... + h.c. \tag{4.4}$$

Here, m_i and A_i are of order m_{susy}. Breaking of the symmetry will arise through terms of the type

$$\delta V_{soft}^2 = \frac{m_{susy}^2}{M_p}(\gamma_1 \Phi_1 Q Q_s^* + ...) + \frac{m_{susy}^2}{M_p^2}(\gamma_1' \Phi_1 Q \Phi_2 Q^* + ...) \tag{4.5}$$

and

$$\delta V_{soft}^3 = \frac{m_{susy}}{M_p} \lambda_5^1 Q \bar{d}_s H_1 (\eta_1 \Phi_1 + \eta_2 \Phi_2 + \eta_3 \Phi_2^*)$$

$$+ \frac{m_{susy}}{M_p^2} \lambda_7^{11} Q \bar{d} H_1 (\eta_1' \Phi_1 \Phi_1 + \eta_2' \Phi_1 \Phi_2 + \eta_3' \Phi_1 \Phi_2^*) + ... \tag{4.6}$$

We have omitted $SU(2)_H$ indices on Q, $\bar{u}$, $\bar{d}$ but terms with all possible contractions should be understood. Here γ, γ', η and η' are dimensionless numbers. By 't Hooft's naturalness criterion,[20] many of these couplings should not be much less than one; the theory does not become any more symmetric if these quantities vanish. As a result, the generic symmetry-violating terms in the first two generations are of order $(\Phi/M_p)^2 \sim 10^{-2}$. Some of these couplings, however, can (and should!) naturally be small. So there is no difficulty with suppressing FCNC's.

This framework is predictive. For example, it suggests that there should be a high degree of degeneracy only in the first two squark and slepton generations. The small value of the neutron electric dipole moment is also readily accomodated here. Recently, Seiberg and Pouliot have considered models in which non-Abelian symmetries play a role both in producing squark degeneracy and in explaining the features of the quark mass matrix. Such models are clearly of great interest.[4]

An alternative approach to the problem of fcnc's in supersymmetry has recently been explored in ref. 3. Here there is no degeneracy at all, but rather a tight alignment between the quark and squark mass matrices. This is done in the context of a more complete theory of fermion masses with many stringy features.

5. Symmetries and String Theory

String theory has much to say about the question of symmetries which might plausibly appear in a low energy effective field theory:

1. Global vs. local continuous symmetries: It has long been argued that it doesn't make much sense to impose global symmetries in field theory. In string theory, this prejudice takes on the status of a theorem: one can show that string theory posseses no global continuous symmetries.[5]

2. Discrete symmetries: String models often exhibit a rich structure of discretesymmetries.[21] These can often be thought of as "gauge symmetries," e.g. coordinate or gauge transformations in some higher dimensional theory.

3. Matter multiplets in string theory do not typically have the structure of conventional grand unified theories.[21] In particular, the transformation properties of fields under discrete symmetries do not correspond to those of GUT multiplets.

Discrete symmetries are of great phenomenological importance in supersymmetric model building: they are needed to forbid proton decay, and perhaps other rare processes ($\mu \to e\gamma$, etc.). They appear in many other possible extensions of the standard model, as well. It has been argued that any discrete symmetries appearing in an effective lagrangian should be gauge symmetries.[22] Otherwise, they are likely to be spoiled by gravitational effects. It has also been stressed that discrete symmetries may be anomalous;[23,24] this suggests anomaly constraints on discrete symmetries. In ref. 23, these constraints were enumerated assuming that the discrete symmetries were embedded in a broken continuous gauge group, and that the charges of the heavy states were integer multiples of those of the light states. This lead to a quite strong set of constraints. In general, however, only a weaker set of constraints hold; these can be understood in terms of instantons in the low energy theory.[23,25]

In ref. 25, it was noted that in string theory there *are* often anomalous discrete symmetries. However, one can always cancel these anomalies by assigning to the (model-independent) axion a non-linear transformation law under the symmetry,

$$a \to a + 2\pi\delta. \tag{5.1}$$

(δ would be a multiple of $1/N$ for a Z_N symmetry). (This possibility had been suggested in ref. 23.) Such a transformation law means, of course, that the discrete symmetry is spontaneously broken (at a scale of order M_p). However, this observation has another consequence: in perturbation theory, in such cases, there is an unbroken, *global*, anomalous discrete symmetry. After all, perturbation theory exhibits an unbroken (spontaneously or explicitly) discrete symmetry .

It is perhaps helpful to give a simple example of this phenomenon (which was discussed in ref. 25). Such an example is provided by the compactification of the heterotic string on the Calabi-Yau manifold associated with the quintic hypersurface in CP^4, discussed, for example, at some length in the textbook of Green, Schwarz and Witten.[26] At certain points in the moduli space, this model possesses a freely-acting $Z_5 \times Z_5$ symmetry. In the textbook treatment, one mods out by this symmetry, including Wilson lines, to obtain a model with a low number of generations and a reasonable gauge group. For our purposes (despite our title) we are not concerned if the number of generations happens to be large. We can mod out by one of the Z_5's, corresponding to rotating the coordinates, Z_a, of CP^4, by phases:

$$Z_a \to \alpha^a Z_a$$

where $\alpha = e^{2\pi i/5}$. This is still freely acting; this means that we don't have to worry about the appearance of massless particles in twisted sectors.

The main virtue of this choice is that it leaves over a set of R-symmetries. For definiteness, consider the symmetry under which $Z_1 \to \alpha Z_1$. Under this symmetry, the gluinos transform by a phase $\alpha^{-1/2}$. Now we can include a Wilson line without breaking this symmetry. For example, we can include a Wilson line in the "second" E_8 (the one which is unbroken in the absence of the Wilson line), described by:

$$a = \frac{1}{5}\left(1, 1, 2, 0, 0, 0, 0, 0\right). \tag{5.2}$$

(I am using the notation which is standard in the orbifold context). By itself, this choice is not modular invariant, but this is easily repaired by including a Wilson line in the first E_8 as well. In the second E_8, there are two unbroken gauge groups. It is easy to determine the effects of instantons by simply examining $SU(2)$ subgroups of these. One finds that instantons of the first group have four gluino zero modes, while instantons of the second have 24. Thus assigning to the axion a transformation law

$$a \to a + \frac{12\pi}{5} \tag{5.3}$$

(the reason for writing 12 rather than 2 will become clear shortly) cancels both anomalies.

This observation is potentially of great importance for model building. It suggests that it is reasonable to impose anomalous, global discrete symmetries. For example, one might impose a symmetry which forces $m_u = 0$ in order to solve the strong CP problem. (Recall that the discrete symmetries of string theory don't have the structure of those of conventional GUT's.)

These anomalous discrete symmetries are interesting from at least one other viewpoint. They should permit one to make *exact non-perturbative* statements about string dynamics.[27] The reason is simple. The axion and dilaton together make up the complex scalar in a supermultiplet, usually denoted S,

$$S = \frac{1}{g^2} + ia. \tag{5.4}$$

Given the axion transformation law of eqn. (5.1), $W(S)$ must take the form

$$W(S) = \sum_n C_n(M) e^{-n/\delta}.$$

This is precisely the behavior one encounters in the gluino condensation scenario. In these cases, one can argue that there can be no stringy non-perturbative effects stronger than the effects observed in the low energy field theory! The main subtlety in this argument lies in the determination of δ. δ cannot be determined unambiguously from the anomaly considerations described above; typically one can add a constant of the form $2\pi r$, where r is an integer. For example, had we chosen 2 rather than 12 in eqn. (5.3), the superpotential of gluino condensation would not have been invariant. This issue will be discussed further in ref. 27.

Can one generalize this? The discussion here is close in spirit to discussions of "dilaton duality" which have appeared recently in the literature.[7] I suspect that some version of this duality is indeed correct; however it is difficult to establish it by the sort of arguments which have been employed here. (See the talks by Joanne Cohn and John Schwarz at this meeting.) These issues will also be discussed in ref. 27.

6. String Theory and the θ Puzzle

There are three known solutions to the strong CP problem.

1. $m_u = 0$. This solution may be consistent with current algebra.[28] As we have noted in the previous section, this is a prediction which could quite plausibly emerge from string theory.

2. Peccei-Quinn symmetries and axions: here one needs a global symmetry to a high degree of approximation in perturbation theory.

3. Spontaneous CP violation: in theories with an exact CP symmetry, the bare θ is zero. The problem is then to arrange that θ be small at tree level after symmetry breaking, and that radiative corrections be sufficiently small.[10,11]

String theory has interesting statements to make about all three possibilities. To consider these, we should first understand the status of CP as a symmetry of string theory. Strominger and Witten were probably the first to comment on CP in string theory.[29] These authors noted that certain Calabi-Yau compactifications possess unbroken CP at points in their moduli spaces. CP, then, is a symmetry of the classical theory which can be spontaneously broken by vev's for moduli. It is not hard to see that in these cases, CP is a good symmetry of perturbation theory, since it corresponds to a two-dimensional symmetry which commutes with the BRS operator.

What about non-perturbatively? For example, there has been speculation on the possibile existance of non-perturbative parameters in string theory. Could some of these exist and violate CP (θ, after all, is the quintessential non-perturbative parameter)? While we don't understand much about non-perturbative string theory, the answer to this question is a definite "no." CP turns out to be a gauge symmetry in string theory, a combination of general coordinate and gauge transformations in the higher dimensional space.[30,31]

To see this, consider the $O(32)$ or $E_8 \times E_8$ heterotic strings in ten dimensions. In ten dimensions, P is not a symmetry of either theory. What one would like to call C has the effect of changing the signs of all of the lattice momenta, while acting trivially on spinors (as a consequence of GSO). This change of signs is a symmetry of the lattice; in fact, it is a gauge transformation. In $O(32)$, this transformation acts on a 32 component vector by the matrix:

$$\Lambda = \begin{pmatrix} i\sigma_2 & 0 & \dots \\ 0 & i\sigma_2 & \dots \\ \dots & \dots & \dots \end{pmatrix}. \tag{6.1}$$

A similar transformation works in E_8.

Now compactify to four dimensions on an ordinary torus. In this theory, it is well-known that P is a symmetry. Where did this come from? Denote the compactified coordinates by y^a, $a = 1 \dots 6$, and the uncompactified coordinates by x^μ. The transformation

$$x^i \to -x^i \quad y^2 \to -y^2 \quad y^4 \to -y^4 \quad y^6 \to -y^6 \tag{6.2}$$

is a good space-time symmetry (it commutes with GSO). From the perspective of ten-dimensions, it is part of the proper Lorentz group. Thus for these compactifications, CP is a product of an ordinary transformation and a general coordinate transformation (with a little work, you can check that this transformation has the expected action on spinors). Other compactifications, such as Calabi-Yau compactifications, typically preserve this product of symmetries at some points in the moduli space.

CP is thus special. It can't be broken by no-perturbative effects (instantons, wormholes, etc.); non-perturbative, CP-violating parameters cannot arise in string theory, and in particular there can be no bare θ (e.g. in the $E_8 \times E_8$ theory). Thus all CP-violating effects in string theory are, in principle, calculable.

An obvious question is whether one can use this to implement the spontaneous CP-violation solution to the strong CP-problem. A few observations are important. First, to accomplish this, one wants CP broken at a scale well below M_p.[30] If Λ is the scale of CP violation, even if one suppresses low dimension operators contributing to θ, there will inevitably be operators of high dimension, whose contribution will be of order $(\Lambda/M_p)^n$. Thus one probably wants to break CP at a scale not larger than 10^{10} GeV, or so (assuming that the leading contributions are due to dimension five operators). Otherwise, the effective θ is almost certainly of order one.[30]

E_6 models, such as those which arise in $(2,2)$ compactifications of string theory, provide a rather natural setting for the Nelson-Barr mechanism.[32] Under $O(10) \times U(1)$, the 27 of E_6 decomposes as

$$27 = 16_{-1/2} + 10_1 + 1_{-2}. \tag{6.3}$$

Here the 16 contains an ordinary generation of quarks and leptons and an additional right-handed neutrino, $\mathcal{N}$, while the ten contains a vectorlike (with respect to the standard model group) set of doublets, H_1 and H_2, and a vectorlike set of charge $-1/3$ quarks, q and $\bar{q}$. We will denote the $O(10)$ singlet by S. The $SU(3) \times SU(2) \times U(1)$ singlet fields, $\mathcal{N}_i$ and S_i can naturally obtain vev's of order[33]

$$M_{INT} = \sqrt{m_W M_p} \approx 10^{11} GeV. \tag{6.4}$$

Moreover, there is typically a range of soft-breaking parameters for which the $< S_i >$ are real, while the $< \mathcal{N}_i >$ are complex. The allowed couplings in the superpotential in these models are of the form

$$W = \mathcal{N} q\bar{d} + S q\bar{q} + H_1 Q\bar{u} + H_2 Q\bar{d}. \tag{6.5}$$

The fermion mass matrix then has the Nelson-Barr structure:

$$m_F = \begin{pmatrix} \Gamma H_2 & \gamma \mathcal{N} \\ 0 & \mu \end{pmatrix}. \tag{6.6}$$

Automatically, $arg\ det\ m_F = 0$.

At tree level, there are many couplings which must be suppressed for all of this to work. For example, it is necessary to avoid phases in the Higgs potential, and to suppress couplings which would lead to phases in the S vev's. This can be accomplished by discrete symmetries. In loops, however, more serious problems arise which cannot be dealt with so easily. Early studies of loop corrections to θ showed that many contributions can be suppresed.[34] However, these analyses assumed that squarks are precisely degenerate. If one relaxes the assumption of exact degeneracy and examines the various corrections one finds that the smallness of θ sets stringent requirements. In particular, diagrams correcting the d quark mass lead to a requirement that $\frac{\delta m^2}{m^2_{susy}} < 10^{-5}$; diagrams contributing to the gluino mass require that certain phases be aligned to one part in 10^7. Neither of the schemes we have discussed above (flavor symmetries or dilaton driven supersymmetry) seem likely to produce anything like this degree of degeneracy. Perhaps these constraints can be satisfied as a consequence of symmetries. But at the least a much more elaborate symmetry structure is required to implement the Nelson-Barr scheme in the framework of supersymmetry than has been considered to date.

7. Axions in String Theory and Elsewhere

The underlying idea to the axion solution to the strong CP problem is to postulate a "Peccei-Quinn symmetry" under which the axion transforms as

$$a \to a + f_a \delta. \tag{7.1}$$

This solution, however, suffers from serious problems, which, like the Barr-Nelson solution discussed above, call its plausibility into question.[9]

The basic problem is simple, and should be "obvious" to string theorists. In string theory, and presumably in any fundamental theory, there are no global, continuous symmetries. Approximate global symmetries, if they exist, might arise accidentally, as an accidental consequence of the structure of low dimension terms in the low energy effective lagrangian (like B and L in the standard model).

To see the difficulty, recall that the usual axion potential has the form

$$V = -Nm_\pi^2 f_\pi^2 \cos(a/f_a).$$ (7.2)

Suppose that $\mathcal{O}_n$ is the lowest dimension operator which breaks the Peccei-Quinn symmetry; its dimension is $n + 4$. Then the symmetry-breaking lagrangian has the form

$$\mathcal{L}_{SB} = \frac{\gamma}{M_p^n}\mathcal{O}_n.$$ (7.3)

This gives rise to an axion potential, on dimensional grounds, of the form

$$\delta V = \frac{\gamma f_a^{n+1}}{M_p^n}a(x).$$ (7.4)

If $f_a \sim 10^{11} GeV$, then requiring $\theta = a/f_a < 10^{-9}$ gives $n > 7$, i.e. it is necessary to suppress operators up to dimension 11 in order to obtain a sufficiently good symmetry!

But string theory always has an axion in perturbation theory.[35] This axion arises from the antisymmetric tensor field, $B_{\mu\nu}$, which in four dimensions is equivalent to a scalar. Moreover, the decay constant of the axion is of order M_p. From the perspective of the argument given above, this is highly suprising. It can be understood in a variety of ways. First, the couplings of the antisymmetric tensor are governed by a gauge principle. This insures that in perturbation theory, $B_{\mu\nu}$ enters the effective lagrangian only through the gauge-invariant field strength $H_{\mu\nu\rho}$.* Alternatively, one can understand this result directly in terms of the string perturbation expansion. At $k = 0$, the vertex operator for the axion is a total derivative.

While the existence of this axion is rather suprising from a field theoretic perspective, it is not at all clear that this field can solve the strong CP problem. First, its decay constant is of order M_p, which poses cosmological difficulties. Second, if there is any sort of hidden sector gauge group, it is likely to give mass to this axion.

Thus, if the string theory solves the strong CP problem by the Peccei-Quinn mechanism, another axion must arise by accident as a consequence of a continuous or a *discrete* gauge symmetry. This latter possibility was already considered some time ago in refs. 36 and 37. These authors noted, as we have mentioned earlier, that it is quite natural to obtain a scale $f_a \sim 10^{11} GeV$ in string theory. For example, the field S of eqn. (6.3) can appear in the superpotential only in a restricted way. Because there is no mass for this field, and because of its gauge quantum numbers, the leading term which can give rise to a potential for the field S and the corresponding field $\bar{S}$ which might arise from the $\overline{27}$, is

$$W = \frac{1}{M_p^2}S^2\bar{S}^2.$$ (7.5)

If the soft breaking terms include terms of the form

$$V_{soft} = -m_{3/2}^2|S|^2 + ...$$ (7.6)

then $S^2 \sim m_{3/2}M_p$. In order that there be an approximate symmetry, one needs to forbid

---

* One might naturally ask how the axion potential due to non-Abelian gauge interactions is consistent with this symmetry. In fact, it is. Moreover, it is not hard to understand why only non-perturbative effects give rise to this mass. I thank Renata Kallosh, Lenny Susskind for discussions of this and related matters, which will appear elsewhere.

many operators in W and V_{soft}. This is possible with rather simple discrete symmetries.[36,37,38]

8. Summary

We have touched here on a large number of topics. There are a few points which I hope you will take away from this talk:

1. Kaplunovsky and Louis have exhibited one is perhaps the first example of something which might cautiously be referred to as a string prediction. In particular, string theory offers a simple option for explaining squark and slepton degeneracy; this framework makes a series of strong predictions.

2. String theory offers several lessons for model building.

   a. Discrete symmetries in string theory can be global and anomalous (relevant for $m_u = 0$?); discrete symmetries in string theory don't look like those from conventional GUT's.

   b. CP is an exact, gauge symmetry of string theory. But even though CP violation is spontaneous and the "bare" θ vanishes, it is difficult to solve the strong CP problem without axions or $m_u = 0$.

3. String theory suggests alternative approaches to the problem of squark degeneracy, based on both continuous and discrete flavor symmetries.

Acknowledgements

I would like to thank A. Kagan, R. Leigh, D. MacIntire, and N. Seiberg for many helpful conversations about these subjects.

REFERENCES

1. V.S. Kaplunovsky and J. Louis, CERN-TH. 6809/93 UTTG-05-93.

2. M. Dine, A. Kagan and R. Leigh, SCIPP-93-05.

3. Y. Nir and N. Seiberg, Phys. Lett. **B309** (1993) 307.

4. N. Seiberg and P. Pouliot, RU-93-39.

5. T. Banks, L. Dixon, D. Friedan and E. Martinec, Nucl. Phys. **B299** (1988) 613.

6. M. Leurer, Y. Nir and N. Seiberg, Phys. Lett. **B398** (1993)319.

7. A. Font, L. Ibanez, D. Lust and F. Quevedo, Phys. Lett. **B249** (1990) 35; S.J. Rey, Phys. Re. **D43** (1991) 526; A. Sen, preprint TIFR-Th-92-41; J. Schwarz, CALT-68-1815; J.E. Cohn and V. Periwal, IASSNS-HEP-93-24.

8. S.H. Shenker, talk given at the Cargese Workshop on Random surfaces, Quantum Gravity and Strings, Cargese, France (1990), Published in the Proceedings.

9. M. Kamionkowski and J. March-Russell, Phys. Lett. **282B** (1992) 137; R. Holman *et al.*, Phys. Lett. **282B** (1992) 132; S.M. Barr and D. Seckel, Bartol preprint BA-92-11.

10. A. Nelson, Phys. Lett. **136B** (1984) 387.

11. S.M. Barr, Phys. Rev. Lett. **53** (1984) 329.

12. M. Dine, R. Rohm, N. Seiberg and E. Witten, Phys. Lett. **156B** (1985) 55; J.P. Derendinger, L.E. Ibanez and H.P. Nilles, Phys. Lett. **155B**(1985) 65

13. M. Dine and N. Seiberg, Phys. Lett. **162B**, 299 (1985). and in *Unified String Theories*, M. Green and D. Gross, Eds. (World Scientific, 1986).

14. N.V. Krasnikov, Phys. Lett. **193B** (1987) 37; L. Dixon, in *Porceedings of the DPF Meeting*, Houston, 1990 : J.A. Casas, Z. Lalak, C. Munoz and G.G. Ross, Nucl. Phys. **B347** (1990) 243; T. Taylor, Phys. Lett. **B252** (1990) 59.

15. F. Gabiani and A. Masiero, Nucl. Phys. **B322** (1989) 235

16. A good overview of the positive features of supersymmetry and supersymmetric unification, with extensive references, is provided by G. Kane, UM-TH-93-10.

17. P. Langacker and M. Luo, Phys. Rev. **D44** (1991) 817; U. Amaldi, W. de Boer and H. Furstenau, Phys. Lett. **B260** (1991) 447.

18. L. Ibanez and D. Lust, Nucl. Phys. **B382** (1992) 305.

19. R. Barbieri, J. Louis and M. Moretti, CERN-TH.6856/93

20. G. 't Hooft, in *Recent Developments in Gauge Theories*, G. 't Hooft et. all, Eds., Plenum (New York) 1980.

21. M. Green, J. Schwarz and E. Witten, *Superstring Theory*, Cambridge University Press, New York, 1986.

22. L. Krauss and F. Wilczek, Phys. Rev. Lett. **62**, 1221 (1989).

23. L. Ibanez and G. Ross, Phys. Lett. **260B** (1991) 291; Nucl. Phys. **B368** (1992) 3.

24. J. Preskill, Sandip Trivedi, F. Wilczek and M. Wise, Nucl. Phys. **B363** (1991) 207.

25. T. Banks and M. Dine, Phys. Rev. **D45** (1992) 424.

26. M. Green, J. Schwarz and E. Witten, *Superstring Theory*, Cambridge University Press, Cambridge (1987).

27. M. Dine, R. Leigh and D. MacIntire, in preparation

28. H. Georgi and I. McArthur, Harvard University Report No. HUTP-81/A011 (unpublished); D.B. Kaplan and A.V. Manohar, *Phys. Rev. Lett.* **56**, 2004 (1986); K. Choi, C.W. Kim and W.K. Sze, *Phys. Rev. Lett.* **61**, 794 (1988); J. Donoghue and D. Wyler, *Phys. Rev.* **D45** (1992) 892; K. Choi, Nucl. Phys. **B383** (1992) 58.

29. A. Strominger and E. Witten, Comm. Math. Phys. **101** (1985) 341.

30. K. Choi, D. Kaplan and A.E. Nelson, Nucl. Phys. **B391** (1992) 515.

31. M. Dine, R. Leigh and D. Macintire, Phys. Rev. Lett. **69** (1992) 2030.

32. P.H. Frampton and T.W. Kephart, Phys. Rev. Lett. **65** (1990) 1549.

33. M. Dine, V. Kaplunovsky, M. Mangano, C. Nappi and N. Seiberg, Nucl. Phys. **B259** (1985) 549.

34. S.M. Barr and A. Masiero, *Phys. Rev.* **D38** (1988) 366;S. Barr and G. Segre, preprint BA-93-01 (1993)..

35. E. Witten, Phys. Lett. **B149** (1984) 359.

36. G. Lazarides, C. Panagiotakopoulos and Q. Shafi, Phys. Rev. Lett. **56** (1986) 432.

37. J. Casas and G. Ross, Phys. Lett. **192B** (1987) 119.

38. M. Dine, SCIPP-92-27.

SYMMETRIES OF THE DISSIPATIVE HOFSTADTER MODEL *

DENISE E. FREED

Center for Theoretical Physics
Massachusetts Institute of Technology
Cambridge, MA 02139, U.S.A.

ABSTRACT

The dissipative Hofstadter model, which describes a particle in 2-D subject to a periodic potential, uniform magnetic field, and dissipation, is also related to open string boundary states. This model exhibits an SL(2,Z) duality symmetry and hidden reparametrization invariance symmetries. These symmetries are useful for finding exact solutions for correlation functions.

Talk presented at the Strings '93 conference in Berkeley, May 1993

1. The Dissipative Hofstadter Model and Open String Theory

The dissipative Hofstadter model describes the quantum mechanics of a particle confined to two dimensions in a periodic potential and transverse magnetic field, subject to a dissipative force. The particle's coordinates are taken to be $\vec{x}(t) = (x(t), y(t))$, and the magnetic field is given by $B\hat{z}$. Classically, the dissipation is described by the $-\eta\dot{\vec{x}}$ term in the equations of motion. To treat the dissipation quantum mechanically, we use the Caldeira-Leggett model.[1] In this model, the Euclidean action is given by

$$\mathcal{S} = S_q + S_\eta + S_V,\qquad(1)$$

where S_q is the usual action of a particle in a constant magnetic field,

$$S_q = \int_{-T/2}^{T/2} dt \left[\frac{M}{2}\dot{\vec{x}}^2 + \frac{ieB}{2c}(\dot{x}y - \dot{y}x)\right].\qquad(2)$$

S_η is a non-local kinetic term that accounts for the friction. It is given by

$$S_\eta = \frac{\eta}{4\pi} \int_{-T/2}^{T/2} \int_{-\infty}^{\infty} dt\, dt' \left(\frac{\vec{x}(t) - \vec{x}(t')}{t - t'}\right)^2.\qquad(3)$$

S_V is the action due to the potential, which we are taking to be

$$S_V = -\int_{-T/2}^{T/2} \left[V_0 \cos\left(\frac{2\pi x(t)}{a}\right) + V_0 \cos\left(\frac{2\pi y(t)}{a}\right)\right].\qquad(4)$$

---

* This work was supported in part by DOE grant #DE-AC02-76ER03069 and NSF grant #87-8447.

The friction term, S_η, is obtained by first introducing a bath of harmonic oscillators which interact linearly with the particle. The integration over the oscillators in the resulting functional integral yields the S_η term in the remaining action for the particle.

It is useful to define the dimensionless parameters describing the friction/unit cell, α, and the flux/unit cell, β. They are given by

$$2\pi\alpha = a^2\eta/\hbar \qquad \text{and} \qquad 2\pi\beta = \frac{eB}{\hbar c}a^2. \tag{5}$$

The action in Eq. (1) is also used to obtain the boundary state in open string theory.[2] This boundary state describes a world sheet with a boundary, where all the fields on the interior of the world sheet are free and all the interactions take place on the boundary.[3] Then the potential corresponds to a tachyon field; the magnetic field corresponds to a guage field; and the mass term acts as a regulator. In order to give a solution to string theory, the theory must be independent of the regulator. Hence it must be scale invariant and lie at a critical point.

2. The Theory at $\alpha = 1$, $\beta = 0$

When $\alpha = 1$ and there is no magnetic field, the theory is expected to be at a critical point.[4] At this point, we can use simple algebraic identities to fermionize the theory as follows:[5,6]

$$e^{ix(t)} = \psi_L{}^\dagger(t)\psi_R(t), \tag{6}$$

$$\dot{x}(t) = i\left[\psi_L{}^\dagger(t)\psi_L(t) - \psi_R{}^\dagger(t)\psi_R(t)\right], \tag{7}$$

and similarly for $y(t)$. The propagator is then given by

$$\langle\psi_L{}^\dagger(t)\psi_L(0)\rangle = \langle\psi_R{}^\dagger(t)\psi_R(0)\rangle = \frac{i}{t} \qquad \text{and} \qquad \langle\psi_L{}^\dagger(t)\psi_R(0)\rangle = 0. \tag{8}$$

The theory is bilinear in ψ, so it can be solved exactly. We find that

$$\langle\dot{x}(t_1)\dot{x}(t_2)\rangle = -\mu\frac{2}{(t_1 - t_2)^2}, \tag{9}$$

where μ is known as the mobility; and all other m-point functions of the $\dot{x}(t)$'s are contact terms. That means they are zero unless at least two points are coincident.

The problem with this treatment is that fermionization is only fine for large-time behavior, since the derivation ignored how the short distance behavior was regulated. One would hope that for the contact terms to be well-defined and independent of the regulator, the symmetries and possibly also the large-time behavior of the system are enough to determine them.

3. Duality Symmetry

The first symmetry we expect this system to have is a duality symmetry in α and β.[5] In Fig. 1, the approximate phase diagram for this system shows

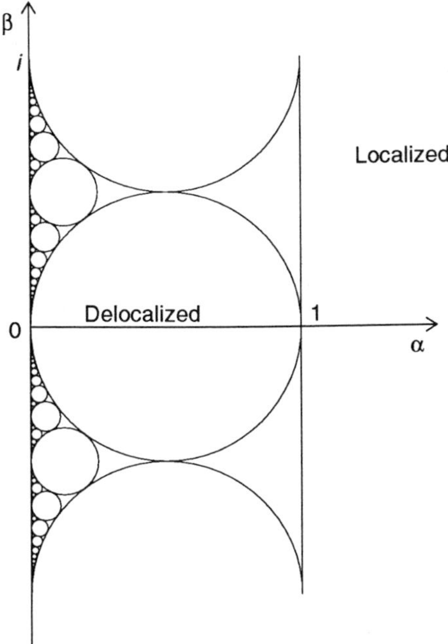

Figure 1: Phase diagram as a function of friction, α, and magnetic flux, β.[5]

the expected transitions between localized states (for $\alpha > 1$), delocalized states (in the interiors of the circles), and unknown states (in the triangular regions between the circles). The diagram exhibits an $SL(2, Z)$ symmetry. This means that if the theory at one value of α and β is critical, then, for any other value of α and β that is related to the first by an $SL(2, Z)$ transformation, it will also be critical. In addition, this symmetry relates theories at different values of flux and friction, so if the correlation functions are known at one value of α and β, then there are simple transformations we can use to obtain them at the other values of α and β related by the $SL(2, Z)$ symmetry. For example, if we know the correlation function $\langle \dot{\tilde{x}}^{\mu_1}(k_1) \ldots \dot{\tilde{x}}^{\mu_m}(k_m) \rangle_0$ at the point $\alpha = 1$, $\beta = 0$, then we can obtain all the correlation functions $\langle \dot{\tilde{x}}^{\mu_1}(k_1) \ldots \dot{\tilde{x}}^{\mu_m}(k_m) \rangle_\beta$ at the other multi-critical points on the large circle centered at $\alpha = 1/2$, $\beta = 0$, as follows:[6]

$$\langle \dot{\tilde{x}}^{\mu_1}(k_1) \ldots \dot{\tilde{x}}^{\mu_m}(k_m) \rangle_\beta = \left[\prod_{i=1}^{m} r^{\mu_i x}(k_i) + \prod_{i=1}^{m} r^{\mu_i y}(k_i) \right] \langle \dot{\tilde{x}}^{\mu_1}(k_1) \ldots \dot{\tilde{x}}^{\mu_m}(k_m) \rangle_0, \tag{10}$$

where $r^{\mu\nu}$ is given by

$$r^{\mu\nu}(k) = \delta^{\mu\nu} - \frac{\beta}{\alpha}\text{sign}(k)\epsilon^{\mu\nu}. \tag{11}$$

It is not difficult to do calculations to $O(V_0^2)$, and, to this order, direct calculations show that Eq. (10) is exact. They also show that for $\beta \neq 0$, not all the

m-point functions are contact terms.[7] This implies that the duality symmetry in Eq. (10) relates contact terms at $\beta = 0$ to functions that are non-zero at large times. We conclude from this that the duality symmetry should be a guiding principle in finding the contact terms.

4. Reparametrization Invariance

The second symmetry the system should have comes from the reparametrization invariance of open string theory. This symmetry implies that the generating function for critical dissipative quantum systems should satisfy "hidden" reparametrization invariance Ward identities.[8] The generating function, $W[\vec{J}]$, is given by

$$e^{W[\vec{J}]} = \int D\vec{x}(t) \exp\left[-\frac{1}{\hbar}(S_q + S_\eta + S_V)\right] \exp\left[-\frac{1}{\hbar}\int \vec{J} \cdot \dot{x}\, dt\right].$$
(12)

For $n \geq 0$, the Ward identity is

$$\sum_{m=1}^{n-1}\left[\frac{1}{2}\frac{\partial W}{\partial \vec{J}_{-m}} \cdot \frac{\partial W}{\partial \vec{J}_{m-n}} - \frac{\partial^2 W}{\partial \vec{J}_{-m} \cdot \vec{J}_{m-n}}\right] + \sum_{\substack{m=-\infty \\ m \neq 0}}^{\infty} m\vec{J}_m \cdot \frac{\partial W}{\partial \vec{J}_{m-n}} = 0,$$
(13)

and there is a similar equation for $n < 0$. This identity is satisfied at all orders in V_0 when $\alpha/(\alpha^2 + \beta^2) = 1$ and $\beta/\alpha \in Z$.[9] The equations with $n = 0, \pm 1$ imply that, inside the correlation functions, the $\ddot{x}(t_i)$ should transform as dimension-one operators under SL$(2, R)$ transformations of time. The only other independent equation is with $n = 2$. One problem with the Ward identities is that they do not give enough information for solving for the correlation functions. However, they do say that any correlation function at $\beta = 0$, $\alpha = 1$ must be SL$(2, R)$ covariant. We can apply the duality transformation to this correlation function to obtain one at another value of β. This new correlation function must once again exhibit the SL$(2, R)$ symmetry. This should give a lot of information about the form of the correlation functions.

One difficulty with this reasoning is that the regulator used to derive the duality transformation is not the same one used to prove the Ward identities. A rigorous derivation of Eq. (10) using the regulator satisfying the Ward identities gives a slightly weaker version of the transformation.

5. Results for Correlation Functions

If we carefully repeat the derivation of fermionization using the regulator that satisfies the Ward identity, then we can obtain additional symmetries and properties of the correlation functions to *all* orders in V_0. We find that, when $\alpha/(\alpha^2 + \beta^2) = 1$ and $\beta/\alpha \in Z$, the m-point functions are given by[6]

$$\langle \dot{\tilde{x}}^{\mu_1}(k_1) \ldots \dot{\tilde{x}}^{\mu_m}(k_m)\rangle_\beta = \left[\prod_{j=1}^{m} r^{\mu_j x}(k_j) + \prod_{j=1}^{m} r^{\mu_j y}(k_j)\right] F(\vec{k}, \beta).$$
(14)

This is the form of the correlation functions predicted by the duality transformation in Eq. (10). The function $F(\vec{k}, \beta) = \vec{a}(\vec{k}) \cdot \vec{k}$ has the following properites: It is finite as the cutoff goes to zero. It is piecewise linear in $\vec{k}$, which suggests a connection with the Duistermaat-Heckman theorem;[10] and it is homogeneous in $\vec{k}$. In real space, this last property gives a new symmetry under non-one-to-one reparametrizations of time. It tells what happens when $z \to z^n$, where $z = e^{2\pi i k t}$. In addition, we find that $F(\vec{k}, \beta) \to 0$ if $k_i = 0$; F is symmetric under $k_i \leftrightarrow k_j$, and also $\vec{k} \leftrightarrow -\vec{k}$; and F is continuous. These results restrict the m-point functions to lie in a finite-dimensional linear space.

We conjecture that these results, combined with the reparametrization invariance Ward identites, determine all the m-point functions when $\alpha/(\alpha^2 + \beta^2) = 1$ and $\beta/\alpha \in Z$. We have found that they do give exact solutions for the two, four and six-point functions, and also for any correlation function with special conditions on the $k_i's$.[11] For example, the 4-point function is proportional to $\min(|k_1|, |k_2|, |k_3|, |k_4|)$. However, it is still an open question whether these symmetries are enough to determine any arbitrary correlation function. In any case, it does appear that these symmetries, combined with the long-time behavior, are enough to obtain exact solutions for all the correlation functions.

6. Acknowledgements

I would like to thank the organizers for arranging such an interesting and stimulating conference.

7. References

1. A. O. Caldeira and A. J. Leggett, *Physica* **121A**(1983) 587; *Phys. Rev. Lett.* **46** (1981) 211; *Ann. of Phys.* **149** (1983) 374.
2. C. G. Callan, L. Thorlacius, *Nucl. Phys.* **B329** (1990) 117.
3. C. G. Callan, C. Lovelace, C. R. Nappi, and S. A. Yost, *Nucl. Phys.* **B293** (1987) 83; *Nucl. Phys.* **B308** (1988) 221.
4. A. Schmid, *Phys. Rev. Lett.* **51** (1983) 1506; F. Guinea, V. Hakim and A. Muramatsu, *Phys. Rev. Lett.* **54** (1985) 263; M. P. A. Fisher and W. Zwerger, *Phys. Rev.* **B32** (1985) 6190.
5. C. G. Callan and D. Freed, *Nucl. Phys* **B374**(1992)543.
6. D. E. Freed, CTP#2170, to appear in *Nucl. Phys.* **B**.
7. C. G. Callan, A. G. Felce and D. E. Freed, *Nucl. Phys.* **B392** (1993) 551.
8. C. G. Callan, L. Thorlacius, *Nucl. Phys.* **B319** (1989) 133.
9. D. E. Freed, CTP#2241, submitted to *Nucl. Phys.* **B**.
10. J. J. Duistermaat, G. J. Heckman, *Invent. Math* **69** (1982)259.
11. D. E. Freed, in preparation.

Duality as a Gauge Symmetry and Topology Change

Amit Giveon [2]

Racah Institute of Physics, The Hebrew University
Jerusalem, 91904, ISRAEL

ABSTRACT

Duality groups as (spontaneously broken) gauge symmetries for toroidal backgrounds, and their role in (∞-dimensional) underlying string gauge algebras are reviewed. For curved backgrounds, it is shown that there is a duality in the moduli space of WZNW sigma-models, that can be interpreted as a broken gauge symmetry. In particular, this duality relates the backgrounds corresponding to axially gauged abelian cosets, $G/U(1)_a$, to vectorially gauged abelian cosets, $G/U(1)_v$. Finally, topology change in the moduli space of WZNW sigma-models is discussed.

[2]e-mail address: giveon@vms.huji.ac.il

1 Introduction

The new results in this talk are based on work with E. Kiritsis [1]. I will discuss two issues in the moduli space of a WZNW sigma-model:

(a) There is a target space duality that can be interpreted, in the effective action, as a broken gauge symmetry. In particular, this duality relates the backgrounds corresponding to axially gauged abelian cosets to vectorially gauged abelian cosets.

(b) There is a topology change, namely, there are deformations lines in the moduli space of sigma-models along which the topology is changed.

The importance of the interpretation of duality as a gauge symmetry is:

(1) It shows that duality is an exact symmetry in string theory. In particular, it shows that axial-vector duality is an exact symmetry; this has implications in the study of black-hole duality in string theory [2], and duality in cosmological string solutions [3, 4].

(2) Dualities appear to be discrete symmetries (some kind of a "Weyl subgroup") of a (infinite-dimensional) universal gauge algebra.

In section 2, I will explain point (2) reviewing known results on duality as a gauge symmetry for toroidal backgrounds. In section 3, I will discuss the $J\bar{J}$ deformation of $SU(2)$ (or "$SL(2)$") WZNW sigma-model, and a duality along this line will be interpreted as a broken gauge symmetry. Finally, in section 4, I will discuss topology change in the moduli space of WZNW sigma-models.

2 Duality as a gauge symmetry and universal gauge algebras

2.1 $R \to 1/R$ circle duality as a gauge symmetry

I will first discuss the simplest case of a single scalar field compactified on a circle with radius R. The moduli space of circle compactifications is the positive real line, describing all possible compactification radii $R > 0$. For each compactification radius, the conformal field theory (CFT) has (at least) a $U(1)_L \times U(1)_R$ affine symmetry, generated by the chiral and antichiral currents J and $\bar{J}$. The truly marginal deformation of the CFT, corresponding to a change of the compactification radius from R to $R + \delta R$, is given by adding $\delta S = \delta(R^2) \int d^2z J\bar{J}$ to the worldsheet action.

A striking property of string theory is that there is a "target-space duality" symmetry relating backgrounds with different geometries, that correspond to the same CFT. In this case, the CFT corresponding to a radius R circle is equivalent to the CFT of a radius $1/R$ circle [5].[3]

The conformal field theory at the self-dual point, $R = 1$, has an extended affine symmetry $SU(2)_L \times SU(2)_R$. At this point there are three chiral currents J^a, $a = 1, 2, 3$, and three antichiral currents $\bar{J}^a$, generating the affine $SU(2)_L$ and $SU(2)_R$ symmetry, respectively. We shall refer to the point $R = 1$ as the $SU(2)$ point. To deform the theory away from the $SU(2)$ point, one may choose to identify $J \equiv J^3$, $\bar{J} \equiv \bar{J}^3$. But this choice is not unique: any deformation of the type

$$(\sum_{a=1}^{3} \alpha_a J^a)(\sum_{b=1}^{3} \beta_b \bar{J}^b) \tag{2.1}$$

is truly marginal. The set of critical points, that can be reached from the $SU(2)$ point by such conformal deformations, span a 5-dimensional surface in the 9-dimensional euclidean space generated by the couplings to $J^a \bar{J}^b$. However, because different truly marginal deformations give rise to the same CFTs, the dimension of the physical moduli space is 1. This can be shown as follows: At the $SU(2)$ point, $SU(2)_L$ and $SU(2)_R$ rotations of the J^a's and $\bar{J}^b$'s are symmetries of the CFT, and as a consequence, any deformation of the type (2.1) give rise to the same CFT given for an appropriate $J^3 \bar{J}^3$ deformation.

In particular, the duality transformation $R \to 1/R$ corresponds to the Weyl transformation in $SU(2)_L$ that takes J^3 to $-J^3$ [6]. Infinitesimaly near the $SU(2)$ point, this corresponds to the identification of the theory given by the deformation $\delta\alpha J^3 \bar{J}^3$ with the theory given by the deformation $-\delta\alpha J^3 \bar{J}^3$.

The discussion, so far, was from the worldsheet point of view. But in string theory, worldsheet properties have a target-space analogue. Indeed, coupling constants to operators in the worldsheet action become space-time fields in the effective action of string theory. In particular, for a string compactified on a circle, we expect to have a scalar field corresponding to the worldsheet coupling to the $J\bar{J}$ deformation. The vacuum expectation values (VEVs) of the scalar field correspond to the compactification radii. At the particular VEV, corresponding to the $R = 1$ point, there is a gauge symmetry $SU(2)_L \times SU(2)_R$. At this point, extra scalar fields and gauge fields become massless. Changing the VEV of scalar fields away from this point (or, equivalently, changing the compactification radius), the gauge symmetry is spontaneously broken to $U(1)_L \times U(1)_R$, and, in addition, there is a symmetry corresponding to $R \to 1/R$ duality. In that sense, the target space duality is a discrete symmetry of the spontaneously broken $SU(2)$ gauge algebra. The Z_2 duality is the Weyl group of $SU(2)$.

2.2 $O(d, d, Z)$ dualities as gauge symmetries

Compactifying the bosonic (heterotic) string on a d–dimensional torus, the circle duality is generalized to a *duality group* isomorphic to $O(d, d, Z)$ ($O(d, d + 16, Z)$) [7, 8, 9].

---

[3]I choose $\alpha' = 1$, where α' is the inverse string tension.

The duality group may be interpreted (in the effective action) as a residual discrete symmetry group of some spontaneously broken (∞–dimensional) universal gauge algebra. Moreover, the $O(d, d, Z)$ dualities are expected to be some kind of a Weyl subgroup of the underlying algebra. ¿From the worldsheet point of view, this can be shown as follows [10]: When $d > 1$ there is an infinite number of points in the moduli space of d–tori backgrounds that have extended affine symmetries. For instance, when $d = 2$ there is an infinite number of points, in the moduli space, where the affine $U(1)^2$ symmetry is extended to $SU(2) \times SU(2)$ or to $SU(3)$. Now, any product of Weyl reflections acting on conformal deformations around points with extended symmetries relate geometrically different toroidal backgrounds that correspond to the same CFT. It turns out that any such "Weyl reflections" product is an element of $O(d, d, Z)$. Moreover, any element of $O(d, d, Z)$ correspond to a product of Weyl reflections.[4]

These worldsheet properties are realized in the target space in an intriguing way. In the effective action, one expects to find a (∞-dimensional) gauge algebra, that is spontaneously broken for any VEV of the scalar fields to an appropriate (finite-dimensional) gauge group, and residual discrete symmetries generating the duality group. This is the interpretation of target space dualities as the residual discrete symmetries of a spontaneously broken gauge algebra.

Such an effective action was constructed in ref. [11] for the $d = 4$, $N = 4$ heterotic string, *i.e.*, the toroidal compactification of the heterotic string to four dimensions. The ∞–dimensional gauge algebra, called the "Duality Invariant String Gauge algebra" (DISG) is isomorphic to

$$\text{Duality Invariant String Gauge algebra} \sim \text{``}LatticeAlg\text{''}(\Gamma^{6,22}), \qquad (2.2)$$

namely, the algebra of dimension 1 operators in the CFT of 28 chiral scalars, compactified on an even-self dual lorentzian lattice with signature (6,22). The infinite-dimensional gauge symmetry is spontaneously broken for *any* VEV of the scalar fields to a finite-dimensional gauge algebra (typically $U(1)^6 \times E_8 \times E_8$), and a group of residual discrete symmetries: the $O(6, 22, Z)$ target-space dualities.

This duality group is some kind of a Weyl subgroup of the underlying gauge algebra. The subgroup of the the duality group, that fixes a point in the moduli space of toroidal backgrounds, is related to the Weyl group of the enhanced (finite-dimensional) gauge symmetry at the fixed point. To get points with large enhanced gauge symmetries and large duality subgroups that fix such points, it is useful to compactify time as well.

2.3 Compactifying time

By compactifying all the coordinates – including timelike dimensions – on a torus, one recovers points in the moduli space of backgrounds with large gauge symmetries; these

---

[4]More precisely, every element of $O(d, d, Z)$ is a product of Weyl reflections in the moduli space of $(d + 1)$–tori [10].

correspond, in the worldsheet, to large on–shell algebras of dimension (1,0) or (0,1) operators. Some backgrounds are fixed points of large duality subgroups. Compactifying time is, therefore, useful in the search for an underlying universal gauge algebra of string theory.

Such a program was initiated for the critical $N = 2$ string, compactified completely on a torus $T^{2,2}$ [12]. A distinguished point in the moduli space of (2,2) toroidal backgrounds is the one where the Narain lattice $\Gamma^{4,4}$ is the direct sum of a $(2,2)_L$ even self-dual lorentzian lattice of left-movers, and a $(2,2)_R$ even self-dual lorentzian lattice of right-movers

$$\text{Maximal Extended Symmetry Point}: \quad \Gamma^{4,4}_{MES} = \Gamma^{2,2}_L \oplus \Gamma^{2,2}_R.$$

At this point the extended on–shell gauge algebra is infinite-dimensional: it is generated by the area-preserving diffeomorphisms of null 2-tori in the $(2,2)_L$ (and $(2,2)_R$) torus.

A universal gauge algebra is a *background independent* algebra that is a minimal Lie algebraic closure which contains *all* the on–shell algebras. Candidates for universal gauge algebras of the $N = 2$ string were presented in ref. [12]. These are:

(1) "Lattice-Algebra"$(\Gamma^{4,4})$.

(2) Volume-Preserving-Diffeomorphisms$(T^{4,4})$.

Here $T^{4,4} \equiv R^{4,4}/\Gamma^{4,4}$ is the Narain torus defined by the Narain lattice $\Gamma^{4,4}$. The first candidate is the analogue of the DISG for the $N = 2$ string; the duality group $O(4,4,Z)$ is some kind of its Weyl subgroup. The second candidate seems more natural for the $N = 2$ string.

In a recent paper [13], Moore has presented a candidate for a universal symmetry algebra of the bosonic string. One starts by compactifying all coordinates on a (1,25) torus. A distinguished point in the moduli space of toroidal backgrounds is the one where the Narain lattice $\Gamma^{26,26}$ is the direct sum of a $(25,1)_L$ even self-dual lorentzian lattice of left-movers, and a $(1,25)_R$ even self-dual lorentzian lattice of right-movers

$$\text{Maximal Extended Symmetry Point}: \quad \Gamma^{26,26}_{MES} = \Gamma^{25,1}_L \oplus \Gamma^{1,25}_R.$$

At this point the extended on–shell gauge algebra is infinite-dimensional: it is the direct sum of the "Fake Monster Lie Algebra" ($\sim$ "Lattice-Algebra"$(\Gamma^{1,25})$) of left-movers and right-movers.

The candidate for a universal gauge symmetry is a modification of the DISG algebra (on a modified Narain lattice, needed to include properly the left-moving enhanced symmetries together with the right-moving enhanced symmetries): Universal Symmetry $\sim D\tilde{I}SG \equiv$ "$LatticeAlg$"$(\tilde{\Gamma}^{26,26})$. For more details see ref. [13].

3 $J\bar{J}$ deformation of $SU(2)$ (or $SL(2)$) WZNW sigma-model, and a duality as a broken gauge symmetry

The discussion in section 2 is restricted to toroidal backgrounds. What about target space dualities in the moduli space of *curved* backgrounds? Can they be interpreted as some (spontaneously broken) gauge symmetries? Here I shall discuss, from the worldsheet point of view, a particular element of the duality group in the moduli space of WZNW models. For simplicity, I will discuss the simplest non-trivial case, namely, duality acting on the conformal deformations line of $SU(2)$ (or "$SL(2)$") WZNW models. In this note I will only present the results; the details appear in [1].

The action for $SU(2)_k$ (in a particular parametrization of the group elements) is given by

$$S[x, \theta, \tilde{\theta}] = \frac{k}{2\pi} \int d^2z (\partial\theta, \partial\tilde{\theta}, \partial x) \begin{pmatrix} \sin^2 x & \cos^2 x & 0 \\ -\cos^2 x & \cos^2 x & 0 \\ 0 & 0 & 1 \end{pmatrix} \begin{pmatrix} \bar{\partial}\theta \\ \bar{\partial}\tilde{\theta} \\ \bar{\partial}x \end{pmatrix}$$
$$-\frac{1}{8\pi} \int d^2z \phi_0 R^{(2)}, \tag{3.1}$$

where $x \in [0, \pi/2)$ and $\theta, \tilde{\theta} \in [0, 2\pi)$. We will refer to the matrix in (3.1) as "the background matrix" E.

The action S (3.1) is manifestly invariant under the $U(1)_L \times U(1)_R$ affine symmetry generated by the currents

$$J = k(-\sin^2 x \partial\theta + \cos^2 x \partial\tilde{\theta}) \qquad \bar{J} = k(\sin^2 x \bar{\partial}\theta + \cos^2 x \bar{\partial}\tilde{\theta}) . \tag{3.2}$$

It is possible to deform the action S to new conformal backgrounds by adding to it any marginal deformation. We will focus on the marginal deformation $J\bar{J}$:

$$S \to S + \alpha \int d^2z J\bar{J}. \tag{3.3}$$

This deformation is equivalent to a particular 1–parameter sub-family of $O(2,2,R)$ rotations acting on the background matrix and dilaton in (3.1). Parametrizing this family by $0 \le R < \infty$, one finds the line of exact CFTs with background matrices:

$$kE_R = k \begin{pmatrix} \tan^2 x/\Delta & 1/\Delta & 0 \\ -1/\Delta & R^2/\Delta & 0 \\ 0 & 0 & 1 \end{pmatrix}, \qquad \Delta = 1 + R^2 \tan^2 x. \tag{3.4}$$

The dilaton also transforms under the deformation (see [1]).

Some special points along the R-line of deformations are:

(1) At $R = 1$, one recovers the original $SU(2)$ point (3.1).

(2) At $R \to \infty$, the background corresponds to a direct product of an axially gauged $SU(2)/U(1)_a$ sigma-model [15], and a non-compact free scalar field.

(3) At $R \to 0$, the background corresponds to a direct product of a vectorially gauged $SU(2)/U(1)_v$ sigma-model, and a non-compact free scalar field.

We now arrive to the important point of this part. The Weyl reflection $J \to -J$ is given by a continuous group rotation at the WZNW point. This implies that $J \to -J$ is a symmetry at the WZNW point, and therefore, the infinitesimal deformation $S_{WZNW} \to S_{WZNW} + \delta\alpha \int d^2z J\bar{J}$ is the same CFT as the one given by the deformation $-\delta\alpha \int d^2z J\bar{J}$.

The points $\delta\alpha$ and $-\delta\alpha$ along the α–modulus (3.3) are *the same CFT*. In string theory we say: $\delta\alpha$ and $-\delta\alpha$ are related by a residual Z_2 symmetry of the broken gauge symmetry (of the enhanced symmetry point); the residual symmetry is the target space duality. This symmetry can be integrated to finite α, giving rise to a $Z_2 \in O(2,2,Z)$ duality[5] along the R–line (3.4). The action of duality on the background E_R is:

$$\text{Duality along the } R-\text{line}: \quad E_R \to E_{1/R} \quad i.e. \quad R \to 1/R. \tag{3.5}$$

In particular, the boundary points $R \to 0$ and $R \to \infty$ are the same CFT. As a consequence, axial-vector duality of $SU(2)/U(1)$ and $SL(2)/U(1)$ is exact, and corresponds to a residual discrete symmetry of the broken gauge algebra.

I shall end this part with few remarks:

(a) In the $SU(2)$ ($SL(2)$) case, the CFT along the R-line is a Z_k orbifoldization of a compact (non-compact) parafermionic theory and a free scalar field with radius $\sqrt{k}R$.

(b) In the $SU(2)$ case, the modular invariant R-dependent genus one partition function can be written.

(c) The results of this section can be extended to *any* group G [1].

4   Topology change in the space of WZNW sigma-models

Along the $0 < R < \infty$ deformations line of the $SU(2)$–WZNW sigma-model, the topology of the background space is of the three sphere. However, at the boundaries ($R = 0, \infty$), the topology is changed to that of a product of a two-disc (corresponding to the $SU(2)/U(1)$ conformal background) with a circle whose radius shrinks to 0. In the "$SL(2)$" case, the background at the boundary is the direct product of a degenerated

---

[5]The $O(d,d,Z)$ group is a duality (sub-)group also for *curved* backgrounds with d abelian isometries [14].

circle with a semi-infinite cigar (at $R \to \infty$), or the infinite trumpet (at $R \to 0$); these correspond to the dual pair of the 2-d euclidean black-hole backgrounds.

A topology change at the boundary of moduli space is not surprizing. However, the R-line of deformations is not the full story in the moduli space of G-WZNW sigma-models. In fact, any conformal sigma-model with d abelian isometries can be transformed to a new conformal background by $O(d, d, R)$ rotations [16, 14].[6] In this bigger moduli space of sigma-models, there are some more interesting deformation lines, along which the topology might change.

For example, in the moduli space of the $SU(2)$–WZNW sigma-model, there is a one-parameter sub-family of $O(2, 2, R)$ rotations that generate backgrounds with metric [1]

$$ds^2(\alpha) = \frac{k}{\Delta(\alpha)}[\sin^2 x d\theta^2 + \cos^2 x d\tilde\theta^2] + kdx^2,$$

$$\Delta = \cos^2 \alpha \cos^2 x + (\cos\alpha + k\sin\alpha)^2 \sin^2 x. \tag{4.1}$$

(There are also non-trivial dilaton and torsion along this α–line). At the point $\alpha = 0$ the background includes a metric of the $SU(2)_k$ group manifold S^3 (as well as an antisymmetric background). Along the line $0 < \alpha < \pi/2$ the background includes the metric (4.1) with the topology of S^3 (as well as an antisymmetric background and a dilaton field). At the point $\alpha = \pi/2$ the background metric is

$$ds^2(\alpha = \pi/2) = \frac{1}{k}d\theta^2 + \frac{1}{k}\cot^2 x d\tilde\theta^2 + kdx^2. \tag{4.2}$$

At this point the manifold has a topology of $D_2 \times S^1_{1/k}$, where D_2 is a two–disc and $S^1_{1/k}$ is a circle with radius $R^2 = 1/k$. One may continue to deform this theory by, for example, changing the compactification radius R of the free scalar field θ.

It is remarkable that (for integer k) the neighborhood of the point $\alpha = \pi/2$ is mapped to the neighborhood of the point $\alpha = 0$ by an element of $O(2, 2, Z)$, namely, a target space generalized duality. Therefore, a region in the moduli space, where a topology change occurs, is mapped to a region where there is no topology change at all. A similar phenomenon happens for more complicated examples in the moduli space of Calabi-Yau compactifications [18].

Finally, let me make few comments:

(a) The sigma-models along the α–line (4.1) have conical singularities. Therefore, to make sense of the CFTs along the α–line one should understand CFTs corresponding to backgrounds with (non-orbifold) conical singularities.

(b) After the topology is changed at (4.2), by deforming the compactification radius of the free scalar field θ, one does not get rid of the curvature singularity encountered at the point where the topology is changed.

---

[6]In the bosonic string, the $O(d, d)$ rotations give only the leading order in α' of the conformal backgrounds; but, there are higher order corrections that make them exact [14, 4, 17].

(c) To cure both problems, one should look at the moduli space of WZNW sigma-models in higher dimensions; for instance, the sigma-model moduli space discussed for cosmological backgrounds in [4].

Acknowledgements

I thank S. Elitzur for his remarks on the manuscript. This work is supported in part by the BSF - American-Israeli Bi-National Science Foundation.

References

[1] A. Giveon and E. Kiritsis, preprint CERN-TH.6816/93, RI-149-93, hep-th/9303016.

[2] A. Giveon, Mod. Phys. Lett. **A6** (1991) 2843;
R. Dijkgraaf, E. Verlinde, and H. Verlinde, Nucl. Phys. **B371** (1992) 269;
E.B. Kiritsis, Mod. Phys. Lett. **A6** (1991) 2871;
I. Bars, preprint USC-91-HEP-B3.

[3] A.A. Tseytlin, Mod. Phys. Lett. **A6** (1991) 1721;
A.A. Tseytlin and C. Vafa, Nucl. Phys. **B372** (1992) 443.

[4] A. Giveon and A. Pasquinucci, Phys. Lett. **B294** (1992) 162.

[5] K. Kikkawa and M. Yamasaki, Phys. Lett. **B149** (1984) 357;
N. Sakai and I. Senda, Prog. Theor. Phys. **75** (1986) 692.

[6] M. Dine, P. Huet and N. Seiberg, Nucl. Phys. **B322** (1989) 301.

[7] A. Giveon, E. Rabinovici and G. Veneziano, Nucl. Phys. **B322** (1989) 167.

[8] A. Shapere and F. Wilzcek, Nucl. Phys. **B320** (1989) 669.

[9] A. Giveon, N. Malkin and E. Rabinovici, Phys. Lett. **220B** (1989) 551.

[10] A. Giveon, N. Malkin and E. Rabinovici, Phys. Lett. **B238** (1990) 57.

[11] A. Giveon and M. Porrati, Phys. Lett. **B246** (1990) 54; Nucl. Phys. **B355** (1991) 422.

[12] A. Giveon and A. Shapere, Nucl. Phys. **B386** (1992) 43.

[13] G. Moore, hep-th/9305139.

[14] A. Giveon and M. Roček, Nucl. Phys. **B380** (1992) 128.

[15] K. Bardakci, M. Crescimanno and E. Rabinovici, Nucl. Phys. **B344** (1990) 344.

[16] K.A. Meissner and G. Veneziano, Phys. Lett. **B267** (1991) 33;
M. Gasperini, J. Maharana and G. Veneziano, Phys. Lett. **B272** (1991) 277;
A. Sen, Phys. Lett. **B271** (1991) 295.

[17] E. Kiritsis, ENS preprint LPTENS-92-29, hepth/9211081; CERN preprint, hepth/9302033.

[18] P.S. Aspinwall, B.R. Greene and D. Morrison, Phys. Lett. **B303** (1993) 249;
E. Witten, Nucl. Phys. **B403** (1993) 159.

THE DUAL FACES OF STRING THEORY

ELIAS KIRITSIS

Theory Division, CERN, CH-1211
Geneva 23, Switzerland

ABSTRACT

Duality symmetries for strings moving in non-trivial spacetime backgrounds are analysed. It is shown that, for backgrounds generated from compact WZW and coset models, such duality symmetries are exact to all orders in string perturbation theory. A global treatment of duality symmetries is given, by associating them to the known symmetries of affine current algebras (affine-Weyl group and external automorphisms). It is argued that self-duality symmetries of WZW and coset models generate the duality symmetries of their moduli space. Some remarks are presented, concerning the survival of such symmetries in the non-compact case. The implications of duality symmetries for string dynamics in non-trivial/singular spacetimes are discussed.

1. Introduction

Strings, being extended objects, sense the target space into which they are embedded, in a different way than point particles. In a compact space this difference appears because, strings, except from their local excitations that mimic point particle behavior ("momentum" modes), have "winding" excitations where the string wraps around non-contractible cycles of the manifold. The masses of momentum modes are inversely proportional to the volume of the manifold, whereas those of the winding modes are proportional to the volume, since it costs energy in order to stretch the string. Moreover, the string contains oscillating modes that respond to background fields differently than the center of mass of the string. In certain cases, the physics of string propagation remains invariant under a reorganization of the one string Hilbert space and a specific change in the background. This symmetry is known as duality. In the simplest possible example, that of a string moving on a circle, it was observed that the spectrum of the theory with radius R and that with radius $1/R$ are identical, once we interchange winding and momentum modes[1].

It turns out that such duality symmetries exist (semi-classically) for all backgrounds with isometries[2]. In CFT, some of these symmetries were identified as different abelian gaugings of a WZW theory[3], and this was generalized to abelian gaugings of arbitrary theories with chiral currents[4], and organized into (semi-classical) $O(d,d,Z)$ type symmetries[5], mimicking the situation for flat backgrounds. Moreover, for coset models, such duality symmetries exist also for backgrounds without any isometries[3, 6]. Non-abelian duality received more attention recently[7] but its status is not yet clear. A careful analysis of the underlying CFT structure, revealed that most of these semiclassical symmetries, pertaining to compact cosets, are indeed exact in string theory[6], and they are intimately related to the affine Weyl symmetries of the "parent" theory, the WZW model.

In the non-compact case, the affine Weyl group is not a manifest symmetry but it can be shown that a particular kind of duality, axial-vector duality[3] is still a symmetry[8]*.

At the semiclassical level, provided there is an abelian isometry, the duality transformation can be effected by gauging this isometry and adding also a langrange multiplier coupled to the field strength of the gauge field[2]. Integrating out the langrange multiplier, forces the gauge field to be pure gauge which can, subsequently, be gauged away, giving back the original model. On the other hand, one can gauge fix to a unitary gauge and then integrate out the gauge field (which appears quadratically in the action). In this way, a different (dual) sigma model action is obtained (the measure can be also taken care off, effectively changing the dilaton). Modulo global properties[9], the original and the dual action describe the same theory.

Duality has important implications for string theory. It can be thought of as an unbroken part of the full string symmetry in a particular background, and it can provide important hints concerning the string physics around that background. In particular, all possible dual backgrounds are relevant for string propagation as each determines the response of some of the string modes. It is also interesting that duality seems to preserves some general relativistic notions as that of a Hawking temperature and entropy of black holes[10].

Explicit studies in specific models[6] (although the result seems to be general) indicate that, in the non-flat case, duality maps zero modes to oscillator modes of the string. In some coordinate system, such oscillator modes resemble winding modes, the only difference being that they are not "topological" (the target manifold has no non-contractible cycles). I will comment more on the possible implications of duality for string physics in non-compact curved backgrounds in the last section.

2. WZW Models

In this section, we will analyze in detail the duality symmetries of WZW models, both from the σ-model and the CFT (affine current algebra) point of view.

It turns out that understanding the simplest group, SU(2), will suffice. In the case of non-simply laced simple groups there are some minor changes due to the short roots that will be dealt with latter on. The case of non-simple groups has further complications that we will not consider here.

The action of the WZW model is

$$I(g) = \frac{k}{4\pi} I_{NS}(g) + \frac{ik}{6\pi} \Gamma_{WZ}(g) \tag{2.1}$$

---

*See also A. Giveon's talk in this volume.

$$I_{NS}(g) = \int d^2x\, Tr[U_\mu U^\mu] \quad , \quad \Gamma_{WZ}(g) = \int_{\substack{B \\ \partial B = S^2}} d^3y\, \varepsilon^{\mu\nu\rho} Tr[U_\mu U_\nu U_\rho] \tag{2.2}$$

where

$$U_\mu = g^{-1}\partial_\mu g \quad , \quad V_\mu = \partial_\mu g g^{-1} \tag{2.3}$$

g is a matrix in the fundamental representation of G and Tr is a properly normalized trace such that

$$\frac{1}{12\pi^2} \int_{S^3} Tr[U \wedge U \wedge U] \in Z \ . \tag{2.4}$$

The action $I(g)$ is invariant under the group $G_R \otimes G_L$, generated by left and right group transformations, $g \to h_1 g h_2$, with associated conserved currents

$$J_R^\mu = \frac{k}{2\pi} P_-^{\mu\nu} U_\nu \quad , \quad J_L^\mu = \frac{k}{2\pi} P_+^{\mu\nu} V_\nu \tag{2.5}$$

with $P_\pm^{\mu\nu} \equiv \delta^{\mu\nu} \pm i\varepsilon^{\mu\nu}$. These currents are conserved and chirally conserved and they generate two copies of the affine $\hat{G}$ current algebra. An important property of the WZW action is that it satisfies the Polyakov-Wiegman formula

$$I(gh) = I(g) + I(h) - \frac{k}{2\pi} \int d^2x\, P_+^{\mu\nu} Tr[U_\mu(g)V_\nu(h)] \tag{2.6}$$

To generate duality transformations in the WZW model, we pick a generator of the Lie algebra of G, T^0, normalized as $Tr[(T^0)^2] = 1$. We can then parametrize $g = e^{i\phi T^0} h$. Using (1.6), the action $I(g)$ takes the form

$$I(g) = I(h) + \frac{k}{4\pi} \int \partial_\mu\phi\partial^\mu\phi - \frac{ik}{2\pi} \int P_+^{\mu\nu} \partial_\mu\phi V_\nu^0(h) \tag{2.7}$$

where $V_\mu^0(h) = Tr[T^0 V_\mu(h)]$. We can now apply the duality map[3] to obtain[†]

$$I^{\text{dual}}(g) = I(h) + \frac{1}{4\pi k} \int \partial_\mu\phi\partial^\mu\phi - \frac{i}{2\pi} \int P_+^{\mu\nu} \partial_\mu V_\nu^0(h) \ . \tag{2.8}$$

The angle ϕ was originally normalized to take values in $[0, 2\pi]$. It is obvious from (2.8) that the effect of the duality transformation is to change the range of values to $[0, 2\pi/k]$. To see how many independent duality transformations exist, we have to explicitly parametrize the Cartan torus dependence of the WZW model. Pick a basis in the Cartan algebra, T^i, $i = 1, 2, \cdots, r$, $[T^i, T^j] = 0$, $Tr[T^i T^j] = \delta^{ij}$ and parametrize,

$$g = e^{i\sum_{i=1}^r \alpha^i T^i} h\, e^{i\sum_{i=1}^r \gamma^i T^i} \ . \tag{2.9}$$

Then using (2.6) the WZW action becomes

$$I(g) = I(h) + \frac{k}{4\pi} \int (\partial_\mu\alpha^i \partial^\mu\alpha^i + \partial_\mu\gamma^i \partial^\mu\gamma^i) - \frac{ik}{2\pi} \int (P_+^{\mu\nu} \partial_\mu\alpha^i V_\nu^i(h) + P_-^{\mu\nu} \partial_\mu\gamma^i U_\nu^i(h)) +$$

$$+ \frac{k}{2\pi} \int P_+^{\mu\nu} \partial_\mu\alpha^i \partial_\nu\gamma^j M^{ij}(h) \tag{2.10}$$

[†]The measure also changes by a finite computable piece[3]

where

$$U^i_\mu(h) = Tr[T^i U_\mu(h)] \, , \ \ V^i_\mu(h) = Tr[T^i V_\mu(h)] \, , \ \ M^{ij}(h) = Tr[T^i h T^j h^{-1}] \, . \tag{2.11}$$

It is obvious from (2.10) that we can apply the duality transformation using any of the α^i, γ^i. Thus, there are $2^{2r} - 1$ non-trivial duality transformations. A duality transformation on α^i effectively makes the substitution $\alpha^i \to \alpha^i/k$ in the action whereas a duality transformation on γ^i makes the substitution $\gamma^i \to -\gamma^i/k$. The new manifold has a similar metric to the group manifold but due to the different angle periodicities, has Taub-NUT type singularities.

In order to identify the underlying property of the WZW model, responsible for the invariance under these duality transformations, we have delve a bit into such elements of the the representation theory of the affine Lie algebras as the affine Weyl group and external automorphisms.

The affine Weyl group $\hat{W}$ is a semidirect product of the Lie algebra Weyl group W times a translation group, $\hat{W} = W \triangleright T$. Apart from the action of finite Weyl group elements, there are Weyl transformations associated to roots which have a component in the direction of the imaginary simple root. The action of such an element $\hat{W}_{\tilde{\alpha}}$ on a finite Lie algebra weight $\vec{\lambda}$ and on the grade n is

$$\hat{W}_{\tilde{\alpha}}(\vec{\lambda}) = W_{\tilde{\alpha}}(\vec{\lambda}) - k\vec{\beta} \tag{2.12a}$$

$$\hat{W}_{\tilde{\alpha}}(n) = n - \vec{\lambda} \cdot \vec{\beta} - \frac{k}{2}\vec{\beta} \cdot \vec{\beta} \tag{2.12b}$$

where $\vec{\beta} = 2\vec{\alpha}/\vec{\alpha} \cdot \vec{\alpha}$ is the coroot associated to the finite Lie algebra root $\vec{\alpha}$, the grade n is basically the mode number[‡] and $W_{\tilde{\alpha}}(\vec{\lambda}) = \vec{\lambda} - \vec{\alpha}(\vec{\lambda} \cdot \vec{\beta})$ is a finite Weyl transformation. It is important to note that affine Weyl transformations, in general, map states inside a representation at different levels.

There are also external automorphisms of the affine algebra which are essentially associated to symmetries of the affine Dynkin diagram. For the $SU(n)$ case, the affine Dynkin diagram consists of n nodes connected around a circle. The external automorphisms are generated by a basic rotation, and a reflection which corresponds to the finite Lie algebra external automorphism (that maps a representation to its complex conjugate). When we write a highest weight $\vec{\Lambda} = \sum_{i=1}^{n-1} m_i \vec{\Lambda}_i$ in terms of the fundamental weights $\vec{\Lambda}_i$, (m_i are non-negative integers), the action of the generating rotation of the affine Dynkin diagram is as follows

$$\sigma(\vec{\Lambda}) = (k - \sum_{i=1}^{n-1} m_i)\vec{\Lambda}_1 + m_1\vec{\Lambda}_2 + \cdots + m_{n-2}\vec{\Lambda}_{n-1} \, . \tag{2.13}$$

σ generates a Z_n group[§] where $\sigma^n = 1$ on the highest weights, but acts as an affine Weyl transformation in the representation. Specializing to SU(2), let $m \in Z/2$ be the weight, and $j \in Z/2$ the highest weight (spin of a representation). Then the finite Weyl group acts as $m \to -m$, and combined with the affine translation $m \to m + k$

[‡] In a highest weight representation where the affine primaries have L_0 eigenvalue Δ, the grade n of a state is the eigenvalue of $L_0 - \Delta$ on that state.

[§] In general this group is isomorphic to the center of the finite Lie group they generate the affine Weyl group. The only nontrivial outer automorphism σ acts as $j \to k - j$ and σ^2 is a Weyl translation.

The non-trivial statement now is: For compact groups, integer level and integrable highest weight representations, both the affine Weyl group and the external automorphisms are symmetries. In particular, in a WZW model the Hilbert space is constructed by tying together (in a modular invariant way) two copies of representations of the affine algebra. Thus, we have invariance under independent affine Weyl transformations acting on left or right representations. Moreover, since the modular transformation properties of the affine characters reflect the external automorphism symmetries, the theory is invariant under external automorphisms that act at the same time on left and right representations. These invariance properties can be verified for correlation functions on the sphere and the torus. This then implies that they hold on an arbitrary Riemann surface since the sphere and torus data are sufficient in order to construct the correlators at higher genus.

As an example, we will present the SU(2) case and focus on the spectrum. We introduce the (affine) $SU(2)_k$ characters

$$\chi_l(q = e^{2\pi i \tau}, w) = Tr_l \left[q^{L_0} e^{2\pi i w J_0^3} \right] = \sum_{m=-k+1}^{k} c_m^l(q) \vartheta_{m,k}(q, w) \qquad (2.14)$$

where l is twice the spin (a non-negative integer) and m is twice the J_0^3 eigenvalue. The trace is in the affine hw representation of spin l,

$$\vartheta_{m,k}(q, w) = \sum_{n \in Z} q^{k(n + \frac{m}{2k})^2} e^{2\pi i w(kn + \frac{m}{2})} \qquad (2.15)$$

and c_m^l are the standard string functions which satisfy $c_m^l = 0$ when $l - m = 1 (mod\ 2)$ (which means that the spin is increased or decreased in units of 1) . For integrable representations (k is a positive integer and $0 \le l \le k$), invariance under the affine Weyl group is equivalent to

$$c_m^l = c_{-m}^l \quad , \quad c_m^l = c_{m+2k}^l \qquad (2.16)$$

The first relation is due to the Weyl group of $SU(2)$ while the second is the generating translation in the affine Weyl group. There is another important relation

$$c_m^l = c_{k-m}^{k-l} \qquad (2.17)$$

which is a consequence of the external affine automorphism.

The duality transformation on α^i amounts to replacing $\bar{J}^i \to -\bar{J}^i$, where $\bar{J}^i$ is the right Cartan current in the T^i basis of the Cartan subalgebra. Similarly the duality transformation on γ^i amounts to the replacement $J^i \to -J^i$ at the level of the Cartan subalgebra. This is not the whole story however. With a bit more effort one can see that they act as Weyl transformations on the left or right SU(2) currents. This identification can be seen clearly by coupling the WZW action to external gauge fields and monitoring the effect of the duality transformation on the currents. It can also be recovered from the twisted partition function via the action of the duality transformation on the gauge field moduli (for the Cartan).

The duality transformations

$$D_i \; : \; J^i \to -J^i \tag{2.18a}$$

$$\bar{D}_i \; : \; \bar{J}^i \to -\bar{J}^i \tag{2.18b}$$

are exact symmetries of the model. This can be verified explicitly, since characters are invariant under the finite Weyl group.

This invariance is similar, but qualitatively different than that present in flat backgrounds. There, one has a family of theories parametrized by G, B and duality is the statement that two theories are equivalent for different values of the parameters. Here, there is no parameter present and, in this sense, this is what we could call self-duality. It corresponds to the self duality (in the flat case) of level 1 WZW models appearing at special values of the moduli. Now we are in a position to discuss the general WZW model for a simple group G. Let M be the root lattice, M_L the long root lattice and M^* the weight lattice. The character of a hw representation of $\hat{G}$ with hw $\vec{\Lambda}$ is defined as

$$\chi_{\vec{\Lambda}}(q, \vec{w}) = Tr[q^{L_0} e^{2\pi i \vec{w} \cdot \vec{J}_0}] \tag{2.19}$$

where $\vec{J}_0$ generates the cartan subalgebra of G. The character admits the string function decomposition[11],

$$\chi_{\vec{\Lambda}} = \sum_{\vec{\lambda} \in M^*/kM_L} c_{\vec{\lambda}}^{\vec{\Lambda}}(q) \Theta_{\vec{\lambda}}(\vec{w}, q) \tag{2.20}$$

with $\Theta_{\vec{\lambda}}$ being the classical ϑ-function of level k of the Lie algebra of G

$$\Theta_{\vec{\lambda}}(\vec{w}, q) = \sum_{\vec{\gamma} \in M_L} q^{\frac{k}{2}\left(\vec{\gamma} + \frac{\vec{\lambda}}{k}\right)^2} e^{2\pi i \vec{w} \cdot (k\vec{\gamma} + \vec{\lambda})} \; . \tag{2.21}$$

The string functions are invariant under the Weyl group and Weyl translations

$$c_{w(\vec{\lambda})}^{\vec{\Lambda}} = c_{\vec{\lambda}}^{\vec{\Lambda}} \; , \quad c_{\vec{\lambda} + k\vec{\beta}}^{\vec{\Lambda}} = c_{\vec{\lambda}}^{\vec{\Lambda}} \tag{2.22}$$

where w is a Weyl transformation and $\vec{\beta} \in M_L$.

The (left) generating duality transformations D_i correspond to Weyl reflections generated by the simple roots $\vec{\alpha}_i$ which implement the transformations (2.18a). The invariance of the spectrum (and partition function) is encoded in the fact, obvious from (2.21,22), that $\chi_{\vec{\Lambda}}$ is invariant under $w_i \to -w_i$. Although $w_{\vec{\alpha}_i}$ do not commute, they do so when applied to the character, thus at the level of the partition function they generate a group isomorphic to Z_2^r. However, at the level of correlation functions the (left) duality group is larger and in fact isomorphic to $W_L \times W_R/W_D$, where $W_{L,R}$ are the left(right) Weyl groups of the (finite) Lie algebra of the WZW model and W_D is the diagonal Weyl group (whose action corresponds to reparametrizations of the action).

We have seen already (from affine algebra representation theory) that the (non-local) symmetry of the (compact and unitary) g-WZW model is larger: It is generated by the left and right affine Weyl groups $\hat{W}^g_{L,R}$ as well as the external affine automorphisms A^g. I conjecture that $\hat{W}^g_L \times \hat{W}^g_R \times A^g/W_D$ is the full self-duality group of the WZW. Afine Weyl transformations act semi-classically as GL(r,Z) rotations, that is, as particular changes of basis in the lattice of weights. This implies, that although semi-classically, for curved backgrounds $GL(r, Z)$ is a symmetry[5], in the exact theory only some part of it survives. Concerning the external affine automorphisms, their action in σ-model language is not known.

3. Compact Coset Models

From the WZW theory, we can built other CFTs by projections. The simplest such projection corresponds to constraining the affine currents of a subalgebra, known as coset construction[13].¶ The σ-model action of coset models is obtained by gauging the appropriate subgroup of the WZW model. Gauging different dual versions of the WZW model, dual versions of the coset model are obtained. It can be shown that the Killing symmetries of a coset model G/H are of two types[6]: Chiral $H'_L \times H'_R$ isometries, when there is a subgroup H' such that $[H, H'] = 0$ and non-chiral abelian isometries in one to one correspondence with U(1) factors of H. In many cases, the aforementioned duality exists for coset actions without isometries[6].

A special form of duality is obtained for coset models where the gauged subgroup contains a U(1) factor. In such a case, one has the option of gauging either the axial or the vector subgroup of the original $U(1)_L \times U(1)_R$ subalgebra. The σ-model actions of these two gauged models are generically different but it can be shown that the models are dual to each other[3]. This type of duality is known as axial-vector duality and at the semiclassical level is powerful in generating different types of backgrounds[4, 5]. It would seem that Weyl symmetry of the affine algebra is enough to guarantee that axial-vector duality is an exact symmetry. The story is more complicated though. It turns out[6] that the underlying symmetry of current algebra responsible for axial-vector duality is affine-Weyl symmetry, $\hat{W}_L \times \hat{W}_R$. For integrable (unitary) representations of (compact) affine algebras the affine Weyl group is a symmetry and so is axial-vector duality.

The full duality symmetry of a coset model g/h is generated by the full duality symmetry of the "parent" g-WZW theory. Let H be a reductive subgroup of G and H^{na} be its non-abelian component, (discard U(1) factors). Denote the Lie algebra of H^{na} by h. Then, $D^h = \hat{W}^h_L \times \hat{W}^h_R \times A^h$ is a normal subgroup of $D^g = \hat{W}^g_L \times \hat{W}^g_R \times A^g$. The full self-duality group of G/H is D^g/D^h. The need to factor D^h comes since it

---

¶There are more general projections though, preserving conformal invariance[12].

acts only on the gauge degrees of freedom and it is thus, invisible in the G/H theory.

4. Duality Symmetries in the Moduli Space of WZW and Coset Models.

So far we have been concerned with the self duality symmetries of WZW and coset models. We will now show how this self duality generates duality symmetries of their moduli space.

The primary example of this is the moduli space of (flat) D-dimensional toroidal models. At special points of the moduli, one finds WZW models (at level one). It can be shown in this case that, the full duality group of the moduli space, $O(D, D, Z)$ can be generated from the self duality symmetries of the WZW points[14].

A similar phenomenon happens in the general case. Consider a WZW (or coset) model and its neighborhood in moduli space which is generated via marginal perturbations by integrable (1,1) operators $O_{1,1}^i$. This set of operators contains dual pairs, that is for each $O_{1,1}^i$ there is a $\tilde{O}_{1,1}^i$ in the set related by a self-duality of the WZW or coset theory. At the level of the (abstract) conformal field theory, O^i and $\tilde{O}^i$ correspond to the same operator. In a σ-model (field theoretic) realization, they are different fields.

Self-duality at the WZW or coset point implies that the line generated by O^i is equivalent as a CFT to that generated by $\tilde{O}^i$. The backgrounds (σ-models) corresponding to the two lines though are different. Thus, in the full moduli space, duality symmetries can be generated by the self-duality symmetries at special points (WZW and cosets)

In the case of a G-WZW model for simple G, at a generic level the only integrable perturbations are generated by $J^i\bar{J}^j$ constructed out of the left and right Cartan currents. Thus (except at special points) the dimension of the moduli space is D^2 where $D = rank_G$. The duality group here is similar to $O(D,D,Z)$, since in such deformations the generalized G-parafermion theory does not change along the moduli space and duality acts on the cartan-torus bosons. When G is semi-simple, then one has extra generic (1,1) operators, however their integrability is an open question.

Similar remarks apply to perturbations by relevant operators. In this case self-duality of the WZW or coset theory implies Krammers-Wanier duality for the off-critical theory. The archetype of this duality exists in the Z_N parafermion models perturbed by the first energy operator, which upon a self-duality transformation in the fixed point theory, changes sign.

It is tempting to conjecture that all duality symmetries of the full moduli space of WZW and coset models are reflections of self-duality at these special points, as it happens in the flat case.

5. Duality in the Non-compact Case.

So far we have dealt with compact WZW models and their cosets. In the non-compact case things can, a priori, be different. At the level of duality symmetries of the G-WZW model itself, those associated with the finite Weyl group are still symmetries (if G is reductive). However the affine Weyl acts differently. For a generic affine representation, affine Weyl transformations map it to different representations. We have seen that affine Weyl symmetry is important for axial-vector duality. The only way that axial-vector duality can survive in the non-compact case is, if the spectrum is organized into complete orbits of the affine Weyl group.

There is a different way of showing that axial-vector duality remains an exact symmetry in the non-compact case[8]. This, in retrospect implies the organization of the spectrum as mentioned above. Important information about the spectrum can be retrieved that way since, (at least for the case of SL(2,R)) non-compact string functions are known[15] as well as their transformation properties under the affine Weyl group. It is an interesting problem to try to obtain the spectrum this way.

6. Comments on the Physics of Duality.

The main lesson from duality and related symmetries is that the background fields do not determine uniquely the spectrum and physics of string theory. Duality can be viewed as a tiny (unbroken) part of the huge string gauge symmetry, whose glory remains obscure to our days. Another way to state this, is, that different modes of the string feel different geometry. Thus, the background interpretation of string vacua should be used with care in order to ascertain the physics. The only cases were the description is reliable is the large volume limit of compact manifolds, and at the asymptotically flat region of non-compact manifolds. Once at a region of finite curvature, the geometrical description of string theory breaks down. Even topology is not preserved under duality and there are continuous families of ground states in string theory where topology changes without the occurrence of anything catastrophic[17, 16, 8].

An example of how this type of symmetry can affect string propagation, can be given (heuristically) as follows[||]. Consider a string background which is highly curved or even singular (semi-classically) in a certain region, (the 2-d black hole[18], is such an example but one needs to add a few extra dimensions in order to have non-trivial massive states). In the asymptotic region, (which is obtained by some spacetime-depended radius becoming very large), one has quantum numbers for asymptotic states that correspond roughly to windings and momenta. Momentum states are the only low energy states in this region. Consider a momentum mode travelling towards the high curvature region. Its effective mass starts growing as it approaches large curvatures. At some point it becomes energetically possible for it to decay to winding states which, in this region, start having effective masses that are lower than momentum modes. In such backgrounds (unlike flat ones) winding and momentum are not separately conserved so that such a transition is possible. The reason for this is that there is a non-trivial dilaton field and thus, winding and momentum conservation is broken by the screening operators which transfer it to

---

[||]This argument is advanced in collaboration with C. Kounnas.

discrete states localized at the high curvature region. An alternative interpretation of this, is that particles interact with such localized states loosing momentum (in discrete steps) and gaining winding number.

Once such a momentum to winding mode transition happens in the strongly curved region, the winding state sees a different geometry, namely the dual one and thus continues to propagate further into the strong curvature region since it feels only the (weak) dual curvature. This phenomenon, implies the need of a novel treatment of the underlying effective field theory approach, which would resemble the effective treatment of particles in condensed matter physics, with position dependent masses.

This type of picture indicates that the physics of string black holes will be qualitatively different that their classical general relativity counterparts.

References

1. K. Kikkawa and M. Yamazaki, *Phys. Lett.* **B149** (1984) 357;
   N. Sakai and I. Senda, *Prog. Theor. Phys.* **75** (1986) 692;
   E. Alvarez and M. Osorio, *Phys. Rev.* **D40** (1989) 1150;
   V. Nair, A. Shapere, A. Strominger and F. Wilczek, *Nucl. Phys.* **B287** (1987) 402;
   A. Shapere and F. Wilczek, *Nucl. Phys.* **B320** (1989) 609;
   A. Giveon, E. Rabinovici and G. Veneziano, *Nucl. Phys.* **B322** (1989) 167.
2. T. Buscher, *Phys. Lett.* **B194** (1987) 51; *ibid.* **B201** (1988) 466; PhD Thesis, unpublished.
3. E. Kiritsis, *Mod. Phys. Lett.* **A6** (1991) 2871.
4. M. Roček and E. Verlinde, *Nucl. Phys.* **B373** (1992) 630.
5. A. Giveon and M. Roček, *Nucl. Phys.* **B380** (1992) 128.
6. E. Kiritsis, CERN preprint, CERN-TH.6797/93, hepth/9302033.
7. X. De la Ossa and F. Quevedo, Neuchâtel preprint, NEIPH92-004, hepth/9210021;
   M. Gasperini, R. Ricci and G. Veneziano, CERN preprint, CERN-TH.6960/93, hepth/9308112;
   A. Giveon and M. Roček, Racah Inst. preprint, RI-152-93, hepth/9308154.
8. A. Giveon and E. Kiritsis, CERN preprint, CERN-TH.6816/93, hepth/9303016.
9. E. Alvarez, L. Alvarez-Gaumé, J. Barbon, Y. Lozano, CERN preprint, CERN-TH.6991/93.
10. G. Horowitz and D. Welch, ITP preprint, NSF-ITP-93-89, hepth/9308077.
11. V. Kač, D. Peterson, Adv. Math. 53 (1984) 125.
12. M. Halpern and E. Kiritsis, *Mod. Phys. Lett.* **A4** (1989) 1373.
13. K. Bardakci and M. Halpern, *Phys. Rev.* **D3** (1971) 2493;

310

M. Halpern, *Phys. Rev.* **D4** (1971) 2398;
P. Goddard, A. Kent and D. Olive, *Phys. Lett.* **B152** (1985) 88.

14. A. Giveon, N. Malkin and E. Rabinovici, *Phys. Lett.* **B238** (1990) 57.

15. I. Bakas and E. Kiritsis, *Int. J. Mod. Phys.* **A7** [**Supp. 1A**] (1992) 55.

16. P. Aspinwall, B. Greene, D. Morrison, *Phys.Lett.* **B303** (1993) 249.

17. E. Witten, *Nucl. Phys.* **B403** (1993) 159.

18. E. Witten, *Phys. Rev.* **D44** (1991) 314.

WHERE IS THE LARGE RADIUS LIMIT?

DAVID R. MORRISON

School of Mathematics, Institute for Advanced Study, Olden Lane
Princeton, NJ 08540, USA

ABSTRACT

By properly accounting for the invariance of a Calabi-Yau sigma-model under shifts of the B-field by integral amounts (analagous to the θ-angle in QCD), we show that the moduli spaces of such sigma-models can often be enlarged to include "large radius limit" points. In the simplest cases, there are holomorphic coordinates on the enlarged moduli space which vanish at the limit point, and which appear as multipliers in front of instanton contributions to Yukawa couplings. (Those instanton contributions are therefore suppressed at the limit point.) In more complicated cases, the instanton contributions are still suppressed but the enlarged space is singular at the limit point. This singularity may have interesting effects on the effective four-dimensional theory, when the Calabi-Yau is used to compactify the heterotic string.

1. Integral Shifts of the B-Field

To write a Lagrangian for the nonlinear sigma-model on a Calabi-Yau manifold X, we must make a choice of metric g_{ij} (Ricci-flat in the one-loop approximation) and "B-field" (a real closed 2-form). This B-field enters into the action only through integration over the world-sheet, in a term proportional to $\int_\Sigma \phi^*(B)$. (The notation refers to a map ϕ from the worldsheet Σ to the target space X, assumed to satisfy $h^{2,0}(X) = 0$.) In fact, this term can naturally be combined with a contribution from the Kähler form J of the metric to produce a term in the action proportional to $\int_\Sigma \phi^*(B + iJ)$. When B, J and the string tension are normalized properly, the partition function and the correlators involve precisely the quantity $\exp\left(2\pi i \int_\Sigma \phi^*(B + iJ)\right)$.

Altering B by adding an exact 2-form to it does not alter any of the quantities $\int_\Sigma \phi^*(B)$, so only the de Rham cohomology class of B matters for specifying the Lagrangian. Furthermore, if we replace B by $B + B_0$ where B_0 represents an *integral* cohomology class, then the crucial quantity $\exp\left(2\pi i \int_\Sigma \phi^*(B + iJ)\right)$ is left unchanged since each $\int_\Sigma \phi^*(B_0)$ is an integer. (The B-field therefore plays a rôle somewhat analogous to that of the θ-angle in QCD.) The importance of the resulting principle of *invariance of physics under integral shifts of the B-field* is not as widely recognized as it ought to be.

This invariance is manifest in the analysis carried out by Candelas et al.[1] of the mirror map for quintic threefolds. Analyzing the behavior of the metric on the moduli space, these authors found that the usual parameter t (proportional to $B + iJ$) on the Kähler moduli space of the quintic-mirror has an asymptotic relationship to the natural parameter $z := \psi^{-5}$ on the complex moduli space of the quintic, of the form

$$t \sim \frac{1}{2\pi i} \log z + \text{constant} + \cdots.$$

Once one has observed that values of t which differ by an integer lead to identical physics, one is led to introduce $e^{2\pi i t}$ as a more natural parameter on the Kähler moduli space. (This effectively modifies the definition of that space by making identifications between points which differ by an integral shift of the B-field.) The new parameter $e^{2\pi i t}$ is then a single-valued function of z, consistent with mirror symmetry. This same idea has led to successful mirror map calculations for other one-parameter families of Calabi-Yau threefolds,[2] and more recently for one-parameter families of Calabi-Yau manifolds in higher dimension.[3]

2. The Large Radius Limit

To analyze the large radius limit in general, we choose a basis $e^1, \ldots, e^r$ of the integral harmonic 2-forms on the target space X, and write $B + iJ = \sum z_j e^j$. The z_j's can be regarded as coordinates on the "complexified Kähler cone," constrained by some inequalities such as $\text{Im}(z_j) > 0$. The identification under integral shifts of the B-field can be implemented by exponentiating these coordinates,[4] introducing $w_j := e^{2\pi i z_j}$. Inequalities such as $0 < \text{Im}(z_j) < \infty$ on the z_j's translate into inequalities such as $0 < |w_j| < 1$ on the w_j's. We partially compactify the space by including points for which some w_j is 0.

To see the large radius limit, we should rescale the metric via $g_{ij} \mapsto \lambda\, g_{ij}$, and take $\lambda \to \infty$. The Kähler form scales as $J \mapsto \lambda J$, and the exponentiated coordinates transform as $w_j \mapsto |w_j|^\lambda \cdot \arg(w_j)$. As $\lambda \to \infty$, all points with $|w_j| < 1$ flow towards the origin $(0, \ldots, 0)$, so we can apparently regard the origin in this coordinate system as the "large radius limit point." This is consistent with the behavior of the instanton expansions of three-point functions, which take the general form[5]

$$\langle \mathcal{O}_A \mathcal{O}_B \mathcal{O}_C \rangle = A \cdot B \cdot C + \sum_\Gamma \frac{w^\Gamma}{1 - w^\Gamma} (A \cdot \Gamma)(B \cdot \Gamma)(C \cdot \Gamma),$$

where the sum is over rational curves Γ on X, and w^Γ is a monomial in $w_1, \ldots, w_r$ determined by the homology class of Γ. All instanton contributions to this correlation function vanish at the origin in the w-coordinates.

In order to ensure that $|w_j| < 1$ for all points in the Kähler moduli space, we must choose the classes $e^1, \ldots, e^r$ to lie in the closure of the Kähler cone of X. When $r = 1$ (i.e., for the mirrors of one-parameter families), this condition completely specifies the integral basis and determines a "large radius limit point" unambiguously. However, when $r > 1$ the freedom to change the basis causes difficulties.

In all examples with $r > 1$ studied in the literature,[4, 6] the Kähler cone has a very simple form and the edges of its closure can be used as the desired basis. However, this is *not* a general feature of Calabi-Yau threefolds: a study from a mathematical perspective[7] reveals some complexities which are not visible in these examples.

3. Blowing Down the Moduli Space

The most concrete way to see the effect of a change of basis is to consider what happens if the moduli space is blown up at the origin in the w-coordinates. We take $r = 2$ for simplicity, and find two new coordinate charts after the blowup, with coordinates $(w_1, \frac{w_2}{w_1})$ and $(\frac{w_1}{w_2}, w_2)$. (The corresponding bases are $\{e^1 + e^2, e^2\}$ and $\{e^1, e^1 + e^2\}$.) Rescaling the metric and taking $\lambda \to \infty$ sends (w_1, w_2) to the origin in the first chart when $|w_2| < |w_1|$, and to the origin in the second chart when $|w_1| < |w_2|$. Both "origins" can thus lay claim to being the "large radius limit" associated to at least *part* of the Kähler moduli space.

Conversely, if we have a partial compactification of the Kähler moduli space which includes more than one large radius limit point (each associated with a different basis $e^1, \ldots, e^r$, and with a different domain inside the moduli space), we should attempt to blow down this space to produce a partial compactification with a *single* large radius limit point for the entire moduli space. These blowdowns are similar to those arising in toric geometry,[8] and will often lead to singularities in the compactified space. The instanton contributions to correlation functions are still suppressed in such a limit, in spite of the singularities—we must accept the possibility that the "true" large radius limit point is not a smooth point.

(Note that all of the large radius limit points under discussion are associated to a *single* Kähler cone. It is also possible to consider other large radius limit points associated to the Kähler cones of *different* birational models of X. This leads to topology-changing transitions,[4] and one would not expect to collapse *those* limit points to a single point by blowing down.)

Even when we expect to be able to blow down and are willing to allow singularities, it may prove to be impossible to perform the desired blowing down, due to the presence of an infinite number of large radius limit points. For example, if the Kähler cone is described as $\frac{2}{1-\sqrt{5}} y < x < \frac{2}{1+\sqrt{5}} y$, then (as shown in figure 1) attempting to cover the cone using integral bases leads to a sequence of rays with slopes $\ldots, -\frac{5}{8}, -\frac{2}{3}, -1, \frac{1}{0}, 2, \frac{5}{3}, \frac{13}{8}, \ldots$ which asymptotically approach the walls[†] of the cone. Each adjacent pair of rays in the sequence gives rise to a distinct large radius limit point.

[†] The figure does not include these walls—the limiting rays with irrational slope $\frac{1 \pm \sqrt{5}}{2}$ —since they are less than a line-width's distance from the outer rays as shown (at the level of resolution of the figure).

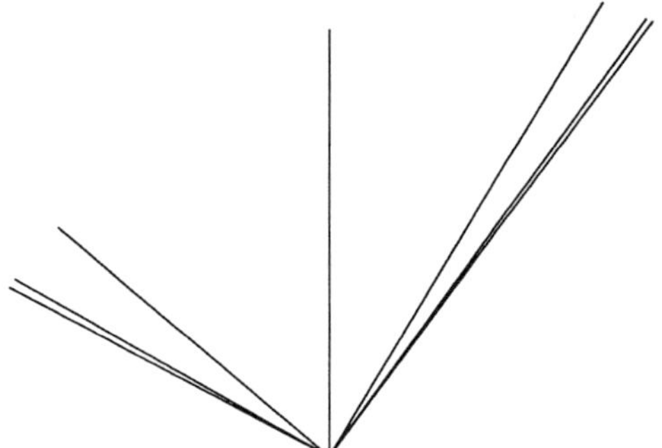

Figure 1. Decomposing the cone $\frac{2}{1-\sqrt{5}}\, y < x < \frac{2}{1+\sqrt{5}}\, y$.

4. Automorphisms

A Calabi-Yau threefold may have holomorphic automorphisms which act nontrivially on the Kähler cone. If we make the identifications on the Kähler moduli space dictated by those automorphisms, it may become possible to do the blowdowns—an infinite number of large radius limit points may turn into a finite number after these identifications.[7, 9]

In the example above, an automorphism acting on the cone as $(x, y) \mapsto (2x + 3y, 3x + 5y)$ leads from an infinite number of large radius limit points on the original Kähler moduli space to two remaining points on the quotient space. The quotient space can then be blown down explicitly,[10] leading to a singular surface with local equation $w^2 = (u^3 - v^2)(u^2 - v^3)$. This is illustrated in figure 2.

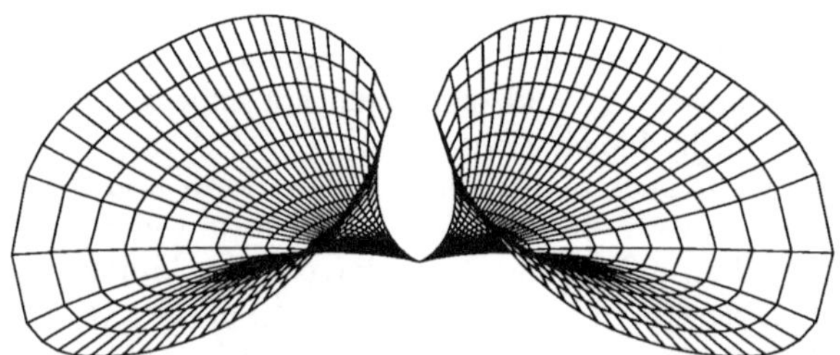

Figure 2. The blown down moduli space $w^2 = (u^3 - v^2)(u^2 - v^3)$

Singularities of this type have a very large "local fundamental group" arising from the holomorphic automorphisms. This allows for the following possibility in a compactification scheme: In the effective 4D theory, there could be varying moduli fields which always remain close to the field theory (large radius) limit, yet which generate global monodromy effects in $M^{3,1}$, somewhat akin to discrete gauge transformations. (Previous examples of such discrete symmetries were only visible upon leaving the sigma-model region of the moduli space, i.e., upon varying the moduli fields to a point far from the field theory limit.) Phenomenological implications of such a scheme are at present unknown.

Acknowledgements

This paper owes much to discussions and collaborations with Paul Aspinwall, Brian Greene, Sheldon Katz, and Ronen Plesser. I thank them all. This research was supported by NSF grant DMS-9103827, and by an American Mathematical Society Centennial Fellowship.

References

1. P. Candelas, X. de la Ossa, P. Green and L. Parkes, *Nucl. Phys.* **B359** (1991) 21.

2. D. R. Morrison, in *Essays on Mirror Symmetry*, ed. S.-T. Yau (Intl. Press, Hong Kong, 1992). See also A. Font, *Nucl. Phys.* **B391** (1993) 358; A. Klemm and S. Theisen, *Nucl. Phys.* **B389** (1993) 153.

3. B. R. Greene, D. R. Morrison and M. R. Plesser, "Mirror manifolds in higher dimension", CLNS-93/1253, YCTP-P31-92.

4. P. S. Aspinwall, B. R. Greene and D. R. Morrison, *Phys. Lett.* **B303** (1993) 249.

5. P. S. Aspinwall and D. R. Morrison, *Comm. Math. Phys.* **151** (1993) 245, and the references therein.

6. P. Candelas, X. de la Ossa, A. Font, S. Katz and D. R. Morrison, "Mirror symmetry for two parameter models – I", CERN-TH.6884/93; S. Hosono, A. Klemm, S. Theisen and S.-T. Yau, "Mirror symmetry, mirror map and applications to Calabi-Yau hypersurfaces", HUTMP-93/0801.

7. D. R. Morrison, "Compactifications of moduli spaces inspired by mirror symmetry", in *Journées de Géometrie Algébrique, Orsay 1992*, ed. A. Beauville, *Astérisque*, in press. (alg-geom/9304007)

8. For a review see P. S. Aspinwall, B. R. Greene and D. R. Morrison, "Calabi-Yau moduli space, mirror manifolds and spacetime topology change in string theory", IASSNS-HEP-93/38, CLNS-93/1236.

9. A. Grassi and D. R. Morrison, *Duke Math. J.* **71** (1993) 831.

10. F. Hirzebruch, in *Lecture Notes in Math.* **627**, (Springer-Verlag, Berlin, 1977), p. 287.

Quantization of the Mirror Symmetry

Hirosi Ooguri

Research Institute for Mathematical Sciences,
Kyoto University, Kyoto 606-01, Japan

1. Introduction

In the 80's, major progress in the string theory had been made on the world–sheet, and various techniques to study correlation functions of conformal field theories in two–dimensions has been developed. However we are still at a rather preliminary stage in understanding the string theory itself. Especially target–space aspects of the string theory are just beginning to be uncovered. Also, to understand even perturbative aspects of the quantum string theory, we need to develop a practical method to perform integrations over the moduli space of Riemann surface to compute perturbative amplitudes explicitly.

In recent years, some progress has been made in this direction. It is realized that some of the superstring computations can be transformed into solvable finite dimensional problems involving topological field theories [1]. A typical example is a computation of Yukawa couplings [2], [3]. In the standard compactification of the heterotic string theory on a Calabi-Yau manifold M, the world–sheet theory is constructed from a supersymmetric sigma–model on M plus four free bosons and fermions corresponding to the uncompactified dimensions. The sigma–model on M has the $N = 2$ superconformal symmetry generated by the energy–momentum tensor T, a pair of supercurrents $G^\pm$ and the $U(1)$ current J. In the minimal scenario, the $E_8 \times E_8$ gauge group is broken on M down to $E_8 \times E_6$ in the low energy effective theory. Each left–handed fermion in the four uncompactified dimensions transforming as $\overline{\mathbf{27}}$ of E_6 corresponds to what is called a (c,c)–primary field with the left and the right $U(1)$ charges (q_L and q_R) equal to 1. A (c,c)–primary field ϕ_i is a primary field of the superconformal field theory, but satisfies an extra–condition that it is annihilated by $G^+_{-1/2,L}$ and $G^+_{-1/2,R}$, i.e.

$$G^+_{-1/2,L}\phi(z,\bar{z}) = \oint_z \frac{dw}{2\pi i} G_L^+(w)\phi(z,\bar{z}) = 0 \tag{1}$$

$$G^+_{-1/2,R}\phi(z,\bar{z}) = \oint_{\bar{z}} \frac{d\bar{w}}{-2\pi i} G_L^+(\bar{w})\phi(z,\bar{z}) = 0. \tag{2}$$

Each (c,c)–primary field ϕ_i is related to an element of the cohomology $H^{1,1}$ of M ($i = 1,...,\dim H^{1,1}$). Similarly each left–handed fermion in $\mathbf{27}$ corresponds to a (c,a)–primary field $\tilde{\phi}_i$ with the $U(1)$ charges $q_L = 1$, $q_R = -1$ associated to an element of

$H^{2,1}(M)$. A (c,a)–primary field is annihilated by $G^-_{-1/2,L}$ and $G^+_{-1/2,R}$,

$$G^-_{-1/2,L}\tilde{\phi}_i = G^+_{-1/2,R}\tilde{\phi}_i = 0. \tag{3}$$

The Yukawa coupling C_{ijk} among $\overline{27}$'s should then be related a three–point amplitude of the (c,c)–primary fields ϕ_i, ϕ_j, ϕ_k on a two–dimensional sphere, and similarly for the Yukawa coupling $\tilde{C}_{ijk}$ among 27's. We should, however, also remember that, in the Yukawa coupling, two of the external states are spacetime fermions. The vertex operators for these spacetime fermions should be in the Ramond sector, and their sigma–model parts are constructed by applying the spectral flow operators on the (c,c) (or (c,a)) primary fields. Thus C_{ijk} (or $\tilde{C}_{ijk}$) is equal to the three–point amplitude of the (c,c) (or (c,a)) primary fields, but with two insertions of the spectral flow operators on the sphere.

The insertion of the two spectral flow operators effectively changes the spins and the weights of operators on the sphere, and this generates what is called the twisting of the $N = 2$ sigma–model into a topological sigma–model [1],[4]. To be precise, the spectral flow operators which appear in C_{ijk} and $\tilde{C}_{ijk}$ are different in their ways in twisting the right–moving sector. Thus the same $N = 2$ sigma–model gives two inequivalent topological models. In the case of C_{ijk} among $\overline{27}$'s, the spectral flow operator changes the spins of the supercurrents G^+_L and G^+_R from 3/2 to 1 and those for G^-_L and G^-_R from 3/2 to 2. Thus we may regard G^+_L and G^+_R as the BRST currents of the topological sigma–model. This procedure is called the A–twist, and the resulting topological sigma–model is called the A–model [5]. On the other hand, the spectral flow operator for $\tilde{C}_{ijk}$ among 27's transforms G^+_L and G^-_R into the spin–1 BRST currents, and this generates what is called the B–twist of the sigma–model. By definition (2), the (c,c)–primary fields commute with the BRST operators of the A–model and are regarded as physical operators in this model. On the other hand, the (c,a)–primary fields are physical operators in the B–model.

In this way, the computations of the Yukawa coupling reduce to the ones in the topological A– and B–models. In the A–model, the BRST symmetries generated by G^+_L and G^+_R imply that the instanton approximation is exact [2], [5]. Thus the Yukawa coupling among $\overline{27}$ is given by a sum over holomorphic maps from S^2 into M as

$$C_{ijk} = \int_M k_i \wedge k_j \wedge k_k + \sum_s N_s^{(0)} s_i s_j s_k \frac{\exp(-\sum_i s_i t^i)}{1 - \exp(-\sum_i s_i t^i)} \tag{4}$$

where k_i, k_j, k_k are elements of $H^{1,1}(M)$ corresponding to the (c,c)–primary fields ϕ_i, ϕ_j, ϕ_k, and $N_s^{(0)}$ is the number of spheres holomorphically embedded in M such that k_i evaluated on them is equal to s_i. The Kähler class k of M can be expressed as a linear combination of the basis of $H^{1,1}(M)$ as $k = \sum_i t^i k_i$, and the expansion coefficient t^i is called the Kähler moduli of M. The combination $\sum_i s_i t^i$ in the exponent in the above is equal to the sigma–model action evaluated on the instanton. It is known that $N_s^{(0)}$ is generically independent of the complex structure of M, and so is the Yukawa coupling C_{ijk} in the A–model.

On the other hand, in the B–model, the Yukawa coupling $\tilde{C}_{ijk}$ does not depend on the Kähler structure of M [3], [5]. This is because any deformation of

318

the Kähler structure on M is BRST trivial in the B–model. Thus, in particular, $\tilde{C}_{ijk}$ is independent of the volume of M. We can then evaluate the Yukawa coupling in the infinite volume limit of M, and the computation reduces to a problem of the classical algebraic geometry. As we mentioned in the above, each (c,a)–primary field corresponds to an element of $H^{2,1}(M)$ which generates an infinitesimal deformation of the complex structure on M. Thus the Yukawa coupling among the (c,a)–primary fields is given by

$$\tilde{C}_{ijk} = \int_M \Omega \wedge \frac{\partial^3}{\partial y^i \, \partial y^j \, \partial y^k} \Omega \tag{5}$$

where Ω is the unique holomorphic 3–form on M, and y^i, y^j, y^k are the complex moduli on M. In many examples, it is possible to compute the Yukawa coupling $\tilde{C}_{ijk}$ among **27** explicitly using the technique of the variation of the Hodge structure.

The Yukawa coupling C_{ijk} among $\overline{\mathbf{27}}$ is much more difficult to evaluate directly. There is a combinatorial method to compute the number $N_s^{(0)}$ for small values of s_i, but we need to know this for all s_i's in order to compute C_{ijk} by using the formula (4). Fortunately, in some cases, the computation of C_{ijk} reduces to that of $\tilde{C}_{ijk}$. It is when the Calabi–Yau manifold M has its mirror partner $\widetilde{M}$. A pair of Calabi–Yau manifold $(M, \widetilde{M})$ is called a mirror pair when the A–model on M is equivalent as a quantum field theory to the B–model on $\widetilde{M}$ (and vice versa). This in particular means that the number of $\overline{\mathbf{27}}$'s for the heterotic string on M is equal to the number of **27**'s for the one on $\widetilde{M}$ and vice versa. This equivalence among the two topological field theories is called the mirror symmetry. The existence of a mirror pair had been conjectured in various places [6],[7],[8], but an explicit construction was first given in [9]. If $(M, \widetilde{M})$ is a mirror pair, the Yukawa coupling among $\overline{\mathbf{27}}$'s on M can be related to the Yukawa coupling among **27**'s on $\widetilde{M}$. In the paper [10], this idea was implemented in the case when M is the quintic 3–fold in $\mathbf{CP}^4$ to compute the Yukawa coupling C_{ijk}. As a by–product, they derived the number $N_s^{(0)}$ of S^2 in M for all values of s by expanding C_{ijk} as in (4). Thus when M has a mirror partner $\widetilde{M}$, it is possible to compute the Yukawa couplings among $\overline{\mathbf{27}}$ as well as among **27**'s.

Since the mirror symmetry is an equivalence of quantum field theories, it should hold whether the world–sheet is S^2 or a surface of higher genus. Thus it should also be useful in studying higher loop amplitudes in the string theory. This is what I would like to discuss in the following.

This talk is based on my recent works with Michael Bershadsky, Sergio Cecotti and Cumrun Vafa [11], [12]. I would like to thank them for the collaboration.

2. Quantum Amplitudes of Topological String

It is realized recently that some higher loop amplitudes in the superstring theory can also be related to the topological string computations [12], [13]. At one–loop, a topological string amplitude is given by

$$\mathcal{F}_1 = \frac{1}{2} \int_{\mathcal{M}_1} \frac{d^2\tau}{Im\tau} trace \left[(-1)^{F_L + F_R} F_L F_R q^{L_0} \bar{q}^{\bar{L}_0} \right], \tag{6}$$

where the trace is taken over the Ramond Hilbert space of the supersymmetric sigma-model on the Calabi-Yau 3–fold, $\mathcal{M}_1$ is the moduli space of a torus, F_L and F_R are the left and right $U(1)$ charges, and $q = e^{2\pi i \tau}$. This can be related to the one–loop correction to the R^2 term in the low energy effective theory in the type II superstring theory compactified on M [12], [13]. It is also found in [12] that the same quantity $\mathcal{F}_1$ can also be related to the one–loop contribution to the threshold correction to the gauge coupling [14] in the heterotic string theory.

At g–loop ($g > 1$), a canonical amplitude in the topological string on the Calabi–Yau 3–fold M is

$$\mathcal{F}_g = \int_{\mathcal{M}_g} d^{3g-3}m\, d^{3g-3}\bar{m} \langle \prod_{a=1}^{3g-3} \int_{\Sigma_g} \mu_a G_L^- \int_{\Sigma_g} \bar{\mu}_a G_R^- \rangle \tag{7}$$

where $\mathcal{M}_g$ is the moduli space of the genus–g surface Σ_g, and μ_a is the Beltrami-differential associated to the modulus m_a. This expression is for the A–model. In the B–model, we simply replace G_R^- in the above by G_R^+. After the twisting [1], [4], the supercurrent G^- becomes an operator of dimension 2, thus the integral $\int \mu_a G_L^-$ is well–defined on Σ_g and so is $\int \bar{\mu}_a G_R^-$. The supersymmetric sigma–model can be defined for a Calabi–Yau of any dimensions, but the 3–fold enjoys a special status for the following reason. Because of the chiral anomaly in the $U(1)$ current, $\bar{\partial}J \sim \hat{c}R$, where $\hat{c} = \dim M$ and R is the Ricci curvature on the world–sheet, the fermion number is violated on Σ_g by the amount $(3g - 3) \cdot \dim M$. This can be compensated by the $(3g - 3)$–insertions of G^- in (7) only if $\dim M = 3$. Thus $\mathcal{F}_g$ at $g > 1$ is non–vanishing only when M is 3–fold.

It is shown in [12], [13] that $\mathcal{F}_g$ for $g > 1$ computes superpotential terms involving the gravi–photon in the effective theory of the type II superstring. The vertex operator for the gravi–photon belongs to the Ramond–Ramond sector. Thus an insertion of the gravi–photon vertex operator twists the sigma–model, and in some case the sigma–model is transformed into the topological model. This is how the superstring amplitudes are related to the topological string amplitudes $\mathcal{F}_g$. In the following, we develop methods to evaluate $\mathcal{F}_g$ and explore its geometric properties. For more details on the superstring interpretation of $\mathcal{F}_g$, we refer the reader to [12].

3. Holomorphic Anomaly

In the A–model, we saw that the Yukawa coupling C_{ijk} depends holomorphically on the Kähler moduli t^i as in (4), but does not depend on the moduli of the complex structure on M. Conversely, in the B–model, the Yukawa coupling $\tilde{C}_{ijk}$ in the B–model depends holomorphically on the complex moduli y^i (5) but does not depend on the Kähler moduli. Some of these properties are inherited by the higher loop amplitudes $\mathcal{F}_g$'s. By using the Ward identity of the $N = 2$ superconformal symmetry, it is shown in [11] and [12] that $\mathcal{F}_g$ in the A–model (B–model) is independent of the complex moduli (Kähler moduli). On the other hand, in these papers, it is also found that their holomorphicities are slightly broken due to contributions from the boundary of the moduli space $\mathcal{M}_g$. Let us briefly explain how this happens.

Let us consider the case of the A–model. For the B–model, we can simply interchange G_R^+ and G_R^-. In the sigma–model, a variation with respect to the modulus parameter t^i is realized by an insertion of the operator

$$\int_{\Sigma_g} G_{-1/2,L}^- G_{-1/2,R}^- \phi_i \tag{8}$$

where ϕ_i is a (c,c)–primary field associated to the modulus t^i. Since the sigma–model itself is a unitary theory with the hermiticity condition $(G^-)^\dagger = G^+$, a variation with respect to $\bar{t}^i$ should be generated by an insertion of

$$\int_{\Sigma_g} G_{-1/2,L}^+ G_{-1/2,R}^+ \bar{\phi}_i, \tag{9}$$

i.e.

$$\frac{\partial}{\partial \bar{t}^i} \mathcal{F}_g = \int_{\mathcal{M}_g} d^{3g-3}m\, d^{3g-3}\bar{m} \langle \int_{\Sigma_g} G_{-1/2,L}^+ G_{-1/2,R}^+ \bar{\phi}_i, \prod_{a=1}^{3g-3} \int_{\Sigma_g} \mu_a G_L^- \int_{\Sigma_g} \bar{\mu}_a G_R^- \rangle. \tag{10}$$

If there were no G^- inserted on Σ_g, we could move $G_{-1/2,L}^+$ and $G_{-1/2,R}^+$ around the surface Σ_g and show that the right–hand side vanishes. However because of $\int_{\Sigma_g} \mu_a G_L^-$ and $\int_{\Sigma_g} \bar{\mu}_a G_R^-$, we must pick up the commutators of G^+ and G^-. They generate insertions of the energy–momentum tensor T, which can be converted into derivatives with respect to the moduli m_a of the Riemann surface. Thus the integrand in the right–hand side becomes a total derivative in the moduli space $\mathcal{M}_g$. We must then evaluate contributions from the boundary of $\mathcal{M}_g$. The non–vanishing contribution to $\bar{\partial}\mathcal{F}_g$ which arises from the boundary of $\mathcal{M}_g$ is called the holomorphic anomaly.

The holomorphic anomaly at $g > 1$ is evaluated in [12] with the result

$$\frac{\partial}{\partial \bar{t}^i} \mathcal{F}_g = \frac{1}{2}\overline{C_{ijk}}e^{2K}G^{j\bar{j}}G^{k\bar{k}}\left(D_j D_k \mathcal{F}_{g-1} + \sum_{r=1}^{g-1} D_j \mathcal{F}_r D_k \mathcal{F}_{g-r} \right) \tag{11}$$

where $G_{i\bar{j}}$ is the Zamolodchikov metric on the moduli space of the sigma–model which appears in the operator product expansion

$$\phi_i(z)\overline{\phi_j}(w) \simeq \frac{G_{i\bar{j}}}{|z-w|^2}, \tag{12}$$

K is the Kähler potential for the metric, $G_{i\bar{j}} = \partial_i \bar{\partial}_{\bar{j}} K$, and D_i is an appropriate covariant derivative ($\mathcal{F}_g$ is a section of some line–bundle over the moduli space of the sigma–model. See [12] for more detail). In the right–hand side, the repeated indices j,k are summed over the (c,c)–primary states with the $U(1)$ charges $q_L = q_R = 1$. In the case of the B–model, we can simply replace $\overline{C_{ijk}}$ in this formula by $\overline{\widetilde{C}_{ijk}}$ and the Zamolodchikov metric $G_{i\bar{j}}$ and the Kähler potential K by the corresponding objects in the B–model. The holomorphic anomaly at $g = 1$ is also derived in [11] as

$$\frac{\partial^2}{\partial t^i \partial \bar{t}^j} \mathcal{F}_1 = \frac{1}{2}tr(C_i \overline{C_j}) - \frac{\chi}{24}G_{i\bar{j}}, \tag{13}$$

where tr is a trace over the (c,c)–primary states of all $U(1)$ charges q_L, q_R. Especially when M is a 3–fold, this becomes

$$\frac{\partial^2}{\partial t^i \partial \overline{t^j}} \mathcal{F}_1 = \frac{1}{2} C_{ikl} \overline{C_{jkl}} e^{2K} G^{k\bar{k}} G^{l\bar{l}} - (\frac{\chi}{24} - 1) G_{i\bar{j}}. \tag{14}$$

Here the sums over the indices j, k are restricted to those with $q_L = q_R = 1$.

Thus, unlike the case of the Yukawa coupling, the higher loop amplitude $\mathcal{F}_g$ is not holomorphic on the moduli space of the sigma–model. Its anti–holomorphic dependence, however, is captured by the anomaly equations (11), (13). In various cases, these equations can be integrated to derive the moduli dependence of $\mathcal{F}_g$. This will be explained in the next sections.

The similar technique can be used to prove that $\mathcal{F}_g$ in the A–model is independent of the complex moduli. In the case of the A–model, the variation with respect to the complex moduli is realized by an insertion of

$$\int_{\Sigma_g} G^{-}_{-1/2,L} G^{+}_{-1/2,R} \phi_i. \tag{15}$$

As in the case of the holomorphic anomaly discussed in the above, we can move G_R^+ around the surface Σ_g, pick up a commutator with G_R^- and generate an insertion of the energy–momentum tensor which is converted into a derivative with respect to the moduli m of Σ_g. Then the computation again reduces to the boundary of $\mathcal{M}_g$. It turned out that the contribution from the boundary vanishes in this case [12]. Thus $\mathcal{F}_g$ in the A–model is independent of the complex moduli. Similarly one can show that $\mathcal{F}_g$ in the B–model is independent of the Kähler moduli of M.

4. One–Loop

Let us examine the property of $\mathcal{F}_1$ in more detail. In the above, we mentioned that $\mathcal{F}_g$ in the A–model (B–model) is independent of the complex moduli y (Kähler moduli t). To be precise, this is the case for $g > 1$. At $g = 1$, $\mathcal{F}_1$ does not distinguish the A–model and the B–model. So the precise statement in this case is that $\partial_t \mathcal{F}_1$ is independent of $y, \bar{y}$ and that $\partial_y \mathcal{F}_1$ is independent of $t, \bar{t}$. Namely $\mathcal{F}_1$ is a sum of two terms, one depends on t and $\bar{t}$ and the other depends on y and $\bar{y}$.

If we are interested in the complex moduli dependence of $\mathcal{F}_1$, we may choose any Kähler structure on M to evaluate $\mathcal{F}_1$. Especially we can take the infinite volume limit. Here we employ the argument by Witten [15]. The infinite volume limit for a fixed and finite world–sheet modulus is dominated by constant maps. But as noted in [15], this is not the full story. The reason is that we are integrating over the modulus τ of the torus. No matter how large a volume of Calabi-Yau we choose if we go close enough to the boundary of the moduli space of τ, we can get a finite action. In other words the world–sheets which will have finite actions are the ones concentrated in long thin tubes, which means that the computation reduces to that of an ordinary field theory. We then find that the complex moduli dependent part of $\mathcal{F}_1$ is expressed in terms of the determinants of the Laplacians $\Delta_{p,q}$ acting on

$\Omega^{(0,p)}(\wedge^q T^{(1,0)})$, the space of $(0,p)$–forms taking values in $\wedge^q T^{(1,0)}$ on M, as

$$\mathcal{F}_1 = \frac{1}{2} \log \left[\prod_{p,q=0}^{n} (\det \Delta_{p,q})^{(-1)^{p+q} pq} \right] + (\text{independent of } y, \bar{y}) \tag{16}$$

where $n = \dim M$.

This particular combination of the determinant can be related to the Ray–Singer holomorphic torsion [16] as

$$\mathcal{F}_1 = \frac{1}{2} \sum_q (-1)^q q \log I(\wedge^q T) + (\text{independent of } y, \bar{y}), \tag{17}$$

where

$$I(V) = \prod_p (\det \Delta_{p,V})^{(-1)^p p}, \tag{18}$$

V is a vector bundle on M and $\Delta_{p,V}$ is the Laplacian acting on $\Omega^{(0,p)}(V)$. One of the important properties of the Ray–Singer holomorphic torsion is that it obeys the Quillen's anomaly formula [17]

$$\partial\bar{\partial} \log I(V) = \partial\bar{\partial} \sum_p (-1)^p \log g_p + 2\pi i \int_M \text{Td}(T)\text{Ch}(V)\Big|_{(1,1)} \tag{19}$$

where g_p is the metric for zero modes of $\Delta_{p,V}$, $\text{Td}(T)$ denotes the Todd class of the tangent bundle over M, and $\text{Ch}(V)$ is the Chern class of the vector bundle V. The symbol $\Big|_{(1,1)}$ in the above formula means that we take a $(n+1, n+1)$–form of the integrand and integrate it over the n–fold M to be left with a $(1,1)$–form on the complex moduli space. Applying this formula to (16), we recover the holomorphic anomaly formula (13) for $\mathcal{F}_1$ [12]. Thus, in this particular case of the B–model at one–loop, the holomorphic anomaly reduces to the Quillen's formula.

The holomorphic anomaly (13) can be easily integrated to obtain

$$\mathcal{F}_1 = \log \left[X(t,\bar{t})|f(t)|^2 \right] + \log \left[\widetilde{X}(y,\bar{y})|\widetilde{f}(y)|^2 \right] \tag{20}$$

where

$$X(t,\bar{t}) = e^{-\frac{\chi}{24}K} \prod_{p,q} (\det g_{p,q})^{(-1)^{p+q}\frac{p+q}{4}}. \tag{21}$$

with $g_{p,q}$ being the metric for the (c,c)–primary fields with the $U(1)$ charges (p,q)

$$X(t,\bar{t}) = e^{-\frac{\chi}{24}\widetilde{K}} \prod_{p,q} (\det \widetilde{g}_{p,q})^{(-1)^{p+q}\frac{p+q}{4}}. \tag{22}$$

with $\widetilde{g}_{p,q}$ being the metric for the (c,a)–primary fields with the $U(1)$ charges $(p,-q)$. Here $f(t)$ and $\widetilde{f}(y)$ is some holomorphic object which cannot be determined from the anomaly equation alone. In order to fix f and $\widetilde{f}$, we need to know the behavior of $\mathcal{F}_1$ near the boundary of the moduli space of M. For $\widetilde{f}$, this is not an easy task since the behavior of the sigma–model near the boundary of the complex moduli space of M is not well–known. On the other hand, the behaviour of $\mathcal{F}_1$ near the boundary of the Kähler moduli space can be derived from the sigma–model path integral as we see below. Thus, $f(t)$ can in principle be derived. However in this case, the computation of the metric $g_{p,q}$ is difficult. Both of these difficulties could be avoided if the manifold M has a mirror partner $\widetilde{M}$. If this is the case, we can compute $g_{p,q}$ of M by using the computation of $\tilde{g}_{p,q}$ on $\widetilde{M}$ (which can be done using the variation of Hodge structure), and at the same time we can fix the boundary condition for $\tilde{f}(y)$ on M by using the Kähler moduli space of $\widetilde{M}$. Thus, if M has a mirror partner $\widetilde{M}$, we can compute $\mathcal{F}_1$ explicitly.

Previously we saw that, in the A–model, the Yukawa coupling C_{ijk} is expressed as a sum over rational curves on M. Thus one may expect that the genus–g amplitude $\mathcal{F}_g$ can also be expanded as a sum over genus–g curves. Similarly the Kähler moduli dependent part of $\mathcal{F}_1$ should be given as a sum over holomorphic elliptic curves on M. This is almost true, but a precise statement is as follows [11]. The fact that the instanton approximation is exact for the Yukawa coupling (an instanton in this case is a holomorphic map $S^2 \to M$) is a consequence of the BRST symmetry on the world–sheet generated by G^+ [2], [5]. Since the BRST symmetry is slightly broken at $g \geq 1$ due to the holomorphic anomaly (11), (13), it is not exactly true that $\mathcal{F}_g$ can be expressed as a sum over holomorphic elliptic curves. In fact, the holomorphic anomaly implies that $\mathcal{F}_g$ depends on the anti–holomorphic Kähler moduli $\bar{t}^i$ as well as t^i. In the sigma–model Lagrangian, the moduli $\bar{t}^i$ appear in the combination

$$\bar{t}^i \int_{\Sigma_g} k_i \bar{\partial} X \partial \bar{X}. \tag{23}$$

Thus if the instanton approximation were exact, $\mathcal{F}_g$ should not depend on $\bar{t}^i$. This clearly is inconsistent with the holomorphic anomaly. However we can recover the sum over holomorphic maps from $\mathcal{F}_g$ as follows. By sending $\bar{t}^i \to \infty$, we can suppress non–holomorphic maps $X : \Sigma_g \to M$ in the sigma–model path integral because of the form (23) of the Lagrangian.

In the case of $g = 1$, the evaluation of the path integral in the limit $\bar{t}^i \to \infty$ gives the following formula [11].

$$\frac{\partial}{\partial t^i} \mathcal{F}_1 \simeq \frac{(-1)^n}{24} \int_M k_i \wedge c_{n-1} + \sum_s s_i N_s^{(1)} \left(\sum_{N=1}^{\infty} \frac{e^{-N \sum s_j t^j}}{1 - e^{-N \sum s_j t^j}} \right) \tag{24}$$

$$+ \frac{1}{12} \sum_s s_i N_s^{(0)} \frac{e^{-\sum s_j t^j}}{1 - e^{-\sum s_j t^j}}. \tag{25}$$

The leading term $(-1)^n \int_M k_i \wedge c_{n-1}$ in the right–hand side is a contribution from the constant map $\Sigma_1 \to M$. This term is multiplied by the factor $\frac{1}{24}$, which is equal a minus–half of the Euler character of the moduli space Σ_1. This is because the constant map exists for any modulus τ of Σ_1, thus we must perform integration over the moduli space of Σ_1. In the second term, $N_s^{(1)}$ is the number of holomorphic elliptic curves on M. The factor

$$\sum_{N=1}^{\infty} \frac{e^{-N \sum s_j t^j}}{1 - e^{-N \sum s_j t^j}} \tag{26}$$

is there in order to count multi–covering of a holomorphic elliptic curve in M by the world–sheet torus Σ_1. It is curious that the number of the rational curves $N_s^{(0)}$ also contribute to $\mathcal{F}_1$ as we see in the third term in the right–hand side. This is due to the following two reasons. One of the origins of the third term is the compactification of the moduli space of curves in M (the instanton moduli space). In general, a curve is not necessarily isolated in M and there may be a continuous family of curves. In this case, $N_s^{(1)}$ represent a computation of the Euler character of a certain vector bundle over the instanton moduli space, the Gromov–Witten invariant. In order to compute the Euler character, we must compactify the instanton moduli space. We must then take into account contributions from the boundary of the moduli space, and they are given by $N_s^{(0)} s_i e^{-\sum s_j t^j}$ which gives a leading term in the Taylor expansion of the third term in $e^{-\sum s_j t^j}$. The rest of the third term is explained as follows. The world–sheet Σ_1 may be represented as a branched double cover of S^2 which may be holomorphically embedded in M. Contributions from such configurations are given by $N_s^{(0)} s_i e^{-2\sum s_j t^j}$ (the factor 2 in the exponent is due to the fact that the world–sheet Σ_1 double–covers S^2), and this gives the second term in the Taylor expansion. We must also take into account multi–coverings of these configurations. Combining all these contributions, we have obtained the formula (25).

To summarize, in the one–loop amplitude $\mathcal{F}_1$, the complex moduli dependence and the Kähler moduli dependence decouple. The complex moduli dependent part is related to the Ray–Singer torsion as in (16) while the Kähler moduli dependent part contains informations on the number of holomorphic elliptic curves $N_s^{(1)}$ on M as in (25). If the manifold M has a mirror partner $\widetilde{M}$, we can combine these two formulae and compute $\mathcal{F}_1$. In the case of the quintic 3–fold in $\mathbf{CP}^4$, we have carried out this procedure explicitly and computed the numbers $N_s^{(1)}$ of holomorphic elliptic curves [11].

Table 1. the number of elliptic curves on the quintic 3–fold

Degree	$N_s^{(1)}$
1	0
2	0
3	609250
4	3721431625
5	12129909700200
6	31147299732677250
7	71578406022880761750
8	154990541752957846986500
9	324064464310279585656399500

5. Higher–Loops

In [12], the formula (25) for $g = 1$ is extended to the case of $g = 2$ as

$$\mathcal{F}_2 \simeq \frac{\chi}{5760} + \sum_s N_s^{(2)} e^{-\sum s_j t^j} + \frac{1}{240} \sum_s N_s^{(0)} \frac{e^{-\sum s_j t^j}}{(1 - e^{-\sum s_j t^j})^2} \tag{27}$$

in the $\bar{t}^i \to \infty$ limit of the A–model. The first term in the right–hand side is the contribution from the constant map $\Sigma_2 \to M$ with the coefficient

$$\frac{1}{5760} = \frac{1}{2} \int_{\mathcal{M}_2} (c_1)^3, \tag{28}$$

and the second term is from holomorphic maps $\Sigma_2 \to M$. Third term is the contributions from the boundary of the instanton moduli space and from the multi–covering of Σ_2 over S^2. The coefficient $\frac{1}{240}$ in front of the third term is a minus of the Euler character of the moduli space of genus–2 curves. In order to extend this result to still higher genera $g \geq 3$, we need to understand better the contributions from the boundary of the instanton moduli space and from the multi–coverings.

On the other hand, in the B–model, the genus–g amplitude $\mathcal{F}_g$ should be related to a g–loop computation of some quantum field theory for the reason we described in the previous section (the infinite–volume limit is exact in the B–model). Such a quantum field theory should be regarded as a string field theory of the B–model. In fact, we have identified such a quantum field theory. It is related to the theory of Kodaira and Spencer [18] which describes the deformation of the complex structure of M. The moduli space of a Calabi–Yau 3–fold could be regarded as the space of solution of the topological string of the B–model since the consistent propagation of this topological string requires the Ricci flatness of the target manifold [5]. It is then natural that the Kodaira–Spencer theory emerges as a string field theory since it described the moduli space of the complex structure of a Calabi–Yau manifold. For more details, we again refer the reader to [12].

As in the case of $g = 1$ (20), there is a systematic procedure to integrate the holomorphic anomaly equation (11) for $\mathcal{F}_g$. To do this, we need to introduce a "canonical" prepotential S satisfying

$$\overline{C_{ijk}} = e^{-2K} \bar{D}_{\bar{i}} \bar{D}_{\bar{j}} \bar{D}_{\bar{k}} S. \tag{29}$$

That we can find such an object S was shown in [12]. We then introduce some notations.

$$S^i = G^{i\bar{j}} \bar{D}_{\bar{j}} S , \quad S^{ij} = G^{i\bar{k}} \bar{D}_{\bar{k}} S^j \tag{30}$$

With these preparations, the holomorphic anomaly (11) can be integrated, for example in the case of $g = 2$, as

$$\mathcal{F}_2 = \frac{1}{2} S^{ij} D_i D_j \mathcal{F}_1 + \frac{1}{2} D_i \mathcal{F}_1 S^{ij} D_j \mathcal{F}_1 - \frac{1}{8} S^{jk} S^{mn} C_{jkmn} \tag{31}$$

$$-\frac{1}{2}S^{ij}C_{ijm}S^{mn}D_n\mathcal{F}_1 + \frac{\chi}{24}S^i D_i\mathcal{F}_1 \qquad (32)$$

$$+\frac{1}{8}S^{ij}C_{ijp}S^{pq}C_{qmn}S^{mn} + \frac{1}{12}S^{ij}S^{pq}S^{mn}C_{ipm}C_{jqn} \qquad (33)$$

$$-\frac{\chi}{48}S^i C_{ijk}S^{jk} + \frac{\chi}{24}(\frac{\chi}{24}-1)S + f_2(t). \qquad (34)$$

As in the case of $g = 1$, there is a holomorphic object $f_2(t)$ which cannot be determined from the anomaly equation alone. This can be fixed when the manifold M has a mirror partner $\widetilde{M}$. In this case, by the mirror symmetry, $\mathcal{F}_g$ is also equal to the g–loop amplitude of the Kodaira–Spencer theory on $\widetilde{M}$, and this geometric interpretation gives a strong restriction on the behavior of $f_2(t)$ near the boundary of the moduli space. We can combine this information with the formula (27) for the A–model on M to determine $f_2(t)$. In the case when M is the quintic 3–fold in $\mathbf{CP}^4$, we again carried out this procedure and derived the number $N_s^{(2)}$ of genus–2 curves on M by using (27).

Table 2. the number of genus–2 curves on the quintic 3–fold

Degree	$N_s^{(2)}$
1	0
2	0
3	0
4	534750
5	75478987900
6	871708139638250
7	5185462556617269625
8	9006736425242367534500
9	3258596871473582660010240500

We have also derived similar formulae for all g by integrating the holomorphic anomaly [12]. It turned out that the resulting expressions can be reproduced by the Feynman rule of some finite dimensional quantum mechanical system. It would be interesting to understand a geometric origin of this system.

References

1. E.Witten, *Commun. Math. Phys.* bf 117 (1988) 353; and 118 (1988) 411.

2. M.Dine, N.Seiberg, X.Wen and E.Witten, *Nucl. Phys.* **B278** (1986) 769 and **B289** (1987) 319.

3. A.Strominger and E.Witten, *Commun. Math. Phys.* **101** (1985) 341.

4. T.Eguchi and S.-K.Yang, *Mod. Phys. Lett.* **A5** (1990) 59.

5. E.Witten, in *Essay on Mirror Manifolds,* ed. by S.-T. Yau, International Press (1992).

6. L.Dixon, in Proceedings of the ICTP Summer Workshop in High Energy Physics and Cosmology, Trieste 1987.

7. T.Eguchi, H.Ooguri, A.Taormina and S.-K.Yang, *Nucl. Phys.* **B315** (1989) 193.

8. W.Lerche, C.Vafa and N.Warner, *Nucl. Phys.* **B324**(1989) 427.

9. B.R.Greene and M.R.Plesser, *Nucl. Phys.* **B338** (1990) 15.

10. P.Candelas, X. de la Ossa, Green and Parkes, *Nucl. Phys.***B359** (1991) 21.

11. M.Bershadsky, S.Cecotti, H.Ooguri and C.Vafa, *Holomorphic Anomalies in Topological Field Theories*, HUTP–93/A008, RIMS–915, hep-th/9302103.

12. M.Bershadsky, S.Cecotti, H.Ooguri and C.Vafa, *Kodaira–Spencer Theory of Gravity and Exact Results for Quantum String Theories*, HUTP–93/A025, RIMS–946, SISSA–142/93/EP.

13. I.Antoniadis, E.Gava, K.S.Narain and T.R.Taylor, *Topological Amplitudes in String Theory*, hep-th/9307158.

14. V.Kaplunovsky, *Nucl.Phys.* **B307** (1988) 36;
L.J.Dixon, V.S.Kaplunovsky and J.Louis, *Nucl.Phys.* **B355** (1991) 649;
S.Ferrara, C.Kounnas, D.Lüst and F.Zwirner, *Nucl. Phys.***B365** (1991) 431;
I.Antoniadis, E.Gava and K.S.Narain, *Nucl. Phys.***B383** (1992) 93 and *Phys. Lett.* **B283** (1992) 209;
J.-P. Derendinger, S. Ferrara, C. Kounnas and F. Zwirner, *Nucl. Phys.***B372** (1992) 145.

15. E.Witten, *Chern–Simons Gauge Theory as a String Theory*, IASSNS-HEP-92/45, hep-th/9207094.

16. D.B.Ray and I.M.Singer, *Ann. Math.***98** (1973) 154.

17. J.M.Bismut and D.S.Freed, *Commun. Math. Phys.* **106** (1986) 159 and 107 (1986) 103.

18. K.Kodaira and D.C.Spencer, *Ann. Math.***67** (1958) 328 and **71** (1960) 43;
K.Kodaira, L.Niremberg and D.C.Spencer, *Ann. Math.***68** (1958) 450;
K.Kodaira and D.C.Spencer, *Acta. Math***100** (1958) 281.

OPEN STRINGS IN CONSTANT ELECTRIC AND MAGNETIC FIELDS

MASSIMO PORRATI*

*Department of Physics, New York University, 4 Washington Place
New York, NY 10003, USA*

ABSTRACT

Various properties of open strings in external constant E.M. fields are reviewed. In particular, the charged-particle pair production rate in an external electric field is evaluated, and shown to reduce to Schwinger's formula in the limit of low-intensity fields. Open strings in external magnetic fields are shown to undergo an infinite number of phase transitions as the strenght of the field increases.

1. Open Strings in a Constant Electric-Field Background

In the presence of an external constant *electric* field, quantum field theory predicts a nonvanishing probability for the creation of charged particle pairs. The rate of particle pair production can be estimated using simple semiclassical arguments. Namely, one represents a virtual charged particle pair in an external electric field of strenght E by a potential well of depth $2m$, where m is the mass of the charged particle, superimposed to a linear potential

$$V = 0, \quad x < 0, \quad V = 2m - eEx, \quad x \geq 0. \tag{1}$$

By denoting with e the charge of the particle and using the standard semi-classical approximation one would get a rate $w \sim \exp(-\mathcal{O}(1)m^2/|eE|)$. This estimate is quite accurate. Indeed, the *exact* rate for a particle of spin s minimally coupled to an external electric field was found long ago by Schwinger[1]:

$$w = \frac{2s + 1}{8\pi^3} \sum_{k=1}^{\infty} (-1)^{(2s+1)(k+1)} (eE/k)^2 \exp(-k\pi m^2/|eE|). \tag{2}$$

A natural question arising at this point is whether it is possible to find an exact formula for the pair-production rate in string theory. The answer to this question is in the affirmative, in the case of the open bosonic and supersymmetric strings[2].

Open bosonic strings in an external *constant* e.m. field are exactly soulble, to lowest order in the string coupling constant. Their world-sheet action reads, in the

---

*On leave of absence from INFN, Sez. di Pisa, Pisa, Italy.

gauge $A_\mu = (1/2)F_{\nu\mu}x^\nu$ and for a state of total electric charge $e_1 + e_2$[2,3,4,5]

$$S = -\frac{1}{2\pi}\int d\sigma d\tau\, \partial_a X^\mu \partial^a X_\mu$$
$$+\frac{1}{2}e_1\int d\tau\, F_{\mu\nu}X^\nu\partial_\tau X^\mu|_{\sigma=0} + \frac{1}{2}e_1\int d\tau\, F_{\mu\nu}X^\nu\partial_\tau X^\mu|_{\sigma=\pi}. \tag{3}$$

Here we set $\alpha' = 1/2$. The world-sheet coordinates are $\sigma \in [0, \pi]$ and the proper time τ. The 26-dimensional target-space metric is $\eta_{\mu\nu} = (-1, +1, .., +1)$. The e.m. field couples only through boundary terms, thus the equations of motion of all coordinates are the same as in the free-string case

$$\partial_a \partial^a X^\mu = 0. \tag{4}$$

The only effect of the e.m. field is to modify the boundary conditions obeyed by the X^μ. Notice that the target-space metric remains flat to lowest order in the string coupling constant $g = e^\phi$. The back reaction due to the presence of a non-zero stress-energy tensor is in fact of order g with respect to the tree-level metric. When the external field is purely electrical one can choose a coordinate system in which only the component F_{01} of $F_{\mu\nu}$ is non-vanishing. By introducing standard light-cone coordinates $X^\pm = (X^0 \pm X^1)/\sqrt{2}$ the boundary conditions read

$$\begin{aligned}\partial_\sigma X^\pm &= \pm\beta_1\partial_\tau X^\pm, & \sigma = 0,\\ \partial_\sigma X^\pm &= \mp\beta_2\partial_\tau X^\pm, & \sigma = \pi.\end{aligned} \tag{5}$$

In Eq. (5) $\beta_i \equiv \pi e_i E$, $E = F_{01}$. The boundary conditions and mode expansion for the transverse coordinates X^i $i = 2, .., 25$ are the same as for the free open string. Equations (4,5) imply instead the following mode expansion for the coordinates $X^\pm$

$$\begin{aligned}X^\pm &= x^\pm + ia_0^\pm\phi_0^\pm + i\left(\sum_{n=1}^\infty a_n^\pm\phi_n^\pm(\sigma,\tau) - h.c.\right),\\ \phi_n^\pm &= (n \mp i\epsilon)^{-1/2}e^{-i(n\mp i\epsilon)\tau}\cos[(n \mp i\epsilon)\sigma \pm \text{arcth}\,\beta_1],\\ (a_0^\pm) &= \pm ia_0^\pm, \quad \epsilon = \frac{1}{\pi}(\text{arcth}\,\beta_1 + \text{arcth}\,\beta_2).\end{aligned} \tag{6}$$

Notice that for small electric fields, that is for fields $e_i E \ll \alpha'^{-1}$, $\epsilon \approx 2\alpha'(e_1 + e_2)E$. The main effect of the non-vanishing electric field is an *imaginary* shift in the free-string frequencies $n \to n \mp i\epsilon$. Moreover, similarly to the case of a particle in an external magnetic field, the center-of-mass coordinates $x^\pm$ do not commute, but obey instead the equation

$$[x^+, x^-] = \frac{-i\pi}{\beta_1 + \beta_2}. \tag{7}$$

By using the above mode expansion one easily finds the Virasoro operators L_n. They can be written as a sum of a "transverse" component $L_n^\perp$, involving only

330

the coordinates $X^2, .., X^{25}$, and identical to the free-string one, and a "longitudinal" one, $L_n^{\parallel}$. The component $L_0^{\parallel}$ reads, for $\epsilon > 0$,

$$L_0^{\parallel} = -\sum_{n=0}^{\infty}(n + i\epsilon)(a_n^+)^* a_n^- - \sum_{n=1}^{\infty}(n - i\epsilon)(a_n^-)^* a_n^+ + \frac{1}{2}i\epsilon(1 - i\epsilon). \tag{8}$$

Besides the imaginary shift in frequencies, the main difference with the free-string case is the shift in the vacuum energy. This shift can be determined either by spectral flow (with a twist $\theta = i\epsilon)^2$, or by imposing that the Virasoro algebra closes in the standard form $[L_m, L_n] = (n - m)L_{n+m} + 1/12cm(m^2 - 1)^{3,4}$.

The vacuum to vacuum transition amplitude $\langle 0| \exp -iTH|0\rangle$ is given in terms of the free-energy density F by a standard field-theory argument as $\exp(-iTVF)$. Thus the total pair-production rate w, equal to the probability of vacuum decay, is $w = -2\Im F$.

In open string theory the free-energy density F is given, at one loop, by a sum of four terms arising from the four different geometries open strings can propagate on: the torus, Klein bottle, annulus, and Möbius strip

$$-iTVF = \frac{1}{2}\mathcal{T} + \frac{1}{2}\mathcal{K} + \frac{1}{2}\mathcal{A} + \frac{1}{2}\mathcal{M} \tag{9}$$

The first two geometries correspond to closed string states: they do not depend on the end-point charges e_1 and e_2, since the torus and the Klein bottle have no boundaries, thus they do not give rise to terms contributing to $\Im F$.

The annulus contribution reads

$$\mathcal{A} = \frac{1}{2}\lim_{\delta \to 0}\int_{\delta}^{\infty} dt\, t^{-1}\mathrm{Tr}\, e^{-\pi t(L_0 - 1)} \approx \frac{1}{2}\mathrm{Tr}\, \log(L_0 - 1). \tag{10}$$

The trace in Eq. (10) is taken over all variables (oscillators, momenta etc.) and over all possible charge sectors of the open string. These charge sectors, in a consistent model, are determined by the embedding of $U(1)_{e.m.}$ into the open string gauge group. The integration variable t is the (real) modular parameter of the annulus.

The Möbius strip has only one boundary, and thus contributes only to sectors where $e_1 = e_2$. Its contribution to the free-energy is

$$\mathcal{M} = \pm\frac{1}{2}\lim_{\delta \to 0}\int_{\delta}^{\infty} dt\, t^{-1}\mathrm{Tr}\, \mathcal{P}e^{-\pi t(L_0 - 1)}. \tag{11}$$

The operator $\mathcal{P}$ implements the change of orientation of the open string $\sigma \to \pi - \sigma$. The Möbius-strip contribution, together with the Klein-bottle one, is needed in order to cancel the small-t divergences arising already in the free-string annulus integral. Only after annulus, Klein-bottle and Möbius-strip contributions are added, and only for a specific choice of the gauge group, do the free-string small-t divergencies cancel. The correct gauge group turns out to be $SO(8192)$ for the bosonic string, and $SO(32)$ for the open superstring[6].

Cancellation of divergencies in the presence of an external field requires to take into proper account the back reaction on the target-space metric induced by the presence of the background field. However, since we are only interested in the imaginary part of the free-energy, which receives no contribution form the small-t region of integration, we can safely ignore this complication.

The annulus contribution to the free-energy can be evaluated straightforwardly, since the Virasoro operators are sums of free oscillators

$$A = \sum_{e_1,e_2 \in Q} \lim_{\delta \to 0} VT \frac{|\beta_1 + \beta_2|}{(2\pi)^2} \int_\delta^\infty dt\, t^{-1} \int \frac{d^{24}p}{(2\pi)^{24}} e^{-\pi t p^2/2} T^\perp T^\parallel. \tag{12}$$

In this equation Q denotes the set of boundary charges determined by the embedding of $U(1)_{e.m.}$ into the open-string gauge group. $T^\perp$ and $T^\parallel$ arise from taking the trace over transverse and longitudinal oscillators, respectively. The integration over transverse momenta is standard. The normalization factor in front of the integral is the only place where the β_i enter explicitly, instead of ϵ, and it is determined by noticing that the commutation relation (7) implies that the $x^\pm$ phase-space integration measure is

$$\frac{|\beta_1 + \beta_2|}{2\pi^2} dx^+ dx^-. \tag{13}$$

Obviously $T^\perp$ has the same form as for the free open string, since it involves only transverse oscillators

$$T^\perp = \eta(it/2)^{-24} \eta(it/2)^2. \tag{14}$$

Notice the contribution η^2 coming from the coordinate-reparametrization ghosts.

$T^\parallel$ depends on the external electric field and reads

$$T^\parallel = e^{-\pi t(\epsilon^2/2 - 1/12)} \left[2i \sin(\pi |\epsilon| t/2) \prod_{n=1}^\infty \left| 1 - e^{-\pi t(n + i\epsilon)} \right|^2 \right]^{-1}. \tag{15}$$

This quantity is imaginary, as imposed by equation (9). The free-energy F though, has a nonzero imaginary part, arising from the integration in t. Indeed, $T^\parallel$ has the form

$$T^\parallel = [2i \sin(\pi |\epsilon| t/2)]^{-1} f(t), \tag{16}$$

with $f(t)$ regular for $t > 0$, and thus it has simple poles for positive t, located at $t = 2k/|\epsilon|$, k integer. When integrating in t one has to deform the contour of integration so as to avoid these poles. The correct prescription on the contour is, as expected, the one allowing for the analytic continuation $t \to it$. Thus, $\Im F$ reduces to a sum over residues at the poles, according to Cauchy's theorem

$$A = \sum_{e_1,e_2 \in Q} VT \frac{\beta_1 + \beta_2}{2(2\pi)^{26} \epsilon} \sum_{k=1}^\infty (-1)^{k+1} e^{-\pi |\epsilon| k} (|\epsilon|/k)^{13} \eta(ik/|\epsilon|)^{-24}. \tag{17}$$

This formula can be recast in the following simple form

$$A = \sum_{e_1,e_2 \in Q} VT \frac{\beta_1 + \beta_2}{2\epsilon} \sum_{k=1}^\infty (-1)^{k+1} Z(2k/|\epsilon|), \tag{18}$$

where $Z(t)$ is the *free-string* partition function on the annulus. Equation (18) holds in any compactification that does not involve the $X^{\pm}$ coordinates. Notice also that the term under the Q-summation sign in Eq. (18) gives the complete pair-production rate for open-string states with $e_1 \neq e_2$, since in this case F receives no contribution from the Möbius strip. To compare Eq. (18) with Schwinger's result (2) we must compactify the bosonic string to four dimensions and recall that four distinct open-string sectors contribute to the production rate of a particle pair of given charge $e = e_1 + e_2$. These are the sectors with boundary charges (e_1, e_2), (e_2, e_1), $(-e_1, -e_2)$, and $(-e_2, -e_1)$. By denoting with M_S the mass of an open-string state S and after a simple computation one finds

$$w = \sum_S \sum_{k=1}^{\infty} (-1)^{k+1} \frac{\beta_1 + \beta_2}{8\pi^4 \epsilon} (|\epsilon|/k)^2 e^{-k\pi(M_S^2 + \epsilon^2)/|\epsilon|}. \tag{19}$$

When $\beta_i \ll 1$, $\epsilon \to (e_1 + e_2)E$, and Eq. (19) reduces to Eq. (2).

When the two boundary charges are equal, only two open-string sectors contribute to the production rate of a given particle pair: (e, e) and $(-e, -e)$. The annulus amplitude in Eq. (17), therefore, would reduce to one-half of Schwinger's result. This discrepancy is resolved by taking into account the contribution of the Möbius strip to equation (9).

For a given state S this contribution is equal to annulus one, up to a sign. This sign is positive or negative according to whether the state S is even or odd under $\mathcal{P}$ (defined in Eq. (11)). The sum of the Möbius and annulus contribution thus reduces to Schwinger's formula for all states which are not eliminated from the physical spectrum by the Möbius projection.

An interesting property of Eqs. (17,18) is that they diverge for a (finite) critical value of the electric field, namely when the force applied by the electric field on either of the boundary charges equals the string tension

$$\min_i |e_i E| = (2\pi\alpha')^{-1}. \tag{20}$$

This instability arises already at the classical level[4]. It is interesting to notice that when $(e_1 + e_2) \to 0$, but $e_1 \neq 0$, the annulus partition function $\mathcal{A}$ does not reduce to the free-string one. In fact, by setting $\beta_1 = -\beta_2 + \Delta$, and taking the limit $\Delta \to 0$, one finds $\epsilon = \Delta/(1 - \beta_1^2)\pi + \mathcal{O}(\Delta^2)$. Using Eqs. (12,14,15) one then finds that the neutral-string annulus amplitude $\mathcal{A}_{neutral}$ and the free-string one $\mathcal{A}_{free}$ are related by

$$\mathcal{A}_{neutral} = (1 - \beta_1^2)\mathcal{A}_{free}. \tag{21}$$

This formula agrees with ref.[3].

Open superstrings in an external field can be solved exactly at lowest order in the string coupling constant by using the same procedure already outlined in the bosonic-string case. The world-sheet action of the critical 10-dimensional superstring reads

$$S = S_{bosonic} + \frac{i}{2\pi} \int d\sigma d\tau \, \bar{\psi}^{\mu} \rho^{\alpha} \partial_{\alpha} \psi_{\mu}$$

$$+\frac{i}{4}\,e_1\int d\tau\,F_{\mu\nu}\bar\psi^\mu\rho^0\psi^\nu\Big|_{\sigma=0}+\frac{i}{4}\,e_2\int d\tau\,F_{\mu\nu}\bar\psi^\mu\rho^0\psi^\nu\Big|_{\sigma=\pi}. \tag{22}$$

The ρ^i are the 2-dimensional Dirac matrices.

The appropriate boundary conditions on the 2-dimensional bosons $X^\pm$ are given in Eq. (5). The boundary conditions on their fermionic partners $\psi^\pm$ are

$$\begin{aligned}
(1\mp\beta_1)\psi_R^\pm &= (1\pm\beta_1)\psi_L^\pm, \quad \sigma=0,\\
(1\pm\beta_2)\psi_R^\pm &= -(-1)^a(1\mp\beta_2)\psi_L^\pm, \quad \sigma=\pi.
\end{aligned} \tag{23}$$

In this equation the subindices R, L denote the 2-dim. handedness of the fermions. The variable a is equal to 0 when the fermions are given anti-periodic (Neveu-Schwarz) boundary conditions and to 1 when they are given periodic (Ramond) boundary conditions. The transverse fermionic coordinates obey standard free-superstring boundary conditions.

All fermionic coordinates obey free equations of motion which, together with Eq. (23), completely determine the fermion mode expansion. Canonical quantization gives rise to the Virasoro operators $L_n = L_n^F + L_n^B$. The bosonic contribution to these operators is denoted by L_n^B and is the same as in the bosonic open string. The fermionic contribution L_n^F can be further decomposed into a sum of two terms: $L_n^{F\perp}$, containing only transverse oscillators, and identical to the corresponding operator for a free open superstring, and $L_n^{F\parallel}$, depending only on the fermions $\psi^\pm$. By introducing the canonical anti-commuting operators d_n^μ obeying $\{d_n^\mu, d_m^\nu\} = \eta^{\mu\nu}\delta_{n+m,0}$ one finds, in particular,

$$L_0^{F\parallel} = -\sum_{n\in Z+1/2+a/2} -(n+i\epsilon):d_{-n}^- d_n^+: +\frac{a}{8}-\frac{i\epsilon}{2}(a-i\epsilon), \quad \epsilon>0. \tag{24}$$

The shift in the vacuum energy can be determined by spectral flow, once a normal-ordering prescription for $d_k^\pm$ is given, or by closure of the Virasoro algebra in exact analogy with the bosonic case. Notice that the shift in the Ramond-sector vacuum energy exactly cancels between bosons and fermions (cfr. Eq. (8)).

The annulus amplitude now reads

$$\mathcal{A}=\sum_{e_1,e_2\in Q}\sum_{a,b}\lim_{\delta\to0}VT\frac{|\beta_1+\beta_2|}{(2\pi)^2}C\begin{bmatrix}a\\b\end{bmatrix}\int_\delta^\infty dt\,t^{-1}\mathrm{Tr}_a\left[(-1)^{bF}e^{-\pi t(L_0-1)}\right]. \tag{25}$$

F is the fermion-number operator commuting with all world-sheet bosons and anti-commuting with the world-sheet fermions. The values of the parameters a and b are 0 or 1, and a is defined as before to be 0 in the Neveu-Schwarz sector and 1 in the Ramond sector. The weights in Eq. (25) are chosen so as to implement space-time supersymmetry in 10 dimensions

$$C\begin{bmatrix}0\\0\end{bmatrix}=-C\begin{bmatrix}1\\0\end{bmatrix}=-C\begin{bmatrix}0\\1\end{bmatrix}=\pm C\begin{bmatrix}1\\1\end{bmatrix}=\frac{1}{2} \tag{26}$$

the trace in Eq. (25) factorizes into the product of a purely bosonic contribution, which is evaluated using Eqs. (12,14,15), and a fermionic one, which reads

$$\mathrm{Tr}\,_a(-1)^{bF}e^{-\pi t L_0^F} = T_F^{ghost}T_F^{\perp}T_F^{\parallel},$$

$$T_F^{ghost} = \mathrm{Tr}\,_a(-1)^{bF}e^{-\pi t L_0^{F\ ghost}} = \eta(it/2)/\Theta\begin{bmatrix} a \\ b \end{bmatrix}(0|it/2), \quad a+b \neq 2,$$

$$T_F^{\perp} = \mathrm{Tr}\,_a(-1)^{bF}e^{-\pi t L_0^{F\perp}} = \left\{\Theta\begin{bmatrix} a \\ b \end{bmatrix}(0|it/2)/\eta(it/2)\right\}^3,$$

$$T^{\parallel} = \mathrm{Tr}\,_a(-1)^{bF}e^{-\pi t L_0^{F\parallel}} = \Theta\begin{bmatrix} a-2i|\epsilon| \\ b \end{bmatrix}(0|it/2)/\eta(it/2). \tag{27}$$

The only term depending on the external electric field in Eq. (27) is, as expected, $T_F^{\parallel}$. Here the conventions on theta functions with characteristics are as in[2]. The contribution T_F^{ghost}, arising from the world-sheet supersymmetry ghosts, takes a special form in the presence of a fermionic zero-mode (when $a+b=2$). We need not to consider it because the contribution of the $a+b=2$ sector to $\mathcal{A}$ gets cancelled by $T_F^{\perp}$, since the theta function with $a=b=1$ vanishes identically.

It is immediate to notice that the poles in the t-integration arise only from the trace over bosons. Thus, the annulus amplitude can be evaluated exactly as in the bosonic case. The residue at the pole $t = 2k/|\epsilon|$ is given by a bosonic term times the fermionic term

$$T_F(t = 2k/|\epsilon|) = (-1)^{ak}e^{\pi k|\epsilon|}\left\{\Theta\begin{bmatrix} a \\ b \end{bmatrix}(0|ik/|\epsilon|)/\eta(ik/|\epsilon|)\right\}^4. \tag{28}$$

Notice that this term contributes an extra $(-1)^k$ to the sum over residues when $a = 1$, that is for space-time fermions, in analogy with Schwinger's formula Eq. (2). After a few manipulations, analogous to the ones performed in the bosonic case, one arrives at the following formula for the vacuum decay rate when $10 - D$ dimensions are compactified

$$w = \frac{1}{(2\pi)^{D-1}}\sum_S \frac{\beta_1 + \beta_2}{\pi\epsilon}\sum_{k=1}^{\infty}(-1)^{(k+1)(a_S+1)}(|\epsilon|/k)^{D/2}e^{-\pi k M_S^2/|\epsilon|}. \tag{29}$$

Here the variable S labels the string states surviving the GSO and Möbius projections, $a_S = 0$ for states in the Neveu-Schwarz sector and $a_S = 1$ for Ramond-sector states. In $D = 4$ and for small fields this expression reduces to Eq. (2). Notice that in Eq. (29) the production rates for each particle species diverge at $\beta_i = 1$. This behavior is different from the bosonic-string one, where the production rates *vanished* species by species, but their sum diverged, thanks to the stronger divergence of the statistical sum over all string states.

Finally, one may notice that equation (5) interpolates between free-string (Neumann) boundary conditions, at $\beta_i = 0$, and reflecting boundary conditions at

$\beta_i = 1$. This latter case corresponds to the critical value of the electric field given in Eq. (20). At $\beta_i = 1$, thus, the world-sheet dynamics of the bosonic string becomes formally equivalent to that of a black-hole solution of 2-dimensional gravity coupled to 24 free massless scalars[7] (see also K. Schoutens's and E. Verlinde's contributions to these proceedings).

2. Open Strings in a Constant Magnetic-Field Background

Open strings in an external purely magnetic field exhibit new interesting features already at tree-level[8]. Let us examine at first the bosonic string. In this case by choosing a coordinate system in which the only non-vanishing component of the magnetic field is $H \equiv H_{12}$, the Virasoro operator L_0 reads[3,8]

$$L_0 = (2b_0^\dagger b_0 + 1)\frac{|h|}{2} - \frac{1}{2}h^2 - |h| \sum_{k=1}^{\infty}(a_k^\dagger a_k - b_k^\dagger b_k) + L_0^{free}. \tag{30}$$

Here L_0^{free} denotes the free-string Virasoro operator of all the transverse coordinates $X^0, X^3, ..., X^{(D-1)}$, and we introduced the complexified creation and annihilation operators

$$\sqrt{2}a_k = \alpha_k^1 + i\,\text{sign}\,(h)\alpha_k^2, \quad \sqrt{2}b_k = \alpha_k^1 - i\,\text{sign}\,(h)\alpha_k^2. \tag{31}$$

Equation (30) differs from the free string expression by terms involving

$$h = \frac{1}{\pi}(\arctan 2\alpha' e_1 \pi H + \arctan 2\alpha' e_2 \pi H). \tag{32}$$

Notice in particular the presence of a magnetic-dipole coupling, non-linear in the magnetic-field strenght

$$|h| \sum_{k=1}^{\infty}(a_k^\dagger a_k - b_k^\dagger b_k) = hS_{12}, \tag{33}$$

where $S_{\mu\nu}$ is the target-space spin operator. In particular, in four dimensions, S_{12} is the spin component along the magnetic field. Since, in the low-field limit ($\alpha' eH \ll 1$) $h \to 2\alpha'(e_1 + e_2)H$, it is apparent that open string states have all a gyromagnetic ratio $g = 2$[9,10]. Moreover, in $D = 4$ and in the low-field limit, Eq. (30) itself reduces to the well-known field-theoretical formula giving the energy levels of a particle of mass M, spin $\vec{S}$, and gyromagnetic ratio g in an external magnetic field (see for instance[11])

$$E^2 = (2n + 1)eH - g_S e\vec{H} \cdot \vec{S} + p_3^2 + M^2. \tag{34}$$

To arrive at this formula, it suffices to recall that the wave equation of a string state Ψ is

$$(L_0 - 1)\Psi \equiv \alpha'[E^2 - (p^0)^2]\Psi = 0, \tag{35}$$

and to notice that the eigenvalues n of $b_0^\dagger b_0$ are the Landau levels. The Virasoro operator L_0 possesses the usual tachionic mode of the free bosonic string, but it also shows additional tachionic modes when $H \neq 0$. By direct examination of Eq. (30)

one may see that only the highest-helicity states belonging to the first ("parent") Regge trajectory can become tachyonic. By denoting with $|0\rangle$ the Fock vacuum for the string oscillators a_k, b_k and α_k, these states read

$$\Psi(m) = \left(a_1^\dagger\right)^m |0\rangle. \tag{36}$$

The value of $\alpha' E^2$ on them is

$$(\alpha' E^2)\Psi(m) = \left[\frac{|h|}{2} - \frac{h^2}{2} + (1 - |h|)m - 1\right]\Psi(m). \tag{37}$$

For fixed value of the magnetic field, and thus of $|h|$, $\alpha' E^2$ is negative on all states with

$$m < \frac{1 + |h|(|h| - 1)/2}{1 - |h|}. \tag{38}$$

In particular, the highest helicity of the open-string massless vector ($m = 1$) is always tachionic when $H \neq 0$. This instability is well-known in Yang-Mills theory[12]. The $m = 0$ instability of open bosonic strings was also noticed in ref.[3]. There is an infinite number of states given by Eq. (37), correspondingly, as the intensity of the external magnetic field increases, open strings may undergo an infinite number of phase transitions. As H approaches infinity $h \to 1$. In this limit all states $\Psi(m)$ would become tachionic, in the standard supersting vacuum, with a common (negative) square energy $\alpha' E^2 = -1$.

The meaning of these phase transitions is that the vacuum where the VEVs of the $\Psi(m)$ vanish is unstable, when the magnetic field passes the threshold value given by Eq. (38): in the stable vacuum $\Psi(m)$-condensates should appear (cfr.[12,13]).

The previous analysis can be straightforwardly extended to the open superstring.

In the superstring case one must introduce fermionic oscillators, either in the Ramond or Neveu-Schwarz sector, besides the bosonic ones. The fermionic oscillators can be complexified in the same way as the bosonic coordinates

$$\sqrt{2}d_k = d_k^1 + i\,\text{sign}\,(h)d_k^2, \quad \sqrt{2}\tilde{d}_k = d_k^1 - i\,\text{sign}\,(h)d_k^2 \tag{39}$$

Using the same notations as before one finds that the Virasoro operator L_0 takes the following form in the Ramond sector

$$L_0 = (2n + 1)\frac{|h|}{2} + |h|d_0^\dagger d_0 - \frac{|h|}{2} - |h|\sum_{k=1}^\infty \left(d_k^\dagger d_k - \tilde{d}_k^\dagger \tilde{d}_k + a_k^\dagger a_k - b_k^\dagger b_k\right) + L_0^{free}. \tag{40}$$

Recalling that the wave equation in the Ramond sector takes the form $L_0\Psi = 0$ one easily finds that all Ramond states have positive square energy. This result follows from the inequality $h \leq 1$. This inequality is due to the fact that in string theory the magnetic-dipole coupling is non-linear (see Eq. (32)). The field-theoretical formula (34) with $g = 2$ would instead give instabilities for *any* spin larger than $1/2$.

The Virasoro operator in the Neveu-Schwarz sector reads

$$L_0 = (2n + 1)\frac{|h|}{2} - |h| \sum_{k=1/2}^{\infty} \left(d_k^\dagger d_k - \tilde{d}_k^\dagger \tilde{d}_k + a_k^\dagger a_k - b_k^\dagger b_k \right) + L_0^{free}. \tag{41}$$

The Neveu-Schwarz wave equation is $(L_0 - 1/2)\Psi = 0$, this implies that there exist Neveu-Schwarz physical states (i.e. states surviving the GSO projection) with $\alpha' E^2 < 0$, when H is sufficiently large. A brief analysis of Eq. (41) shows that these states have the form

$$(a_1^\dagger)^m d_{1/2}^\dagger |0\rangle_{NS}, \tag{42}$$

and that they become tachionic when

$$m < \frac{|h|}{2(1 - |h|)}. \tag{43}$$

They are, again, highest-helicity states belonging to the parent Regge trajectory of the Neveu-Schwarz sector. When $h \to 1$ the negative square energy of all these states becomes $\alpha' E^2 = -1/2$.

An interesting and difficult problem still to be solved is to find an ansatz for the scalar potential of the fields given in Eqs. (36,42). Knowledge of this potential would allow a detailed study of high magnetic field phase transitions in string theory.

References

1. J. Schwinger, *Phys. Rev.* **82** (1951) 664.

2. C. Bachas and M. Porrati, *Phys. Lett.* **296B** (1992) 77.

3. A. Abouelsaood, C.G. Callan, C.R. Nappi and S.A. Yost, *Nucl. Phys.* **B280** (1987) 599.

4. V.V. Nesterenko, *Int. J. Mod. Phys.* **A4** (1989) 2627.

5. C.P. Burgess, *Nucl. Phys.* **B294** (1987) 427.

6. M.B. Green and J.H. Schwarz, *Phys. Lett.* **149B** (1984) 117; **151B** (1985) 21; M.J. Douglas and B. Grinstein, *Phys. Lett.* **183B** (1987) 52; S. Weinberg, *Phys. Lett.* **187B** (1987) 278; N. Marcus and A. Sagnotti, *Phys. Lett.* **188B** (1987) 58.

7. K. Schoutens, H. Verlinde and E. Verlinde, *preprint* **PUPT-1395, IASSNS-HEP-93-25, hep-th 9304128** (1993).

8. S. Ferrara and M. Porrati, *preprint* **CERN-TH.6915/93, UCLA/93/TEP/17, NYU-TH.93/06/01, hep-th 9306048** (1993).

9. S. Ferrara, M. Porrati and V.L. Telegdi, *Phys. Rev.* **D46** (1992) 3529.

10. E. Del Giudice, P. Di Vecchia and S. Fubini, *Ann. Phys.* **70** (1972) 378; S. Matsuda and T. Saido, *Phys. Lett.* **43B** (1973) 123; M. Ademollo *et al.*, *Nuovo Cim.* **A21** (1976) 77.

11. J. Schwinger, in *Particles, Sources and Fields* Vol. III (Advanced Book Classics, New York, 1989).

12. P.K. Nielsen and P. Olesen, *Nucl. Phys.* **B144** (1978) 376.

13. J. Ambjørn and P. Olesen, *Nucl. Phys.* **B315** (1989) 606; **B330** (1990) 193.

DOES STRING THEORY HAVE A DUALITY SYMMETRY RELATING WEAK AND STRONG COUPLING?[*]

JOHN H. SCHWARZ[†]

California Insitute of Technology, Pasadena, CA 91125

ABSTRACT

The heterotic string theory, compactified to four dimensions, has been conjectured to have a duality symmetry (S duality) that transforms the dilaton nonlinearly. If valid, this symmetry could provide an important means of obtaining information about nonperturbative features of the theory. Even though it is inherently nonperturbative, S duality exhibits many similarities with the well-established target-space duality symmetry (T duality), which does act perturbatively. These similarities are manifest in a new version of the low-energy effective field theory and in the soliton spectrum obtained by saturating the Bogomol'nyi bound. Curiously, there is evidence that the roles of the S and T dualities are interchanged in passing to a five-brane formulation.

---

[*] Supported in part by the U.S. Dept. of Energy under Grant No. DE-FG03-92-ER40701.
[†] JHS@THEORY3.CALTECH.EDU

Introduction

This talk reports on work done recently in collaboration with Ashoke Sen [1] [2], which investigated various issues concerning two types of duality symmetries that have been considered in string theory. One, which is well-established to all orders in string perturbation theory, is known as "target space duality," or more succinctly as "T duality." This discrete symmetry group is a generic feature of theories with compactified spatial dimensions. It is actually a discrete gauge group, so it should remain valid when nonperturbative effects are taken into account. The simplest example is the Z_2 symmetry that arises when one dimension is compactified with the topology of a circle. In this case, it can be described as the equivalence of a circle of radius R and one of radius α'/R. In fact, it can be realized as a field transformation, since R corresponds to the classical value of a scalar field. More generally, for the four-dimensional heterotic string, compactified on a torus that is dual to an even self-dual lattice in the manner proposed by Narain [3], the corresponding T duality group turns out to be O(6,22;Z). This example has N=4 supersymmetry, and is therefore certainly unrealistic, but it is a particularly nice example to study. Many, but not all, of its properties are expected to hold in more realistic settings, but that question will not be explored in detail here.

The second kind of duality, which is our main focus, is much more speculative. It is an SL(2,Z) group that was discovered many years ago as a symmetry of the classical field equations of N=4 supergravity [4]. (This paper actually identified an SL(2,R) symmetry. Instanton effects are expected to break the symmetry to the discrete subgroup.) N=4 supergravity contains a dilaton ϕ and an axion χ, which can be combined in a complex scalar field λ as follows

$$\lambda = \lambda_1 + i\lambda_2 = \chi + ie^{-\phi}. \tag{1}$$

This field transforms nonlinearly under SL(2,Z)

$$\lambda \to \frac{a\lambda + b}{c\lambda + d}, \tag{2}$$

where a, b, c, d are integers satisfying $ad - bc = 1$. Since the value of the field λ determines the coupling constant g and the vacuum angle θ according to

$$< \lambda >= \frac{\theta}{2\pi} + \frac{8\pi i}{g^2}, \tag{3}$$

such a symmetry is necessarily nonperturbative. In particular, the transformation $\lambda \to -1/\lambda$, for $\theta = 0$, inverts the coupling constant. The complex field λ is present in the massless spectrum even for compactifications that only leave N=1 supersymmetry. It is therefore possible to speculate that it might be a symmetry of the full nonperturbative string theory in that case, which is what was done in 1990 by Font *et al.*[5]. This was, to the say the least, a very bold conjecture. The proposed symmetry could be called "dilaton–axion duality" or "weak coupling–strong coupling duality." These are rather cumbersome, so we propose to refer to it as "S duality." While S duality is still far from established, we find that all our studies support it. Indeed, when analyzed in the proper way, S duality and T duality have a great deal in common.

4D Effective Field Theory

This section summarizes material I presented at Strings '92 in Rome [6]. The low-energy effective field theory that describes the massless bosonic fields associated with Narain compactification of the heterotic string has a T duality symmetry group $G_T = \mathrm{O}(6,22;Z)$ and scalar fields (moduli) that parametrize the moduli space $\mathrm{O}(6,22)/\mathrm{O}(6)\times\mathrm{O}(22)\times G_T$. These fields are conveniently described by a 28×28 matrix-valued scalar field M^{ab} satisfying the constraints

$$M^T = M, \quad M^T L M = L, \tag{4}$$

where L is the O(6,22) metric

$$L = \begin{pmatrix} 0 & I_6 & 0 \\ I_6 & 0 & 0 \\ 0 & 0 & -I_{16} \end{pmatrix}. \tag{5}$$

For generic values of the moduli fields, the spectrum also contains 28 massless abelian gauge fields (so that the gauge group is $[\mathrm{U}(1)]^{28}$). For special values of the moduli there are additional massless gauge fields and enhanced gauge symmetry. However, to keep things as simple as possible, we will only include the 28 gauge fields A_μ^a that are massless for all values of the moduli. The other massless bosons are the graviton (described by a metric tensor $g_{\mu\nu}$), the dilaton ϕ, and an antisymmetric tensor $B_{\mu\nu}$. The action for this theory can be obtained in a variety of ways, one of the easiest of which is dimensional reduction from ten dimensions [7]. In terms of the "string metric," the result is

$$S = \int_M dx\sqrt{-g}\, e^{-\phi}(\mathcal{L}_1 + \mathcal{L}_2 + \mathcal{L}_3 + \mathcal{L}_4 + \mathcal{L}_5)\,, \tag{6}$$

$$\mathcal{L}_1 = R$$
$$\mathcal{L}_2 = g^{\mu\nu}\partial_\mu\phi\partial_\nu\phi$$
$$\mathcal{L}_3 = -\frac{1}{12}H_{\mu\nu\rho}H^{\mu\nu\rho}$$
$$\mathcal{L}_4 = \frac{1}{8}g^{\mu\nu}\mathrm{tr}\big(\partial_\mu ML\partial_\nu ML\big)$$
$$\mathcal{L}_5 = -\frac{1}{4}F^a_{\mu\nu}(LML)_{ab}F^{b\mu\nu}, \tag{7}$$

where

$$F^a_{\mu\nu} = \partial_\mu A^a_\nu - \partial_\nu A^a_\mu . \tag{8}$$

$$H_{\mu\nu\rho} = \partial_\mu B_{\nu\rho} + \frac{1}{2}A^a_\mu L_{ab}F^b_{\nu\rho} + (\text{cyc. perms.}) . \tag{9}$$

This result has manifest T duality since the metric and dilaton are invariant under T transformations, and the gauge fields transform by the vector representation of O(6,22).

This theory also has S duality symmetry, though this is not at all apparent in the form given above. To exhibit this symmetry, it is convenient to replace the string metric by the canonical metric by means of the Weyl rescaling $g_{\mu\nu} \to e^\phi g_{\mu\nu}$, since the canonical metric will be invariant under S duality, but the dilaton field is not. Also, to exhibit the axion χ, it is necessary to make a duality transformation

$$\sqrt{-g}\, e^{-2\phi}H^{\mu\nu\rho} \to \epsilon^{\mu\nu\rho\lambda}\partial_\lambda\chi, \tag{10}$$

which (as usual) interchanges the role of a field equation and a Bianchi identity. Then one can introduce the complex field λ defined in eq. (1) and write a "dual action," whose classical field equations are equivalent to those obtained from the original action.

$$S_{\text{dual}} = \int_M dx\sqrt{-g}(\mathcal{L}'_1 + \mathcal{L}'_{2,3} + \mathcal{L}'_4 + \mathcal{L}'_5) , \tag{11}$$

$$\mathcal{L}'_1 = R$$
$$\mathcal{L}'_{2,3} = -\frac{1}{2\lambda_2^2}g^{\mu\nu}\partial_\mu\lambda\partial_\nu\bar\lambda$$
$$\mathcal{L}'_4 = \frac{1}{8}g^{\mu\nu}\mathrm{tr}\big(\partial_\mu ML\partial_\nu ML\big)$$
$$\mathcal{L}'_5 = -\frac{\lambda_2}{4}F^a_{\mu\nu}(LML)_{ab}F^{b\mu\nu} + \frac{\lambda_1}{4}F^a_{\mu\nu}L_{ab}\tilde{F}^{b\mu\nu}, \tag{12}$$

where

$$\tilde{F}^{a\mu\nu} = \frac{1}{2\sqrt{-g}}\epsilon^{\mu\nu\rho\sigma}F^a_{\rho\sigma}. \tag{13}$$

All terms in the action $S_{\rm dual}$ are invariant under S duality, except for $\mathcal{L}'_5$. However, the equations of motion do transform covariantly under S duality provided that when $\lambda \to \frac{a\lambda+b}{c\lambda+d}$ (and $ad - bc = 1$),

$$F^a_{\mu\nu} \to c\lambda_2(ML)_{ab}\tilde{F}^b_{\mu\nu} + (c\lambda_1 + d)F^a_{\mu\nu}. \tag{14}$$

Note that in terms of the gauge fields themselves this is a nasty nonlocal transformation.

Manifest S Duality

The construction of the action $S_{\rm dual}$ is a significant step towards exhibiting S duality, but because of the noninvariance of $\mathcal{L}'_5$, it is not the final form. We would like to have a third form of the action with both dualities (S and T) made manifest, thereby putting them on an equal footing. The main problem in realizing S duality in $S_{\rm dual}$ is attributable to the gauge fields. The basic idea for overcoming this difficulty is to replace each gauge field A_μ by a pair of independent gauge fields $A_\mu^{(\alpha)}$, $\alpha = 1, 2$ and to obtain the relation $F_{\mu\nu}^{(2)} = \tilde{F}_{\mu\nu}^{(1)}$ as an equation of motion. The formulas will not have manifest Lorentz invariance, though they will have manifest rotational symmetry. Accordingly, it is convenient to introduce separate "electric" and "magnetic" fields

$$E_i^{(\alpha)} = \partial_0 A_i^{(\alpha)} - \partial_i A_0^{(\alpha)}, \quad B^{(\alpha)i} = \epsilon^{ijk}\partial_j A_k^{(\alpha)} \quad 1 \leq i, j, k \leq 3. \tag{15}$$

Let us first describe a two-potential version of free Maxwell theory, and explain how that theory is coupled to gravity, before applying the results to the problem of making S duality manifest. Our action for free Maxwell theory is

$$S = -\frac{1}{2}\int d^4x \left(B^{(\alpha)i}\mathcal{L}_{\alpha\beta}E_i^{(\beta)} + B^{(\alpha)i}B^{(\alpha)i}\right), \tag{16}$$

where

$$\mathcal{L} = \begin{pmatrix} 0 & 1 \\ -1 & 0 \end{pmatrix}. \tag{17}$$

is the metric for SL(2,R) = Sp(2). The symbols $\mathcal{L}$ and $\mathcal{M}$ (introduced below) to describe S duality are chosen to emphasize the analogy with L and M used in the description of T duality. The action (16) has the following gauge invariances

$$\delta A_0^{(\alpha)} = \Psi^{(\alpha)}, \quad \delta A_i^{(\alpha)} = \partial_i \Lambda^{(\alpha)}. \tag{18}$$

It is easy to show that it describes a single propagating photon with two physical polarizations. To understand how it works, note that up to total derivatives the first term is proportional to $\epsilon^{ijk}\partial_0 A_i^{(1)}\partial_j A_k^{(2)}$. Therefore the fields $A_0^{(\alpha)}$ do not give any classical equations of motion. Cast in this form, the fields $A_i^{(2)}$ have no time derivatives and can be treated as auxiliary. Thus the (integrated) equation of motion $B_i^{(2)} = E_i^{(1)}$ can be used to eliminate $A_i^{(2)}$ from the action. This gives rise to the standard Maxwell action in the $A_0^{(1)} = 0$ gauge. Gauss's law ($\partial_i E_i^{(1)} = 0$) is implied by the Bianchi identity for $B_i^{(2)}$. The action (16) is manifestly invariant under the electric–magnetic duality symmetry

$$A_\mu^{(\alpha)} \to \mathcal{L}_{\alpha\beta} A_\mu^{(\beta)}, \tag{19}$$

which corresponds to $F_{\mu\nu} \to \tilde{F}_{\mu\nu}$. This is to be contrasted with the usual Maxwell action, $\frac{1}{2}\int(E^2 - B^2)d^4x$, which goes to its negative. While not *manifestly* Lorentz invariant, the action (16) is in fact invariant under the global transformations

$$\delta A_i^{(\alpha)} = x^0 v^k \partial_k A_i^{(\alpha)} + \vec{v} \cdot \vec{x} \mathcal{L}_{\alpha\beta} \epsilon^{ijk} \partial_j A_k^{(\beta)}. \tag{20}$$

On the mass shell, this is identical to the usual Lorentz transformation formula with boost parameter v^i.

Let us now consider how to couple (16) to gravity. Since we do not have manifest Lorentz invariance before coupling to gravity, we do not expect manifest general coordinate invariance after coupling to gravity. Nonetheless it is not difficult to figure out which action gives the desired generally covariant field equations. The result is

$$S_g = -\frac{1}{2}\int d^4x \left[B^{(\alpha)i}\mathcal{L}_{\alpha\beta}E_i^{(\beta)} - \frac{g_{ij}}{\sqrt{-g}g^{00}}B^{(\alpha)i}B^{(\alpha)j} + \epsilon^{ijk}\frac{g^{0k}}{g^{00}}B^{(\alpha)i}\mathcal{L}_{\alpha\beta}B^{(\beta)j} \right]. \tag{21}$$

As usual, $\sqrt{-g} = \sqrt{-\det(g_{\mu\nu})}$ and $g^{\mu\nu}$ is the inverse of $g_{\mu\nu}$, the ordinary four-dimensional metric. These conventions are retained even when space and time components are enumerated separately. If one eliminates the fields $A_\mu^{(2)}$ by using their field equations as before, one obtains the standard action $-\frac{1}{4}\int d^4x\sqrt{-g}F^{(1)\mu\nu}F_{\mu\nu}^{(1)}$. Equation (21) is invariant under general coordinate transformations with the metric transforming in the standard way and

$$\delta A_i^{(\alpha)} = \xi^j \partial_j A_i^{(\alpha)} + (\partial_i \xi^j) A_j^{(\alpha)} + \xi^0 \left\{ -\frac{g_{ij}}{\sqrt{-g}g^{00}} \mathcal{L}_{\alpha\beta} B^{(\beta)j} - \frac{g^{0k}}{g^{00}} \epsilon^{ijk} B^{(\alpha)j} \right\}. \tag{22}$$

As in the case of Lorentz transformations, which corresponds to setting $\xi^0 = \vec{v} \cdot \vec{x}$, $\xi^i = v^i x^0$, and $g_{\mu\nu} = \eta_{\mu\nu}$, this differs from the standard transformation formula by an amount that vanishes when the classical equations of motion are satisfied.

Let us now return to the case of flat space-time and consider the generalization of (16) that includes the coupling to the axion–dilaton field λ. The appropriate formula is

$$S_\lambda = -\frac{1}{2} \int d^4x [B^{(\alpha)i} \mathcal{L}_{\alpha\beta} E_i^{(\beta)} + B^{(\alpha)i} (\mathcal{L}^T \mathcal{M} \mathcal{L})_{\alpha\beta} B^{(\beta)i}], \tag{23}$$

where

$$\mathcal{M}(\lambda) = \frac{1}{\lambda_2} \begin{pmatrix} 1 & \lambda_1 \\ \lambda_1 & |\lambda|^2 \end{pmatrix}. \tag{24}$$

The matrix $\mathcal{M}$ is a symmetric SL(2,R) matrix, which therefore satisfies $\mathcal{M}^T = \mathcal{M}$, $\mathcal{M}\mathcal{L}\mathcal{M}^T = \mathcal{L}$, and $\mathcal{L}$ is given in eq. (17). Under the SL(2,R) transformation (2) of the field λ, $\mathcal{M} \to \omega^T \mathcal{M} \omega$, where $\omega = \begin{pmatrix} d & b \\ c & a \end{pmatrix}$. The action (23) is manifestly invariant provided that at the same time $A_i^{(\alpha)} \to (\omega^T)_{\alpha\beta} A_i^{(\beta)}$. If we eliminate $A_i^{(2)}$ from (23) by using its clasical equation of motion, then we obtain the covariant expression $\mathcal{L}_5'$ given in eq. (12).

We are now in a position to give a third version of our theory that is classically equivalent to the two versions given earlier, but with both S duality and T duality realized as manifest symmetries. The formula that does the job is

$$S = \int d^4x \left[\sqrt{-g} \{ R - \frac{1}{4} g^{\mu\nu} tr(\partial_\mu \mathcal{M} \mathcal{L} \partial_\nu \mathcal{M} \mathcal{L}) + \frac{1}{8} g^{\mu\nu} Tr(\partial_\mu M L \partial_\nu M L) \} \right.$$
$$- \frac{1}{2} \left\{ B^{(a,\alpha)i} \mathcal{L}_{\alpha\beta} L_{ab} E_i^{(b,\beta)} + \epsilon^{ijk} \frac{g^{0k}}{g^{00}} B^{(a,\alpha)i} \mathcal{L}_{\alpha\beta} L_{ab} B^{(b,\beta)j} \right. \tag{25}$$
$$\left. \left. - \frac{g_{ij}}{\sqrt{-g}g^{00}} B^{(a,\alpha)i} (\mathcal{L}^T \mathcal{M} \mathcal{L})_{\alpha\beta} (LML)_{ab} B^{(b,\beta)j} \right\} \right].$$

In the above equation Tr denotes trace over the indices a, b and tr denotes trace over the indices α, β. In this expression we have recast the kinetic term of the λ field ($\mathcal{L}_{2,3}'$ in eq. (12)) in terms of the matrix $\mathcal{M}$.

Written in the form (25), it is clear that the S and T duality are realized in quite analogous ways in the low energy effective action, despite their profound difference from the point of view of compactified string theory. As we have seen, the price for making both of them manifest is the loss of manifest general coordinate invariance. The first form we presented (with manifest T duality) was derived by dimensional reduction of the ten-dimensional N=1 supergravity theory containing a two-form potential $B_{\mu\nu}$. However, there is also a dual version of that theory in which the two-form potential is replaced by its dual, which is a six-form potential $B_{\mu_1\mu_2\cdots\mu_6}$ [8]. Dimensional reduction of this dual theory gives a version of the four-dimensional theory that has manifest S duality but not T duality. This fact gives an interesting insight into the possible significance of S duality.

The two-form potential couples in a natural way to the world volume of the string and is therefore the natural choice from that point of view. However, in similar fashion the six-form potential couples to the world volume of a 'five-brane,' a five-dimensional extended object discovered as a soliton solution of the heterotic string [9] [10]. It has been conjectured in these references that there is a dual formulation of the heterotic string theory in which the five-brane is fundamental and the string is the soliton. It is an interesting possibility that in five-brane perturbation theory S duality is true order-by-order, whereas T duality becomes a non-perturbative symmetry. In other words, the roles of the S and T dualities are interchanged in passing from strings to five-branes, a phenomenon that we call "duality of dualities." As evidence in support of this conjecture, it was shown that S duality describes the interchange of five-brane Kaluza–Klein modes and winding modes, just as T duality does for the compactified string theory.

Before continuing with our main theme, let us pause to describe briefly generalizations of the two-potential formalism we have used for recasting the Maxwell action. What we have done is to introduce a pair of independent unconstrained potentials whose field strengths are dual to one another as a consequence of the equations of motion. This construction is easily generalized to m-forms in d dimensions. Let A denote an m-form potential and F its $(m+1)$-form field strength, so that

$$F_{\mu_1\ldots\mu_{m+1}} = \partial_{[\mu_1}A_{\mu_2\ldots\mu_{m+1}]}, \quad 0 \le \mu_l \le d-1. \tag{26}$$

Similarly, let B denote a $(d-m-2)$-form potential and G its $(d-m-1)$-form field strength. Then form the action

$$S = \int d^d x \left\{ \frac{\epsilon^{i_1\ldots i_m j_1\ldots j_{d-m-1}}}{m!(d-m-1)!} F_{0i_1\ldots i_m} G_{j_1\ldots j_{d-m-1}} + \frac{1}{2\cdot(m+1)!} F_{i_1\ldots i_{m+1}} F_{i_1\ldots i_{m+1}} \right.$$
$$\left. + \frac{1}{2\cdot(d-m-1)!} G_{i_1\ldots i_{d-m-1}} G_{i_1\ldots i_{d-m-1}} \right\}, \quad 1 \le i_l, j_l \le d. \tag{27}$$

This action is invariant under the following gauge transformations:

$$\delta A_{0 i_1 \ldots i_{m-1}} = \Psi^{(1)}_{i_1 \ldots i_{m-1}}, \quad \delta B_{0 i_1 \ldots i_{d-m-3}} = \Psi^{(2)}_{i_1 \ldots i_{d-m-3}}$$
$$\delta A_{i_1 \ldots i_m} = \partial_{[i_1} \Lambda^{(1)}_{i_2 \ldots i_m]}, \quad \delta B_{i_1 \ldots i_{d-m-2}} = \partial_{[i_1} \Lambda^{(2)}_{i_2 \ldots i_{d-m-2}]}. \tag{28}$$

The field equations of (27) imply that G is dual to F. These equations provide a natural generalization of eqs. (16) and (18), which correspond to the case $d = 4$ and $m = 1$. When $d = 4n + 2$ and $m = 2n$ it is possible to have a self-dual field strength (which amounts to equating F and G). An example of this occurs in type 2B supergravity in ten dimensions, which contains a self-dual five-form field strength. In this case our formulas reduce to ones considered previously by Henneaux and Teitelboim [11]. In the particular case of a self-dual boson in two dimensions, they agree with those of ref.[12]. This case will be utilized in the next section. These formulas also can be used to reformulate N=1 supergravity in ten dimensions in a version containing *both* the two-form $B_{\mu\nu}$ and the six-form $B_{\mu_1 \mu_2 \cdots \mu_6}$. This requires a slight generalization of the formulas given above to accommodate the Chern-Simons terms that are present.

T-Duality Symmetric World-Sheet Theory

Let us now consider the dynamics of strings propagating in the presence of arbitrary background values of the fields in the low-energy effective action. Specifically, they are the string metric $g_{\mu\nu}(x)$ (which includes the dilaton as a factor), the two-form potential $B_{\mu\nu}(x)$, the moduli $M^{ab}(x)$, and the 28 abelian gauge fields $A^a_\mu(x)$. As is well-known, this world-sheet theory is conformally invariant when the backgrounds satisfy the appropriate equations of motion. The dynamical 'fields' of the world-sheet theory are the space-time coordinates $x^\mu(\sigma, \tau)$, $\mu = 0, 1, 2, 3$ and 28 internal coordinates $y^a(\sigma, \tau)$, $a = 1, 2, \cdots, 28$, describing six compactified right-movers and 22 compactified left-movers. The y^a's are periodically identified, *i.e.*, $y^a \sim y^a + 2\pi$. The fact that they are chiral bosons of the world-sheet theory is reflected in their world-sheet equations of motion

$$D_0 y^a = -(ML)^a_b D_1 y^b, \tag{29}$$

where

$$D_\alpha y^a = \partial_\alpha y^a + A^a_\mu \partial_\alpha x^\mu. \tag{30}$$

Note that $(ML)^2 = 1$, and the matrix ML has 22 eigenvalues that are -1 and 6 eigenvalues that are $+1$. In addition, the x^μ equations of motion are

$$g_{\mu\nu}\partial^\alpha\partial_\alpha x^\nu + \Gamma_{\mu\nu\rho}\partial^\alpha x^\nu\partial_\alpha x^\rho = -\frac{1}{2}D_1 y^a(L\partial_\mu ML)_{ab}D_1 y^b - \epsilon^{\alpha\beta}\partial_\alpha x^\nu F^a_{\mu\nu}L_{ab}D_\beta y^b$$
$$+\frac{1}{2}\epsilon^{\alpha\beta}H_{\mu\nu\rho}\partial_\alpha x^\nu\partial_\beta x^\rho, \tag{31}$$

where Γ is the usual Christoffel symbol and H is defined in eq. (9). These equations are written in a form having manifest O(6,22,Z) symmetry (T duality). The restriction to integers arises from the periodicity properties of the y^a.

At this point it is natural to wonder whether these T duality symmetric equations can be obtained from a world-sheet action that has this symmetry. This is certainly not true for the usual formulation. In fact, it is easy to write down such an action. The answer is

$$S = \frac{1}{4\pi}\int d^2\sigma\left\{g_{\mu\nu}\eta^{\alpha\beta}\partial_\alpha x^\mu\partial_\beta x^\nu - D_0 y^a L_{ab}D_1 y^b - D_1 y^a(LML)_{ab}D_1 y^b\right.$$
$$\left.+\epsilon^{\alpha\beta}[B_{\mu\nu}\partial_\alpha x^\mu\partial_\beta x^\nu - A^a_\mu\partial_\alpha x^\mu L_{ab}D_\beta y^b]\right\}. \tag{32}$$

This result contains chiral bosons in the manner discussed earlier. It is a generalization of a result given previously by Tseytlin [13].

To understand this theory better, it is important to exhibit the coupling to a world-sheet metric $h_{\alpha\beta}$ that gives 2D Weyl invariance and reparametrization invariance. This is achieved by replacing the first term (as usual) by $\sqrt{-h}h^{\alpha\beta}g_{\mu\nu}(x)\partial_\alpha x^\mu\partial_\beta x^\nu$, and the third term by

$$\frac{1}{\sqrt{-h}h^{00}}D_1 y^a(LML)_{ab}D_1 y^b + \frac{h^{01}}{h^{00}}D_1 y^a L_{ab}D_1 y^b, \tag{33}$$

which is analogous to the B^2 terms in eq. (25). Having this form of the world-sheet action, it is straightforward to deduce the (traceless) energy-momentum tensor $T_{\alpha\beta}$ and the corresponding Virasoro constraints. It is also possible to eliminate the world sheet metric to obtain a "Nambu form" that maintains all the symmetries (including T duality).

The existence of the version of the world-sheet theory presented above is actually quite remarkable. It has manifest T duality, which is a symmetry that relates Kaluza–Klein excitations of the string (which can be regarded as elementary 'particles' of the world-sheet theory) to winding-mode excitations (which are solitons of the world-sheet theory). In terms of a compactification scale R and the string scale α', the corresponding masses are $M_{KK} \sim 1/R$ and $M_{\text{winding}} \sim R/\alpha'$. Thus T duality is a nonperturbative symmetry of the world-sheet theory, just as S duality is conjectured to be for the space-time theory. A world-sheet theory that makes such a nonperturbative symmetry manifest is necessarily strongly coupled. We know it is correct, however, since the y coordinates only appear quadratically and can therefore be treated exactly. The analogy with the space-time theory raises the very interesting question whether it is possible to formulate a string field theory with manifest S duality. (It is already known how to implement T duality in string field theory [14].)

Discussion

The toroidally compactified heterotic string has an infinite spectrum of excitations carrying electric and magnetic charges (with respect to the 28 abelian gauge fields). The states that carry electric charges only are elementary in the sense that they have a perturbative description in the space-time theory. States that carry magnetic charge, on the other hand, must be regarded as solitons of the string theory. The electric and magnetic charges of a state can be defined by the asymptotic behavior of the gauge fields

$$F_{0i}^a \sim q_{\text{el}}^a \frac{x^i}{r^3} \qquad \tilde{F}_{0i}^a \sim q_{\text{mag}}^a \frac{x^i}{r^3}. \tag{34}$$

The allowed values of the electric and magnetic charges are determined by the asymptotic values of the moduli fields ($M_{ab}^{(0)}$ and $\lambda^{(0)}$). In terms of these one has [15]

$$q_{\text{el}}^a = \frac{1}{\lambda_2^{(0)}} M_{ab}^{(0)}(\alpha_0^b + \lambda_1^{(0)} \beta_0^b), \qquad q_{\text{mag}}^a = L_{ab}\beta_0^b, \tag{35}$$

where both α_0^a and β_0^a are 28-component vectors belonging to a reference lattice P_0, which is even and self-dual with respect to the metric L. These charges automatically incorporate the Dirac quantization condition suitably generalized to allow for dyons and a vacuum angle [16]. The appearance of the electric and magnetic charges as central charges in the supersymmetry algebra allows one to deduce a Bogomol'nyi lower bound on the masses of states with specified electric and magnetic charges [17]. In the present context this bound is given by

$$(m_0)^2 = \frac{1}{16} \left(\alpha_0^a \quad \beta_0^a \right) \mathcal{M}^{(0)}(M^{(0)} + L)_{ab} \begin{pmatrix} \alpha_0^b \\ \beta_0^b \end{pmatrix}. \tag{36}$$

Remarkably, this formula turns out to be symmetric under both S and T dualities, providing yet further evidence that at a fundamental level they should operate in much the same way.

The perturbative string spectrum contains all states with electric charges only, in other words all states with $\beta_0^a = 0$. This is a T duality invariant (but not S duality invariant) subset of the complete spectrum. The perturbative five-brane spectrum, on the other hand, contains all states for which the last 22 components of α_0^a and β_0^a vanish. This is an S duality invariant (but not T duality invariant) subset of the complete spectrum. In view of these facts, it is tempting to speculate that the classical five-brane theory has S duality symmetry much as the string world-sheet theory has T duality symmetry. However, the severe nonlinearities of the five-brane theory have so far prevented us from proving this.

Since the S duality group SL(2,Z) relates electrically charged states to magnetically charged states, it relates perturbative states and nonperturbative states of the space-time theory, just as the T duality group O(6,22;Z) did for the world-sheet theory. Specifically, the SL(2,Z) group element $\begin{pmatrix} a & b \\ c & d \end{pmatrix}$ maps states with charges $(\vec{\alpha}_0, 0)$ to ones with charges $(a\vec{\alpha}_0, c\vec{\alpha}_0)$. It is possible to find group elements for any pair of relatively prime integers a and c. If we simultaneously allow the transformations to act on the background fields of the world-sheet action to give transformed background fields, whose relationship to the original ones are nonlocal in general, then we obtain a "dual" world-sheet action that is isomorphic to the original one. From the point of view of this dual theory states with charges of the form $(a\vec{\alpha}_0, c\vec{\alpha}_0)$ arise perturbatively and all others are solitons. Thus we have a generalization of Olive–Montonen duality [18] to a situation where there are an infinite number of isomorphic dual descriptions of the same theory labeled by pairs of relatively prime integers.

To conclude, we have seen that S duality and T duality have many similarities. To the extent that a fundamental five-brane formulation of the heterotic string theory makes sense, S duality should be a fundamental symmetry. However, even if five-branes turn out to only make sense as solitons, the symmetry could still be true.

I wish to acknowledge the hospitality of the Institute of Theoretical Physics at U.C. Santa Barbara, where this work was done.

REFERENCES

[1] J. Schwarz and A. Sen, "Duality Symmetric Actions," preprint NSF-ITP-93-46 (hep-th/9304154).

[2] J. Schwarz and A. Sen, "Duality Symmetries of 4D Heterotic Strings," preprint NSF-ITP-93-64 (hep-th/9305185).

[3] K. Narain, Phys. Lett. **B169** (1986) 41; K. Narain, H. Sarmadi and E. Witten, Nucl. Phys. **B279** (1987) 369.

[4] E. Cremmer, J. Scherk and S. Ferrara, Phys. Lett. **B74** (1978) 61; M. Gaillard and B. Zumino, Nucl. Phys. **B193** (1981) 221; M. De Roo, Nucl. Phys. **B255** (1985) 515.

[5] A. Font, L. Ibáñez, D. Lust and F. Quevedo, Phys. Lett. **B249** (1990) 35; S.J. Rey, Phys. Rev. **D43** (1991) 526; A. Shapere, S. Trivedi and F. Wilczek, Mod. Phys. Lett. **A6** (1991) 2677.

[6] J. Schwarz, preprint CALT-68-1815 (hep-th/9209125).

[7] J. Maharana and J. Schwarz, Nucl. Phys. **B390** (1993) 3.

[8] A. Chamseddine, Phys. Rev. **D24** (1981) 3065; S. Gates and H. Nishino, Phys. Lett. **B173** (1986) 52; A. Salam and E. Sezgin, Physica Scripta **32** (1985) 283.

[9] A. Strominger, Nucl. Phys. **B343** (1990) 167; C. Callan, J. Harvey and A. Strominger, Nucl. Phys. **B359** (1991) 611; **B367** (1991) 60; preprint EFI-91-66 (hep-th/9112030).

[10] M. Duff, Class. Quantum Grav. **5** (1988) 189; M. Duff and J. Lu, Nucl. Phys. **B354** (1991) 129, 141; **B357** (1991) 534; Phys. Rev. Lett. **66** (1991) 1402; Class. Quantum Grav. **9** (1991) 1; M. Duff, R. Khuri and J. Lu, Nucl. Phys. **B377** (1992) 281; J. Dixon, M. Duff and J. Plefka, Phys. Rev. Lett. **69** (1992) 3009.

[11] M. Henneaux and C. Teitelboim, p. 79 in Proc. *Quantum Mechanics of Fundamental Systems 2* (Santiago 1987); Phys. Lett. **B206** (1988) 650.

[12] R. Floreanini and R. Jackiw, Phys. Rev. Lett. **59** (1987) 1873.

[13] A. Tseytlin, Phys. Lett. **B242** (1990) 163; Nucl. Phys. **B350** (1991) 395.

[14] T. Kugo and B. Zwiebach, Prog. Theor. Phys. **87** (1992) 801.

[15] A. Sen, Phys. Lett. **B303** (1993) 22; A. Sen, preprint NSF-ITP-93-29 (hep-th/9303057).

[16] P. Dirac, Proc. R. Soc. **A133** (1931) 60; J. Schwinger, Phys. Rev. **144** (1966) 1087; **173** (1968) 1536; D. Zwanziger, Phys. Rev. **176** (1968) 1480, 1489; E. Witten, Phys. Lett. **86B** (1979) 283.

[17] D. Olive and E. Witten, Phys. Lett. **78B** (1978) 97; G. Gibbons and C. Hull, Phys. Lett. **B109** (1982) 190.

[18] C. Montonen and D. Olive, Phys. Lett. **B72** (1977) 117; P. Goddard, J. Nuyts and D. Olive, Nucl. Phys. **B125** (1977) 1; H. Osborn, Phys. Lett. **B83** (1979) 321.

MANIFEST DUALITY IN
LOW-ENERGY SUPERSTRINGS

W. SIEGEL*

Institute for Theoretical Physics
State University of New York, Stony Brook, NY 11794-3840

ABSTRACT

String theories inspire a new formalism for their low-energy limits. In this approach to these field theories, spacetime duality and stringy left/right handedness are manifest. Enlarged tangent-space symmetries allow the different fields (graviton, axion, Yang-Mills) to be treated as a single multiplet, even in the bosonic case, except for the dilaton (multiplet), which appears as the measure.

1. Introduction

Duality[1] is the global symmetry relating the fields g_{mn} (metric), $b_{[mn]}$ (axion), and $A_{m\hat{a}}$ (abelian vectors) in string theory when these fields are constant in some directions. This symmetry is preserved in the low-energy limit. (In fact, it becomes larger: The elements of the group matrices generalize from integral to continuous.)

Our main result is that duality can be made manifest (linear) in a new formalism for (super) gravity + axion.[2] Fields are defined in a (2D+n)-dimensional space with manifest O(D,D+n) symmetry. This space is reduced to D dimensions by an O(D,D+n)-covariant constraint, whose solution "spontaneously" breaks this symmetry to O(n) (the global symmetry of the n abelian gauge vectors). However, for solutions that are independent of d of the dimensions (or by reducing the theory to D−d dimensions) the symmetry is partially restored, to O(d,d+n). The fields and their gauge symmetry are unified in the (2D+n)-dimensional space; the graviton and axion (and abelian gauge vectors) are parts of the same (2D+n)-dimensional field. (In the supersymmetric case there are also anticommuting coordinates, and thus a larger, graded, spontaneously broken symmetry.)

---

* Work supported by National Science Foundation grant PHY 9211367.
  Internet address: siegel@max.physics.sunysb.edu.

The major consequences are: (1) The field theory that represents the low-energy limit of string theory, when written in this formalism, more closely resembles string theory: The left/right handedness of string theory is built in, as the local tangent-space symmetry. A gauge can be chosen where the global part of this symmetry is preserved, so Feynman diagrams factorize into left- and right-handed parts. (2) Duality-related solutions to the field equations can be treated together. (3) The formalism may suggest new ways of understanding gravity (using larger tangent spaces), σ-models (with duality-covariant vertex operators), and string field theory (with an enlarged set of fields, corresponding to the new tangent-space gauge invariances).

The outline of this talk is: (1) We first consider the algebra of the string oscillators, which is manifestly duality covariant. As a result, duality is manifest in the hamiltonian (operator) formalism of string mechanics, but not in the lagrangian one. This affine Lie algebra defines the gauge group. (2) The (massless) fields are introduced as backgrounds for these operators. By restricting to the classical mechanics of strings, we obtain a field theory formalism which can be applied to the bosonic theory in all dimensions, and to the supersymmetric theories in (at least) D=3,4,6,10. (3) As usual, in the supersymmetric case some torsions can be defined without connections. They differ from the usual ones because the corresponding Lie algebras are affine, and they are all constrained to be trivial (vanishing or pure gauge). (4) Connections can be introduced to define covariant derivatives. (5) In constructing the complete set of torsions and curvatures, divergences play as important a role as curls. We explicitly construct curvatures and actions for the bosonic sector. The lagrangian for the metric, axion, and vectors is all contained within the (unique) curvature scalar, with the dilaton as the measure. (In the first-order formalism, the dilaton appears *only* as the measure.) (6) We consider the supersymmetric case in more detail for D=4. We find the (unconstrained) prepotential superfields, on which left/right handedness and duality are manifest. (In a final section added after the talk, we discuss modifications for nonabelian Yang-Mills.)

2. String oscillator algebra

We work in the hamiltonian (operator) formalism for the string because duality can be made manifest there. The oscillators of the string (including the Green-Schwarz formulation of the superstring) form an (abelian) affine Lie algebra:

$$[Z_M(1), Z_N(2)\} = i\delta'(2-1)\eta_{MN}$$

where "1" and "2" label the σ coordinate of the string, Z_M includes both the (super)spacetime momentum (σ-)density and the σ derivative ($'$) of the spacetime coordinates, and η_{MN} is the constant metric in the space of vectors that carry the superindices M:

$$Z_M = (P_{\mathbf{M}}, X'^{\mathbf{M}}, P_{\hat{m}} - \tfrac{1}{2}X'_{\hat{m}}) \quad \Rightarrow \quad \eta_{MN} = \begin{pmatrix} 0 & \delta_{\mathbf{M}}{}^{\mathbf{N}} & 0 \\ \delta^{\mathbf{M}}{}_{\mathbf{N}} & 0 & 0 \\ 0 & 0 & -\delta_{\hat{m}\hat{n}} \end{pmatrix}$$

(We have included the heterotic directions $\hat{m}$ by using only the right-handed oscillators.) Although η_{MN} is constant, it is still generally covariant in the sense of ordinary spacetime, since it consists of Kronecker δ's. Duality is simply the global symmetry acting on the superindices M that preserves the metric η_{MN}.

The local symmetry group elements $e^{-i\Lambda}$ are generated by the affine Lie algebra Z_M:

$$\Lambda = \int \lambda^M Z_M$$

(integrated over σ, where λ is a function of $X^M(\sigma)$). The gauge parameters include those for general (super)coordinate transformations (λ^M), axion (2-form) gauge transformations (λ_M), and abelian vector(-multiplet) gauge transformations ($\lambda^{\hat{m}}$). There is also the gauge invariance of the gauge invariance:

$$\delta\lambda^M = \partial^M \lambda \quad \Rightarrow \quad \delta\Lambda = 0$$

The affine Lie algebra defines a new Lie derivative:

$$[\Lambda_1, \Lambda_2] = i\Lambda_{[1,2]} \quad \Rightarrow \quad \lambda^M_{[1,2]} = \lambda^N_{[1}\partial_N\lambda^M_{2]} - \tfrac{1}{2}\lambda^N_{[1}\partial^M \lambda_{2]N}$$

(where brackets on indices indicate antisymmetrization).

We have used the fundamental identities (expressing the triviality of the winding modes):

$$A' = Z^M \partial_M A, \quad (\partial^M A)(\partial_M B) = \partial^M \partial_M A = 0$$

The Klein-Gordon-like equation $\partial^M \partial_M = 0$ (not the usual d'Alembertian because η is off-diagonal) acts as a partner to the commutation of partial derivatives $[\partial_M, \partial_N\} = 0$. This condition is weaker than the vanishing of $\partial^M, \partial_{\hat{m}}$ in

$$\partial_M = (\partial_M, \partial^M, \partial_{\hat{m}})$$

However, the most general solution to $(\partial^M A)(\partial_M B) = \partial^M \partial_M A = 0$ is of this form: The choice of a solution just chooses which derivatives, labeled $\partial^M, \partial_{\hat{m}}$, are set to vanish. (I.e., the fields now depend on only X^M.) Thus the solution of these conditions spontaneously breaks the manifest O(D,D+n) symmetry to O(n). However, if we choose fields such that an additional d of the D ∂_M vanish, the symmetry is partially restored to O(d,d+n). (In the supersymmetric case the original symmetry is a graded one; e.g., for the heterotic string we have OSp(D,D+n|2D') for D' fermionic coordinates.)

3. Background fields

Background fields $e_A{}^M$ are introduced by covariantizing Z_M:

$$\Pi_A = e_A{}^M Z_M, \quad g_{AB} = e_A{}^M e_B{}^N \eta_{MN}$$

The algebra of Π_A defines λ^M-covariant objects F:

$$[\Pi_A(1), \Pi_B(2)\} = i\delta'(2-1)\tfrac{1}{2}[g_{AB}(1) + g_{AB}(2)] + i\delta F_{AB}{}^C \Pi_C$$

g and F thus represent a background-dependent metric and structure functions for the new algebra. The first-class (Virasoro) constraints restrict the tangent-space metric and the gauge group to GL(D)$\otimes$GL(D+n) on $A = (\mathcal{A}, \tilde{\mathcal{A}})$ (or in the supersymmetric case a graded symmetry, such as GL(D$|$D$'$)$\otimes$GL(D+n) for the heterotic string):

$$g_{A\tilde{B}} = 0$$

$$\Rightarrow \quad L_+ = \tfrac{1}{2}g^{AB}\Pi_B\Pi_A, \quad L_- = \tfrac{1}{2}g^{\tilde{A}\tilde{B}}\Pi_{\tilde{B}}\Pi_{\tilde{A}}; \quad L_+ + L_- = \tfrac{1}{2}Z^M Z_M$$

L_+ and L_- are the generators of left- and right-handed conformal transformations, respectively, so Π_A is left-handed while $\Pi_{\tilde{A}}$ is right-handed. (Others have considered SO(d)$\otimes$SO(d+n)[3] and GL(d)[4] as gauge groups; here those groups can be obtained by partial gauge fixing. General linear groups have also been considered as Yang-Mills groups in ordinary gravity;[5] this is merely a reinterpretation of the early work by Cartan.[6]) Note that the generators of σ reparametrizations ($\tfrac{1}{2}Z^M Z_M$) are background independent ($X'^M P_M$ is already generally covariant); the background couples only to τ reparametrizations.

The gauge transformations of the background fields are given by:

$$\delta\Pi_A = [-i\Lambda, \Pi_A] + \lambda_A{}^B \Pi_B, \quad \lambda_A{}^{\tilde{B}} = \lambda_{\tilde{A}}{}^B = 0$$

$$\Rightarrow \quad \delta e_A{}^M = (\lambda^N \partial_N e_A{}^M + e_{AN}\partial^{[M}\lambda^{N]}) + \lambda_A{}^B e_B{}^M$$

$$\delta g_{AB} = \lambda^M \partial_M g_{AB} + \lambda_{(AB]}$$

The vielbein $e_A{}^M$ thus transforms as a cross between (but not a product of) a covariant and a contravariant world-vector. This is to preserve the super spacetime metric η, which is constant. On the other hand, the tangent-space metric g is not constant (but can be gauged to constant by $\lambda_{(AB]}$). This is the opposite of the situation in the usual vierbein formulation of gravity. (The usual spacetime metric can be found in $L_\pm$ as part of the coefficients of $Z_M Z_N$.)

In the conventional gauge for the tangent-space group $\lambda_A{}^B$, $e_A{}^M$ reduces to the usual fields $e_A{}^M$ (graviton), b_{MN} (axion), and $A_{M\hat{a}}$ (vectors).

The dilaton appears as the integration measure, which can't be constructed covariantly from $e_A{}^M$ ($\det e_A{}^M$ is not a density):

$$\delta\Phi^2 = \partial_M(\lambda^M \Phi^2)$$

4. Torsions and constraints

We now specialize to the (heterotic) superstring, in the Green-Schwarz formalism. (Background fields for the Green-Schwarz formalism were first considered by Witten,[7] and in the string hamiltonian formalism by Shapiro and Taylor.[8]) The indices take the values

$$M = (\mathbf{M},{}^{\mathbf{M}},\hat{m}) = ({}_{\mu},{}_{m};{}^{\mu},{}^{m};\hat{m}), \quad A = (\mathcal{A},\tilde{\mathcal{A}}) = (\alpha, a, \tilde{\alpha}; \tilde{a}, \hat{a})$$

X^M now has fermionic parts Θ^μ as well as the bosonic parts X^m. (Similarly, the new tangent-space indices $\alpha, \tilde{\alpha}$ are fermionic.) We choose Π_α as the second-class constraints. This leads to further restrictions on the tangent-space metric and gauge group:

$$g_{\alpha\beta} = 0, \quad \lambda_\alpha{}^b = \lambda_\alpha{}^{\tilde{\beta}} = \Pi_\alpha{}^\beta{}_\delta{}^\gamma \lambda_\gamma{}^\delta = 0$$

where $\Pi_\alpha{}^\beta{}_\delta{}^\gamma$ is a projection operator constraining $\lambda_\alpha{}^\beta$ to be Lorentz$\oplus$scale$\oplus$internal.

The connection-independent torsions are:

$$F_{\alpha\beta}{}^{\tilde{c}}, \quad F_{\alpha\beta}{}^{c,\tilde{\gamma}}, \quad \Pi_{\alpha\beta}{}^\gamma{}_\delta{}^{\epsilon\zeta} F_{\epsilon\zeta}{}^\delta, \quad F_{\alpha\tilde{\beta}}{}^{c,\tilde{\gamma}}, \quad \Pi_\alpha{}^\gamma{}_\delta{}^\epsilon F_{\epsilon\tilde{\beta}}{}^\delta$$

($\Pi_{\alpha\beta}{}^\gamma{}_\delta{}^{\epsilon\zeta}$ is another projection operator serving the same purpose. "$c,\tilde{\gamma}$" means c or $\tilde{\gamma}$.) By requiring that the constraint algebra takes the usual form,

$$[L_-, \Pi_\alpha] \approx 0, \quad \{\Pi_\alpha, \Pi_\beta\} \approx i\delta F_{\alpha\beta}{}^c \widehat{\Pi}_c$$

(where "$\approx$" means modulo the constraints) for some $\widehat{\Pi}$, we find the constraints on the background fields:

$$F_{\alpha\beta}{}^{\tilde{c}} = F_{\alpha\beta\gamma} = F_{\alpha\tilde{\beta}}{}^{c,\tilde{\gamma}} = 0, \quad F_{\alpha\beta}{}^{c,\tilde{\gamma}} = \gamma^d_{\alpha\beta} F_d{}^{c,\tilde{\gamma}}$$

(for some $F_d{}^{c,\tilde{\gamma}}$), where $\gamma^d_{\alpha\beta}$ are the usual γ (or σ) matrices. The Bianchi identities then constrain the remaining connection-independent torsions:

$$\Pi_{\alpha\beta}{}^\gamma{}_\delta{}^{\epsilon\zeta} F_{\epsilon\zeta}{}^\delta = \Pi_\alpha{}^\gamma{}_\delta{}^\epsilon F_{\epsilon\tilde{\beta}}{}^\delta = 0$$

In the conventional gauge these constraints give the usual result:[9]

$$H_{\alpha\beta\gamma} = 0; \quad \mathcal{F}_{\alpha\beta\hat{c}} = 0, \quad c_{\alpha\beta}{}^c = -H_{\alpha\beta}{}^c = \gamma^c_{\alpha\beta};$$

$$H_{abc} = c_{\alpha(bc)} = 0, \quad \mathcal{F}_{ab}{}^{\hat{c}} = \gamma_{b\alpha\beta} W^{\beta\hat{c}}$$

where c, H, and $\mathcal{F}$ are the usual duality-noncovariant superfield strengths for the supergraviton (connection-free torsion), superaxion, and superMaxwell fields. (W, which originally appeared as the gauge field $e_{\tilde{a}}{}^{\hat{m}}$, now becomes the usual spinor superfield strength.)

5. Covariant derivatives

We first define the covariant derivative in the naive way:

$$\nabla_A = e_A + \omega_{AB}{}^C G_C{}^B + \omega_{A\tilde{B}}{}^{\tilde{C}} G_{\tilde{C}}{}^{\tilde{B}}, \quad e_A = e_A{}^M \partial_M$$

$G_C{}^B$ and $G_{\tilde{C}}{}^{\tilde{B}}$ are the generators of the left- and right-handed GL symmetries, which act on all tangent-space indices. However, there is a new definintion of the torsion, determined by the new Lie derivative:

$$\lambda^A_{[1,2]} = \lambda^B_{[1} \nabla_B \lambda^A_{2]} - \tfrac{1}{2} \lambda_{[1B} \nabla^A \lambda^B_{2]} + \lambda^B_1 \lambda^C_2 T_{BC}{}^A$$

$$\Rightarrow \quad T_{AB}{}^C = F_{AB}{}^C + (\omega_{[AB)}{}^C + \tfrac{1}{2}\omega^C{}_{[AB)})$$

Consequently, the connection constraints don't fix all the connections:

$$\omega\text{-dependent } T_{AB}{}^C = \nabla_A g_{BC} = 0$$

$$\Rightarrow \quad \omega_{A\tilde{B}}{}^{\tilde{C}} = -F_{A\tilde{B}}{}^{\tilde{C}}, \quad \omega_{A(BC]} = -e_A g_{BC}, \quad \omega_{[ABC)} = -\tfrac{1}{3} F_{[ABC)}$$

(In the bosonic case, all the torsions are ω-dependent; in the supersymmetric case, the other torsions are fixed for other reasons, as described in the previous section.) Integration by parts for the measure Φ^2 defines another new torsion and connection constraint:

$$\tilde{T}_A \equiv \Phi^2 \overleftarrow{\nabla}_A \Phi^{-2} = 0$$

$$\Rightarrow \quad \omega_{BA}{}^B = -\tilde{F}_A \equiv -\Phi^2 \overleftarrow{e}_A \Phi^{-2}$$

(The backward arrow means the partial derivatives and GL generators act backwards, even on the fields contained in ∇.) As the analog of the usual relation

$$[\nabla_A, \nabla_B\} A = T_{AB}{}^C \nabla_C A$$

from $[\partial_M, \partial_N\} = 0$, we have from $\partial_M \partial^M = 0$

$$\nabla_A \nabla^A A = -\tilde{T}_A \nabla^A A$$

Since not all connections are defined, only certain kinds of differentiation are covariant: These include the new Lie derivative, divergences of left- and right-handed vectors, gradients of scalars, and gradients of opposite-handed vectors (e.g. $\nabla_A V_{\tilde{B}}$), as well as combinations of these, and certain spinor derivatives in the supersymmetric case.

6. Curvatures and actions (bosonic)

Because of the condition $\partial^M \partial_M = 0$, traces are now a fundamental part of the definition of curvatures as well as torsions:

$$R_{A\bar{B}} = 2R_{\bar{C}A\bar{B}}{}^{\bar{C}} = 2R_{C\bar{B}A}{}^{C}$$
$$= (e_{\bar{B}}\widetilde{F}_A - F_{\bar{B}A}{}^C \widetilde{F}_C) - (e_C F_{\bar{B}A}{}^C - F_{C\bar{B}}{}^{\bar{E}} F_{\bar{E}A}{}^C)$$
$$R = -\tfrac{1}{2}R_{AB}{}^{AB} = \tfrac{1}{2}R_{\bar{A}\bar{B}}{}^{\bar{A}\bar{B}}$$
$$= e_A \widetilde{F}^A + \tfrac{1}{2}\widetilde{F}_A{}^2 + \tfrac{1}{2}e_A e_B g^{AB} - \tfrac{1}{2}F_{AB\bar{C}}{}^2 - \tfrac{1}{432}F_{[ABC)}{}^2 + \tfrac{1}{8}(e^A g^{BC})(e_B g_{AC})$$

These satisfy the Bianchi identities:

$$\nabla^{\bar{B}} R_{A\bar{B}} + \nabla_A R = \nabla^B R_{B\bar{A}} - \nabla_{\bar{A}} R = 0$$

The bosonic action is then simply:

$$S = -\int d^D x \; \Phi^2 R$$

This action implies the field equations:

$$-\delta S = \int R \, \delta\Phi^2 + \Phi^2 R^{A\bar{B}} e_{\bar{B}}{}^M \delta e_{AM} \quad \Rightarrow \quad R = R_{A\bar{B}} = 0$$

The action can also be expressed in 1st-order form:

$$S = \int d^D x \; \Phi^2 \tfrac{1}{4}(R_{AB}{}^{AB} - R_{\bar{A}\bar{B}}{}^{\bar{A}\bar{B}})$$

$$-R_{AB}{}^{AB} = \tfrac{1}{2}(\tfrac{1}{2}\omega_B{}^{[AB)2} - 2e_A\omega_B{}^{[AB)} - \omega_B{}^{[AB)}e^C g_{AC})$$
$$- \tfrac{1}{2}(\tfrac{1}{2}\omega_B{}^{(AB]2} + \omega_B{}^{(AB]}e^C g_{AC}) + \tfrac{1}{12}(\tfrac{1}{2}\omega^{[ABC)2} + \tfrac{1}{3}\omega^{[ABC)}F_{[ABC)})$$
$$+ \tfrac{1}{12}(\tfrac{1}{2}\omega^{(ABC]2} + \tfrac{1}{2}\omega^{(ABC]}e_{(A}g_{BC]}) + (\tfrac{1}{2}\omega^{\bar{A}BC2} + \omega^{\bar{A}BC}F_{\bar{A}BC})$$

Note that in the 1st-order form the dilaton appears only as the measure. (We can also add a cosmological term $\int d^D x \; \Phi^2$.)

For purposes of perturbation theory we can write

$$e_A{}^M \equiv \langle e_A{}^M \rangle + h_A{}^B \langle e_B{}^M \rangle, \quad \Phi^2 \equiv 1 + \phi$$

Using the tangent-space gauge invariance and the constraint $g_{A\bar{B}} = 0$, we can choose $h_{A\bar{B}}$ as the only independent field (its indices raised and lowered with the flat metric $\eta_{AB} \equiv \langle g_{AB} \rangle$). Using the gauge-fixing functions

$$f_A \equiv d^{\bar{B}} h_{A\bar{B}} + \tfrac{1}{2}d_A \phi, \quad f_{\bar{A}} \equiv -d^A h_{A\bar{B}} + \tfrac{1}{2}d_{\bar{B}}\phi \quad (d_A \equiv \langle e_A{}^M \rangle \partial_M)$$

the kinetic terms from $L - \tfrac{1}{2}f_A{}^2 + \tfrac{1}{2}f_{\bar{A}}{}^2$ are simply

$$-\tfrac{1}{4}\phi \Box \phi - \tfrac{1}{2}h^{A\bar{B}} \Box h_{A\bar{B}}, \quad \Box = d^A d_A = -d^{\bar{A}} d_{\bar{A}}$$

so the propagators are string-like. We can also write the rest of the action in terms of just $h_{A\bar{B}}$, ϕ, and d_A. We then see that the Feynman diagrams involve only d_A's and η_{AB}'s (i.e., η_{AB}'s and $\eta_{\bar{A}\bar{B}}$'s): no objects which mix left- and right-handed indices. As a result, the (integrand of) each Feynman diagram factorizes into left- and right-handed parts (even though d_A and $d_{\bar{A}}$ aren't independent).

7. D=4

In D=3 the constraints determine the vielbein in terms of just the spinor part, as usual,[10] $e_A{}^M \to e_\alpha{}^M$, which is now extended to include the axion (multiplet) by virtue of its extended index M. In D=4 the constraints further determine the vielbein in terms of a prepotential:

$$\Pi_\alpha = A_\alpha{}^\mu e^W Z_\mu e^{-W}, \quad \Pi_{\dot\alpha} = A_{\dot\alpha}{}^{\dot\mu} e^{-\overline{W}} Z_{\dot\mu} e^{\overline{W}}, \quad W = \int W^M Z_M$$

$$e_\alpha = A_\alpha{}^\mu e^w \partial_\mu e^{-w}, \quad w = W^M i \partial_M$$

Again this result is similar to pure supergravity in ordinary superspace,[11] except that the index on the prepotential W^M is extended to include the additional fields, and the expansion of the exponentials as multiple commutators contains extra terms due to the use of the new Lie bracket. (A similar result has been obtained by Berkovits,[12] but duality was not manifest[13] because the solution to the constraints was expressed differently.[14])

The generalization of the gauge group is also analogous to the pure supergravity case, with modifications from the extended indices and new Lie bracket:

$$e^{W'} = e^{-i\Lambda} e^W e^{i\Xi}, \quad \Xi = \int \xi^M Z_M, \quad \partial_\mu \xi^M = 0 \; except \; \partial_\mu \xi^\nu \neq 0$$

We thus have the analog of the gauge invariance of the gauge invariance:

$$W'^M = W^M + \partial^M \zeta$$

As in pure supergravity, these new transformations replace the old when using the Λ-invariant combination of the fields:

$$e^U \equiv e^{\overline{W}} e^W$$

The original gauge invariances can be used to eliminate all fields except the prepotential U^M, which transforms under only the new invariances Ξ and ζ. (As for the bosonic case, the dilaton multiplet appears in a separate superfield.)

The (superfield) curvatures are now R (chiral) and $R_{\dot A}$ (real). The component Ricci scalar R appears at order θ^2 in the superfield R (as in pure supergravity), while the component $R_{a\bar B}$ (which contains the remaining field equations) appears at order $\theta\bar\theta$ in the superfield $R_{\dot A}$. The latter superfield exhibits the direct-product (left⊗right) structure of the states: $R_{\dot A}(\theta,\bar\theta) \leftarrow V(\theta,\bar\theta) \otimes A_{\dot A}$ (vector multiplet⊗(vector⊕scalars)).

8. Epilog

The true low-energy limit of the heterotic superstring (at least for interesting compactifications of the 16 extra dimensions) includes nonabelian Yang-Mills. After this talk was given we generalized the above methods to this case. It requires only the corresponding generalization of the affine Lie algebra:

$$[Z_M(1), Z_N(2)\} = i\delta'(2-1)\eta_{MN} + i\delta(1-2)f_{MN}{}^P Z_P$$

The group structure constants f satisfy the usual Jacobi identities. If we define the unbroken duality group still to be the invariance O(D,D+n) (with n now the number of generators of the nonabelian Yang-Mills group) of the metric η, then f transforms covariantly: After an O(D,D+n) transformation, it still has the algebraic properties of the structure constants of the Yang-Mills group (crossed with 2D U(1)'s). The solution of $\partial^M \partial_M = 0$ that spontaneously breaks this symmetry can be chosen such that the only nonvanishing components of f are $f_{\hat{m}\hat{n}}{}^{\hat{p}}$. Then, when choosing the fields to be independent of d coordinates, the restored O(d,d+n) symmetry includes the nonabelian generators.

We then find the corresponding modification in the definition of the new Lie bracket:

$$\lambda^M_{[1,2]} = \lambda^N_{[1} \partial_N \lambda^M_{2]} - \tfrac{1}{2}\lambda^N_{[1} \partial^M \lambda_{2]N} + \tfrac{1}{2}\lambda^N_1 \lambda^P_2 f_{NP}{}^M$$

F (and thus the torsions) are modified simply by $F \to F + f$, and the vielbein transformation law by $\delta e_A{}^M \to \delta e_A{}^M + e_A{}^P \lambda^N f_{NP}{}^M$. The low-energy action then has the same form as before, the nonabelian terms being automatically included in the redefinition of F. All these changes follow directly from the change in $[Z, Z\}$. Effectively, we have just changed our choice of vacuum, writing $\Pi_A = e_A{}^M \langle \Pi_M \rangle$, where $\langle \Pi \rangle$ is our new Z, which can be expressed in terms of some $\langle e \rangle$ and abelian Z. However, working directly in terms of the nonabelian Z simplifies algebra, and allows the use of arbitrary representations (e.g., bosonic or fermionic) of this affine Lie algebra. If Lorentz generators are introduced as additional generators in the affine Lie algebra instead of as an ordinary[15] (but local in σ) Lie algebra (i.e., they then have a δ' term in their algebra), the low-energy action will then include Lorentz Chern-Simons terms.

These results can also be applied to the infinite-dimensional (spontaneously broken) Yang-Mills groups ($n=\infty$) which can occur when considering directly all the possible low-energy limits of a specific compactified string.[16]

Acknowledgments

I thank Amit Giveon and Martin Roček for discussions related to the topic of the last section.

REFERENCES

[1] K. Kikkawa and M. Yamasaki, *Phys. Lett.* **149B** (1984) 357;
N. Sakai and I. Senda, *Prog. Theor. Phys.* **75** (1986) 692;
V.P. Nair, A. Shapere, A. Strominger, and F. Wilczek, *Nucl. Phys.* **B287** (1987) 402;
B. Sathiapalan, *Phys. Rev. Lett.* **58** (1987) 1597;
R. Dijkgraaf, E. Verlinde, and H. Verlinde, *Comm. Math. Phys.* **115** (1988) 649;
K.S. Narain, M.H. Sarmadi, and E. Witten, *Nucl. Phys.* **B279** (1987) 369;
P. Ginsparg, *Phys. Rev.* **D35** (1987) 648;
P. Ginsparg and C. Vafa, *Nucl. Phys.* **B289** (1987) 414;
S. Cecotti, S. Ferrara, and L. Girardello, *Nucl. Phys.* **B308** (1988) 436;
R. Brandenberger and C. Vafa, *Nucl. Phys.* **B316** (1988) 391;
A. Giveon, E. Rabinovici, and G. Veneziano, *Nucl. Phys.* **B322** (1989) 167;
A. Shapere and F. Wilczek, *Nucl. Phys.* **B320** (1989) 669;
M. Dine, P. Huet, and N. Seiberg, *Nucl. Phys.* **B322** (1989) 301;
J. Molera and B. Ovrut, *Phys. Rev.* **D40** (1989) 1146;
K.A. Meissner and G. Veneziano, *Phys. Lett.* **267B** (1991) 33;
A.A. Tseytlin and C. Vafa, *Nucl. Phys.* **B372** (1992) 443;
M. Roček and E. Verlinde, *Nucl. Phys.* **B373** (1992) 630;
J.H. Horne, G.T. Horowitz, and A.R. Steif, *Phys. Rev. Lett.* **68** (1992) 568;
A. Sen, *Phys. Lett.* **271B** (1992) 295;
A. Giveon and M. Roček, *Nucl. Phys.* **B380** (1992) 128.

[2] W. Siegel, *Phys. Rev.* **D47** (1993) 5453; Superspace duality in low-energy superstrings, Stony Brook preprint ITP-SB-93-28 (May 1993), to appear in *Phys. Rev.* **D.**

[3] M.J. Duff, *Nucl. Phys.* **B335** (1990) 610.

[4] J. Maharana and J.H. Schwarz, *Nucl. Phys.* **B390** (1993) 3.

[5] E.A. Lord, *Phys. Lett.* **65A** (1978) 1;
A. Trautman, *Czech. J. Phys.* **29** (1979) 107;
A. Komar, *Phys. Rev.* **D30** (1984) 305;
R. Percacci, Role of soldering in gravity theory, *in* Differential geometric methods in theoretical physics, Aug. 20-24, 1984 (World Scientific, Singapore, 1986) p. 250; Geometry of nonlinear field theories (World Scientific, Singapore, 1986) p. 157.

[6] É. Cartan, *Leçons sur la géométrie des espaces de Riemann* (Gauthier-Villars, Paris, 1963) 2nd ed., pp. 177-184.

[7] E. Witten, *Nucl. Phys.* **B266** (1986) 245;
M.T. Grisaru, P. Howe, L. Mezincescu, B. Nilsson, and P.K. Townsend, *Phys. Lett.* **162B** (1985) 116;
E. Bergshoeff, E. Sezgin, and P.K. Townsend, *Phys. Lett.* **169B** (1986) 191.

[8] J.A. Shapiro and C.C. Taylor, *Phys. Lett.* **181B** (1986) 67, **186B** (1987) 69; *Phys. Reports* **191** (1990) 221.

[9] M.T. Grisaru, H. Nishino, and D. Zanon, *Phys. Lett.* **206B** (1988) 625, *Nucl. Phys.* **B314** (1989) 363.

[10] S.J. Gates, Jr., *Phys. Rev.* **D17** (1978) 3188;
M. Brown and S.J. Gates, Jr., *Annals Phys.* **122** (1979) 443;
S.J. Gates, Jr., M.T. Grisaru, M. Roček, and W. Siegel, Superspace, *or* One thousand and one lessons in supersymmetry (Benjamin/Cummings, Reading, 1983), p. 35.

[11] W. Siegel, *Nucl. Phys.* **B142** (1978) 301;
S.J. Gates, Jr. et al., *loc. cit.*, p. 280.

[12] N. Berkovits, *Phys. Lett.* **304B** (1993) 249.

[13] M. Roček, private communication.

[14] V.I. Ogievetskii and E. Sokatchev, *Phys. Lett.* **79B** (1978) 222; *Sov. J. Nucl. Phys.* **31** (1980) 140, 424, **32** (1980) 447, 589.

[15] E. Bergshoeff, P.S. Howe, C.N. Pope, E. Sezgin, and E. Sokatchev, *Nucl. Phys.* **B354** (1991) 113.

[16] A. Giveon and M. Porrati, *Nucl. Phys.* **B355** (1991) 422.

STATUS OF FRACTIONAL SUPERSTRINGS

S.-H. Henry Tye

Newman Laboratory of Nuclear Studies
Cornell University
Ithaca, NY 14853-5001

Abstract

We argue why it is important to search beyond the superstring for consistent string theories that have space-time supersymmetry with critical space-time dimensions less than 10. We give a lengthy introduction on the motivation and approach. We discuss briefly some promising possibilities, namely, fractional superstrings. More specifically, we consider the $K = 4$, or spin-4/3 fractional superstring, which has a 3-dimensional flat Minkowski space-time representation with bosons and fermions in its spectrum. Its central charge of 5 is less than the critical value of 10. The no-ghost theorem for the bosonic sector is proved, using the recently constructed Kac determinants. We summarize the present status and point out some open questions.

1 Introduction

String theory is the only known theory with the potential for describing all matter and forces in nature in a unified way. In particular, the superstring theory and the closely related heterotic string theory entail many structures, including gravity, gauge fields and chiral fermions, that are central to the present understanding of our universe. However, their critical space-time dimension is ten, and there are numerous mechanisms to reduce the number of observable space-time dimensions. Although the number of models in ten dimensions is very limited, namely the $SO(32)$ and the $E_8 \otimes E_8$ models[1], the number of models with four flat space-time dimensions is huge. Furthermore, there is no known compelling reason why the superstring theory should have only four large space-time dimensions. While it is important to search for dynamical and/or symmetry reasons explaining how and why our universe can be realized in the heterotic/superstring framework, it is also important to search for a new string theory that gives four large space-time dimensions naturally. Whether such a string theory exists or not is a question that we should be able to answer definitively.

For self-consistency, it is reasonable to request that the new string theory have space-time supersymmetry(SUSY). Let us recall the superstring case. ensional superstring model with spacetime SUSY automatically has a zero cosmological constant to all orders in perturbation theory. On the other hand, a non-supersymmetric string model has, generically, a non-zero cosmological constant through string loop corrections. Such a cosmological constant is typically over a hundred orders of magnitude too big. One may argue that this does not happen in superstring theory, where space-time SUSY is broken spontaneously. The argument is based on two observations: (1) in string theory, a non-zero cosmological constant will destroy the factorization property in multi-loop string diagrams, thus invalidating the consistency of string perturbation expansions; (2) starting with a consistent quantum theory, spontaneous/dynamical symmetry breaking will occur only in a self-consistent way. These observations imply that, if a superstring model with spacetime SUSY is a consistent quantum theory, then, after SUSY breaking, it must remain a consistent theory, i.e., the resulting cosmological constant in flat Minkowski spacetime must remain precisely zero, since a consistent string perturbative expansion must exist in the new non-supersymmetric vacuum. Of course, this argument sheds no light on whether or not (or how) SUSY breaking takes place in string theory.

So our goal is to search for the existence of new consistent string theories that have space-time SUSY and lower critical space-time dimensions. Our approach starts from the string world-sheet. String theories are characterized by the local symmetries of two-dimensional conformal field theories on the string world-sheet. The bosonic string is invariant under diffeomorphisms and local Weyl rescalings on the world-sheet; whereas the superstring is characterized by a locally supersymmetric version of these symmetries. This involves a spin-3/2 supercurrent, in addition to the energy-momentum tensor, on the string world-sheet. It is natural to ask whether other symmetries on the world-sheet can give rise to consistent string theories, which have lower space-time dimensions. To obtain a string theory from the world-sheet symmetry, the symmetry algebra is treated as a constraint algebra on the string Fock space. This approach have been successfully applied to the bosonic string and the superstring. It is natural to conjecture that, given any world-sheet symmetry algebra, there is a corresponding string theory with that algebra as the constraint algebra. Recently, a lot of progress have been made on the W-string[2], supporting this conjecture.

The first step of this approach is to understand the properties of world-sheet symmetries. I give an overall view of the possible world-sheet symmetry algebras. Specific examples[3] are given to illustrate the richness of possibilities. In Section 2, I discuss some specific possibilities of new string theories, namely, fractional superstrings. Since the properties of fractional superstrings were described in detail in the literature[4-13], I shall restrict myself to a brief summary of the present status and some remarks. Then I shall discuss a specific example, namely, the $K = 4$, or spin-4/3 fractional superstring[10]. This theory is the simpliest non-trivial example, and is characterized by a chiral algebra involving a pair of spin-4/3 currents on the world-sheet in addition to the energy-momentum tensor[14]. Scattering amplitudes of this theory are briefly described. They satisfy both spurious state decoupling and cyclic symmetry (duality). Examples of such amplitudes are calculated using an explicit $c = 5$ realization of the spin-4/3 current algebra[10]. This representation has a 3-dimensional flat Minkowski space-time interpretation, with fermions and bosons only in its spectrum[11]. The no-ghost theorem for the bosonic sector is proved using the recently constructed Kac determinants[12]. Again, I shall restrict myself to a brief introduction and a summary of the present status. Section 3 contains some remarks.

The structure of two-dimensional conformal field theories(CFT) is in large part determined by their underlying conformal symmetry algebra, which organizes all the fields in a CFT into sets of primary fields of definite conformal dimensions, and their associated infinite towers of descendant fields. The fundamental conformal symmetry is the Virasoro algebra. The most general extended conformal symmetry algebra can be written as follows. Consider a set of currents $J_i(z)$, primary with respect to $T(z)$, with conformal dimensions h_i, where $i \in \{1, 2, \ldots, N\}$:

$$T(z)T(w) = \frac{c/2}{(z-w)^4} + \frac{2T(w)}{(z-w)^2} + \frac{\partial T(w)}{(z-w)} + \ldots , \tag{1}$$

$$T(z)J_i(w) = \frac{h_i J_i(w)}{(z-w)^2} + \frac{\partial J_i(w)}{(z-w)} + \ldots.$$

The operator product expansions (OPEs) among the currents have the following generic form:

$$J_i(z)J_j(w) = q_{ij}(z-w)^{-h_i-h_j}(1 + \ldots) + \tag{2}$$
$$\sum_k f_{ijk}(z-w)^{-h_i-h_j+h_k}\left(J_k(w) + \ldots\right).$$

where q_{ij} and f_{ijk} are structure constants. The ellipses stand for current algebra descendants, whose dimensions differ from those of the identity and the J_k by positive integers. Since the algebra is chiral, i.e., independent of $\bar{z}$, the conformal dimensions are the spins of the fields. The parameters c, h_i, q_{ij} and f_{ijk} are not free and must be chosen such that the algebra(2) is associative. This condition places strong constraints on the set of consistent conformal dimensions h_i and restricts the structure constants q_{ij} and f_{ijk} as functions of the central charge c.

Extended conformal algebras naturally fall into three classes:

(i) Local algebras: the simplest type, where all powers of $(z-w)$ appearing in (2) are integers. This class includes the most familiar extended algebras, such as the superconformal, Kač-Moody and W_n algebras. These examples are unitary and therefore consist of currents with only integer and half-integer spins. Additionally, there exist non-unitary algebras such as the BRST ghost systems where arbitrary spins may be present.

When fractional powers of $(z-w)$ appear in some of the OPEs in (2), some of the currents will necessarily have fractional spins. In this case, the algebra is non-local, due to the presence of Riemann cuts in the complex plane. Such algebras are more complicated to construct and analyze than the local ones. Among non-local algebras there is a further division, again along lines of complication.

(ii) Abelian non-local algebras: also known as parafermion (PF) or generalized parafermion current algebras, were first constructed by Zamolodchikov and Fateev[16]. They are the simplest type of non-local algebras, involving at most one fractional power of $(z-w)$ in each OPE in (2). Any two currents in a PF algebra obey abelian braiding relations, i.e., upon braiding the two currents (analytically continuing one current along a path encircling the other), any correlation function involving these two currents only changes by a phase. The analysis of the associativity conditions for PF theories can be carried out using algebraic methods.

(iii) Non-abelian non-local algebras: or non-abelian algebras for short, since they are necessarily non-local. This is the most general class of extended algebras. Their characteristic feature is that their OPEs involve multiple cuts, i.e., there are terms in at least one of the OPEs in (2) with different fractional powers of $(z - w)$. Any two fractional spin currents appearing in one of these OPEs will in general obey non-abelian braiding properties. The analysis of the associativity conditions for non-abelian algebras requires more powerful methods.

In general, the holomorphic n-point correlation functions of the currents, $\langle J_i(z_1) J_j(z_2)....J_k(z_n) \rangle$, can be expressed as a linear combination of some set of conformal blocks. The relative coefficients of the various conformal blocks are fixed by the closure condition and the associativity condition. The closure condition is simply the requirement that no new currents beyond the currents of the algebra should appear. Associativity is the condition that the particular linear combination of conformal blocks that appears in the n-point function is invariant under fusion transformations (i.e., duality). For the local algebras, each conformal block involves only integer powers of $(z_i - z_j)$; for the abelian non-local algebras, each correlation function of the parafermion currents has exactly one conformal block, even though it involves fractional powers of $(z_i - z_j)$. Of course, for the most general case we expect each correlation function to have multiple conformal blocks, and the conformal blocks to involve different fractional powers of $(z_i - z_j)$. This general case corresponds to the non-abelian non-local algebras. From this point of view we see that upon braiding the currents, the correlation function is, in general, transformed into an independent linear combination of conformal blocks. This reflects the different phases that are picked up upon analytically continuing the different fractional powers of $(z_i - z_j)$.

The different braiding properties of the different types of extended algebras described above are reflected in the moding of their currents. On a given state in any representation of the algebra, we can obtain new states by acting with current modes

$$J^{i_1}_{-n_1-r_1} J^{i_2}_{-n_2-r_2} ... J^{i_m}_{-n_m-r_m} |\Phi\rangle, \tag{3}$$

where the n_j are integers and the r_j are fractional in general. For local unitary algebras, the half-integer spin currents can have only integer or half-integer modings, and $r_1 = r_2 = \cdots = r_m = r$ where either $r = 0$ or $r = \frac{1}{2}$, depending on the state $|\Phi\rangle$. For PF theories, the situation is slightly more complicated. Generically the r_i are different, determined by the state $|\Phi\rangle$ and the currents that preceded it. For non-abelian algebras, the moding of a particular current in (3) is not unique; it depends both on the state it operates on as well as on the state we want it to create.

Another useful way to classify symmetry algebras is their linearity property. An algebra is said to be linear if tensoring any two of its representations automatically gives a representation of the same algebra. The Virasoro and the superconformal algebras are linear in this sense. Most other algebras are non-linear, i.e., they do not have this property. The construction of representations for these algebras are rather non-trivial.

The first set of non-abelian algebras was conjectured in Ref.[15] and explicitly constructed in Ref.[3]. Consider a fractional supercurrent G with conformal dimension (i.e., spin) $\Delta = (K + 4)/(K + 2)$. The following OPEs define the level K FSCA with arbitrary central charge c:

$$T(z)T(w) = \frac{(c/2)}{(z - w)^4} + \frac{2T(w)}{(z - w)^2} + \frac{\partial T(w)}{(z - w)} + \dots ,$$

$$T(z)G(w) = \frac{\Delta}{(z - w)^2}G(w) + \frac{1}{(z - w)}\partial G(w) + \dots , \qquad (4)$$

$$G(z)G(w) = (z - w)^{-2\Delta}\left\{\frac{c}{\Delta} + 2(z - w)^2 T(w) + \dots\right\} +$$

$$\lambda(c)(z - w)^{-\Delta}\left\{G(w) + \frac{1}{2}(z - w)\partial G(w) + \dots\right\} .$$

The associativity condition for this algebra fixes the structure constant $\lambda^2(c)$:

$$\lambda^2(c) = \frac{2K^2 c_{111}^2}{3(K + 4)^2}(c_1 - c) . \qquad (5)$$

Here c_{111} is the $SU(2)_K$ structure constant for the OPE of two chiral spin-1 primary fields to give a chiral spin-1 field. It depends only on K and is explicitly given in Ref.[3]; $c_1 = 24/K + c_0$, and $c_0 = 3K/(K + 2)$.

For $K = 1$, $G(z)$ is absent and we recover the Virasoro algebra. For $K = 2$, c_{111} is zero and we recover the usual superconformal algebra, which has no cuts in the compex plane in its OPEs. The absence of cuts is an enormous simplification, and is the reason why the superconformal algebra is substantially easier to study than the $K > 2$ cases. We call these algebras the fractional superconformal algebras (FSCAs), since they are the natural generalization of the superconformal algebra. The closed algebra generated by the currents $T(z)$ and $G(z)$ is the level K FSCA. It is closely associated with the $SU(2)_K$ WZW model, in the sense that a representation of level K FSCA can be constructed from the currents and the $j = 1$ fields of $SU(2)_K$. We can also construct representations using $\mathbf{Z}_K$ parafermions. We note that there exist other level K FSCAs with additional fractional supercurrents [3]. It is not hard to convince oneself that, for each affine Lie algebra, there is at least one such corresponding extended conformal symmetry algebra; some examples are W_N algebra associated with $SU(N)_1$[17] and the PF current algebras associated with $so(N)_2$[18]. The usefulness of these symmetry algebras in organizing CFTs is illustrated by the application of FSCA to the $SU(2)$ WZW coset models[15, 19].

2 Fractional superstrings

It is natural to ask whether symmetries other than the superconformal symmetry on the world-sheet can give rise to consistent string theories, which have lower space-time dimensions. Recently, we find strong evidence for new string theories based on FSCAs, called fractional superstrings (FSS), which have interesting phenomenologies in space-times with critical dimensions less than 10. The evidence come from the following:

(1) Critical central charge and null states; any consistent string theory is expected to have extra sets of physical null states at a special value of the central charge, known as the critical central charge. This property is presumably a reflection of the underlying gauge symmetry of the string theory. Recently the Kac determinants for the level K FSCA have been constructed using the BRST cohomology techniques developed in the conformal field theory[12]. In particular, we reproduce the Kac determinants for the Virasoro ($K = 1$) and the superconformal ($K = 2$) algebras. The Kac determinant formulae allow us to examine the null state structure of plausible critical FSS with an arbitrary level K world-sheet fractional supersymmetry. The critical central charge is found to be[4, 12]

$$c_{\text{critical}} = c_1 + c_0 = \frac{6K}{K+2} + \frac{24}{K} . \tag{6}$$

Note that this recovers the well-known results for the bosonic ($K = 1$, with $c_{\text{critical}} = 26$) and the superstring ($K = 2$, with $c_{\text{critical}} = 15$) theories. This result also agrees with the more explicit determination of the spin-4/3 ($K = 4$, with $c_{\text{critical}} = 10$) FSS given in Appendix E of Ref.[8].

(2) One-loop modular invariant partition function; since the fractional superstring involves world-sheet fractional spin fields, i.e., parafermions, we search for closed string modular invariant partition functions that: (i) involve parafermion characters, (ii) contain the massless graviton, and (iii) are tachyon-free. Rather unique solutions are obtained, and they are consistent with space-time supersymmetry[5, 6]. The FSS with spin-3/2 ($K = 2$), 4/3 ($K = 4$), 6/5 ($K = 8$) and 10/9 ($K = 16$) fractional supercurrent on the world-sheet seem to have maximum flat dimension 10, 6, 4 and 3 respectively. Of course, the spin-3/2 case is simply the usual superstring. It is natural to view the other cases as non-trivial generalizations of the superstring case.

(3) Tree-level scattering amplitudes; we construct tree-level scattering amplitudes for the spin-4/3 fractional superstring[10]. Despite the non-local and non-linear structure of the world-sheet symmetry, we construct the N-point scattering amplitude for bosons and demonstrate its cyclic symmetry (duality) and the decoupling of spurious states to the physical states. In a particular representation which has a three dimensional Minkowski space-time, we also argue that only space-time bosons and fermions are present in this model[10, 11]. The closed string version contains the massless graviton.

(4) No-ghost theorem; we apply the Kac determinant formula for the K=4 fractional superconformal algebra to the above three-dimensional Minkowski space-time representation of the spin-4/3 fractional superstring[12]. We prove the no-ghost theorem for the space-time bosonic sector of this model; that is, at the tree-level scattering, its physical spectrum is free of negative-norm states.

Here, a few remarks on the above properties are in order:

(a) In the construction of the modular invariant partition functions, we found the critical dimension to be, for $K > 1$, $D_{\text{crit}} = 2 + 16/K$. There, we tensor ($D_{\text{crit}} - 2$) copies of Z_K parafermions plus boson X^μ in the light-cone gauge, each copy with central charge c_0. Naively, this would imply that the critical central charge is $c_{\text{crit}} = D_{\text{crit}} c_0 = 6(K + 8)/(K + 2)$, which disagrees with (6). However, a remarkable cancellation happens in the modular invariant partition functions, within the characters of the sectors containing the massless fields. This delicate cancellation, referred to as "internal projection" [6], reduces the central charge in the light-cone gauge from $c_{lc} = (D_{crit} - 2)c_0 = 48/(K + 2)$ to an effective value $c_{lc} = 24/K$. This can agree with (6) if we add back the central charge $2c_0$ for the time and longitudinal dimensions[9]. This "internal projection" feature is not surprising, for the following reason. The level K FSCA are non-linear for $K > 2$, so its representations cannot be obtained from tensoring copies of Z_K parafermion plus boson X^μ. Inside the space of tensored copies, one can imagine a projection to a subspace of states which does form a representation of the algebra. The explicit construction discussed in the next section is motivated by this observation. Note that there is no internal projection for the superstring case. This is expected, since the $K = 2$ FSCA (the superconformal algebra) is linear.

(b) It was argued that the critical dimensions for the $K = 4$ and $K = 8$ FSS are 6 and 4 respectively. However, a closer examination[7] of the modular invariant partition functions for these two cases indicates that they do not have the necessary Lorentz symmetry: the $K = 4$ modular invariant partition function allows at most a four-dimensional Poincare symmetry while the $K = 8$ case allows at most a three-dimensional Poincare symmetry. This seems to imply that the two of the flat directions in the $K = 4$ FSS cannot be normal space-time dimensions. To avoid a continuous spectrum, they must be compactified. This maximum possible space-time dimension was referred to as the "natural dimension", as opposed to the critical dimension. In this sense, the $K = 4$ FSS is the phenomenologically interesting theory to pursue, since its natural space-time dimension is 4. As we shall see in the next section, the $K = 4$ FSS is also by far the easiest case (among all FSS with non-linear FSCA) to study.

(c) In uncompactified superstrings, the spectrum does not contain chiral fermions in fundamental representations of gauge groups. They appear only after appropriate compactification. On the other hand, it is the compactification that removes the uniqueness of the superstring theory, and raises the non-trivial dynamical questions of why and how does it happen. So it is intriguing to see that the $K = 4$ FSS has natural dimension of 4, which is less than its critical dimension of 6; in this situation, one can hope the compactification of the extra dimensions to appear more naturally[7]. Of course, an explicit realization of this feature remains to be found.

To be concrete, let us consider the $K = 4$ FSCA. This is the simpliest non-trivial case because one can split the G current into two pieces, $G(z) =$

$G^+(z) + G^-(z)$, which then satisfy abelian braid relations:

$$G^\pm(z)G^\mp(w) = \frac{1}{(z-w)^{8/3}}\left\{\frac{3c}{8} + (z-w)^2 T(w) + ...\right\}, \qquad (7)$$

$$G^\pm(z)G^\pm(w) = \frac{\lambda}{(z-w)^{4/3}}\left\{G^\mp(w) + \frac{1}{2}(z-w)\partial G^\mp(w) + ...\right\}.$$

Here, $\lambda^2 = (8-c)/6$. In this split form, this algebra is the spin-4/3 parafermionic current algebra constructed by Zamolodchikov and Fateev [14]. Appendix C of Ref.[8] gives a detailed construction of this algebra. To construct a string theory based on this algebra, we use an approach that mimics the original construction of the superstring[20]. By analogy with the superconformal gauge of the superstring, the stress-energy tensor and fractional supercurrents are assumed to generate the physical state conditions. In particular, physical states, taken to be annihilated by all the positive modes L_n and $G_r^\pm$ of T and $G^\pm$ respectively, are the highest-weight states of the FSC algebra. This FSC algebra has a $\mathbf{Z}_3$ symmetry. In particular, the currents G^+ and G^- can be assigned $\mathbf{Z}_3$ charges $q = 1$ and -1, respectively, while the energy-momentum tensor T (as well as the identity) has charge $q = 0$. So the highest-weight modules of this algebra are organized by the $\mathbf{Z}_3$ symmetry of the algebra. Highest-weight states with $\mathbf{Z}_3$ charge ± 1 are said to belong to D-modules, while those with $\mathbf{Z}_3$ charge 0 are in S-modules. The consistency of the FSS can be demonstrated by defining tree scattering amplitudes and showing that they are consistent with the assumed physical state conditions following from the spin-4/3 FSCA. In other words, in tree scattering, physical states never scatter to unphysical states, and null states can also be consistently decoupled from scattering of other physical states. This property is commonly referred to as *spurious state decoupling*, and can be shown in a representation-independent way. The argument for spurious state decoupling follows closely that used in the "old covariant formalism" [21] for ordinary superstring amplitudes;

Scattering of D-module states can be written in three physically equivalent "pictures," reflecting the $\mathbf{Z}_3$ symmetry of the fractional superconformal algebra, in which the vertex operators for scattering can be one of $W^\pm$ of conformal dimension 1/3 and $\mathbf{Z}_3$ charge ± 1 or V of conformal dimension 1 and $\mathbf{Z}_3$ charge 0. In the old covariant formalism, the N-point scattering

374

amplitude can be written in three pictures

$$\begin{aligned}
\mathcal{A}_N &= 2\langle W_N^+|V_{N-1}(1)\Delta....\Delta V_2(1)|W_1^-\rangle \\
&= 2\langle W_N^-|V_{N-1}(1)\Delta....\Delta V_2(1)|W_1^+\rangle \\
&= \langle V_N|V_{N-1}(1)\tilde{\Delta}....\tilde{\Delta}V_2(1)|V_1\rangle.
\end{aligned} \tag{8}$$

Here the propagators are $\Delta = (L_0 - \frac{1}{3})^{-1}$ and $\tilde{\Delta} = (L_0 - 1)^{-1}$; and the vertex operators are related by

$$[G_r^\pm, V(1)] = \left(L_0 + r - \frac{1}{3}\right) W^\pm(1) - W^\pm(1)\left(L_0 - \frac{1}{3}\right) \tag{9}$$

for all $r \in \mathbf{Z}/3$. This is closely analogous to the two different pictures for scattering of Neveu-Schwarz sector states in the old covariant formalism for the ordinary superstring, in which vertex operators can be either G-parity even dimension-1/2 operators or G-parity odd dimension-1 operators (this property follows from the $\mathbf{Z}_2$ symmetry of the superconformal algebra). Scattering of S-module states is problematic due to the absence of an appropriate dimension-1 vertex operator in that sector. Since the S-module typically contains the tachyonic state, we are fortunate that they can be decoupled from tree scattering amplitudes of D-module states by a $\mathbf{Z}_3$ analog of the GSO projection [22]. A separate issue that can be addressed at tree level is the unitarity of scattering amplitudes. In particular, spurious state decoupling implies unitarity only if one can prove that the space of physical states has non-negative norm. This latter property is called the *no-ghost theorem*. Using the recently constructed Kac determinants for FSCAs, such a theorem is proven in Ref.[12] for the three-dimensional model that we shall now discuss.

This model is a particular conformal field theory representation of the spin-4/3 FSC algebra with central charge $c = 5$. It is made up of three free coordinate boson fields X^μ on the world-sheet and a two-boson representation of the $so(2,1)_2$ WZW model, which is closely related to the $SU(3)_1$ WZW model; (note that, conformally, $SU(3)_1 = SU(2)_4 = so(3)_2$). This model has a global three-dimensional Poincare invariance. The non-linear nature of the spin-4/3 FSC algebra makes the existence of such a representation non-trivial. Also, the states in the model are found to be space-time bosons or fermions, showing that the existence of fractional-spin constraints on the world-sheet need not imply fractional spins in space-time. The untwisted sectors of the fractional superconformal algebra describe space-time bosonic physical states in this representation. In particular, the lowest-mass D-module states describe massless gauge fields for the open string and a graviton for closed fractional superstrings.

Since three-dimensional Minkowski space-time is too small to decribe nature, it is encouraging that the central charge of the three-dimensional model is less than the critical value of 10, allowing the possibility of a critical spin-4/3 fractional superstring containing four-dimensional Minkowski space-time.

3 Discussion and outlook

Of course, the existence of FSS remains to be demonstrated. There are many open questions remain to be answered. More specifically, we are still steps away from showing that

(1) consistent fractional superstring theory exists and

(2) it is phenomenologically interesting.

However, we are encouraged by some non-trivial "coincidences" that are worth emphasizing. Let me mention two:

(1) In the construction of modular invariant partition functions, typically one finds there are more constraints than the number of parameters available in finding a solution. A priori, solutions should not exist, but they do.

(2) In the construction of scattering amplitudes for D-module states in the $K = 4$ FSS, which includes the massless vector particle as the ground state, the constraints necessary for spurious state decoupling and for picture-changing to work seems to be tighter than the freedom allowed by the parameters present. Naively, the picture-changing formula (9) should not exist a priori, but it does.

We believe that these "coincidences" are really a reflection of an underlying consistent theory being looked at from a funny angle. They clearly suggest that FSS is tighter than the usual superstring theory. This can be seen in a number of ways:

(1) the number of possible modular invariant partition functions of both the FSS type and the heterotic type are a lot fewer than in the corresponding cases in usual superstring, at any fixed space-time dimensions[7]. This comes about because of (i) the initial critical dimension of FSS starts out to be smaller than that of the superstring; so there are less dimensions to compactify; and (ii) the FSCAs underlying the FSS are non-linear, hence

376

there are fewer representations for fixed central charge.

(2) the tachyonic state (or any state in the same sector) in the Neveu-Schwarz sector is completely consistent in the superstring theory, even though it is undesirable phenomenologically; the same tachyonic state (or any state in the same sector) in the $K = 4$ FSS is not consistent: the four-point scattering amplitude of such states does not exist in the old covariant formalism. So, the GSO-like projection is necessary in the FSS case even at the tree-level.

At the moment, we are trying to construct the BRST ghosts and the BRST invariance of the $K = 4$ FSS. Here, the ghosts are expected to couple to the matter fields, much like the Yang-Mills theory. In this case, one should seek a $c = 0$ representation of the FSC algebra that contains both the matter and the ghost fields and permits the construction of a nilpotent BRST charge. Within this $c = 0$ representation, the goal is to find a $c = 10$ sub-representation that has a four-dimensional Poincare invariance. It is interesting to note that, since $so(4) = so(3) \otimes so(3)$, the world-sheet symmetry algebra corresponding to $so(4)_2$ is simply two copies of the spin-4/3 algebra; however, the coordinate bosons coupled to the $so(3)_2$ model in our $c = 5$ representation do not transform in the vector representation of $so(4)$, and so cannot give a flat space-time interpretation.

In general, other fractional superstrings will be technically more difficult to work with than the spin-4/3 FSS. The main reason is that the general FSC algebra are non-abelian. Since conformally, $su(2)_{2L} = so(3)_L$, we expect all of them (at least the ones with even K) to have a representation (similar to the $c = 5$ one in the $K = 4$ FSS) with a three-dimensional Poincare invariance. However, like the $K = 4$ case, none of them will be at the critical central charge except when K is infinite. In this limit, the fractional supercurrent has conformal dimension 1, with $c_{\text{critical}} = 6$ for the corresponding FSS. The representation is made up of three free coordinate boson fields X^μ on the world-sheet and a three-boson representation of the $so(2, 1)_\infty$ WZW model, so the central charge equals c_{critical}. This string theory has critical space-time dimension 3, and it may be used to describe the three-dimensional Ising model. Since the OPEs in this $K = \infty$ FSCA has only poles, the analysis may be quite tractable.

Acknowledgements

This talk reports on joint work with Stephen Chung, Keith Dienes, Jim Grochocinski, Zurab Kakushadze, Andre LeClair, Ed Lyman, and especially Philip Argyres. It is a pleasure to thank them all for fruitful collaborations and many useful discussions. This work was supported in part by the National Science Foundation.

References

[1] D.J. Gross, J.A. Harvey, E. Martinec, and R. Rohm, *Nucl. Phys.* **B256** (1985) 253.

[2] See e.g., Talk by C. Pope in this conference.

[3] P.C. Argyres, J. Grochocinski and S.-H.H. Tye, *Nucl. Phys.* **B367** (1991) 217, hep-th/9110052; *Nucl. Phys.* **B391** (1993) 409, hep-th/9202007.

[4] P.C. Argyres, A. LeClair and S.-H.H. Tye, *Phys. Lett.* **253B** (1991) 306.

[5] P.C. Argyres and S.-H.H. Tye, *Phys. Rev. Lett.* **67** (1991) 3339, hep-th/9109001.

[6] P.C. Argyres, K.R. Dienes and S.-H.H. Tye, *Commun. Math. Phys.* **154** (1993) 471, hep-th/9201078.

[7] K.R. Dienes and S.-H.H. Tye, *Nucl. Phys.* **B376** (1992) 297, hep-th/9112015.

[8] P.C. Argyres, E. Lyman and S.-H.H. Tye, *Phys. Rev.* **D46** (1992) 4533, hep-th/9205113.

[9] P.C. Argyres and K.R. Dienes, *Phys. Rev. Lett.* **71** (1993) 819, hep-th/9305093; K.R. Dienes, *The Worldsheet Conformal Field Theory of the Fractional Superstring*, McGill preprint McGill-93-01, to appear in *Nucl. Phys.* **B**, hep-th/9305094.

[10] P.C. Argyres and S.-H.H. Tye, *Tree scattering amplitudes of the spin-4/3 fractional superstring I: The untwisted sectors*, IASSNS-HEP-93/57, CLNS 92/1176, hep-th/9310131.

[11] P.C. Argyres and S.-H.H. Tye, *Tree scattering amplitudes of the spin-4/3 fractional superstring II: The twisted sectors,* IASSNS-HEP-93/58, CLNS 93/1251, to appear.

[12] Z. Kakushadze and S.-H.H. Tye, *Kač and new determinants of fractional superconformal algebras,* CLNS 93/1243, hep-th/9310160.

[13] P. Frampton and J. Liu, *Phys. Rev. Lett.* **70** (1993) 130, hep-th-/9210049; G.B. Cleaver and P.J. Rosenthal, *Aspects of Fractional Superstrings,* Caltech preprint CALT-68-1756, hep-th/9302071.

[14] A.B. Zamolodchikov and V.A. Fateev, *Theor. Math. Phys.* **71** (1987) 451.

[15] D. Kastor, E. Martinec and Z. Qiu, *Phys. Lett.* **200B** (1988) 434; J. Bagger, D. Nemeschansky and S. Yankielowicz, *Phys. Rev. Lett.* **60** (1988) 389; F. Ravanini, *Mod. Phys. Lett.* **A3** (1988) 397;

[16] A.B. Zamolodchikov and V.A. Fateev, *Sov. Phys. J.E.T.P.* **62** (1985) 215; *Sov. Phys. J.E.T.P.* **63** (1986) 913.

[17] A.B. Zamolodchikov, *Theo. Math. Phys.* **63** (1985) 1205; V.A. Fateev and S.L. Lykyanov, *Int. J. of Mod. Phys.* **A3** (1988) 507.

[18] P. Goddard and A. Schwimmer, *Phys. Lett.* **206B** (1988) 62.

[19] S.-w. Chung, E. Lyman and S.-H.H. Tye, *Int. J. Mod. Phys.* **7A** (1992) 3339.

[20] P. Ramond, *Phys. Rev.* **D3** (1971) 2415; A. Neveu and J. H. Schwarz, *Nucl. Phys.* **B31** (1971) 86.

[21] A. Neveu, J.H. Schwarz and C.B. Thorn, *Phys. Lett.* **B35** (1971) 529; for a detailed explanation of this formalism, see, *e.g.*, Chapter 7 of M.B. Green, J.H. Schwarz and E. Witten, *Superstring Theory,* Cambridge University Press (1987).

[22] F. Gliozzi, J. Scherk and D. Olive, *Nucl. Phys.* **B122** (1977) 253.

SPACE-TIME TRANSITIONS IN STRING THEORY

Edward Witten[†]

School of Natural Sciences,
Institute for Advanced Study,
Olden Lane,
Princeton, N.J. 08540

ABSTRACT

Simple mean field methods can be used to describe transitions between different space-time models in string theory. These include transitions between different Calabi-Yau manifolds, and more exotic things such as the Calabi-Yau/Landau-Ginzburg correspondence.

---

† Research supported in part by NSF Grant PHY-92-45317.

Today I will be talking about transitions between different space-time models in string theory. In string theory, space-time is represented by a two dimensional quantum field theory. We will therefore study transitions among such two dimensional theories that occur as the parameters are varied. I will describe very simple mean field methods that can be used [1] to relate Calabi-Yau models with different target spaces (as recently developed in a different way by P. Aspinwall, D. Morrison, and B. Greene [2]). The same methods also give a new explanation of the familiar relation between certain Calabi-Yau models and certain Landau-Ginzburg models.

In fact, I believe that in this way we get the generalization of the Calabi-Yau/Landau-Ginzburg correspondence to arbitrary Calabi-Yau sigma models. The general story involves an extension of the sigma model moduli space beyond the classical region. If X is a Calabi-Yau manifold, the Kahler class of a Ricci-flat Kahler metric defines a point in $H^2(X, \mathbb{R})$. This point lies in a conical region of $H^2(X, \mathbb{R})$ called the Kahler cone. In classical field theory, the region of $H^2(X, \mathbb{R})$ outside the Kahler cone does not have much use. In string theory, however, (or in a world-sheet quantum field theory), nothing is wasted. The quantum moduli space is all of $H^2(X, \mathbb{R})$ – and in fact, $H^2(X, \mathbb{C})$, to allow for theta angles. $H^2(X, \mathbb{R})$ is divided into cones, of which the classical Kahler cone of X is only one. In each cone, the theory has a different "geometrical" description. For instance, the traditional Calabi-Yau/Landau-Ginzburg correspondence arises when $H^2(X, \mathbb{R})$ is one dimensional. There are then only two cones: the positive half-line, which is the Kahler cone of X; and the negative half-line, which is the Landau-Ginzburg region.

Let me illustrate these ideas in the Calabi-Yau/Landau-Ginzburg case. Let $z_1, \ldots, z_5$ be complex variables, and let $F(z_1, \ldots, z_5)$ be a homogeneous quanta polynomial, with the transversality property that the equations

$$0 = \frac{\partial F}{\partial z_1} = \ldots = \frac{\partial F}{\partial z_5} \tag{1}$$

have a common solution only at the origin. A generic homogeneous F has that property.

Consider the hypersurface X of solutions of $F = 0$ in $\mathbb{CP}^4$. It is a smooth Calabi-Yau manifold by virtue of the transversality of F. The sigma model with target space X is an $N = 2$ superconformal field theory.

On the other hand, we can introduce chiral superfields $\Phi_1, \ldots, \Phi_5$ in $N = 2$ superspace in two dimensions, with superpotential

$$W(\Phi_1, \ldots, \Phi_5) = F(\Phi_1, \ldots, \Phi_5). \tag{2}$$

With this superpotential, we can make an $N = 2$ model that is not conformally invariant

$$\mathcal{L} = \int d^2 x \, d^4 \theta \sum_i \Phi_i \bar{\Phi}_i - \left(\int d^2 x \, d^2 \theta F(\Phi_i) + h.c. \right) \tag{3}$$

The ordinary potential is then

$$V = \sum_i \left| \frac{\partial F}{\partial \Phi^i} \right|^2 . \tag{4}$$

This model has (by virtue of transversality of F) an isolated vacuum at $\Phi_i = 0$ with massless particles; it is called a Landau-Ginzburg model.

It is claimed (by Martinec and by Greene, Vafa and Warner) that a suitable orbifold of this Landau-Ginzburg model is "equivalent" in the infrared to the Calabi-Yau model determined by the same F.

To obtain a new insight about this relationship, we consider a U(1) gauge theory in N=2 superspace, described by a vector superfield. The gauge invariant field strength is $\Sigma = \{\overline{\mathcal{D}}_+, \mathcal{D}_-\}/2\sqrt{2}$ and the gauge kinetic energy is

$$\mathcal{L}_{\text{gauge}} = -\frac{1}{4e^2} \int d^2 x \, d^4 \theta \, \overline{\Sigma}\Sigma. \tag{5}$$

Such a $U(1)$ gauge theory in two dimensions has two additional interactions possible, the θ angle and the Fayet-Iliopoulos D term. These can be written as

$$L_{D,\theta} = \frac{it}{2\sqrt{2}} \int d^2 x \, d\theta^+ \, d\bar{\theta}^- \Sigma + h.c. \tag{6}$$

with

$$t = ir + \frac{\theta}{2\pi} \tag{7}$$

θ is the usual θ angle, and r is the coefficient of the D term.

To this we add chiral superfields – the five superfields Φ_i of the earlier discussion, which we now take to have charge 1, and a new superfield P of charge -5. Their kinetic energy is

$$\mathcal{L}_{kin} = \int d^2x\, d^4\theta \left\{ \sum_i \bar{\Phi}_i\, \Phi_i + \bar{P}\, P \right\}. \tag{8}$$

Finally, we introduce the superpotential, which we take to be the gauge invariant function $W = PF(\Phi_i)$. The corresponding piece of the Lagrangian is hence

$$\mathcal{L}_W = -\int d^2x\, d^2\theta \, PF(\Phi_i) - h.c. \tag{9}$$

After performing the θ integrals and eliminating the auxiliary fields, the ordinary potential of the model is

$$V = \frac{e^2}{2}\left(\sum \bar{\varphi}_i\varphi_i - 5\bar{p}p - r \right)^2 + |F(\varphi_i)|^2 + \sum_i \bar{p}p\left|\frac{\partial F}{\partial \varphi_i}\right|^2 \tag{10}$$

Here φ_i, p are the bosonic components of superfields Φ_i, P. What remains is to study the vacuum structure as a function of r.

For $r \gg 0$, vanishing of $V_1 = \frac{e^2}{2}(\sum \bar{\varphi}_i\,\varphi_i - 5\bar{p}p - r)^2$ requires $\varphi_i \neq 0$. Vanishing of $V_2 = \bar{p}p\sum_i \left|\frac{\partial F}{\partial \varphi_i}\right|^2$ then implies (given transversality of F) that $p = 0$. Vanishing of V_1 then requires

$$\sum \bar{\varphi}_i\varphi_i = r \tag{11}$$

and after dividing by the gauge group $U(1)$, this gives a copy of $\mathbb{CP}^4$ with Kahler class proportional to r. Finally, vanishing of $V_3 = |F(\varphi_i)|^2$ shows that the space of classical ground states is the Calabi-Yau hypersurface $F = 0$ in $\mathbb{CP}^4$.

Now for $r \ll 0$, vanishing of V_1 gives $p \neq 0$. Vanishing of V_2 gives then $\varphi_i = 0$, so there is up to gauge transformation a unique classical ground state with $p = \sqrt{-r/5}$. In expanding around this ground state, the φ_i are massless and governed by an effective superpotential $\widetilde{W} = \langle p \rangle F(\varphi_i)$. The expectation value of p breaks the gauge group to $\mathbb{Z}_5$, and the low energy theory is a $\mathbb{Z}_5$ orbifold of a Landau-Ginzburg theory with superpotential W.

So Calabi-Yau and Landau-Ginzburg arise as two different "phases" of one system. To probe more deeply, one must understand what happens near $r = 0$. For this I refer to my paper [1]. Suffice it to say that as long as the θ angle is non-zero and at least for quantities such as Yukawa couplings, there is a smooth continuation from $r > 0$ to $r < 0$. The transition involves passing through a situation where stringy effects are big and field theory is not a good approximation.

So at least to this extent, Calabi-Yau and Landau-Ginzburg are two different limits of the same system, rather than two different phases in the strict sense. They are related like water to steam. This is the sense in which Calabi-Yau and Landau-Ginzburg are "equivalent."

By changing the gauge group and the quantum numbers of the chiral superfields, one can work out many generalizations of this. Among other things one gets transitions among space-times of different topology. For more detail, I refer to [1, 2].

I find these results fascinating because along with phenomena such as the $R \leftrightarrow 1/R$ duality, they are among relatively few examples of really "stringy" phenomena that we know. I am sure there is much more lurking under the surface, waiting to be unearthed once we have understood the geometrical tools and language appropriate to string theory. For the time being, not very much of this is accessible, though we can hopefully do more even with the methods we already have.

The phenomenon of topology change makes me wonder about the relation of the diffeomorphism group $diff(X)$ of a classical space-time to the corresponding symmetry group – call it $str(X)$ – of string theory with target space X. (It could be that $str(X)$ is an equivalence relation not generated by a group action, and conceivably well-defined only on shell. I will ignore such questions here.) Can $diff(X)$ be embedded in $str(X)$? One might suppose so, but there is no evident way to make this embedding in any formalism I know of. I would tend to believe that $diff(X)$ cannot be embedded in $str(X)$; such an embedding is certainly possible in the long distance limit, but there may be deviations of order α'. The topology-changing phenomena show that under certain conditions $str(X) = str(X')$ for distinct space-times X and X'. So if $diff(X)$ can be embedded in $str(X)$, so can be $diff(X')$, presumably, as X and X' seem to be on an entirely equal footing.

One can ask the opposite question: is there a homomorphism from $str(X)$ onto $diff(X)$? I would presume the answer is no; the existence of such a homomorphism would more or less give a way to recover field theoretic physics from string theoretic physics, and any such attempt ought to be thwarted by terms of order α'.

At any rate, the topology-changing phenomena ought to mean that if there is a homomorphism from $str(X)$ onto $diff(X)$, there should also be one onto $diff(X')$. This seems even less plausible.

The topology-changing phenomena really ought to mean that the relation between $str(X)$ and $diff(X)$ is purely a long-distance approximation that is relevant in a particular limit of the string theory moduli space. It is tantalizing to think that the deviation of the symmetry group of the world from $diff(X)$ might have some observable effect in the low energy world; but it is hard to see how that would be.

REFERENCES

1. E. Witten, "Phases of $N = 2$ Models in Two Dimensions," (IASSNS-HEP-93/3), to appear in Nucl. Phys. B.

2. P. Aspinwall, D. Morrison, and B. Greene, "Multiple Mirror Manifolds and Topology Change in String Theory" (IASSNS-HEP-93/4), Phys. Lett. **303B**, 249 (1993.)

5. 2-D Gravity

TWO OBSTACLES IN QUANTUM GRAVITY

JAN AMBJØRN

Niels Bohr Institute, University of Copenhagen, Blegdamsvej 17
Copenhagen, DK-2100 Ø, Denmark

1. c > 1

In the last four years there has been some progress in our understanding of 2d quantum gravity and non-critical strings. However, we have still not understood 2d quantum gravity coupled to matter with a central charge larger than one. If we use the discretized approach[1-5] there seems to be no formal problems with such a coupling. But two theorems established soon after the invention of these models made it unlikely that interesting theories would exist for $c > 1$. Recent results of numerical nature seem to indicate loopholes related to the theorems[6-12] In the following I will review the theorems and explain my present understanding of the loopholes.

1.1. The Theorems

The following theorems were proven in the context of 2d gravity coupled to multiple Gaussian fields[2,5,13,21], but the proofs can presumably be generalized to other matter fields:

theorem 1 $\gamma \leq \frac{1}{2}$ *and* $\gamma > 0 \Rightarrow \gamma = \frac{1}{2}$.

theorem 2 $\sigma(\mu_c) > 0$.

In the first theorem γ denotes the string susceptibility. In the second theorem $\sigma(\mu)$ denotes the string tension as a function of the cosmological constant μ, and μ_c is the critical value of the cosmological constant, where the continuum limit is taken. Theorem 2 is only relevant for multiple Gaussian fields since these can be given the interpretation as target space coordinates of the string. The string tension is defined by the exponential decay of the one-loop function, the loop being a fixed planar boundary ∂A enclosing an area A. Let $G_\mu(\partial A)$ denote the one-loop Greens function. Then $\sigma(\mu)$ is defined by

$$G_\mu(\partial A) \sim e^{-\sigma(\mu)A}, \qquad \sigma(\mu) \equiv \lim_{A \to \infty} \frac{\log G_\mu(\partial A)}{A}. \tag{1}$$

According the the KPZ formula γ is an increasing function of c, increasing to 0 as c increases to 1, while the formula leads to complex γ's for $c > 1$. It is natural to interpret the first theorem as an indication that the random surfaces degenerate to so-called branched polymers as soon as $c > 1$, since the value $\gamma = 1/2$ is the generic value for branched polymer surfaces. This interpretation is further corroborated by theorem 2 in the case of multiple Gaussian fields provided we define the relation between the bare (dimensionless) string tension $\sigma(\mu)$ and the "continuum" string tension σ_{phys} having dimension $(\text{mass})^2$ in the usual way. Let a be a scaling parameter with dimension of length, usually identified as a typical lattice cut-off in the regularized theory. We imagine that a is a function of the cosmological constant which scales to 0 for $\mu \to \mu_c$, and defined with reference to some mass parameter of the theory which scales to zero. One choice could e.g. be the mass defined by the exponential decay of the two-point function in target space (i.e. the "tachyon" mass in ordinary string theory):

$$m(\mu) \sim const. \cdot (\mu - \mu_c)^\nu = m_{phys} a(\mu). \tag{2}$$

If it is not possible to find a bare mass which scales to zero, one would be tempted to conclude that the theory has no continuum interpretation in target space. If we *assume* the existence of a bare mass which scales according to (2) we get $a(\mu) \sim (\mu - \mu_c)^\nu$, and the physical mass m_{phys} is simply *our* choice of ratio between $m(\mu)$ and $a(\mu)$. The relation between the physical string tension σ_{phys} and the bare string $\sigma(\mu)$ tension has to be defined in a similar way:

$$\sigma(\mu) = \sigma_{phys} a^2(\mu). \tag{3}$$

Since $a(\mu)$ by assumption scales to zero we conclude that $\sigma_{phys} \to \infty$ for $\mu \to \mu_c$, due to theorem 2. The only excitations of a surface with very large string tension are those which do not increase minimal the area dictated by the boundary conditions. The only possible outgrowths will be thin tubes which are allowed to branch, i.e. precisely branched polymer surfaces.

1.2. Theorem 1

Let me mention possible loopholes in the proof of this theorem[*] Let us for notational simplicity consider pure 2d gravity. In that case we know of course that $c = 0$ and $\gamma = -1/2$ but we can apply the general arguments to the model which is defined as follows:

$$Z(\mu) = \sum_{T \in \mathcal{T}} e^{-\mu|T|} \tag{4}$$

where $\mathcal{T}$ denotes a suitable class of triangulations. We will here restrict the topology to be spherical. Let us further define the n-puncture function as[†]

$$G_n(\mu) = (-1)^n \frac{d^n}{d\mu^n} Z(\mu). \tag{5}$$

---

[*]This analysis was first performed by B. Durhuus in the case of multiple Ising spins, and most of the following discussion of theorem 1 is inspired by discussions with him.

[†]In order not to make the discussion too technical we have chosen not to specify certain symmetry factors in eq. (4) and eq. (5).

We have

$$G_n(\mu) \geq G_2^n(\mu), \qquad n > 2. \tag{6}$$

The geometrical interpretation of eq. (6) is as follows: The rhs of eq. (6) can be viewed as a surface made out of n elongated balloons each with two marked points: one at the "root" and one at the top of the balloon, all tied together at their "roots", this common point being viewed as an interior point. In this way the rhs can be viewed as a surface with n marked points and this restrict sum over surfaces with n marked points is smaller that the sum over all surfaces with n marked points, i.e. the lhs of eq. (6). These arguments can be made rigorous in specific models (see for instance ref. 13).

By definition of the string susceptibility we have

$$G_n(\mu) \sim a.p. + (\mu - \mu_c)^{2-n-\gamma} + \cdots \tag{7}$$

where $a.p.$ denotes an analytic part at μ_c while $\cdots$ denote less singular terms. From (6) we get

$$\gamma < 1 - \frac{1}{n-1} \tag{8}$$

and the strongest bound on γ comes from $n = 3$ and is $\gamma \leq 1/2$. It is tempting to use (8) for $n = 2$, in which case we would get $\gamma \leq 0$. However, (6) is not correct in that case due to a non-unique decomposition of surfaces in a sequence of "balloons"-like surfaces.

We can derive an equation in the case where $n = 2$ as follows: Let us consider the one-point function. To be more precise we will consider the function where one triangle is considered the external boundary. This function has the same scaling behaviour as the one-point function where only a vertex is kept fixed. Assume that the smallest allowed loops in our class of triangulations are of length 3, i.e. are triangles. Any 3-loop of links in the surface is either a triangle belonging to the surface or it can be viewed as the boundary of a surface which grows out from our the rest of the surface and which can be separated from the rest by cutting along the 3-loop. If we still use the notation "one-point" function for the surfaces with a triangle (a 3-loop) as boundary we can write:

$$G_1(\mu) = \sum_{T \in \tilde{T}} e^{-\mu|T|} (1 + G_1(\mu))^{|T|} \equiv \sum_{T \in \tilde{T}} e^{-\bar{\mu}|T|} \tag{9}$$

where $\tilde{T}$ denotes the class of triangulations which cannot be separated in two by cutting along a loop of length 3. Clearly this class of triangulations is a perfectly good one by its own right and we can define a partition function and n-point functions similar to (4)-(5), only with $\tilde{T}$ instead of T. Let us denote the corresponding n-point functions with $\bar{G}_n(\bar{\mu})$, where $\bar{\mu}$ denotes the cosmological constant of the modified model. By definition we have

$$\bar{\mu} = \mu - \log(1 + G_1(\mu)), \qquad \bar{G}_1(\bar{\mu}) = G_1(\mu). \tag{10}$$

We can differentiate these equations with respect to μ:

$$\frac{d\bar{\mu}}{d\mu} = 1 + \phi(\mu), \qquad \phi(\mu) \equiv \frac{G_2(\mu)}{1 + G_1(\mu)} \tag{11}$$

and

$$\phi(\mu) = \frac{\bar{\phi}(\bar{\mu})}{1 - \bar{\phi}(\bar{\mu})}. \tag{12}$$

ϕ is essentially the two-point function G_2. They have the same critical behaviour and we will not distinguish between them. $\bar{\phi}$ is defined as ϕ, just using $\bar{G}_1$, $\bar{G}_2$ and $\bar{\mu}$ instead of G_1, G_2 and μ. It can be viewed as the sum over surfaces which connect the two boundary points (or boundary 3-loops) and where the surfaces in the sum have the property that they cannot be separated in two components, each of which contains one of the boundaries, by a cut along a 3-loop. By expanding the denominator eq. (12) gets the obvious graphical interpretation of the two-point function $\phi(\mu)$ as the sum over all surfaces (of trivial topology) made up by successive gluing of "irreducible" $\bar{\phi}$-surfaces along their boundaries, such that only two boundaries are left as external.

If we assume that $\gamma > 0$, we know from (7) that $\phi(\mu)$ will diverge for $\mu \to \mu_c$. From (12) this is impossible for $\bar{\phi}(\bar{\mu}(\mu))$ since it follows that

$$\bar{\phi}(\bar{\mu}_0) = 1, \qquad \bar{\mu}_0 \equiv \bar{\mu}(\mu_c) \ (\geq \bar{\mu}_c). \tag{13}$$

We conclude that either $\bar{\mu}_0 > \bar{\mu}_c$, the critical point for the modified model defined by the class of triangulations $\bar{\mathcal{T}}$, or $\bar{\gamma}$ corresponding to $\bar{\mathcal{T}}$ has to satisfy $\bar{\gamma} \leq 0$ in order that $\bar{\phi}(\bar{\mu}_c)$ is bounded (in fact $\bar{\phi}(\bar{\mu}_c) = 1$ by (12)).

In the first case we can expand $\bar{\phi}(\bar{\mu})$ in a Taylor series around $\bar{\mu}_0$ and use that due to (11) we have

$$\bar{\mu}_0 - \bar{\mu} \approx \left. \frac{d\bar{\mu}}{d\mu} \right|_{\mu} (\mu_c - \mu) \approx -(\mu - \mu_c)^{1-\gamma} \tag{14}$$

$$\bar{\phi}(\bar{\mu}) \approx 1 + \bar{\phi}'(\bar{\mu}_0)(\bar{\mu} - \bar{\mu}_0) \approx 1 + \bar{\phi}'(\bar{\mu}_0)(\mu - \mu_c)^{1-\gamma}. \tag{15}$$

If we insert $\phi(\mu) \approx (\mu - \mu_c)^{-\gamma}$ and (15) for $\bar{\phi}$ in eq. (12) we get the promised result $\gamma = 1/2$, i.e. the exponent corresponding to branched polymers.

In the Taylor expansion around $\bar{\mu}_0$ we used that $\bar{\phi}'(\bar{\mu}_0) < 0$. This follows from the fact that the two-point function is the sum over surfaces with positive weight $\exp(-\bar{\mu}|T|)$. If we however lift the restriction that the surfaces should enter with positive weight, by adding to the definition (4) a weight $\rho(T)$ which can take positive and negative values according to the triangulation, it is possible to arrange that

$$\bar{\phi}'(\bar{\mu}_0) = \cdots = \bar{\phi}^{(n-1)}(\bar{\mu}_0) = 0, \qquad \bar{\phi}^{(n)}(\bar{\mu}_0) \neq 0. \tag{16}$$

In this case we get instead of eq. (15)

$$\bar{\phi}(\bar{\mu}) \approx 1 + \bar{\phi}^{(n)}(\bar{\mu}_0)(\bar{\mu} - \bar{\mu}_0)^n \approx 1 + \bar{\phi}^{(n)}(\bar{\mu}_0)(\mu - \mu_c)^{(1-\gamma)n}. \tag{17}$$

If we insert this in eq. (12) we get

$$\gamma = 1 - \frac{1}{n+1}.$$ (18)

We can clearly arrange that $\gamma > 1/2$. But it is unlikely that the theory is different from a theory of branched polymers. In fact there exists a theory of so-called multicritical branched polymers, where the probability for branching can be negative[14]. The possible values of γ for such branched polymers are precisely given by (18).

The only possibility to get something non-trivial for $\gamma > 0$ seems to be the situation where $\bar{\mu}_0 = \bar{\mu}_c$. In this case we know that $\bar{\gamma} \neq \gamma$. In fact we have to have $\bar{\gamma} \leq 0$ since $\bar{\phi}(\bar{\mu}_c) = 1$. We can no longer make a Taylor expansion as in (15). We can however use (14) and the fact that $\bar{\phi}(\bar{\mu})$ (at least for $\bar{\gamma} \geq -1$) by the definition of $\bar{\gamma}$ behaves like

$$\bar{\phi}(\bar{\mu}) \approx 1 + (\bar{\mu} - \bar{\mu}_c)^{-\bar{\gamma}} \approx 1 + (\mu - \mu_c)^{-\bar{\gamma}(1-\gamma)}.$$ (19)

If we insert this in (12) we get

$$\gamma = -\frac{\bar{\gamma}}{1-\bar{\gamma}}, \qquad \bar{\gamma} = -\frac{1}{n} \Rightarrow \gamma = \frac{1}{n+1}$$ (20)

which gives a remarkable relation between γ and $\bar{\gamma}$, first obtained by Durhuus[15]. The interpretation of this relation is however a little disappointing. We see that it might be possible to obtain a γ between 0 and 1/2. However, it does not really reflect a new theory. It seems rather to be a "polymer-like" or "bubble-like" manifestation of an underlying $c < 0$ theory.

The above analysis only opened for the possibility that $0 < \gamma < 1/2$. It did not provide any examples. The treatment was also simplified in the sense that we used the notation of pure gravity where we know by explicit calculation that $\gamma < 0$. There is an example of a model which realizes precisely the scenario mentioned above[16,17,18]. The partition function (in the sense of generating connected random surfaces) is given by

$$Z(\mu, g) = \log \left\{ \int dM \exp \left[-\mu N \mathrm{Tr}\, V(M) + g\mu^2 (\mathrm{Tr}\, M^2)^2 \right] \right\}.$$ (21)

where M denotes an $N \times N$ Hermitian matrix and $V(M)$ is a potential corresponding the nth multicritical matrix model. If $g = 0$ this matrix model will generate random surfaces with $\gamma = -1/n$. We have a critical line in the (μ, g)-coupling constant plane and at a certain point along this line the critical behaviour changes and γ jumps from $-1/n$ to $1/(n+1)$. The geometrical interpretation of the model is that the term $(\mathrm{Tr}\, M^2)^2$ introduces a new kind of vertex in the model, such that to leading order in $1/N$ ordinary spherical surfaces dominates, except that two (or more) such surfaces can touch each other by means of the new vertices, the topology of the total surface still being spherical. The jump in γ is related to a change in behaviour the effective "touching" coupling constant $N\bar{g} = -g\mu < \mathrm{Tr}\ M^2 >$ as a function of μ, from being analytic to non-analytic, i.e. a situation precisely as described above in general terms.

The other class of models where one could imagine that $0 < \gamma < 1/2$ is multiple Ising models coupled to gravity. Durhuus has analysed this situation along the lines indicated above[15]. Things are slightly more complicated since we will now have several coupling constants and the scalar equations written above will be replaced by matrix equations. However, the result is unchanged: The only possible alternative to $\gamma = 1/2$ seems to be that $\gamma = 1/(n+1)$, and this model is somehow a shadow of an underlying unitary model with $\bar{\gamma} = -1/n$ and a central charge $\bar{c} < 1$.

1.3. Theorem 2

Theorem 2 is more specific for discretized string models, and the loopholes not obvious, if any. One cure, which is inspired by the geometrical picture of spiky polymers emerging from a minimal surface, attempts to add by hand extrinsic curvature terms. The model was first formulated in[19,20], partly inspired by theorem 2 (proven in ref. 21), partly by the fact that one could get extrinsic curvature terms by integrating out the fermionic part of the superstring. It seems difficult to obtain analytic results for the model (see however ref. 22 for a renormalization group approach), and most work has been of numerical nature. From large scale Monte Carlo simulations the following have been established (with the usual reservation of numerical simulations, which have a lot in common with experimental physics, including the ability to produce results which have to be retracted eventually): As a function of the extrinsic curvature coupling constant λ we have two phases : One phase where the physics is the same as for $\lambda = 0$, i.e. the string tension $\sigma(\mu, \lambda)$ is not scaling to zero as $\mu \to \mu_c(\lambda)$ (the critical value of the cosmological constant μ now depends on λ), and, for $\lambda > \lambda_c$, a now phase where the surfaces essentially are "smooth" embeddings in target space with Hausdorff dimension $d_H = 2$. At the transition point λ_c we have according to the numerical simulations[23-25] that the string tension scales to zero as $\mu \to \mu_c(\lambda_c)$. In addition the mass gap $m(\mu, \lambda_c)$ of the two-point function scales to zero for $\mu \to \mu_c(\lambda_c)$ in such a way that

$$\frac{m^2(\mu, \lambda_c)}{\sigma(\mu, \lambda_c)} \approx constant. \tag{22}$$

According to the discussion above this is precisely what is needed in order to achieve an interesting continuum limit which allow the interpretation as a string theory. It is, however, at present unclear which continuum theory could serve as a candidate for the continuum limit of the regularized string theory.

2. Quantum Gravity for D > 2

Attempts to quantize quantum gravity in dimensions $D > 2$ differ radically from the attempts described above for $D = 2$. For $D > 2$ we have not a well understood theory when the rotation to Euclidean space is performed: Due to the conformal mode the Einstein-Hilbert action is unbounded from below. The Euclidean path integral is not well defined. In addition to the problems with the conformal mode we are in quantum gravity faced with the question of topology. Should all topologies be included in the quantum theory, and, if this is the case, how can we imagine this should be done, since it is impossible to classify 4-geometries?

Simplicial quantum gravity is an attempt to give a non-perturbative definition of Euclidean gravity. Its main advantage is that it has been shown to work well (at least as well as any method based on a continuum formulation) in $D = 2$. In addition it provides a very natural geometric discretization of the Einstein-Hilbert action. A drawback is that it deals in a superficial way with the conformal mode problem[‡] since the regularization is such that for a fixed lattice volume the action is bounded from below. Of course this just postpone the conformal mode problem until the scaling limit has to be taken. In addition it has little to say about the summation over topologies, but at least it highlights the problem, since the model is simply not well defined unless we restrict the topology. In the following we will always assume the simplest topology, i.e. S^4 in $D = 4$.

1.2. The Model

As in $D = 2$ we imagine the manifolds which enters in the regularized path integral to be constructed by gluing together identical simplexes, in this case 4-simplexes rather than the equilateral triangles used in $D = 2$. The curvature of such a piecewise linear manifold is given by Regge calculus, which in this case becomes extremely simple[26,27]: Suppose that our manifold is constructed out of N_4 simplexes and that it has N_2 two-dimensional sub-simplexes. Using the construction of Regge a little algebra leads to to the following identifications (which we write down in D dimensions for the sake of generality):

$$\int d^D \xi \sqrt{g} \; \propto \; N_D \tag{23}$$

$$\int d^D \xi \sqrt{g} R \; \propto \; \frac{2\pi}{\arccos(1/D)} N_{D-2} - \frac{D(D+1)}{2} N_D. \tag{24}$$

For a given four-dimensional triangulation T we can therefore write the cosmological term plus the Einstein-Hilbert action as

$$S[T] = \kappa_4 N_4(T) - \kappa_2 N_2(T) \tag{25}$$

where κ_2 is proportional to the inverse of the bare gravitational coupling constant and κ_4 involves a combination of the bare cosmological coupling constant and the bare gravitational coupling constant.

The formal recipe for going from the continuum functional integral to the discretized one is now:

$$\int \mathcal{D}[g_{\mu\nu}] \; \to \; \sum_{T \in \mathcal{T}} \tag{26}$$

[‡]This is however true for almost all regularizations.

$$\int \mathcal{D}[g_{\mu\nu}]e^{-S[g]} \quad \rightarrow \quad \sum_{T \in \mathcal{T}} e^{-S[T]} \tag{27}$$

where $[g_{\mu\nu}]$ denotes equivalence classes of metrics, i.e. the volume of the diffeomorphism group is divided out, and it is understood that the functional integral is only over connected four-manifolds of topology S^4.

The rhs of eq. (27) shares a number of properties with its two-dimensional analogue: The number of triangulations of topology S^4 and a given volume N_4 is exponentially bounded[§]. This implies that the partition function for a given κ_2 has a critical κ_4 where the infinite volume limit can be taken. It does not imply that we necessarily have an interesting continuum theory. One has to scan κ_2 to look for divergent correlation lengths, at least if we want to use the standard intuition from statistical mechanics and the theory of critical phenomena. Unfortunately the only tool which seems useful at the moment in the analysis of the scaling limit is numerical simulations.

1.2. Observables and Numerical Simulations

Due to the invariance under diffeomorphism and the fact that we in quantum gravity have to integrate over all Riemannian manifolds, the observables which are most readily available are averages of invariant local operators like the curvature $R(x)$. A non-local observable is the integrated curvature-curvature correlation. In a continuum formulation it would be

$$\chi(\kappa_2) \equiv \langle \int d^4\xi_1 \, d^4\xi_2 \, \sqrt{g(\xi_1)}R(\xi_1) \, \sqrt{g(\xi_2)}R(\xi_2) \, \rangle - \langle \int d^4\xi \sqrt{g(\xi)}R(\xi) \rangle^2. \tag{28}$$

In a lattice regularized theory one would expect that away from the critical points short range fluctuations will prevail, while approaching the critical point long range fluctuation might be important and would result in an increase of $\chi(\kappa_2)$. The observable $\chi(\kappa_2)$ is the second derivative of the free energy $F = -\ln Z$ with respect to the inverse gravitational coupling constant. In the case where the volume N_4 is kept fixed we have

$$\chi(\kappa_2, N_4) \sim \langle N_2^2 \rangle_{N_4} - \langle N_2 \rangle_{N_4}^2 \;\; = \;\; -\frac{d^2 \ln Z(\kappa_2, N_4)}{d\kappa_2^2} \tag{29}$$

From the above discussion it follows that we have to search for values of κ_2 where $\chi(\kappa_2, N_4)/N_4$ diverges in the infinite volume limit $N_4 \to \infty$.

Another observable is the Hausdorff dimension. One can define the Hausdorff dimension in a number of ways, which are not necessarily equivalent. Here we will simply measure the average volume $V(r)$ contained within a radius r from a given point:

$$V(r) \sim r^{d_h}. \tag{30}$$

By the use of Regge calculus it is straightforward to convert these continuum formulas to our piecewise flat manifolds. Let me now summarize some of the numerical results obtained with the last $1\frac{1}{2}$ year by various groups[28-29].

[§]This is only verified by numerical methods. It would be interesting to have an analytic proof of this.

The measurements of $\chi(\kappa_2)$ show a peak at $\kappa_2 \approx 1.1$. We will denote this value κ_2^c. At this point there indeed seems to be some kind of phase transition in geometry. For $\kappa_2 < \kappa_2^c$ the radius of the typical universes generated by computer is very small, and the radius grows very slowly with the volume N_4, indicating a large fractal dimension of space-time. For $\kappa_2 > \kappa_2^c$ the situation is the opposite: The radius of the typical universe is large, in fact much larger than one would naively expect, and there is a strong dependence on the volume N_4. A closer analysis of the distribution of distances reveals that the geometrical structures generated are of low fractal dimension (between 1 and 2). If we approach k_2^c it seems that we have some chance of generate structures with a fractal dimension which is not that different from 4, but it is difficult to measure the Hausdorff dimension precisely. At a first glance κ_2^c seems to be the obvious candidate for a critical point where we can take the continuum limit. There is one obstacle, however. It turns out that the scalar curvature does not scale at the transition point. Since the continuum curvature has the dimensions of $(\text{mass})^2$, standard scaling arguments will tell us that that the bare curvature has to scale to zero at the critical point. This seems not to be the case in the numerical simulations and the manifolds which are generated can therefore not in any simple way be identified with smooth continuum manifolds! This problem may be similar in nature to the non-scaling of the string tension mentioned above and if this conjecture can be substantiated it might give some hints to possible cures.

5. Acknowledgements

It is a pleasure to thank Bergfinnur Durhuus for explaining to me his ideas about $c > 1$ Ising spin models. They formed the foundation for section 1.2.

6. References

1. F. David, *Nucl. Phys.* **B257** (1985) 45.
2. J. Ambjørn, B. Durhuus and J. Fröhlich, *Nucl. Phys.* **B257** (1985) 433.
3. F. David, *Nucl. Phys.* **B257** (1985) 543.
4. V. A. Kazakov, I. K. Kostov and A. . Migdal, *Phys. Lett.* **157B** (1985) 295.
5. J. Ambjørn, B. Durhuus, J. Fröhlich and P. Orland, *Nucl. Phys.* **B270** (1986) 457.
6. E. Brezin and S. Hikami, *Phys. Lett.* **B295** (1992) 209.
7. S. Hikami, *Phys. Lett.* **B305** (1993) 327.
8. J. Ambjørn, S. Jain and G. Thorleifsson, *Phys. Lett.* **B307** (1993) 34.
9. J. Ambjørn, B. Durhuus, T. Jonsson and G. Thorleifsson, *Nucl. Phys.* **B398** (1993) 568.
10. M. Wexler, *Matrix models on large graphs.*, PUPT-1398 and PUPT-1384.
11. C.A. baillie and J.D. Johnston, *Phys. Lett.* **B286** (1992) 44.

12. S. M Catterall, J.B. Kogut and R.L. Renken, *Phys. Lett.* **B292** (1992) 277.

13. J. Ambjørn, B. Durhuus and J. Fröhlich, *Nucl. Phys.* **B275** [FS17] (1986) 161.

14. J. Ambjørn, B. Durhuus, T. Jonsson, *Phys. Lett.* **B244** (1990) 403

15. B. Durhuus, unpublished.

16. S.R. Das, A. Dhar, A.M. Sengupta and S.R. Wadia, *Mod. Phys. Lett.* **A5** (1990) 1041.

17. G.P. Korchemsky, *Phys. Lett.* **B296** (1992) 323; *Matrix models perturbed by higher curvature terms*, UPRF-92-334.

18. L. Alvarez-Gaume, J.L.F. Barbon, and C. Crnkovic, *Nucl. Phys.* **B394** (1993) 383.

19. J. Ambjørn, B. Durhuus, T. Jonsson, *Phys. Rev. Lett.* **58** (1987) 2619.

20. J. Ambjørn, B. Durhuus, J. Fröhlich and T. Jonsson, *Nucl. Phys.* **B290** [**FS20**] (1987) 480.

21. J. Ambjørn and B. Durhuus, *Phys. Lett.* **188B** (1987) 253.

22. J. Ambjørn, B. Durhuus, J. Fröhlich and T. Jonsson, *J. Stat. Phys.* **55** (1989) 29.

23. S.M. Catterall, *Phys. Lett.* **B243** (1990) 121.

24. J. Ambjørn, A. Irbäck, J. Jurkiewicz and B. Petersson, *Nucl. Phys.* **B393** (1993) 571.

25. J. Ambjørn, A. Irbäck, J. Jurkiewicz, B. Petersson and S. Varsted, *Phys. Lett.* **B275** (1992) 295.

26. J. Ambjørn, J. Jurkiewicz and C.F. Kristjansen, *Nucl. Phys.* **B393** (1993) 601.

27. J. Ambjørn, Z. Burda, J. Jurkiewicz and C.F. Kristjansen, *Acta Physica Polonica* **B23** (1992) 991.

28. J. Ambjørn and J. Jurkiewicz, *Phys. Lett.* **B278** (1992) 42.

29. M.E. Agishtein and A.A. Migdal, *Nucl. Phys.* **B385** (1992) 395; *Mod. Phys. Lett* **A7** (1992) 1039.

INTRODUCTION TO DIFFERENTIAL W-GEOMETRY

Jean-Loup GERVAIS

Laboratoire de Physique Théorique de l'École Normale Supérieure[1],
24 rue Lhomond, 75231 Paris CEDEX 05, France.

Abstract

Ideas recently put forward by Y. Matsuo and the author are summarized on the example of the simplest (W_3) generalization of two-dimensional gravity. These notes are based on lectures given at the workshop " Strings, Conformal Models and Topological Field Theories", Cargese 12-21 May 1993; and the meeting "String 93", Berkeley, 24-29 May 1993.

---

[1]Unité Propre du Centre National de la Recherche Scientifique, associée à l'École Normale Supérieure et à l'Université de Paris-Sud.

1   Introduction

In many ways, W-algebras are natural generalizations of the Virasoro algebra. They were first introduced as consistent operator-algebras involving operators of spins higher than two[1]. Moreover, the Virasoro algebra is intrinsically related with the Liouville theory which is the Toda theory associated with the Lie algebra A_1, and this relationship extends to W-algebras which are in correspondence with the family of conformal Toda systems associated with arbitrary simple Lie algebras[2]. Another point is that the deep connection between Virasoro algebra and KdV hierarchy has a natural extention[3] to W-algebras and higher KdV (KP) hierachies[4]. On the other hand, W symmetries exhibit strikingly novel features. First, they are basically non-linear algebras. Since the transformation laws of primary fields contain higher derivatives, product of primaries are not primaries at the classical level. Naive tensor-products of commuting representations do not form representations. A related novel feature is that W-algebras generalize the diffeomorphisms of the circle by including derivatives of degree higher than one. Going beyond linear approximation (tangent space) is a highly non-trivial step. Taking higher order derivatives changes the shape of the world-sheet in the target-space, thus W-geometry should be related to the extrinsic geometry of the embedding. Finally, Virasoro algebras are notoriously related to Riemann surfaces. The W-generalization of the latter notion is a fascinating problem.

In a series of recent papers, we have developed a geometrical framework for the class of Conformal field Theory CFT mentioned above, where these features emerge from the standard Riemannian geometry of particular manifolds which we called W surfaces[5]–[7]. These references cover quite a lot of material, and the present lecture will go in the opposite way. Leaving the description of the general scheme to refs[5]–[7], we shall, instead, illustrate the ideas by two explicit examples: the 2D gravity case (section 2), and its simplest generalization to the W_3 gravity (section 3). In both cases, we discuss the two current approaches, namely, the conformal one where W-gravity is identified with a conformal Toda (or Liouville theory for 2D gravity) theory, and the light-cone approach.

Before starting, however, let us recall how our general scheme goes. One basic point is to make use of the fact that one deals with integrable models, but geometrically there are two aspects. The first uses extrinsic geometry. In ref.[6], we showed that the A_n-W–geometry corresponds to the embedding of holomorphic two-dimensional surfaces in the complex projective space CP^n. These (W) surfaces are specified by embedding equations of the form $X^A = f^A(z)$, $\bar{X}^{\bar{A}} = \bar{f}^{\bar{A}}(\bar{z})$, where z, and $\bar{z}$ are the two surface-parameters. The fact that they are functions of a single variable is equivalent to the Toda field-equations, so that this describes W gravity in the conformal gauge. These functions have a natural extension to CP^n using the higher variables z^k, $\bar{z}^k$ of the Toda hierarchy of integrable flows, and this provides a local parametrization of CP^n. The original variables z and $\bar{z}$ are identified with z^1, $\bar{z}^1$, respectively. For the embedding functions the extension is such that they become functions of half of the variables noted $f^A([z]) = f^A(z^0, \cdots z^n)$, and $\bar{f}^{\bar{A}}([\bar{z}]) = \bar{f}^{\bar{A}}(\bar{z}^0, \cdots \bar{z}^n)$ such that

$$\frac{\partial f^A([z])}{\partial z^k} = \frac{\partial^k f^A([z])}{(\partial z)^k}, \qquad \frac{\partial \bar{f}^{\bar{A}}([\bar{z}])}{\partial \bar{z}^k} = \frac{\partial^k \bar{f}^{\bar{A}}([\bar{z}])}{(\partial \bar{z})^k} \tag{1.1}$$

One main virtue of the coordinates z^k, $\bar{z}^k$ is that, due to the last equations, higher derivaties in z and $\bar{z}$ are changed to first-order ones, and this is how our geometrical scheme gets rid of the troublesome higher derivatives of the usual approachs.

The second aspect[5][6] only makes use of intrinsic geometries, but introduces a family of associated surfaces in the standard Grassmannians associated with CP^n. This is useful to discuss global aspects by using the fact developed in ref.[6], that W surfaces are instantons of non-linear σ models. We shall not dwell into this aspect in these lectures.

So far this is only for the conformal gauge. Concerning the light-cone approach, our recent insight[7] is that, for any Kähler manifold, there exist changes of coordinates such that metric tensors of the light-cone type come out. This allows us to relate conformal and light-cone descriptions by diffeomorphisms.

Finally, let us stress that we shall remain entirely at the level of classical field theory, thus describing only the $C \to \infty$ limit of the problem. The quantum approach to Toda theories is making steady progress[9]–[15], [18]–[20] but its connection with the present geometrical scheme remains as a fascinating problem for the future.

2   Two-dimensional Gravity

In this part we discuss the case of two-dimensional gravity in some details, using methods that will be later on generalized to W gravity. With general world-sheet parameters ξ, the Weyl anomaly takes the form

$$S = \frac{1}{\gamma} \int d_2\xi (\frac{1}{2}\mathcal{R}\frac{1}{\Delta}\mathcal{R} + \mu\sqrt{-\mathcal{G}}) \tag{2.1}$$

In this expression, Δ is the Laplacian with the 2D-gravity metric $\mathcal{G}_{\alpha,\beta}$, γ is the coupling constant, $\mathcal{R}$ the scalar curvature, and μ the cosmological constant.

2.1   Conformal gauge

First, choose conformal coordinates z and $\bar{z}$, so that the arc length takes the form[2]

$$ds^2 = 2e^{2\Phi}dzd\bar{z}, \tag{2.2}$$

the Weyl anomaly becomes the Liouville action

$$S = \frac{1}{\gamma} \int d_2z (-\frac{1}{2}\partial\Phi\bar{\partial}\Phi + \mu e^{2\Phi}). \tag{2.3}$$

In general ∂ and $\bar{\partial}$ denote $\partial/\partial z$, and $\partial/\partial\bar{z}$ respectively. At the level of the present discussion we may always set $\mu = 1$ by a shift of ϕ, and we shall do so from now on. The general solution of the Liouville equation is given by the holomorphic decomposition

$$e^{-\Phi} = \sum_{j=1}^{2} \chi_j(z)\bar{\chi}_j(\bar{z}). \tag{2.4}$$

The functions $\chi_j(z)$, and $\bar{\chi}_j(\bar{z})$ are pairs of solutions of the differential equations

$$-\frac{d^2\chi_j(z)}{(dz)^2} + T(z)\chi_j(z) = 0, \quad -\frac{d^2\bar{\chi}_j(\bar{z})}{(d\bar{z})^2} + \bar{T}(\bar{z})\bar{\chi}_j(\bar{z}) = 0, \tag{2.5}$$

---

[2]This may not be possible globally, but we shall only consider the local aspects

where T and $\bar{T}$ are the two non-vanishing components of the stress-energy tensor. They are normalized so that

$$\chi_1(z)\frac{d\chi_2(z)}{dz} - \chi_2(z)\frac{d\chi_1(z)}{dz} = \bar{\chi}_1(\bar{z})\frac{d\bar{\chi}_2(\bar{z})}{d\bar{z}} - \chi_2(\bar{z})\frac{d\chi_1(\bar{z})}{d\bar{z}} = 1. \qquad (2.6)$$

At this point, and in the following, we need a simple mathematical lemma concerning differential equations, which we state once for all next.

Lemma 1 *Automatic differential equation.*
Consider N functions f^A of a single variable x, whose Wronskian does not vanish. They satisfy a differential equation of the form

$$f^{(N)A}(x) = \sum_{\ell=1}^{N} \kappa_\ell(x) f^{(N-\ell)A}(x). \qquad (2.7)$$

Proof: Recall the definition of the Wronskian:

$$\mathrm{Wr}(f^1,\cdots,f^N) \equiv \begin{vmatrix} f^1 & \cdots & f^N \\ f^{(1)\,1} & \cdots & f^{(1)\,N} \\ \vdots & \cdots & \vdots \\ f^{(N-1)\,1} & \cdots & f^{(N-1)\,N} \end{vmatrix}. \qquad (2.8)$$

In this formula, and hereafter, upper indices in between parentheses denote the order of derivatives. Clearly, any of the functions f^A may be trivially written as a linear combination of the type $\sum_B c_B^A f^B$. It is then clear that

$$\begin{vmatrix} f^1 & \cdots & f^N & f \\ f^{(1)\,1} & \cdots & f^{(1)\,N} & f^{(1)} \\ \vdots & \cdots & \vdots & \\ f^{(N)\,1} & \cdots & f^{(N)\,N} & f^{(N)} \end{vmatrix} = 0.$$

Expanding this determinant with respect to the last column immediately gives Eq.2.7. **Q.E.D.** Moreover, one immediately sees that

$$\kappa_1 = \frac{d\ln[\mathrm{Wr}(f^1,\cdots,f^n)]}{dx} \qquad (2.9)$$

Returning to our main line, we see that, in the differential equations Eq.2.5 the first order term vanishes, so that the Wronskians $\mathrm{Wr}(\chi)$ and $\mathrm{Wr}(\bar{\chi})$

are constant, and may be chosen equal to one (see Eq.2.6). For our geometrical description this is not appropriate, however. The basic reason is as follows. Under conformal transformations $\delta z = \epsilon(z)$, the χ_j fields[3] transform as primary fields of weight $-1/2$ (that is such that $\chi_j(z)\,(dz)^{-1/2}$ is invariant). This is consistent, since it follows that the Wronskian $\mathrm{Wr}(\chi)$ transforms with weight zero, so that condition Eq.2.6 is conformally invariant. In our geometrical description, these functions will become geometrical objects whose form should not change under conformal transformation, so that they should transform with weight zero. We may change the conformal weights by using a projective description. For this we define

$$f^A = \sqrt{w(z)}\,\chi_{A+1}, \quad \bar{f}^A = \sqrt{\bar{w}(\bar{z})}\,\bar{\chi}_{A+1}, \quad A = 0, 1, \tag{2.10}$$

so that

$$\mathrm{Wr}(f^0, f^1) = w(z), \quad \mathrm{Wr}(\bar{f}^0, \bar{f}^1) = \bar{w}(z) \tag{2.11}$$

where w and $\bar{w}$ are arbitary functions of a single variable. Now the f's satisfy a more general differential equation of the type Eq.2.7, where $\kappa_1 \neq 0$. Substituting into Eq.2.4, we derive the arc length

$$ds^2 = dz\,d\bar{z}\,\frac{\mathrm{Wr}(f^0, f^1)\mathrm{Wr}(\bar{f}^0, \bar{f}^1)}{(f^0\bar{f}^0 + f^1\bar{f}^1)^2}, \tag{2.12}$$

and, re-arranging the terms, we may write

$$ds^2 = dz\,d\bar{z}\left(\frac{-(f^{(1)0}\bar{f}^{(1)0} + f^{(1)1}\bar{f}^{(1)1})(f^0\bar{f}^0 + f^1\bar{f}^1)}{(f^0\bar{f}^0 + f^1\bar{f}^1)^2} + \right.$$
$$\left.\frac{(f^{(1)0}\bar{f}^0 + f^{(1)1}\bar{f}^1)(f^0\bar{f}^{(1)0} + f^1\bar{f}^{(1)1})}{(f^0\bar{f}^0 + f^1\bar{f}^1)^2}\right). \tag{2.13}$$

As is well-known $\mathrm{Wr}(\rho f^0, \rho f^1) = \rho^2\mathrm{Wr}(f^0, f^1)$, for arbitrary function ρ, so that[4] Eq.2.12 is invariant under $f^A \to \rho f^A$, and $\bar{f}^A \to \bar{\rho}\bar{f}^A$. We arrive at a projective structure, following our general scheme[6] where A_n-W geometries are described in CP^n (we will have $n = 1$ in the present case). Let us recall some useful definitions at this point.

---

[3]In the conformal gauge χ and $\bar{\chi}$ are on the same footing. When we discuss properties of chiral component we some times talk about the χ fields as an example. Clearly, the $\bar{\chi}$ fields are analogous.

[4]this is obviously true by construction.

Definition 1 *Kähler manifolds*

A Kähler manifold of real dimension $2n$ is complex manifold with a special class of coordinates X^A, $X^{\bar A}$, $1 \le A, \bar A \le n$, such that the only components of the metric are $G_{A\bar B} = G_{\bar B A}$, and

$$G_{A\bar B} = \partial_A \partial_{\bar B} K, \qquad (2.14)$$

where K is the Kähler potential.

In general we denote the differential operators $\partial / \partial X^A$, and $\partial / \partial X^{\bar B}$ by ∂_A, and $\partial_{\bar B}$.

Definition 2 *Complex projective space CP^n*

The complex projective space CP^n is defined from the trivial complex space C^{n+1} with coordinates X^A, $\bar X^{\bar A}$, $A, \bar A = 0, \cdots n$, by identifying any two points $X, \bar X$, and $Y, \bar Y$ related by the scale transformation

$$X^A = Y^A \rho, \quad \text{and} \quad \bar X^{\bar A} = \bar Y^{\bar A} \bar\rho. \qquad (2.15)$$

This is a simplest non-trivial example of a Kähler manifold. The metric on this space is given by the Fubini-Study equation

$$G_{A\bar A} = \left(\delta_{A\bar A} \left(\sum_{B=0}^{n} X^B \bar X^B \right) - X^{\bar A} \bar X^A \right) \Big/ \left(\sum_{B=0}^{n} X^B \bar X^B \right)^2, \qquad (2.16)$$

whose Kähler potential is given by

$$K = \ln \sum_{A=0}^{n} X^A \bar X^{\bar A}. \qquad (2.17)$$

Eq.2.16 is invariant under the scale transformation Eq.2.15. Using this freedom, it is customary to parametrize CP^n, by letting one component (say, X^0) equal 1.

Closing the parenthesis, we return to Eq.2.13. Consider the space CP^1. The equations

$$X^A = e^\zeta f^A(z), \quad \bar X^A = e^{\bar\zeta} \bar f^A(\bar z), \quad A = 0, 1, \qquad (2.18)$$

define a holomorphic change of coordinates in C^2 from X^A to ζ, z. It is easy to see that Eq.2.13 is equivalent to

$$G_{z\bar z} \equiv e^{2\Phi} = f^{(1)A} \bar f^{(1)\bar B} G_{A\bar B} \Big|_{X^A = f^A, \bar X^A = \bar f^A} \qquad (2.19)$$

By construction, Eq.2.13 is invariant under the rescaling $f \to \rho(z)f$, $\bar{f} \to \bar{\rho}(\bar{z})\bar{f}$. Thus we may let $\zeta = \bar{\zeta} = 0$, and the geometrical meaning of Eqs.2.18, Eq.2.19 is that $z\,\bar{z}$ are parameters of CP^1, such that the Liouville exponential $G_{z,\bar{z}}$ is equivalent to the metric tensor of Fubini-Study. Thus any solution of Liouville equation defines a local holomorphic parametrization of CP^1. Note that, since the change of coordinate is holomorphic, the Kähler condition is preserved. Indeed, one has

$$G_{z\bar{z}} = \partial\bar{\partial}\mathcal{K}, \quad \mathcal{K} = \ln(f^0\bar{f}^0 + f^1\bar{f}^1). \tag{2.20}$$

2.2 Light-cone gauge

2.2.1 The Weyl anomaly

First, rederive the ideas of ref.[8] is the present context. At the level of Eq.2.1, one changes coordinates from $z\,\bar{z}$ to $u\,\bar{u}$, such that

$$e^{2\Phi}dzd\bar{z} = dud\bar{u} + hdu^2 \tag{2.21}$$

The Weyl anomaly Eq.2.1 becomes

$$S = \frac{1}{2\gamma}\int d_2 u \left((\bar{D}^2 h)\frac{1}{D\bar{D} - \bar{D}h\bar{D}}(\bar{D}^2 h) + 2\right), \tag{2.22}$$

where D and $\bar{D}$ stand for $\partial/\partial u$, and $\partial/\partial\bar{u}$, respectively. Here the second term is a constant and can be forgotten. Using the obvious relation

$$\frac{1}{D\bar{D} - \bar{D}h\bar{D}} = \frac{1}{D - h\bar{D}}\frac{1}{\bar{D}}$$

we may write after partially integrating

$$S = \frac{1}{\pi}\int d_2 u\, h\, \mathcal{T}, \quad \mathcal{T} \equiv \frac{\pi}{2\gamma}\bar{D}^2 \frac{1}{D - h\bar{D}}\bar{D}h \tag{2.23}$$

Starting from this expression, we compute

$$\left(D - h\bar{D} - 2(\bar{D}h)\right)\mathcal{T} = \frac{\pi}{2\gamma}\bar{D}^2 \frac{D - h\bar{D}}{D - h\bar{D}}\bar{D}h$$

so that we get

$$\left(D - h\bar{D} - 2(\bar{D}h)\right)\mathcal{T} = \frac{\pi}{2\gamma}\bar{D}^3 h \tag{2.24}$$

Thus the deformed conservation law are satisfied. Minimizing S, under the form Eq.2.23, with respect to h, one finds

$$0 = \frac{\pi}{2\gamma}\bar{D}^2 \frac{1}{D - h\bar{D}}\bar{D}h = \mathcal{T}$$

so that Polyakov's anomaly equation comes out:

$$\bar{D}^3 h = 0 \tag{2.25}$$

2.2.2 Liouville dynamics

Next, following ref.[7], we connect the results just recalled with the conformal-gauge Liouville appproach recalled above. Since the Liouville equations also follow from minimizing S, they are equivalent to Polyakov's anomaly equations. This may be seen as follows. Making use of Eq.2.21, one sees that

$$h = -\frac{\partial \bar{u}}{\partial z}, \quad e^{2\Phi} = \frac{\partial \bar{u}}{\partial \bar{z}}, \tag{2.26}$$

The last equation is immediately solved, thanks to Eq.2.20, and we find that the change of variables form conformal to light-cone explicitly reads

$$u = z, \quad \bar{u} = \partial K = \frac{f^{(1)0}\bar{f}^0 + f^{(1)1}\bar{f}^1}{f^0\bar{f}^0 + f^1\bar{f}^1} \tag{2.27}$$

Computing h in terms of f^1, and $\bar{f}^1$ one finds.

$$h = -\frac{f^{(2)0}\bar{f}^0 + f^{(2)1}\bar{f}^1}{f^0\bar{f}^0 + f^1\bar{f}^1} + \left[\frac{f^{(1)0}\bar{f}^0 + f^{(1)1}\bar{f}^1}{f^0\bar{f}^0 + f^1\bar{f}^1}\right]^2$$

The first term is retransformed by using the lemma which gives $f^{(2)A} = \kappa_1 f^{(1)A} + \kappa_0 f^A$. One gets

$$h = -\kappa_0 - \kappa_1\bar{u} + \bar{u}^2 \tag{2.28}$$

For fixed $z = u$, it is quadratic in $\bar{u}$, so that Eq.2.25 is indeed verified. In terms of the $sl(2, R)$ light-cone currents we have $J^+ = 1$, $J^0 = \kappa_1/2$, and $J^- = -\kappa_0$.

3 W_3 gravity

Unfortunately no formula similar to the Weyl-anomaly term Eq.2.1 is known at present. Our starting point will be the generalization of the Liouville equation known as the A_2 Toda equation, where the W_3 algebra is realized by Poisson bracket, and which should thus describe W_3 gravity in the conformal gauge.

3.1 Conformal gauge

3.1.1 W surface

In the A_2 Toda theory there are two fields ϕ_1, ϕ_2. The field equations read

$$\partial\bar\partial\phi_1 = -e^{2\phi_1-\phi_2}, \qquad \partial\bar\partial\phi_2 = -e^{2\phi_2-\phi_1} \tag{3.1}$$

with solution

$$e^{-\phi_1} = \sum_{j=1}^{3} \chi_j(z)\bar\chi_j(\bar z), \qquad e^{-\phi_2} = \sum_{i<j=1}^{3} \chi_{ij}(z)\bar\chi_{ij}(\bar z) \tag{3.2}$$

There are now three pairs of arbitary functions of a single variable χ_j, $\bar\chi_j$, the functions χ_{ij}, and $\bar\chi_{ij}$ are defined by

$$\chi_{ij} = \mathrm{Wr}(\chi_i, \chi_j), \quad \bar\chi_{ij} = \mathrm{Wr}(\bar\chi_i, \bar\chi_j). \tag{3.3}$$

The Toda equations 3.1 are satisfied if

$$\mathrm{Wr}(\chi_1, \chi_2, \chi_2) = 1, \quad \mathrm{Wr}(\bar\chi_1, \bar\chi_2, \bar\chi_2) = 1, \tag{3.4}$$

One sees that the generalization of the Liouville case is very direct. The lemma, combined with the last Wronskian condition shows that the χ_j satisfy a differential equation of the form

$$\chi_j^{(3)} = T\chi_j^{(1)} + W\chi_j \tag{3.5}$$

The potentials are the generators of the W transformations: T is the stress-energy tensor, and W the W_3 charge. Again we relax the condition 3.4 by letting

$$f^A = (w(z))^{1/3}\,\chi_{A+1}, \quad \bar f^A = (\bar w(\bar z))^{1/3}\,\bar\chi_{A+1}, \quad A = 0, 1, 2 \tag{3.6}$$

so that

$$\mathrm{Wr}(f^0, f^1, f^2) = w(z), \quad \mathrm{Wr}(\bar{f}^0, \bar{f}^1, \bar{f}^2) = \bar{w}(z) \tag{3.7}$$

$$
\begin{aligned}
e^{-\phi_1} &= \frac{\sum_{A=0}^{2} f^A(z)\bar{f}^A(\bar{z})}{\left[\mathrm{Wr}(f^0, f^1, f^2)\mathrm{Wr}(\bar{f}^0, \bar{f}^1, \bar{f}^2)\right]^{1/3}} \\
e^{-\phi_2} &= \frac{\sum_{A<B=0}^{2} \mathrm{Wr}(f^A, f^B)\mathrm{Wr}(\bar{f}^A, \bar{f}^B)}{\left[\mathrm{Wr}(f^0, f^1, f^2)\mathrm{Wr}(\bar{f}^0, \bar{f}^1, \bar{f}^2)\right]^{2/3}}
\end{aligned}
\tag{3.8}
$$

Our task is now to give a meaning for these formulae in Riemannian geometry. There are two new features as compared with the Liouville case. First we now have three pairs of functions, but we still only have two variables z and $\bar{z}$. Second, derivatives of second degree appear so that one does not have a direct tangent-space interpretation. These two problems are solved simultaneously according to our general scheme. First, we introduce CP^2 as target space where the equations

$$X^A = f^A(z), \quad \bar{X}^A = \bar{f}^A(\bar{z}), \quad A = 0, 1, 2, \tag{3.9}$$

define a holomorphic (W) surface. Second, we will also consider the extrinsic geometry of these surfaces where higher derivatives naturally appear when one defines normals. It is convenient to define

$$\tilde{G}_{a\bar{b}} = \sum_{A,\bar{B}} f^{(a)A}(z)\bar{f}^{(\bar{b})\bar{B}}(\bar{z})G_{A\bar{B}}\Big|_{X^A=f^A, \bar{X}^{\bar{A}}=\bar{f}^{\bar{A}}} = \tilde{G}_{\bar{b}a} \tag{3.10}$$

Consider first the intrinsic geometry. It is straightforward to see that, according to Eq.3.8, the arc-length on the surface is given by

$$ds^2 \equiv \tilde{G}_{1\bar{1}}dzd\bar{z} = e^{2\phi_1 - \phi_2}dzd\bar{z} \tag{3.11}$$

so that this particular exponential of the Toda fields gives the induced metric on the W surface. The complete geometrical interpretation may be obtained following the usual method of differential geometry, namely by considering normals to the surface. There are two of them, noted e_1, and $e_{\bar{1}}$. Their components may be written compactly by the generalized Frenet-Serret formulae

$$e_1^A = \frac{1}{\sqrt{\tilde{G}_{1\bar{1}}\Delta_2}}\begin{vmatrix} \tilde{G}_{1\bar{1}} & \tilde{G}_{2\bar{1}} \\ f^{(1)A} & f^{(2)A} \end{vmatrix}, \quad e_1^{\bar{A}} = 0, \tag{3.12}$$

408

$$e_{\bar{1}}^A = 0, \quad e_{\bar{1}}^{\bar{A}} = \frac{1}{\sqrt{\widetilde{G}_{1\bar{1}}\Delta_2}} \begin{vmatrix} \widetilde{G}_{1\bar{1}} & \widetilde{G}_{1\bar{2}} \\ \bar{f}^{(1)\bar{A}} & \bar{f}^{(2)\bar{A}} \end{vmatrix}. \tag{3.13}$$

where

$$\Delta_2 \equiv \begin{vmatrix} \widetilde{G}_{1\bar{1}} & \widetilde{G}_{1\bar{2}} \\ \widetilde{G}_{2\bar{1}} & \widetilde{G}_{2\bar{2}} \end{vmatrix} \tag{3.14}$$

It is easy to see that, if we denote by $(\mathbf{X}, \mathbf{Y})$ the inner product $G_{A\bar{B}}(X^A Y^{\bar{B}} + Y^A X^{\bar{B}})$ in CP^2, we have $(e_1, f^{(1)}) = 0$, $(e_1, \bar{f}^{(1)}) = 0$, $(\bar{e}_1, f^{(1)}) = 0$, $(\bar{e}_1, \bar{f}^{(1)}) = 0$, $(e_1, e_1) = (e_{\bar{1}}, e_{\bar{1}}) = 0$, $(e_1, e_{\bar{1}}) = 1$. So that the last equations define orthonormalized normals. With the present notations, the second fundamental form Q is obtained by writing the equivalent expressions

$$f^{(2)A} - f^{(1)A}\frac{\widetilde{G}_{2\bar{1}}}{\widetilde{G}_{1\bar{1}}} = e_1^A Q, \quad \bar{f}^{(2)\bar{A}} - \bar{f}^{(1)\bar{A}}\frac{\widetilde{G}_{2\bar{1}}}{\widetilde{G}_{1\bar{1}}} = e_{\bar{1}}^{\bar{A}} Q \tag{3.15}$$

with $Q = \sqrt{\Delta_2/\widetilde{G}_{1\bar{1}}}$. By explicit computation[5] one finds that $\Delta_2 = \exp(3\phi_1)$, so that

$$Q = e^{(\phi_1+\phi_2)/2} = \sqrt{\frac{\mathrm{Wr}(f^0, f^1, f^2)\mathrm{Wr}(\bar{f}^0, \bar{f}^1, \bar{f}^2)}{\sum_{A=0}^2 f^A(z)\bar{f}^A(\bar{z}) \sum_{C<B=0}^2 \mathrm{Wr}(f^C, f^B)\mathrm{Wr}(\bar{f}^C, \bar{f}^B)}}. \tag{3.16}$$

3.1.2 Associated holomorphic change of coordinates

Although the geometrical meaning of the Toda fields is now clear, the description is not covariant under change of parametrization of CP^2, since it involves second derivatives. This problem is solved by using the fact that the Toda field equations are a subsystem of the so-called Toda hierarchy of integrable flows that makes use of 6 real parameters. These parameters provide a parametrization of C^3 where the formulae we summarized may be re-expressed in terms of first order derivatives only, so that their covariance properties become clear. This part is best handled by means of the fermionic method and Hirota equations[4] [5] [6]. We shall nevertheless remain at the same elementary simple-minded level for illustration. In terms of f and $\bar{f}$, the Toda hierarchy equations become extremely simple (see Eq.1.1). The

---

[5]this is done more powerfully using the fermionic method as explained in ref.[6]

variables are[6] z^0, $z^1 \equiv z$, z^2 denoted collectively as $[z]$; and $\bar{z}^0$, $\bar{z}^1 \equiv \bar{z}$, $\bar{z}^2$ denoted collectively as $[\bar{z}]$. The functions $f^A(z)$ and $\bar{f}^{\bar{A}}(\bar{z})$ are extended as holomorphic or antiholomorphic functions $f^A([z])$ and $\bar{f}^{\bar{A}}([\bar{z}])$ solutions of the equations

$$\partial_0 f^A([z]) = f^A([z]), \quad \partial_2 f^A([z]) = (\partial_1)^2 f^A([z])$$
$$\bar{\partial}_0 \bar{f}^{\bar{A}}([\bar{z}]) = \bar{f}^{\bar{A}}([\bar{z}]), \quad \bar{\partial}_2 \bar{f}^{\bar{A}}([\bar{z}]) = (\bar{\partial}_1)^2 \bar{f}^{\bar{A}}([z]) \qquad (3.17)$$

with the initial conditions

$$f^A([z]) = f^A(z), \quad \bar{f}^{\bar{A}}(\bar{z}) = \bar{f}^{\bar{A}}([\bar{z}]), \quad \text{for } z^0 = z^2 = \bar{z}^0 = \bar{z}^2 = 0 \qquad (3.18)$$

Of course the differential equations in z^0 and $\bar{z}^0$ are trivial and give simple factors $\exp(z^0)$ or $\exp(\bar{z}^0)$. Since we work homogeneously we may forget these variables completely and keep z^0 and $\bar{z}^0$ equal to zero. The point of the differential equations in z^2 is to turn second derivatives in z into a first order derivatives. Thus the matrix of inner product of derivatives Eq.3.10 is extended to CP^2 as

$$G([z], [\bar{z}])_{a\bar{b}} = \sum_{A,\bar{B}} \partial_a f^A([z]) \partial_{\bar{b}} \bar{f}^{\bar{B}}([\bar{z}]) G_{A\bar{B}}\Big|_{X^A = f^A([z]), \bar{X}^{\bar{A}} = \bar{f}^{\bar{A}}([\bar{z}])} \qquad (3.19)$$

where now **only first order derivatives appear**. As a matter of fact, this last equation is but the Riemannian metric obtained by making the holomorphic change of variables $X^A = f^A(z, z^2)$, $\bar{X}^{\bar{A}} = \bar{f}^{\bar{A}}(\bar{z}, \bar{z}^2)$. Thus again, we associate, with any solution of the Toda equation, a conformal parametrization of the complex projective space. Note again that, since the change is conformal, the Kähler condition is preserved. The metric $G([z], [\bar{z}])_{a\bar{b}}$ may be written as

$$G([z], [\bar{z}])_{a\bar{b}} = \partial_a \partial_{\bar{b}} \ln \left[\sum_A f^A([z]) \bar{f}^{\bar{A}}([\bar{z}]) \right] \qquad (3.20)$$

Our next topic is the W transformations of f. We shall use the parametrization such that $X^0 = \bar{X}^0 = 1$, so that $f^0([z]) = \bar{f}^0([\bar{z}]) = 1$. Due to these

---

[6]Upper indices are always covariant indices. When we want to indicate powers such as z square, we put parentheses, and write $(z)^2$.

conditions, and according to the lemma, the functions $f^A(z)$ are solution of a differential equation of the form

$$f^{(3)A}(z) = \lambda(z)f^{(2)A}(z) + \mu(z)f^{(1)A}(z) \tag{3.21}$$

there is no term without derivative since $f^0 = 1$ is a solution. The most general infinitesimal transformation that leaves the condition $f^0 = 1$ invariant is

$$\delta f^A(z) = (\alpha(z)\partial + \beta(z)\partial^2)f^A(z), \tag{3.22}$$

Higher derivatives may be re-expressed in terms of the first two by means of Eq.3.21. Consider, now the extension to CP^2. To simplify the notation we will let $t = z^2$, so that, according to Eqs.3.17,

$$\partial_t f^A(z,t) = (\partial)^2 f^A(z,t). \tag{3.23}$$

This is used in practice as follows. Consider

$$\delta_n f(z) = \epsilon(z)(\partial)^n f(z)$$

Clearly one has

$$f^A(z,t) = e^{t(\partial)^2} f^A(z,0) \tag{3.24}$$

One writes

$$\delta_n f(z,t) = e^{t(\partial)^2}\delta f(z) = \epsilon(z + 2t\partial)e^{t(\partial)^2}\partial^n f(z) = \epsilon(z + 2t\partial)(\partial)^n f(z,t)$$

where $\epsilon(z + 2t\partial)$ is defined by

$$\epsilon(z + 2t\partial) = \sum_r \frac{\epsilon_r}{r!}(z + 2t\partial)^r, \quad \epsilon_r = (\partial)^r \epsilon(z)|_{z=0} \tag{3.25}$$

In general, one sees that an arbitrary W transformation leads to

$$\delta f(z,t) = \alpha(z + 2t\partial)\partial f(z,t) + \beta(z + 2t\partial)\partial_t f(z,t) \tag{3.26}$$

Eq.3.21 is straightforwardly extended to CP^2 where it takes the form

$$\partial_z \partial_t f^A([z]) = L([z])\partial_t f^A([z]) + M([z])\partial_z f^A([z]). \tag{3.27}$$

Using this relation, and Eq.3.23, we may re-express all higher derivetives in Eq.3.26 in terms of the first-order ones, obtaining a transformation of form

$$\delta f(z,t) = A(z,t)\partial f(z,t) + B(z,t)\partial_t f(z,t) \tag{3.28}$$

which correspond to the change of parametrization

$$\delta z = A(z,t), \quad \delta t = B(z,t), \text{ with } \delta f(z,t) = f(z + \delta z, t + \delta t) - f(z,t). \tag{3.29}$$

Thus W transformations are extended as particular diffeomorphisms of CP^2.

3.2 Light-cone description

Finally, we connect the above discussion with the light-cone approach to W_3 gravity, which so far was developed independently of the conformal gauge[16],[17]. The basic point, displayed in ref.[7], is that, for any Kähler manifold, there exist parametrizations such that the metric takes a block-form identical to the light-cone metric Eq.2.21 introduced by Polyakov for two-dimensional gravity.

3.2.1 Light-cone parametrization for Kähler manifolds

Consider an arbitrary Kähler manifold, introduced by Definition 1. There exists another class of prefered coordinates denoted U^A, and $U^{\bar{A}}$, such that the new metric tensor denoted by the letter H takes a form which is the generalization of the right-hand side of Eq.2.21. The change of coordinates is of the form

$$U^{\bar{A}} = U^{\bar{A}}(X^1, \cdots, X^n; X^{\bar{1}} \cdots, X^{\bar{n}}), \quad U^A = X^A. \tag{3.30}$$

The functions $U^{\bar{A}}(X; \bar{X})$ will be determined next, in such a way that the light-cone metric takes the form

$$H_{\bar{A}\bar{B}} = 0, \quad H_{AB} = 2h_{AB}$$

$$H_{A\bar{B}} = H_{\bar{B}A} = \delta_{A,\bar{B}}, \tag{3.31}$$

where h_{AB} will be related to the Kähler potential. Before going on, let us remark that the determinant of H is equal to -1, and that its inverse is given by

$$H^{AB} = 0, \quad H^{\bar{A}\bar{B}} = -2h_{AB}$$

$$H^{A\bar{B}} = H^{\bar{B}A} = \delta_{A,\bar{B}}. \tag{3.32}$$

On the contrary, no general form may be given for the determinant of G or its inverse. This is a very nice point of the light-cone parametrization[7]. Going back to our main line, one finds, by standard computations, that conditions Eqs.3.31 are fulfilled if one has

$$G_{A\bar{B}} = H_{A\bar{C}}\partial_{\bar{B}}U^{\bar{C}} \tag{3.33}$$

---

[7]inverting the metric tensor is often a pain in the neck.

$$2h_{AB} = -H_{B\bar{C}}\partial_A U^{\bar{C}} - H_{A\bar{C}}\partial_B U^{\bar{C}} \tag{3.34}$$

These equations are easily solved by using the Kähler potential, obtaining

$$U^{\bar{A}} = H^{\bar{A}C}\partial_C K, \tag{3.35}$$

$$h_{AB} = -\partial_A \partial_B K. \tag{3.36}$$

Thus we reach the important conclusion that, for any Kählerian manifold there is a choice of coordinates such that the metric tensor takes the block-form

$$H = \begin{pmatrix} 2h & 1 \\ 1 & 0 \end{pmatrix} \tag{3.37}$$

which is the same as the one introduced by Polyakov to describe 2D gravity in the light-cone gauge.

3.2.2 Light-cone coordinates for W_3 gravity

We finally apply this change of coordinates to CP^2, calling $\bar{u}^1$, and $\bar{u}^2$ the two light-cone coordinates. Combining Eq.3.21 with Eq.3.35, 3.36, one sees that one has

$$\bar{u}^\ell = \frac{\sum_{A=0}^2 \partial_\ell f^A([z])\bar{f}^A([\bar{z}])}{\sum_{A=0}^2 f^A([z])\bar{f}^A([\bar{z}])} \tag{3.38}$$

$$h_{\ell m} = -\frac{\sum_{A=0}^2 \partial_\ell \partial_m f^A([z])\bar{f}^A([\bar{z}])}{\sum_{A=0}^2 f^A([z])\bar{f}^A([\bar{z}])} + \bar{u}^\ell \bar{u}^m$$

Making use of Eq.3.27, this gives

$$h_{11} = -\bar{u}^2 + (\bar{u}^1)^2, \quad h_{12} = -L\bar{u}^2 - M\bar{u}^1 + \bar{u}^1\bar{u}^2$$

$$h_{22} = -\bar{u}^2(\partial L + L^2 + M) - \bar{u}^1(\partial M + LM) + (\bar{u}^2)^2 \tag{3.39}$$

so that h is quadratic in the $\bar{u}$'s. On the other hand, Polyakov's equation for light-cone W_3 gravity takes the form $(\bar{D})^3 h = 0$, and $(\bar{D})^5 A = 0$. This contradiction may be resolved as follows: **The W surface where light-cone physics takes place is not at $z^{(2)} = \bar{z}^{(2)} = 0$, contrary to conformal physics.** Indeed, let us assume that the light-cone W surface has equations

$$z^{(2)} = 0, \quad \bar{u}^2 = \alpha(z)(\bar{u}^1)^2 + \beta(z)\bar{u}^1 + \gamma(z) \tag{3.40}$$

Then, on this surface, $h_{\ell m}$ is a polynomial in $\bar{U}^1$ of degree $\ell + m$. Thus it seems natural to let

$$h_{11} = h, \quad h_{22} = A \tag{3.41}$$

obtaining

$$(\bar{D})^3 h = 0, \quad (\bar{D})^5 A = 0, \quad \text{with } \bar{D} = \bar{D}_1 \tag{3.42}$$

Clearly, α, β, and γ are arbitrary. They may be precisely changed by an $sl(3)$ transformation. Thus, one is led to conjecture that **the geometrical origin of the $sl(3)$ invariance is a change of the light-cone W surface so that Eq.3.39 keeps its form.**

4   Outlook

These lecture notes left out many aspects, some of which have been developed elsewhere. In particular, quantization has led to many interesting progress. For the Liouville case, it was realized[12, 9] that there exists a quantum-group structure where the mathematical parameter of the quantum-group deformation coincides with Planck's constant. In this theory, the non-commutativity which is inherent in the quantum group – since the co-product in non-symmetric – is brought about by the actual quantization of the system. This appearance of quantum group seems to be very natural geometrically, since one deals with a gravity theory, where the space-time metric is quantized, which seems tantamount to quantizing the 2D space-time itself. Thus an object like a quantum plane should appear, and quantum groups are natural transformations of such "quantum manifolds". For recent progress in this direction, see refs. [18, 19, 20]. The quantum group picture was extended to the quantum A_n Toda theories in ref.[11]. The classical geometry setting summarized in these notes indicates that at the quantum level one should be dealing with quantum complex projective spaces. This remains to be clarified.

Classically, what we describe is also incomplete. In the light-cone connection, the ansatz Eq.3.41 remains somewhat ad hoc, and its Kac-Moody structure should be clarified. Since we use the Toda equation throughout, we stayed on the mass shell. Thus we only get Polyakov equations (Eqs.2.25, 3.42), and not the full anomaly equations that characterize light-cone W gravities. Besides the particular change of gauge we have discussed, the most basic problem is to formulate W gravity without gauge fixing, and to couple it to some appropriate kind of matter from first principle.

References

[1] A.B.Zamolodchikov, *Theor. Math. Phys.* **65** (1985) 1205; V.A. Fateev and A.B.Zamolodchikov, *Nucl. Phys.* **B280 [FS18]** (1987) 644.

[2] A. Bilal and J.-L. Gervais, *Phys. Lett.* **B206** (1988) 412; *Nucl. Phys.* **B314** (1989) 646; *Nucl. Phys.* **B318** (1989) 579.

[3] J.-L. Gervais, *Phys. Lett.* **B160** (1985) 277; K. Yamagishi, *Phys. Lett* **B205** (1988) 406; I. Bakas, *Phys. Lett.* **B213** (1988) 313.

[4] M. Sato, *RIMS Kokyuroku* **439** (1981) 30; E. Date, M. Jimbo, M.Kashiwara and T.Miwa, "Transformation Groups for Soliton Equations" in *Proc. of RIMS Symposium on Non-linear Integrable Systems-Classical Theory and Quantum Theory* (Kyoto, Japan, May 1981) (World Scientific Publication Co. Singapore, 1983); G.Segal and G.Wilson, *Pub. Math. IHES* **61** (1985) 5.

[5] J.-L. Gervais, Y. Matsuo: *Phys. Lett.* **274B** (1992) 309;

[6] J.-L. Gervais, Y. Matsuo: Comm. in Math. Phys. **152** (1993) 317.

[7] J.-L. Gervais, Y. Matsuo: *Phys. Lett.* **312B** (1993) 285.

[8] A. Polyakov: Mod. Phys. Lett. A **2, 11** (1987) 893.

[9] J.-L. Gervais, Comm. in Math. Phys. **130**, 257, (1990) .

[10] J.-L. Gervais, Phys. Lett. **B243**, 85, (1990) .

[11] E. Cremmer, J.-L. Gervais, Comm. in Math. Phys. **134**, 619, (1990).

[12] O. Babelon, Phys. Lett. **B215**, (1988) 523.

[13] J.-L. Gervais, Comm. in Math. Phys. **138** 138, 301, (1991) .

[14] J.-L. Gervais, "Quantum group derivation of 2D gravity-matter coupling" Invited talk at the Stony Brook meeting *String and Symmetry 1991*, LPTENS preprint 91/22, Nucl. Phys. **B391**, 287, (1993).

[15] J.-L. Gervais, J. Schnittger, Phys. Lett. **B315** , 258, (1993).

[16] Y. Matsuo: Phys. Lett. **B227** (1989) 209.

[17] see, for instance, the related reviews in "String and symmetries 91", *in* Proceeding of the Stony Brook Meeting, World Scientific ed.

[18] E.Cremmer, J.-L. Gervais. J.-F. Roussel, "The genus-zero bootstrap of chiral vertex operators in Liouville theory", preprint LPTENS-93/29, Nucl. Phys. to be published.

[19] E.Cremmer, J.-L. Gervais. J.-F. Roussel, "The quantum group structure of 2D gravity and minimal models II: The genus zero chiral bootstrap" preprint LPTENS-93/02, hep-th 9302035, Comm. Math. Phys. to be published.

[20] J.-L. Gervais, J. Schnittger, "The many faces of the quantum Liouville exponentials" preprint LPTENS-93/30, hep-th 9308134, Nucl. Phys. to be published.

CHIRAL RINGS AND INTEGRABLE SYSTEMS
FOR MODELS OF TOPOLOGICAL GRAVITY

W. LERCHE

CERN, CH 1211 Geneva 23, Switzerland

ABSTRACT

We review the superconformal properties of matter coupled to $2d$ gravity, and W-extensions thereof. We show in particular how the $N = 2$ structure provides a direct link between certain matter-gravity systems and matrix models. We also show that much, probably all, of this can be generalized to W-gravity, and this leads to an infinite class of new exactly solvable systems. These systems are governed by certain integrable hierarchies, which are generalizations of the usual KdV hierarchy and whose algebraic structure is given in terms of quantum cohomology rings of grassmannians.

1. Introduction

There has been some recent progress in understanding theories describing conformal matter coupled to $2d$ gravity. Matter-plus-gravity systems are interesting to study because they are, for certain choices of matter theories, supposed to be exactly solvable. More precisely, they are supposed to be equivalent to matrix models[1], which are exactly solvable by themselves as a consequence of an underlying structure of KdV-type integrable hierarchies. To deduce this equivalence directly from Liouville theory is quite difficult, largely due to technical complications. We will review how the $N = 2$ superconformal structure helps to provide a manifest and direct relationship of (certain of) such models to matrix models, by making use of a connection to topological Landau-Ginzburg theory,[2] and to integrable hierarchies.

Since theories of W-gravity coupled to matter appear by now[3,4] to be on a footing similar to ordinary gravity, one might suspect, by analogy, that there should exist a corresponding infinite sequence of new types of matrix models that describe these theories, governed by certain integrable hierarchies. In a recent paper[5], we made some progress in understanding these new integrable systems in terms of chiral rings, and we will briefly review the main ingredients of this construction as well.

2. $N\!=\!2$ superconformal symmetry of the matter-gravity system

We like to briefly recapitulate ordinary gravity coupled to conformal $2d$ matter. For simplicity, we will consider mainly minimal matter models, but this is not really important. These matter models, denoted by $M_{p,q}$, where $p, q = 1, 2, \ldots$ are coprime

Talk given at Strings '93, Berkeley and at XXVII Internationales Symposium über Elementarteilchentheorie, Wendisch-Rietz.

integers, have central charges $c_M = 13 - 6(t + \frac{1}{t})$, where $t \equiv q/p$. We thus consider tensor products

$$M_{p,q}^{\text{matter}} \otimes M_{p,-q}^{\text{Liouville}} \otimes \{b, c\} \,, \tag{2.1}$$

where $M_{p,-q}^{\text{Liouville}}$ denotes a Liouville theory with appropriate central charge, and $\{b, c\}$ denotes the fermionic ghost system with spins $\{2, -1\}$.

In $BRST$ quantization the physical states of the combined matter-gravity system are given by the non-trivial cohomology classes of a $BRST$ operator,

$$\mathcal{Q}_{BRST} = \oint \frac{dz}{2\pi i} \mathcal{J}_{BRST} \,, \qquad \mathcal{J}_{BRST} = c\,[T_M + T_L + \tfrac{1}{2}T_{gh}] \,, \tag{2.2}$$

which is nilpotent for $c_L + c_M = 26$. The most prominent physical states correspond to the tachyon operators[6]:

$$T_{r,s} = c\,V_{r;-s}^L\,V_{r;s}^M \,, \tag{2.3}$$

where $V_{r;s}$ denotes exponential vertex operators in the usual notation. By convention, the tachyons have bc-ghost number equal to one. In addition, there exist[7] extra physical states whose number and precise structure depends on the specific value of t. For unitary minimal models, where $t = (p + 1)/p$, there exist infinitely many of such extra states for each matter primary, whereas for generic t, there exists basically only one extra sort of states besides the tachyons: these are the operators with vanishing ghost charge. They form what is called[8] the ground ring, which we will denote by $\mathcal{R}^{\text{gr}}$. It is precisely because these operators have zero ghost charge (and zero dimension like all physical operators), that the set of ground ring operators closes into itself under operator products. In fact, even though this ring is in general infinite, it is finitely generated, ie., it has two generators by whose action all other ring elements can be generated[8]:

$$
\begin{aligned}
x &= \left[bc - \frac{t}{\sqrt{2t}}(\partial\phi_L - i\partial\phi_M)\right] V_{1,2}^L V_{1,2}^M \\
\gamma^0 &= \left[bc - \frac{1}{\sqrt{2t}}(\partial\phi_L + i\partial\phi_M)\right] V_{2,1}^L V_{2,1}^M \,.
\end{aligned} \tag{2.4}
$$

(Here, $\phi_{L,M}$ denotes the Liouville field and the matter free field, respectively). The properties of the ground ring elements remind very much to the typical features of chiral fields in $N = 2$ superconformal theories. The whole point is, of course, that the matter-gravity-ghost system is essentially nothing but a (twisted) $N = 2$ superconformal theory. More precisely, it is known[9,10] that one can improve the $BRST$ current (2.2) by a total derivative piece,

$$G^+ = \mathcal{J}_{BRST} - \partial\left(\sqrt{\tfrac{2}{t}}(c\partial\phi_L) + \tfrac{1}{2}(1 - \tfrac{2}{t})\partial c\right) \,, \tag{2.5}$$

such that G^+ together with

418

$$G^- = b\,, \qquad T = T_L + T_M + T_{gh}\,, \qquad J = cb + \sqrt{\tfrac{2}{t}}\,\partial\phi_L\,, \qquad (2.6)$$

indeed generates the (topologically twisted[11,12]) $N=2$ superconformal algebra,

$$T(z){\cdot}T(w) \sim \frac{2T(w)}{(z-w)^2} + \frac{\partial T(w)}{(z-w)}\,,$$

$$T(z){\cdot}G^{\pm}(w) \sim \frac{\tfrac{1}{2}(3\mp 1)G^{\pm}(w)}{(z-w)^2} + \frac{\partial G^{\pm}(w)}{(z-w)}\,,$$

$$T(z){\cdot}J(w) \sim \frac{\tfrac{1}{3}c^{N=2}}{(z-w)^3} + \frac{J(w)}{(z-w)^2} + \frac{\partial J(w)}{(z-w)}\,,$$

$$J(z){\cdot}J(w) \sim \frac{\tfrac{1}{3}c^{N=2}}{(z-w)^2}\,, \qquad J(z){\cdot}G^{\pm}(w) \sim \pm\frac{G^{\pm}(w)}{(z-w)}\,, \qquad (2.7)$$

$$G^+(z){\cdot}G^-(w) \sim \frac{\tfrac{1}{3}c^{N=2}}{(z-w)^3} + \frac{J(w)}{(z-w)^2} + \frac{T(w)+\partial J(w)}{(z-w)}\,,$$

$$G^{\pm}(z){\cdot}G^{\pm}(w) \sim 0\,,$$

with anomaly

$$c^{N=2} = 3\left(1 - \tfrac{2}{t}\right)\,. \qquad (2.8)$$

Upon untwisting, $T \to T - \tfrac{1}{2}\partial J$, $c^{N=2}$ becomes the central charge of an ordinary $N=2$ algebra. Note that the free-field realization (2.5), (2.6) of the $N=2$ algebra is different from the usual one[13]. This is however irrelevant, and one can show[10] that the above realization can be obtained by hamiltonian reduction[14] from a $SL(2|1)$ WZW model in a way that is analogous and equivalent to the way of deriving the usual free-field realization of the $N=2$ algebra. Alternatively, one can show that the two free field realizations of the $N=2$ algebra can be obtained as two different gauge choices in a topological gauged WZW model[15].

Actually, the construction of the twisted $N=2$ algebra is a priori not unique[16,17]. Indeed, one may replace the Liouville field ϕ_L in (2.5), (2.6) by some appropriate combination of ϕ_L with the matter free field, ϕ_M. However, for describing minimal models coupled to gravity, the above choices for G^+ and J are the unique, correct ones[*]. Namely, if G^+ and J depended on ϕ_M, then the various different vertex operator representatives $V_{r,s}^M$ that describe the same given physical state of the minimal model would have different properties under the $N=2$ algebra, which clearly would not make any sense. However, for non-minimal models, where these vertex operators describe distinct physical states, there are other possible choices. For example, in order to describe black holes in $N=2$ language, one chooses[16] the $N=2$ currents to depend only on the matter field, ϕ_M.

---

[*] Our choice for J implies that it is not holomorphically conserved if the theory is perturbed by the cosmological constant[16], and one might get the impression that this is not desirable. For our application to minimal models, there is however nothing wrong with that.

An immediate question is about the significance of the twisted $N = 2$ superconformal symmetry. For generic t, the mere presence of an $N = 2$ algebra doesn't really seem to provide any important new insights, since the representation theory for arbitrary $c^{N=2}$ is not very restrictive. On the other hand, for integer $t \equiv k + 2$, $k \geq 0$, a lot can be learned: namely then the anomaly (2.8) becomes equal to the anomaly of the twisted $N = 2$ minimal models, A_{k+1}^{top}: $c^{N=2} = \frac{3k}{k+2}$. This is a powerful statement, since minimal models tend to be easily solved entirely by representation theory. (For $t = -1$, which describes $c = 1$ matter coupled to gravity and which can be related to the $2d$ black hole[16], the theory is solvable as well, but we will focus on $t \geq 2$ in the following.)

However, this does not yet imply that the minimal models $M_{1,2+k}$ coupled to gravity are the same as the topological minimal models A_{k+1}^{top}. What we have shown is simply that these theories have the same free field realization with the same central charges. A priori, they don't have even the same spectra. The spectrum of a topological $N = 2$ model is well known[12]: it is given by the chiral ring[18], which is the finite set of primary chiral fields. For A_{k+1}^{top}, this is a nilpotent, polynomial ring generated by one element x:

$$\mathcal{R}^{A_{k+1}^{\text{top}}} = \frac{P(x)}{[x^{k+1} \equiv 0]} = \left\{ 1, x, x^2, \ldots, x^k \right\} . \tag{2.9}$$

One can check that powers of the ground ring generator x in (2.4) are indeed primary and chiral with respect to the $N = 2$ currents (2.5) and (2.6), and that $\mathcal{R}^{A_{k+1}^{\text{top}}}$ is identical to the subring of the ground ring $\mathcal{R}^{\text{gr}}$ that is generated by x. (For $t = k + 2$, it turns out that the corresponding tachyons (2.3) have the same $N = 2$ quantum numbers as the ground ring elements, so that they can be viewed as different representatives of the same set of physical fields.) On the other hand, the full ground ring $\mathcal{R}^{\text{gr}}$ of the matter-gravity system contains infinitely many more operators[19]:

$$\mathcal{R}^{\text{gr}} = \mathcal{R}^{A_{k+1}^{\text{top}}} \otimes \left\{ (\gamma^0)^n, \ n = 0, 1, 2, \ldots \right\} . \tag{2.10}$$

These extra operators simply do not exist in the topological minimal models. The difference between the spectra (2.9) and (2.10) can be accounted for as follows: it turns out that the extra operators are exact with respect to an additional *BRST* like operator, $\widetilde{Q}$:

$$\gamma^0 = -\left\{ \widetilde{Q}, \left(\tfrac{t+1}{t} \partial c + \tfrac{1}{\sqrt{2t}} c \, \partial \phi_L \right) \right\} , \quad \text{where} \quad \widetilde{Q} = \oint \frac{dz}{2\pi i} b \, e^{-\frac{t}{\sqrt{2t}}(\phi_L - i\phi_M)} . \tag{2.11}$$

One can show[10] that $\widetilde{Q}$ is one of the Felder-like screening operators that arise in our particular free field realization of the minimal models. That is, *by definition* the full *BRST* operator of the topologically twisted $N = 2$ minimal models is the sum

420

of $\mathcal{Q}_{BRST}$, $\widetilde{\mathcal{Q}}$ and the other screening operators, such that it maximally truncates the infinite free field spectrum precisely to the finite set of physical operators (2.9).

We thus see that the full $BRST$ operator of the topological minimal models is not the correct one if we wish to describe the minimal models $M_{1,2+k}$ coupled to gravity. The correct operator obtains if we drop $\widetilde{\mathcal{Q}}$ as an extra piece of the full $BRST$ operator, and it can be shown that then indeed the "missing" operators $(\gamma^0)^n$ become physical. This can be actually be better formulated in terms of equivariant cohomology. Roughly speaking, imposing equivariant cohomology means that one restricts the Hilbert space to states $|X\rangle$ that satisfy $b_0|X\rangle = 0$. The operator $(\frac{t+1}{t}\,\partial c + \frac{1}{\sqrt{2t}}\,c\,\partial\phi_L)$ in (2.11) does not obey this condition, and this means that the ground ring generator γ^0 is not the $BRST$ variation of a physical operator – hence, it is physical. It is actually well-known[20,21,22] that in pure topological gravity one has to require equivariant cohomology, in order to obtain a non-empty theory.

The situation can be summarized in Fig.1. What we have discussed so far corresponds to step A.

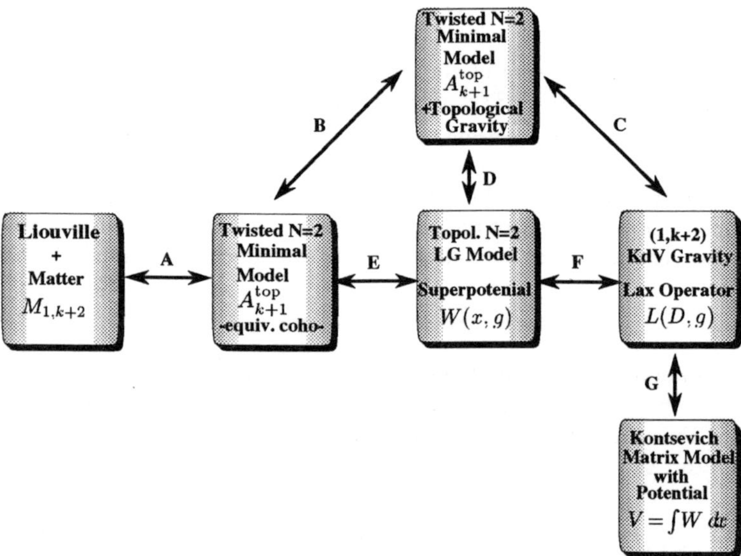

Fig. 1 *Models describing gravity coupled to conformal minimal matter of type $(1, k+2)$.*

It can be shown[23] that the *modified* minimal topological models, which contain the operators $(\gamma^0)^n$ and which are equivalent to the minimal models $M_{1,2+k}$ coupled to gravity, are in fact also equivalent to the *un-modified* models A^{top}_{k+1} coupled to *topological* gravity[20]. A priori, the building blocks of A^{top}_{k+1} and of the same models coupled[24] to topological gravity appear to be quite different. There is, however, the remarkable fact that the total $BRST$ operator of the topological matter plus topological gravity system obeys[23]

$$Q^{tot} \; \equiv \; Q^{N=2} + Q_{BRST} \; = \; U^{-1} Q^{N=2} U \; , \tag{2.12}$$

where $U = e^{\oint c [G_M + G_L + \frac{1}{2} G_{gh}]}$ is a homotopy operator. The upshot is that the cohomologies of A^{top}_{k+1} coupled to topological gravity and of the modified minimal topological models are isomorphic, so that at least at the level of Fock spaces the theories are equivalent. This refers to step B in Fig.1. Step C is an expression of the fact[24] that the recursion relations of $[A^{top}_{k+1} \otimes$ topological gravity$]$ are the same as those of the corresponding matrix models[25].

3. Relation to dispersionless KdV hierarchy

The relationship between the matter-gravity system and matrix models can be exhibited also via a more direct route. One can make use of the fact that the Landau-Ginzburg realization[2] (step E in Fig.1) of the topological matter models can be directly related[26] to the KdV integrable structure of the matrix models (step F). More precisely, one considers the dependence of correlators on perturbation parameters t, defined by* $\langle \ldots e^{\int d^2 z \, d^2 \theta \sum_{i=0}^{k} t_{i+2} x^{k-i}} \rangle$. It was shown[26] that the effect of such perturbations can be described in terms of a Landau-Ginzburg superpotential of the form

$$W(x, g(t)) \; = \; \frac{1}{k+2} x^{k+2} - \sum_{i=0}^{k} g_{i+2}(t) \, x^{k-i} \; . \tag{3.1}$$

The coupling constants $g_{i+2}(t_j)$ are certain, in general non-trivial functions of the perturbation parameters. Since the correlation functions can easily be computed[26] once one knows $W(x, g(t))$, solving the theory just amounts to determining these functions. This can be done by making use of the fact that t are very particular, namely flat[27] coordinates on the LG deformation space. Requiring that the appropriate Gauß-Manin connection vanishes, leads to the following differential equations for $g(t)$:

$$-\partial_{t_{i+2}} W(x, g(t)) \; = \; \partial_x \, \Omega_{k+1-i}(x, g(t)) \; , \tag{3.2}$$

---

* Note that such perturbations lead away from the conformal point. We restrict here to the "small phase space", ie., to perturbations generated by the primary fields.

422

$(i = 0, \ldots, k)$, which involve the hamiltonians

$$\Omega_i(x, g(t)) \;=\; \tfrac{1}{i} \left((k+2) W \right)_+^{\frac{i}{k+2}} (x, g(t)) \; . \tag{3.3}$$

(Here, the subscript "+" denotes, as usual, the truncation to non-negative powers of x.) The crucial observation[26,28] is that under the substitutions $x \to D$ and $W(x, g) \to L(D, g)$, these equations are nothing but the dispersionless limit of the KdV flow equations

$$\partial_{t_{i+2}} L(D, g(t)) \;=\; \left[(L^{\frac{k+1-i}{k+2}})_+, L \right](D, g(t)) \; . \tag{3.4}$$

if one imposes as boundary condition the string equation: $D\, g_{k+2} = 1$. These equations describe[1,29] (step G in the figure) the dynamics of the matrix models of type $(1, k+2)$. This immediately proves the equality of correlation functions (as functions of the small phase space variables t) of the primary fields with the corresponding correlators of the matrix models (step F). These arguments, which involve only $N = 2$ Landau-Ginzburg theory, can also be extended to the gravitational descendants and to some of the recursion relations they obey[30,23] (in the small phase space). In fact, (2.12) implies that all states of the matter-gravity system have $BRST$ representatives in the matter sector alone. That is, the gravitational descendants can be expressed in terms of the LG field x (in equivariant cohomology) as well:[23]

$$\sigma_n(\phi_i)(x, t) \;=\; \partial_x\, \Omega_{i+(k+2)n+1}(x, g(t)) \tag{3.5}$$

(where $\sigma_n(\phi_i)(x, 0) \equiv (\gamma^0)^n x^i$ in previous notation). This is precisely in the spirit of what was said above: the ingredients of the coupling of A_{k+1}^{top} to topological gravity are already build in the structure of the models A_{k+1}^{top} themselves[30]. All what is necessary to describe the coupling of these models to topological gravity is to modify their cohomological definition. The fact that topological gravity coupled to A_{k+1}^{top} can be described purely in terms of LG theory corresponds to step D in Fig.1.

Of particular interest is the perturbation of these models by the "cosmological constant" term. In our language[10,16], it is the perturbation by the top element of $\mathcal{R}^{A_{k+1}^{\text{top}}}$,

$$S_{\text{cosm}} = \mu \int d^2 z \, e^{\sqrt{\frac{2}{t}}\phi_L} \;\equiv\; t_2 \int d^2 z \, d^2\theta \, x^k \; . \tag{3.6}$$

It is known[31] that this perturbation is integrable and leads to the massive quantum $N = 2$ sine-Gordon model; although not invariant under the full (twisted) $N = 2$

superconformal symmetry, it is supersymmetric, and the corrected supercharge $\oint G^+$ still serves as a $BRST$ operator. Under the perturbation, the $N = 2\ U(1)$ current J ceases to be holomorphically conserved, which is a typical feature of perturbations leading away from the conformal point. The effective superpotential is given by a Chebyshev polynomial:

$$W(x, t_2 = \mu) \;=\; \tfrac{2}{k+2}\mu^{\frac{k+2}{2}}\,T_{k+2}(\tfrac{1}{2}\mu^{-\frac{1}{2}}x) \;=\; \tfrac{1}{k+2}\,x^{k+2} - \mu\,x^k + O(\mu^2)\;. \tag{3.7}$$

At $\mu = 1$, the deformed chiral ring becomes identical[32] to the fusion ring of the $SU(2)_k$ WZW model, which it is also the same as the operator product algebra of the $SU(2)_k/SU(2)_k$ topological field theory. This observation then allows to finally make contact to the formulation of matter-plus-gravity models in terms of topological G/G theories[33]: it is known[33,16] that at the level of Fock space cohomology, the (suitably defined) $SU(2)/SU(2)$ model is indeed equivalent to the matter-gravity system. We thus have the relation:

$$\left[M_{1,2+k} \otimes \text{Liouville gravity}\right]\Big|_{\mu=1} \;\cong\; \left[\frac{SU(2)_k}{SU(2)_k}\bigg|_{\substack{\text{modified}\\ \text{cohomology}}}\right] \tag{3.8}$$

4. Extension to W-gravity

One obvious motivation for investigating generalizations is the wish to step beyond the $c_M = 1$ barrier of ordinary gravity. This can be achieved by considering matter theories with extended symmetries, coupled to the corresponding extended geometry. The prime candidates for such models are those related to W-algebras. (One may also consider supersymmetric versions: it turns out[10] that $N = 1$ matter coupled to $N = 1$ supergravity yields $N = 3$ superconformal models, etc.).

For a given theory of W_n-gravity coupled to matter, there is a barrier at $c_M = n-1$, below of which there is a finite number of (dressed) primary fields and below of which the theory should be solvable. In analogy to ordinary gravity, one would expect that such theories should be solvable also at the accumulation points $c_M = n-1$ (where there exists an extra $SU(n)$ current algebra symmetry). At these points, such models are presumably related to black hole type of objects in spacetimes with signature $(n-1, n-1)$ and are characterized by topological field theories based on non-compact versions of $\mathrm{CP}^{\mathrm{top}}_{n-1,k}$.

The physical models in question are tensor products

$$W_n^{\mathrm{matter}} \otimes W_n^{\mathrm{Liouville}} \otimes_{j=1}^{n-1} \{b_j, c_j\}\;, \tag{4.1}$$

which might be called "non-critical W-strings"[34]. Above, W_n^{matter} denotes conformal field theories that have a W-algebra as their chiral algebra, which can be for example W_n minimal models $M_{p,q}^{(n)}$ with central charges $c_M = (n-1)[1 - n(n+$

424

$1) \frac{(t-1)^2}{t}]$, $t \equiv q/p$. Furthermore, $W_n^{\text{Liouville}}$ denotes a $(n-1)$-component generalization of Liouville theory (Toda theory), and $\{b_j, c_j\}$ denotes the Hilbert space of a ghost system with spins $j+1$ and $-j$, respectively. As it turns out, the structure of these theories for arbitrary n is very similar to $n = 2$, which corresponds to ordinary gravity. However, only for $n = 3$ the generalization* of the $BRST$ current is explicitly known[3]:

$$
\begin{aligned}
\mathcal{J}_{BRST} = {} & c_2\left[\tfrac{1}{b_L}W_L + \tfrac{i}{b_M}W_M\right] + c_1\left[T_L + T_M + \tfrac{1}{2}T_{gh}^1 + T_{gh}^2\right] \\
& + \left[T_L - T_M\right]b_1 c_2(\partial c_2) + \mu(\partial b_1)c_2(\partial^2 c_2) + \nu b_1 c_2(\partial^3 c_2) ,
\end{aligned}
\tag{4.2}
$$

where $b_{L,M}^2 \equiv \frac{16}{5c_{L,M}+22}$ and $\mu = \frac{3}{5}\nu = \frac{1}{10b_L{}^2}(1 - 17b_L{}^2)$. In this equation, $T_{L,M}$ and $W_{L,M}$ denote the usual stress tensors and W-generators of the Liouville and matter sectors, and T_{gh}^i are the stress tensors of the ghosts.

Using this $BRST$ current, one can study the spectrum of physical operators of W_3 matter coupled to W_3 gravity, and one finds[10,35,4,36] that the analogs of ground ring elements and tachyons are states with ghost numbers equal to $0, 1, 2, 3$ (the first number corresponds to ground ring elements, and the last one to tachyons). The explicit expressions are however too complicated to be written down here.

The interesting point is that there appears an $N = 2$ superconformal symmetry for all n. For example, for W_3 gravity one finds that

$$
\begin{aligned}
G^+ = {} & \mathcal{J}_{BRST} + \partial\Big[-c_1 J + 2i\sqrt{\tfrac{t}{3}}b_1 c_1 c_2 J + i\tfrac{(1+t)}{2}\sqrt{\tfrac{3}{t}}b_1 c_1(\partial c_2) \\
& - i\tfrac{(3+2t)}{\sqrt{3t}}b_1(\partial c_1)c_2 - \tfrac{(7t^2-10t-15)}{4t}b_1(\partial^2 c_2)c_2 + i\tfrac{(t-9)}{\sqrt{3t}}b_2(\partial c_2)c_2 \\
& - i\tfrac{(3+4t)}{\sqrt{3t}}(\partial b_1)c_1 c_2 - \tfrac{3(4t^2-2t-3)}{2t}(\partial b_1)(\partial c_2)c_2 + \tfrac{(t-3)}{t}(\partial c_1) \\
& + i\tfrac{1}{2\sqrt{3t}}c_2[2tJ^2 - 3(t-5)T_L - 3(t-1)T_M - 6(1+t)\partial J] \\
& + i\tfrac{(1+t)}{2}\sqrt{\tfrac{3}{t}}(\partial c_2)J - i\tfrac{(t^2-4t-1)}{2t}\sqrt{\tfrac{3}{t}}(\partial^2 c_2) + tb_1(\partial c_2)c_2 J \Big] ,
\end{aligned}
\tag{4.3}
$$

together with

$$
\begin{aligned}
G^- &= b_1 , \qquad\qquad T = T_L + T_M + T_{gh} , \\
J &= c_1 b_1 + c_2 b_2 + \tfrac{3}{\sqrt{t}}(\lambda_1 \cdot \partial\phi_L) + \tfrac{i}{2}\sqrt{\tfrac{3}{t}}(t-1)\partial[b_1 c_2]
\end{aligned}
\tag{4.4}
$$

gives a non-standard free field realization of the topological algebra (2.7) with

$$
c^{N=2} = 6\left(1 - \tfrac{3}{t}\right) .
\tag{4.5}
$$

Since we are dealing here with theories with an extended symmetry, coupled to an extended "W-geometry", it is perhaps not too surprising to find that these

---

$\star$ The existence of $BRST$ currents for arbitrary n can be inferred from indirect arguments[10,3].

topological algebras actually extend to topologically twisted $N=2$ W-algebras. For $t = n + k$, which corresponds to W_n-minimal matter models $M^{(n)}_{1,n+k}$, the anomaly indeed becomes equal to the central charges of the minimal $N=2$ W_n models at level k: $c^{N=2} = 3\frac{(n-1)k}{n+k}$. These models are just the well-known Kazama-Suzuki models[37] based on cosets $\frac{SU(n)_k}{U(n-1)}$, which are known to have an $N=2$ W_n chiral algebra[38]. The models that arise here are of course the topologically twisted versions, which we will denote by $\mathrm{CP}^{\mathrm{top}}_{n-1,k}$; $n = 2$ corresponds to ordinary gravity coupled to matter: $\mathrm{CP}^{\mathrm{top}}_{1,k} \equiv A^{\mathrm{top}}_{k+1}$.

The chiral rings of these topological minimal W_n matter models are well understood[18,39], and are described further below. They are generated by primary chiral fields x_i, $i = 1, \ldots, (n - 1)$ (with $U(1)$ charges equal to $i/(n + k)$), and have elements

$$\mathcal{R}^{\mathrm{CP}^{\mathrm{top}}_{n-1,k}} = \left\{ \prod_{i=1}^{n-1} (x_i)^{n_i} , \sum n_i \leq k \right\} . \tag{4.6}$$

The full ground rings of the minimal models $M^{(n)}_{1,n+k}$ coupled to W_n-gravity contain in addition generators γ^0_i, $i = 1, \ldots, (n - 1)$ (with $U(1)$ charges equal to i) and are the "W-gravitationally extended" chiral rings of the Kazama-Suzuki models:

$$\mathcal{R}^{\mathrm{gr}} = \mathcal{R}^{\mathrm{CP}^{\mathrm{top}}_{n-1,k}} \otimes \left\{ \prod_{i=1}^{n-1} (\gamma^0_i)^{n_i} , n_i = 0, 1, 2, \ldots \right\} . \tag{4.7}$$

These rings have an obvious interpretation in terms of topological minimal W_n-matter $\mathrm{CP}^{\mathrm{top}}_{n-1,k}$ coupled to topological W_n-gravity[40]. Like for ordinary gravity, the ground ring generators x_i can be interpreted as the fields of topological LG models, with superpotentials given in refs.[18,41] and in eq. (5.20) below. It would be very interesting to investigate as to what extent also the W-gravitational descendants (γ^0_i) can be expressed in terms of LG fields. Ideally, the whole topological W_n-matter-gravity system can be described in terms of Landau-Ginzburg theory.

Although this has not yet been thoroughly investigated for general n, our considerations seem so far to indicate that the structure for general n is indeed very much parallel to the one of $n = 2$. Accordingly, one would have for W_n-matter models of type $(1, k+n)$ coupled to W_n-gravity a scheme that is analogous to Fig.1. It would be exciting to verify the remaining links in the figure for W-gravity. In particular, by analogy to step F in Fig.1 one would expect the existence of an infinite sequence of new integrable systems, whose Lax operators are given in terms of the Kazama-Suzuki superpotentials, and in analogy to step G one would expect the existence of an infinite class of new matrix models. While this latter assertion is more difficult to prove, we were so far indeed successful[5] to get an idea about the structure of the new integrable systems, and this is what we like to briefly outline next.

5. Quantum rings and integrable systems for W-gravity

The issue is to find a multi-variable generalization[5] of the dispersionless KdV hierarchy, which describes the models $M_{1,n+k}^{(n)}$ coupled to W_n-gravity, as well as the models $\mathrm{CP}_{n-1,k}^{\mathrm{top}}$ coupled to topological W_n-gravity. Our strategy is inspired by a general relationship between topological LG theory, chiral rings and Drinfeld-Sokolov types of integrable systems. At the heart of our construction is the generalization of the above-mentioned relationship between LG superpotential and dispersionless Lax operator to many variables x_i.

We like first to reformulate the ordinary, dispersionless[28] KdV hierarchy[*] (pertaining to the LG models $\mathrm{CP}_{1,k}^{\mathrm{top}} \equiv A_{k+1}^{\mathrm{top}}$) in matrix language, because it is this form of the hierarchy that is most suitable for our generalization. One starts with the linear Drinfeld-Sokolov system[42]

$$\left[D\mathbb{1} - \mathcal{L}_1 \right] \cdot \Psi = 0 , \tag{5.1}$$

where the "Lax operator" $\mathcal{L}_1$ is given by the $(k+2) \times (k+2)$ dimensional matrix

$$\mathcal{L}_1(g) = \Lambda_1^{(z)} + Q_1(g) , \tag{5.2}$$

where $\Lambda_1^{(z)}$ has the familiar form

$$\Lambda_1^{(z)} = \begin{pmatrix} 0 & 1 & 0 & \ldots & 0 \\ 0 & 0 & 1 & \ldots & 0 \\ \vdots & \vdots & \vdots & \ddots & \vdots \\ 0 & 0 & 0 & \ldots & 1 \\ z & 0 & 0 & \ldots & 0 \end{pmatrix} , \tag{5.3}$$

with z representing the spectral parameter. In (5.2), Q_1 is usually taken to be a lower triangular matrix that is determined only up to gauge transformations belonging to the nilpotent subgroup N^-. Upon recursively solving for the components of Ψ in favor to the first component Ψ_0, the system (5.1) is equivalent to the gauge invariant, scalar spectral equation

$$L(D,g) \Psi_0 = \tfrac{1}{k+2} z \Psi_0 . \tag{5.4}$$

In the dispersionless limit, where $D \to x$ and $L(D,g) \to W(x,g)$ (cf., (3.1)), this is precisely the characteristic equation of the Lax operator $\mathcal{L}_1$, which therefore must satisfy

$$W(\mathcal{L}_1(g),g) = \tfrac{1}{k+2} z \mathbb{1} . \tag{5.5}$$

---

[*] With "KdV hierarchy" we will always mean the $(k+1)$th generalized KdV hierarchy.

This "superpotential spectral equation" can be taken as the definition of $\mathcal{L}_1$ in terms of the Landau-Ginzburg superpotential W, and (non-uniquely) determines $Q_1(g)$. The gauge freedom can be fixed by going to any particular gauge. The choice that is most appropriate for us is however not given by taking Q_1, as usual, to be a lower triangular matrix, but by taking Q_1 to belong[43] to the Heisenberg subalgebra generated by $\Lambda_1^{(z)}$ (Q_1 is then lower triangular only up to $\mathcal{O}(1/z)$). That is, we have an infinite expansion

$$\mathcal{L}_1(g) \ = \ \Lambda_1^{(z)} + \sum_{l=1}^{\infty} q_l(g)(\Lambda_1^{(z)})^{-l} \ , \tag{5.6}$$

whose coefficients can be computed from (5.5) in a recursive way.

The KdV flow equations (3.2) that determine the LG couplings $g(t)$ take the form

$$\partial_{t_{i+2}} \Omega_{k+1-j}(g(t)) \ = \ \partial_{t_{j+2}} \Omega_{k+1-i}(g(t)) \ , \tag{5.7}$$

and involve the following, matrix-valued hamiltonians:

$$\begin{aligned}
\Omega_i(\mathcal{L}_1(g),g) &= \tfrac{1}{i}(\Lambda_1^{(z)}(\mathcal{L}_1(g),\mathcal{L}_1^{-1}(g),g))_+^i \ , \\
\text{with } \left[\Omega_i\,,\,\Omega_j\right] &\equiv 0 \ , \qquad \Omega_1 \equiv \mathcal{L}_1 \ ,
\end{aligned} \tag{5.8}$$

where the subscript "+" denotes the truncation to positive powers of $\mathcal{L}_1$ in the expansion of the constant matrix $(\Lambda_1^{(z)})^i$. It is clear that the constant flows associated with $\Omega_{n(k+2)} = \frac{1}{n(k+2)} z^n \mathbb{1}$ are trivial and correspond to perturbations by the null operators $\sigma_n(\phi_{k+1})$; the hamiltonians Ω_i with $i > k+1$ correspond to the gravitational descendants (cf., (3.5)).

It is well-known[42,43] that the basic underlying structure of the KdV integrable system is the algebra generated by $\hat{\Lambda}_1^{(z)}$, which is the principal Heisenberg subalgebra of $\widehat{s\ell}(k+2)$. Its positive part,

$$\mathcal{H}^+ \ \equiv \ \left\{ (\Lambda_1^{(z)})^m, m \in \mathbb{Z}_+ \right\} \ , \tag{5.9}$$

is precisely what determines the hamiltonians, $\Omega = (\mathcal{H}^+)_+$. In view of our later generalization, it is very helpful to note that $\Lambda_1^{(z)}$ is identical to the chiral ring structure constant $C_1(z)$ that pertains to the following LG potential "at one level higher":

$$W^{A_{k+2}^{\text{top}}}(x, t_{k+2} = z, t_l = 0) \ = \ \tfrac{1}{k+3} x^{k+3} - z\,x \ . \tag{5.10}$$

This means that the underlying algebraic structure of the A_{k+1} type matter-gravity system is that of a specifically deformed chiral ring pertaining to the LG theory A_{k+2}:

$$\mathcal{H}^+ \ \cong \ \mathcal{R}^{A_{k+2}^{\text{top}}}(t_{k+2} = z, t_l = 0) \ . \tag{5.11}$$

For $g = 0$, the superpotential spectral equation (5.5) represents a specific relation in this ring, and can be viewed as the equation of motion associated with the LG potential (5.10):

$$W^{A^{top}_{k+1}}(x,0) - \frac{1}{k+2}z = \frac{1}{k+2}\,\partial_x\,W^{A^{top}_{k+2}}(x,z)$$
$$= 0\,.$$

$$(5.12)$$

This important fact, namely that $\Lambda_1^{(z)} = C_1(z)$ so that the matrix-valued spectral equation $W(\Lambda_1^{(z)}) \equiv \frac{1}{k+2}(\Lambda_1^{(z)})^{k+2} = \frac{1}{k+2}z\,\mathbb{1}$ can be interpreted as some chiral ring vanishing relation (associated to a *different* LG theory), is our starting point of the generalization to many variables. More precisely, our plan is to use appropriate ring structure constants $C_i(z)$ to construct hamiltonians and Lax operators for the models $CP^{top}_{n-1,k}$ coupled to gravity. This is motivated by the fact that their chiral rings have a common underlying structure for all n: it is the structure of principal embeddings[44] of $s\ell(2)$. Such kind of embeddings is also precisely what underlies the construction of the W_n-algebras.[45]

Specifically, it is well-known that the Heisenberg algebra generator $\Lambda_1^{(z)}$ that figures in the Drinfeld-Sokolov matrix system[42,43] is nothing but an $s\ell(2)$ step generator I_+ (principally embedded in $s\ell(k+2)$),

$$\Lambda_1^{(z)} = \Lambda_1 + z\,\Lambda_{-(k+1)}\,, \qquad \text{where}$$

$$\Lambda_1 = I_+ \equiv \sum_{\substack{\text{simple} \\ \text{roots }\alpha}} E_\alpha\,, \qquad \Lambda_{-(k+1)} = E_{-\psi}\,,$$

$$(5.13)$$

perturbed by the spectral parameter z (ψ denotes the highest root). The point is, as mentioned above, that this is also the structure of the perturbed chiral ring $\mathcal{R}^{CP^{top}_{1,k+1}}(z)$: it is known[18,39] that the unperturbed ring $\mathcal{R}^{CP^{top}_{1,k+1}}$ is isomorphic to the cohomology ring $H^*(CP_{k+1})$, and there is a theorem by Kostant[46] that says that $H^*(CP_{k+1})$ is generated by an $s\ell(2)$ step generator I_+. The deformation by the spectral parameter is then precisely what deforms the cohomology ring H^* into the quantum cohomology ring QH^*, whence

$$\mathcal{H}^+ \cong \mathcal{R}^{CP^{top}_{1,k+1}}(z) \cong QH^*_{\bar{\partial}}(CP_{k+1},\mathbb{R})\,.$$

$$(5.14)$$

(The word "quantum" indicates that the deformation of the classical cohomology ring by the spectral parameter z is precisely the effect of the instanton corrections in a supersymmetric CP_{k+1} σ-model[47]).

For the more general models $\mathrm{CP}^{\mathrm{top}}_{n-1,k}$ that are related to W_n-gravity, the story is very similar: it is known[18,39] that the chiral rings are isomorphic to the Dolbeault cohomology rings of certain grassmannians:

$$\mathcal{R}^{\mathrm{CP}^{\mathrm{top}}_{n-1,k}} \;\cong\; H^*_{\bar{\partial}}\Big(\frac{SU(n+k-1)}{SU(n-1)\times SU(k)\times U(1)} , \mathbb{R} \Big) . \qquad (5.15)$$

These chiral rings are generated by ring structure constants C_i, $i = 1,\ldots,(n-1)$, which represent the LG fields x_i. The important point is that these ring structure constants are determined by principal embeddings of $s\ell(2)$ as well !

More precisely, consider the following matrices:

$$\Lambda_p \;=\; \sum_{\{\alpha:\rho_G\cdot\alpha=p\}} a^{(p)}_\alpha E_\alpha , \qquad \text{for each } p \in \{1,2,\ldots,(n+k-2)\} , \qquad (5.16)$$

where the coefficients $a^{(p)}_\alpha$ are determined trough $[\Lambda_p, \Lambda_{p'}] = 0$, with $\Lambda_1 \equiv I_+$ as in (5.13). Kostant's theorem[46] now tells that when taken in the $(n-1)$th fundamental representation of $s\ell(n+k-1)$, the matrices Λ_i, $i = 1,\ldots,(n-1)$, generate $H^*\big(\frac{SU(n+k-1)}{SU(n-1)\times SU(k)\times U(1)}\big)$, and this means that they are precisely the ring structure constants of the Kazama-Suzuki models:

$$C_i \;=\; \Lambda_i . \qquad (5.17)$$

Our idea is to employ the Λ_i to construct Lax operators and hamiltonians for generalized Drinfeld-Sokolov systems. The relevant objects are of course matrices $\Lambda_i^{(z)}$ that are perturbed by a spectral parameter; they are uniquely defined by requiring $[\Lambda_i^{(z)}, \Lambda_{i'}^{(z)}] = 0$ where $\Lambda_1^{(z)}$ is as in (5.13) (but now in the $(n-1)$th fundamental representation of $s\ell(n+k-1)$). They generate precisely the quantum deformation[47] of the grassmannian cohomology rings, which are the same as specifically perturbed chiral rings of the models $\mathrm{CP}^{\mathrm{top}}_{n-1,k}$:

$$\mathcal{R}^{\mathrm{CP}^{\mathrm{top}}_{n-1,k}}(t_{k+n-1} = z) \;\cong\; QH^*_{\bar{\partial}}\Big(\frac{SU(n+k-1)}{SU(n-1)\times SU(k)\times U(1)} , \mathbb{R} \Big) . \qquad (5.18)$$

These perturbed chiral rings are associated with the LG superpotentials

$$W^{\mathrm{CP}^{\mathrm{top}}_{n-1,k}}(x_i, z) \;=\; W^{\mathrm{CP}^{\mathrm{top}}_{n-1,k}}(x_i, 0) - z\,x_1 , \qquad (5.19)$$

which were investigated previously[48] in the context of integrable perturbations of the models $\mathrm{CP}^{\mathrm{top}}_{n-1,k}$. Such superpotentials were first explicitly written down in refs.[18,49], and have the form:

$$W^{\mathrm{CP}^{\mathrm{top}}_{n-1,k}}(x_i, 0) \;=\; \sum_{l=1}^{k} (\xi_l)^{n+k}(x_i) , \qquad (5.20)$$

$$\text{where} \qquad x_i \;=\; \sum_{1\leq l_1\leq\ldots\leq l_i\leq k} \xi_{l_1}\xi_{l_2}\cdots\xi_{l_k}$$

430

are the elementary symmetric polynomials. This formula was obtained by making use of the fact that, in the Borel-Weil picture, the cohomology of the grassmannian G/H is generated by Chern classes c_i of certain H-valued vector bundles, which satisfy relations of the form

$$\mathrm{Ch}_{(n-1,1)}(c_i, t) \cdot \mathrm{Ch}_{(1,k)}(c_i, t) \;=\; 1 \,, \tag{5.21}$$

where

$$\mathrm{Ch}_v(c_i, t) \;=\; \sum_{j=0}^{\dim v} c_j(\xi)\, t^j \tag{5.22}$$

is the total graded Chern form associated with the H-representation v. The relations among the $c_i \cong x_i$ generated by (5.21) lead precisely to the vanishing relations associated with the potentials (5.20). The formula (5.20) for the superpotentials was subsequently used in in ref.[41], where the following generating function was found:

$$-\log\Big[\sum_{i=1}^{n-1}(-t)^i x_i\Big] \;=\; \sum_{k=-n+1}^{\infty} t^{n+k}\, W^{\mathrm{CP}^{\mathrm{top}}_{n-1,k}}(x_i, 0) \,. \tag{5.23}$$

From this it is easy to prove that

$$W^{\mathrm{CP}^{\mathrm{top}}_{n-1,k}}(x_i, 0) \;=\; \tfrac{1}{n+k}\Big(\sum_{i=1}^{n-1}(n-i)\,x_{i-1}\partial_{x_i}\Big) W^{\mathrm{CP}^{\mathrm{top}}_{n-1,k+1}}(x_i, 0) \,, \tag{5.24}$$

which means that a given superpotential can be written as a vanishing relation of the superpotential "at one level higher". This is the key point which makes the whole construction fly. Namely, (5.19) and (5.24) imply that

$$W^{\mathrm{CP}^{\mathrm{top}}_{n-1,k}}(x_i, 0) - \tfrac{1}{n+k}z \;=\; \tfrac{1}{n+k}\Big(\sum_{i=1}^{n-1}(n-i)\,x_{i-1}\partial_{x_i}\Big) W^{\mathrm{CP}^{\mathrm{top}}_{n-1,k+1}}(x_i, z) \,, \tag{5.25}$$

and this means that the ring structure constants $\Lambda_i^{(z)} = C_i$ of the models $\mathrm{CP}^{\mathrm{top}}_{n-1,k+1}$ satisfy $W^{\mathrm{CP}^{\mathrm{top}}_{n-1,k}}(\Lambda_i^{(z)}, 0) = \tfrac{1}{n+k} z\, \mathbb{1}$. This is precisely what we have been looking for: namely we can take for the Lax operators of the integrable hierarchies just the perturbed versions of these $\Lambda_i^{(z)}$,

$$\mathcal{L}_i(g) \;=\; \Lambda_i^{(z)} + Q_i(g) \;\equiv\; \Lambda_i^{(z)} + \sum_{l_j} q^i_{l_1,\dots,l_{n-1}}(g)\,(\Lambda_1^{(z)})^{-l_1}\dots(\Lambda_{n-1}^{(z)})^{-l_{n-1}} \,, \tag{5.26}$$

whose coefficients $q(g)$ are such that

$$W^{\mathrm{CP}^{\mathrm{top}}_{n-1,k}}(\mathcal{L}_1(g), \dots, \mathcal{L}_{n-1}(g), g) \;=\; \tfrac{1}{n+k} z\, \mathbb{1} \,. \tag{5.27}$$

This is the desired generalization of the matrix-valued superpotential spectral equation (5.5).

To obtain a hierarchy of differential equations, we need to construct appropriate hamiltonians. By analogy to (5.8), we simply take the commuting matrices

$$\Omega_{l_1,\ldots,l_{n-1}}(\mathcal{L}_i,g) \;=\; \big((\Lambda_1^{(z)})^{l_1}\ldots(\Lambda_{n-1}^{(z)})^{l_{n-1}}\big)_+ \,, \qquad l_i \geq 0 \,, \tag{5.28}$$

where "+" denotes projection to positive grade. That is, we take as relevant Heisenberg algebra

$$\mathcal{H}^+ \;\cong\; QH_{\bar{\partial}}^*\Big(\frac{SU(n+k)}{SU(n-1)\times SU(k+1)\times U(1)}\,,\mathbb{R}\Big) \,, \tag{5.29}$$

which just means, like previously for $n = 2$, *that the underlying algebraic structure of the* $\mathrm{CP}^{\mathrm{top}}_{n-1,k}$ *matter-gravity integrable system is given by the quantum ring associated with the matter model "at one level higher"*, $\mathrm{CP}^{\mathrm{top}}_{n-1,k+1}$. The perturbation by the spectral parameter z deforms the finite, nilpotent ring $\mathcal{R}^{\mathrm{CP}^{\mathrm{top}}_{n-1,k+1}}$ into an infinite dimensional, affine algebra, which reflects the extension of the matter ring (4.6) to the gravitationally extended ground ring (4.7) of the matter-gravity system. It would be very interesting to study in more detail the structure $\mathcal{H}^+$ in relation with the W-gravity descendants of (4.7). How this precisely works is not so clear because the number of hamiltonians (5.28) per grade does not grow indefinitely with increasing grade, since there are relations between polynomials of the $\Lambda_i^{(z)}$ (for example the superpotential spectral equation). These relations are just the multi-generator analogs of the well-known condition that reduces the KP to the KdV hierarchy.

Strictly speaking, $\mathcal{H}^+$ in (5.29) is the enveloping algebra of the principal Heisenberg algebra. That is, since the generators $\Lambda_i^{(z)}$ are in general in a higher fundamental representation of $s\ell(n+k)$, powers of the $\Lambda_i^{(z)}$ will in general not belong to the principal Heisenberg subalgebra of $\widehat{s\ell}(n+k)$, but to its enveloping algebra. This is precisely how this construction makes it possible to have more commuting hamiltonians at a given grade as compared to the usual KdV type of systems[42,43], where one considers only the flows associated with the algebra, which are representation-independent.

The flow equations that determine the small phase space couplings $g(t)$ have then supposedly the generic form

$$\big[D_{l_1,\ldots,l_{n-1}}, D_{l'_1,\ldots,l'_{n-1}}\big] \;=\; 0 \,,$$

$$D_{l_1,\ldots,l_{n-1}} \;\equiv\; \frac{\partial}{\partial t_{l_1,\ldots,l_{n-1}}} - \sum_{k_j} Z^{k_1,\ldots,k_{n-1}}_{l_1,\ldots,l_{n-1}}\,\Omega_{k_1,\ldots,k_{n-1}}(\mathcal{L}_i(g(t)),g(t)) \tag{5.30}$$

(where Z are normalization constants), but whether these equations really determine the correct LG couplings $g(t)$ in terms of the flat coordinates t, is a problem that we don't know how to answer yet in general. All what we have done so far was to check these equations for a couple of examples, where they indeed produced the correct results[5].

But these results as well as the general structure strongly suggest that that the kind of integrable systems we proposed makes sense and correctly describes the quasi-classical dynamics of the models $M^{(n)}_{1,n+k}$ coupled to W_n-gravity, which are supposedly equivalent to the models $\mathrm{CP}^{\mathrm{top}}_{n-1,k}$ coupled to topological W_n-gravity. This would correspond to the completion of step F in Fig.1 for W-gravity, and make step G in the figure appear feasible.

6. Acknowledgements

I would like to thank the organizers of the conference for their hard work, and for providing an opportunity for me to present this material.

7. References

1. D.J. Gross and A.A. Migdal, *Phys. Rev. Lett.* **64** (1990) 717; M. Douglas and S. Shenker, *Nucl. Phys.* **B235** (1990) 635; E. Brezin and V. Kazakov, *Phys. Lett.* **236B** (1990) 144.

2. C. Vafa, *Mod. Phys. Let.* **A6** (1991) 337.

3. M. Bershadsky, W. Lerche, D. Nemeschansky and N.P. Warner, *Phys. Lett.* **B292** (1992) 35; E. Bergshoeff, A. Sevrin and X. Shen, *Phys. Lett.* **B296** (1992) 95; J. de Boer and J. Goeree, *Nucl. Phys.* **B405** (1993) 669; E. Bergshoeff, H. Boonstra, M. de Roo S. Panda and A. Sevrin, *Phys. Lett.* **B308** (1993) 34.

4. P. Bouwknegt, J. McCarthy and K. Pilch, *Semi-infinite cohomology of W-algebras*, preprint USC-93/11 and ADP-23-200/M15.

5. W. Lerche, *Generalized Drinfeld-Sokolov Hierarchies, Quantum Rings, and W-Gravity*, preprint CERN-TH.6988/93.

6. J. Distler and T. Kawai, *Nucl. Phys.* **B321** (1989) 509.

7. B. Lian and G. Zuckerman, *Phys. Lett.* **254B** (1991) 417; P. Bouwknegt, J. McCarthy and K. Pilch, *Comm. Math. Phys.* **145** (1992) 541; E. Witten, *Nucl. Phys.* **B373** (1992) 187.

8. E. Witten, *Nucl. Phys.* **B373** (1992) 187.

9. B. Gato-Rivera and A.M. Semikhatov, *Phys. Lett.* **B293** (1992) 72.

10. M. Bershadsky, W. Lerche, D. Nemeschansky and N.P. Warner, *Nucl. Phys.* **B401** (1993) 304.

11. E. Witten, *Comm. Math. Phys.* **117** (1988) 353; *Comm. Math. Phys.* **118** (1988) 411; *Nucl. Phys.* **B340** (1990) 281.

12. T. Eguchi and S. Yang, *Mod. Phys. Let.* **A4** (1990) 1693.

13. G. Mussardo, G. Sotkov, M. Stanishkov, *Int. J. Mod. Phys.* **A4** (1989) 1135; N. Ohta and H. Suzuki, *Nucl. Phys.* **B332** (1990) 146.

14. M. Bershadsky and H. Ooguri, *Comm. Math. Phys.* **126** (1989) 49; M. Bershadsky and H. Ooguri, *Phys. Lett.* **229B** (1989) 374.

15. T. Nakatsu and Y. Sugawara, *Nucl. Phys.* **B385** (1992) 276.

16. S. Mukhi and C. Vafa, *Nucl. Phys.* **B407** (1993) 667.

17. S. Panda and S. Roy, *Phys. Lett.* **B317** (1993) 533, and preprint IC-93-307.

18. W. Lerche, C. Vafa and N.P. Warner, *Nucl. Phys.* **B324** (1989) 427.

19. H. Kanno and M .Sarmadi, *BRST Cohomology Ring in 2-d Gravity coupled to Minimal Models*, preprint IC/92/150; S. Govindarajan, T. Jayamaran and V. John, *Nucl. Phys.* **B402** (1993) 118.

20. E. Witten, *Nucl. Phys.* **B340** (1990) 281; E. and H. Verlinde, *Nucl. Phys.* **B348** (1991) 457.

21. P. Bouwknegt, J. McCarthy and K. Pilch, *On physical states in 2d (topological) gravity*, preprint CERN-TH.6645/92.

22. E. Getzler, *Two-dimensional topological gravity and equivariant cohomology*, MIT preprint 1993.

23. T. Eguchi, H. Kanno, Y. Yamada and S.-K. Yang, *Phys. Lett.* **B305** (1993) 235.

24. K. Li, *Nucl. Phys.* **B354** (1991) 711; *Nucl. Phys.* **B354** (1991)725.

25. R. Dijkgraaf and E. Witten, *Nucl. Phys.* **B342** (1990) 486; R. Dijkgraaf and E. and H. Verlinde, *Nucl. Phys.* **B348** (1991) 435; For a review, see: R. Dijkgraaf, *Intersection theory, integrable hierarchies and topological field theory*, preprint IASSNS-HEP-91/91.

26. R. Dijkgraaf, E. Verlinde and H. Verlinde, *Nucl. Phys.* **B352** (1991) 59.

27. K. Saito, J. Fac. Sci. Univ. Tokyo Sec. IA.28 (1982) 775; M. Noumi, Tokyo. J. Math. 7 (1984) 1; B. Blok and A. Varchenko, preprint IASSNS-HEP-91/5; E. Verlinde and N.P. Warner, *Phys. Lett.* **269B** (1991) 96; Z. Maassarani, *Phys. Lett.* **273B** (1991) 457; S. Cecotti and C. Vafa, *Nucl. Phys.* **B367** (1991) 359; W. Lerche, D. Smit and N. Warner, *Nucl. Phys.* **B372** (1992) 87; A. Klemm, M. G. Schmidt and S. Theisen, *Int. J. Mod. Phys.* **A7** (1992) 6215.

28. I. Krichever, *Comm. Math. Phys.* **143** (1992) 415; B. Dubrovin, *Nucl. Phys.* **B379** (1992) 627, *Comm. Math. Phys.* **145** (1992) 195, *Comm. Math. Phys.* **152** (1993) 539.

29. M. Douglas, *Phys. Lett.* **238B** (1990) 176.

30. A. Lossev, *Descendants constructed from matter field and K. Saito higher residue pairing in Landau-Ginzburg theories coupled to topological gravity*, preprint TPI-MINN-92-40-T.

31. P. Fendley, W. Lerche, S. Mathur and N.P. Warner, *Nucl. Phys.* **B348** (1991) 66; S. Cecotti and C. Vafa, *Nucl. Phys.* **B367** (1991)359; D. Nemeschansky and N.P. Warner, *Nucl. Phys.* **B380** (1992) 241.

32. M. Spiegelglas and S. Yankielowicz, *Nucl. Phys.* **B393** (1993) 301; D. Gepner,*Comm. Math. Phys.* **141** (1991) 381; M. Spiegelglas, *Phys. Lett.* **274** (1992) 21.

33. E. Witten, *Nucl. Phys.* **B371** (1992)191; O. Aharony, O. Ganor, J. Sonnenschein, S. Yankielowicz and N. Sochen, *Nucl. Phys.* **B399** (1993) 527; O. Aharony, J. Sonnenschein and S. Yankielowicz, *Phys. Lett.* **B289** (1992) 309; J. Sonnenschein, *Physical states in topological coset models*, preprint TAUP-1999-92; V. Sadov, *on the spectra of* $sl(N)_k/sl(N)_k$ *cosets and* W_N *gravities I*, preprint HUTP-92/A055.

34. A. Bilal and J. Gervais, *Nucl. Phys.* **B326** (1989) 222; P. Mansfield and B. Spence, *Nucl. Phys.* **B362** (1991) 294; S. Das, A. Dhar and S. Kalyana Rama, preprint TIFR/TH/91-20; C.N. Pope, L.J. Romans and K.S. Stelle, *Phys. Lett.* **268B** (1991) 167; H. Lu, C.N. Pope, S. Schrans and K. Xu, *Nucl. Phys.* **B385** (1992) 99; C.N. Pope, *A Review of W-strings*, preprint CTP-TAMU-30/92; *W Strings 93*, preprint CTP-TAMU-55-93.

35. For extra states at $c_M = 100$, see: C.N. Pope, E. Sezgin , K.S. Stelle and X.J. Wang, *Nucl. Phys.* **B299** (1993) 247; H. Lu, B.E.W. Nilsson, C.N. Pope, K.S. Stelle, and P.C. West, Int. J. Mod. Phys. A8 (1993) 4071; H. Lu, C.N. Pope, S. Schrans and X. Wang, *Nucl. Phys.* **B408** (1993) 3

36. E. Bergshoeff, J. de Boer, M. de Roo and T. Tjin, *On the cohomology of the non-critical W-string*, preprint UG-7/93.

37. Y. Kazama and H. Suzuki, *Nucl. Phys.* **B321** (1989) 232.

38. K. Ito, *Phys. Lett.* **B259** (1991) 73; *Nucl. Phys.* **B370** (1992) 123; D. Nemeschansky and S. Yankielowicz, *N=2 W-algebras, Kazama-Suzuki models and Drinfeld-Sokolov reduction*, preprint USC-91-005A; L.J. Romans, *Nucl. Phys.* **B369** (1992) 403; W. Lerche, D. Nemeschansky and N.P. Warner, unpublished.

39. D. Gepner, *A comment on the chiral algebras of quotient superconformal field theories*, preprint PUPT-1130; S. Hosono and A. Tsuchiya, *Comm. Math. Phys.* **136** (1991) 451.

40. K. Li, *Phys. Lett.* **B251** (1990) 54, *Nucl. Phys.* **B346** (1990) 329; H. Lu, C.N. Pope and X. Shen, *Nucl. Phys.* **B366** (1991) 95; S. Hosono, *Algebraic definition of topological W-gravity*, preprint UT-588; H. Kunitomo, Prog. Theor. Phys. 86 (1991) 745.

41. D. Gepner, *Comm. Math. Phys.* **141** (1991) 381.

42. Drinfel'd and V. G. Sokolov, Jour. Sov. Math. **30** (1985) 1975.

43. M. de Groot, T. Hollowood and J. Miramontes, *Comm. Math. Phys.* **145** (1992) 57; N. Burroughs, M. De Groot, T. Hollowood and J. Miramontes, *Phys. Lett.* **B277** (1992) 89; T. Hollowood, J. Miramontes and J. Guillen, *Generalized integrability and two-dimensional gravitation*, preprint CERN-TH-6678-92.

44. B. Kostant, Am. J. Math. 81 (1959) 973.

45. For reviews on *W*-algebras, see: P. Bouwknegt and K. Schoutens, Phys. Rep. 223 (1993) 183; J. de Boer, *Extended conformal symmetry in non-critical string theory*, Ph.D. thesis, 1993; J. Goeree, *Higher spin extensions of two-dimensional gravity*, P.D. thesis, 1993; T. Tjin, *Finite and infinite W-algebras*, P.D. thesis, 1993.

46. B. Kostant, private communication via N. Warner.

47. E. Witten, *Comm. Math. Phys.* **118** (1988) 411, *Nucl. Phys.* **B340** (1990) 281; K. Intriligator, *Mod. Phys. Let.* **A6** (1991) 3543; C. Vafa, *Topological mirrors and quantum rings*, in: Essays in Mirror Symmetry, ed. S.T. Yau, 1992; E. Witten, *The Verlinde algebra and the cohomology of the grassmannian*, preprint IASSNS-HEP-93/41.

48. W. Lerche and N.P. Warner, *Nucl. Phys.* **B358** (1991) 571.

49. P. Fendley, W. Lerche, S. Mathur and N.P. Warner, *Nucl. Phys.* **B348** (1991) 66.

CONFORMALLY INVARIANT OFF-SHELL STRINGS

ROBERT C. MYERS

Physics Department, McGill University, Ernest Rutherford Building,
Montréal, Québec, H3A 2T8, Canada

and

VIPUL PERIWAL
The Institute for Advance Study
Princeton, New Jersey, 08540-4920, U.S.A.

ABSTRACT

Recent advances in non-critical string theory allow a unique continuation of critical Polyakov string amplitudes to off-shell momenta, while preserving conformal invariance. These continuations possess unusual, apparently stringy, characteristics, as we illustrate with our results for three-point functions.

1. The Polyakov functional integral

The Polyakov path integral[1] appears only a recipe for perturbative computations of scattering amplitudes, yet it serves as the foundation for our entire understanding of string theory. It is natural to ask whether this framework might yield any insight into the off-shell properties of strings. We report on how the recent progress in understanding non-critical strings provides a new technique to calculate off-shell Polyakov amplitudes[2]. This approach preserves conformal invariance, and so these amplitudes lend themselves to analysis by essentially the standard techniques of conformal field theory.

Space-time scattering amplitudes of string excitations are calculated as correlation functions of vertex operators in a functional integral over the metric on the string world-sheet and the space-time string configurations:

$$\left\langle \prod_i \int \mathrm{d}^2 z_i \sqrt{g}\, V_i(z_i) \right\rangle \equiv \int \frac{\mathrm{D}g\,\mathrm{D}X}{\mathrm{vol.(Diff \times Weyl)}}\, e^{-S[g,X]} \prod_i \int \mathrm{d}^2 z_i \sqrt{g}\, V_i(z_i) \ . \quad (1)$$

The measure is divided by the 'volume' of the symmetries of the classical action $S \equiv 1/(8\pi) \int \mathrm{d}^2 z \sqrt{g} g^{ab} \partial_a X^\mu \partial_b X_\mu$, with $\mu = 1, \ldots, D$ — namely, diffeomorphisms and local Weyl rescalings on the world-sheet. Choosing conformal gauge, $g_{ab} \equiv e^{2\phi} \hat{g}_{ab}(m)$, and fixing diffeomorphisms à la Faddeev-Popov, one finds that these functional integrals reduce to

$$\int \mathrm{d}m\, \frac{\mathrm{D}\phi}{\mathrm{vol.(Weyl)}} \frac{\mathrm{D}X\ \mathrm{Det}'_{\mathrm{FP}}}{\mathrm{vol.(c.k.v.)}}\, e^{-S[\hat{g},X]} \prod_i \int \mathrm{d}^2 z_i \sqrt{\hat{g}(m)} V_i(z_i) \ . \quad (2)$$

Here, c.k.v. stands for the conformal Killing vectors which arise if the world-sheet is a sphere or a torus, and $\mathrm{d}m$ denotes the measure for integrating over the moduli of surfaces with one or more handles. In Eq. (2), the Weyl factor should decouple from the theory, and the integration $\mathrm{D}\phi$ would cancel against the volume of the group of Weyl rescalings in the denominator. This decoupling is only actually achieved if Weyl rescaling survives as a symmetry of the quantum path integral. This requires[1] that $D = 26$ in order to cancel the anomalous dependences on the Weyl field in the measure factor, $\mathrm{D}X \, \mathrm{Det}'_{\mathrm{FP}}/\mathrm{vol.(c.k.v.)}$. Also, one must impose various *space-time* mass-shell and polarization/gauge conditions on the external string states to avoid any anomalous Weyl dependences from normal-ordering the vertex operators. Combined, these restrictions ensure that ϕ is decoupled from *on-shell* correlation functions in *critical* string theory, and the Weyl factor simply disappears from the functional integral.

Therefore the mass-shell conditions are obtained from requiring Weyl invariance. It follows, in the Polyakov approach, that the calculation of amplitudes for *off-shell* string states requires the ability to compute correlation functions of vertex operators with an anomalous Weyl dependence, in the normalized measure $\mathrm{D}\phi/\mathrm{vol.(Weyl)}$. Why are such computations difficult? The problem resides in the non-linearity of the Riemannian metric that defines $\mathrm{D}\phi$. The norm on infinitesimal changes of the conformal factor is constructed with the full world-sheet metric g_{ab}

$$(\delta\phi, \delta\phi) = \int \mathrm{d}^2x \sqrt{g}(\delta\phi)^2 = \int \mathrm{d}^2x \sqrt{\hat{g}}\, \mathrm{e}^{2\phi}(\delta\phi)^2, \tag{3}$$

which then explicitly depends on ϕ. The functional integral over ϕ would be a standard quantum field theory with the measure, $\mathrm{D}_0\phi$, defined by the translation-invariant norm, $(\delta\phi, \delta\phi)_0 = \int \mathrm{d}^2x \sqrt{\hat{g}}(\delta\phi)^2$. The relation between these two measures is remarkably simple,

$$\mathrm{D}\phi = \mathrm{D}_0\phi \, \exp\left(S_L - \frac{\mu}{\pi} \int \mathrm{d}^2z \, \mathrm{e}^{\alpha\phi} \right), \tag{4}$$

where $S_L \equiv \int \frac{\mathrm{d}^2z}{6\pi}\left[\partial\phi\bar{\partial}\phi + \frac{1}{4}\sqrt{\hat{g}}\hat{R}\phi\right]$. The 'cosmological constant' μ is the coefficient of a local counterterm, and remains undetermined in this computation. The constant α in this interaction is explicitly fixed (see below). This relation (4) originally arose in the study of two-dimensional gravity coupled to conformal matter[3]. It is important to note that the derivation of Eq. (4) is entirely independent of the rest of the functional integrals involved. Thus it remains valid for our studies of off-shell amplitudes.

2. Correlation functions

To proceed further, we assume that the correlation functions of interest may be calculated with conformal field theory methods. For non-critical strings, this approach has been verified by comparison with results determined by matrix model techniques. The stress tensor deduced from S_L is $T_L = \frac{1}{6}\left[(\partial\phi)^2 - \partial^2\phi\right]$, which makes no contribution to the total central charge. Off-shell vertex operators receive exponential Weyl dressings, $e^{\beta\phi}V$ (just as matter operators in non-critical string theory do), where

$$\beta = \frac{1}{6}\left[\sqrt{1 - 12\delta} - 1\right]. \tag{5}$$

Here, $\delta = k^2 + 2n$ (with $n = -1, 0, 1, \ldots$) is the mass-shell operator of the particular external string state. Explicitly, off-shell tachyon vertex operators are $e^{\beta\phi}e^{ik\cdot X}$ with $\delta = k^2 - 2$, and hence $\beta = (\sqrt{25 - 12k^2} - 1)/6$. The dressing exponents are chosen to vanish when the vertex operator goes on-shell (*i.e.*, $\delta = 0$), which insures that the off-shell amplitudes reduce precisely to the usual on-shell amplitudes. Rather puzzling is the non-analyticity in this prescription at $\delta = \frac{1}{12}$. Further insight into quantum Liouville theory is needed to overcome this barrier, but for the present time, we will limit our attention to $\delta \leq \frac{1}{12}$ or $\beta \geq -\frac{1}{6}$. Treating the cosmological constant term within the conformal field theory context fixes $\alpha = \frac{2}{3}$ in Eq. (4).

Explicit calculations may be carried out for tree-level scattering amplitudes. We omit the details here, but as an example present the result for an off-shell tachyon three point amplitude $\mathcal{A} = \langle\prod_{i=1}^{3}\int d^2 z_i \, e^{\beta_i\phi}\, e^{ik_i\cdot X}\rangle$. The final result, valid for arbitrary external momenta subject only to the constraint $\delta_i \leq \frac{1}{12}$ or $k_i^2 \leq \frac{25}{12}$, is

$$\mathcal{A} = X \, \Gamma(\tfrac{1+3\gamma}{2})\Gamma(\tfrac{1-3\gamma}{2}) \prod_{\ell=0}^{3} \lambda(\gamma + \tfrac{1}{3}, 1 - 3\beta_\ell) \, \lambda(-\gamma - \tfrac{1}{3}, \tfrac{3}{2} + 3\beta_\ell) \tag{6}$$

where we have defined $X \equiv [\mu\,\Gamma(\tfrac{1}{3})/\Gamma(\tfrac{2}{3})]^{-(3\gamma+1)/2}(\tfrac{2}{3})^{\gamma-2/3}$, $\gamma \equiv \sum_{i=1}^{3}\beta_i$ and $\beta_0 \equiv -\frac{1}{6}$, as well as

$$\lambda(z, b) \equiv (2\pi)^{z/2} \exp[-3z + (\log 3 - E)f(z)] \prod_{\ell=0}^{\infty}\left(\exp\left[-3z + \frac{f(z)}{\ell+1}\right]\right.$$
$$\left.\left[\prod_{k=0}^{2} \frac{(\tfrac{3}{2}z + b + 3\ell + k)\,(\tfrac{3}{2}z + b + 3\ell + k + \tfrac{3}{2})}{(b + 3\ell + k)\,(b + 3\ell + k + \tfrac{3}{2})}\right]^{\ell+1}\right). \tag{7}$$

Here $f(z) = \frac{3}{4}z^2 + (b - \frac{5}{4})z$, and $E = .577\ldots$ is Euler's constant. The λ-function in Eq. (8) satisfies $\lambda(z + 1, b) = \Gamma(\frac{3}{2}z + b)\lambda(z, b)$, amongst other properties.

One of the most interesting, and presumably most physically significant features of this amplitude is the infinite set of poles which it contains. Keeping in mind the restriction $\beta_i \geq -\frac{1}{6}$, the λ-functions introduce two sequences of poles of order $\ell + 1$ at $\beta_i = \ell + \frac{m}{3} + \frac{1}{3}$, and $\beta_i = \ell + \frac{m}{3} + \frac{5}{6}$ for $\ell = 0, 1, 2, \ldots$ and $m = 0, 1, 2$. These are leg poles depending on the external momenta of the individual tachyons. From the relation $k_i^2 = 2 - \beta_i - 3\beta_i^2$, one finds that the poles in the first sequence with $m = 0$, and all of the poles in the second sequence, do *not* correspond to the physical mass shell of external string states. The prefactor $\Gamma(\frac{1+3\gamma}{2})\Gamma(\frac{1-3\gamma}{2})$ also contains a sequence of simple poles at $\gamma = -1/3$ and $1/3 + 2n$ with n, an non-negative integer. In fact, the latter are cancelled by zeroes appearing in the product of λ-functions, in particular the $\ell = 0$ term. There remains a single simple pole in the amplitude at $\gamma = -1/3$. In contrast to the previous leg poles, this pole depends collectively on all of the external momenta. Note that $\gamma = -1/3$ is precisely the value of γ at which the amplitude is independent of the cosmological constant, or equivalently the world-sheet area. This scale independence appears then to be the physical reason for the pole, but its occurrence is an entirely stringy feature.

3. Conclusions and prospects

It has been our aim here to show that the effort expended on the study of non-critical strings has important physical consequences in critical string theories. Any future progress in non-critical string physics, or in quantum Liouville theory, will be of use in understanding off-shell critical string physics. In particular, advances in the conformal field theory treatment of the Liouville correlators are clearly needed. There are no conceptual barriers to the extension of our results to supersymmetric strings, or to open string theories.

A striking feature of the amplitudes is the presence of poles that do not correspond to excitations in the matter sector (even if combined with the ghost sector). They may indicate the presence of excitations that are entirely stringy in nature. In particular, a pole which depends collectively on all of the external momenta, such as that at $\gamma = -1/3$, is entirely unknown in the amplitudes one obtains from a field theory. In field theories, the off-shell character of the amplitude is a function of individual external states. It is difficult then to imagine how this γ dependence could be reproduced in a string field theory. Thus our results may indicate that some fundamentally new framework, other than string field theory, will be required to extend our understanding of critical string theory beyond the Polyakov path integral.

It is a pleasure to thank the organizers of the Strings '93 for their hospitality. R.C.M. was supported by NSERC of Canada, and Fonds FCAR du Québec. V.P. was supported by D.O.E. grant DE-FG02-90ER40542.

References

1. A.M. Polyakov, *Phys. Lett.* **103B** (1981) 207, 211.
2. R.C. Myers and V. Periwal, *Phys. Rev. Lett.* **70** (1993) 2841.
3. N.E. Mavromatos and J.L. Miramontes, *Mod. Phys. Lett.* **A4** (1989) 1847; E. D'Hoker and P.S. Kurzepa, *Mod. Phys. Lett.* **A5** (1990) 1411; E. D'Hoker, *Mod. Phys. Lett.* **A6** (1991) 745; F. David, *Mod. Phys. Lett.* **A3** (1988) 1651; J. Distler and H. Kawai, *Nucl. Phys.* **B321** (1989) 509.

W-Strings '93

C.N. Pope[*]

Center for Theoretical Physics, Texas A&M University,
College Station, TX 77843-4242, USA.

ABSTRACT

We present a review of the status of W string theories, their physical spectra, and their interactions.

1. Introduction

W algebras have received a considerable amount of attention since their discovery [1] in the middle of the 1980's. These efforts have broadly concentrated in two directions, namely the construction and classification of W algebras on the one hand, and their application to W gravity and W strings on the other. Since W algebras are higher-spin extensions of the Virasoro algebra, it is natural to expect that they could be associated with some kind of extensions of the usual bosonic string. The idea of building W-string theories was first developed in [2]. One of the hopes was that they might have massless physical states with spins greater than 2 [2], but so far all the explicit examples that have been constructed do not give rise to such higher-spin massless states. Nonetheless, the structures that emerge in W-string theories are not without interest, revealing intriguing connections with Virasoro and W minimal models.

Most of the efforts so far in constructing W-string theories have been concentrated on the case of the W_3 string, since W_3 is the simplest non-trivial W algebra. It has a primary current of spin 3 in addition to the energy-momentum tensor. Like all W algebras, it is non-linear. The starting point for building a W string is to construct an anomaly-free quantum theory of the associated W gravity. By far the simplest way to do this is by BRST methods [3]. The essential requirement, therefore, is to find the BRST operator Q_B for the W algebra in question. From this, one can write down the full Lagrangian, with ghosts and gauge-fixing terms. The requirement of anomaly-freedom translates into the requirement that the BRST operator Q_B be nilpotent. This is achieved provided that the matter currents satisfy a closed (albeit non-linear) algebra at the full quantum level, with the correct value of central charge to cancel the anomaly from the ghosts for the gauge fields.

---

[*]   Supported in part by the U.S. Department of Energy, under grant DE-FG05-91ER40633.

Having arrived at an anomaly-free quantum W gravity, the next step in building a W string is to find an appropriate W matter system, with the critical value of the central charge, that admits some sort of a spacetime interpretation. Thus some of the matter fields, which we may suggestively call X^μ, should be worldsheet scalars that enter the matter currents in a Lorentz-invariant way. It is non-trivial that this can be done, since the W algebra is non-linear and one cannot, by contrast with ordinary string theory, simply take tensor products of elementary realisations to obtain new ones with larger central charge.

The basic elementary realisations of W algebras are those that come from Hamiltonian reduction and the Miura transformation. In the case of W_3, for example, this is a two-scalar realisation [1]. The central charge can be adjusted to its critical value ($c = 100$ for W_3) by an appropriate choice of a free background-charge parameter. This realisation gives rise to what one may call a theory of pure W_3 gravity, and can be thought of as a generalisation of a one-scalar realisation of the Virasoro algebra, or pure Liouville gravity.

In [4], the observation was made that one of the two-scalars of the Miura realisation of W_3 appears in the spin-2 and spin-3 currents only through its energy-momentum tensor. Thus one can obtain more general realisations of W_3 by replacing this energy-momentum tensor by that for an arbitrary matter system T^{eff} with the same central charge c^{eff}, the only other requirement being that it should commute with the other original scalar field, which we shall call φ. The value of c^{eff}, which is dictated by the details of the Miura realisation and the required critical value ($c = 100$) for the total matter central charge, turns out to be $c^{\text{eff}} = \frac{51}{2}$. It is no coincidence, as we shall see later, that this is 26 minus the central charge of the Ising model. By realising T^{eff} in terms of worldsheet scalars X^μ, the desired goal of obtaining a W_3 realisation with a string-like spacetime interpretation is achieved. The first investigations of the W_3 string based on this realisation were carried out in [5].

Although these multi-scalar realisations enable one to construct W_3 strings that admit a sensible-looking multi-dimensional spacetime interpretation, in some sense the way in which the W_3 symmetry is realised on the fields X^μ is a little trivial, and this ultimately shows up when one calculates the spectrum and scattering of physical states in the theory. As we shall see later, the multi-scalar W_3 string behaves, at least at tree level, precisely like a bosonic string in which criticality is achieved by tensoring a $c^{\text{eff}} = \frac{51}{2}$ energy-momentum tensor for X^μ (with a background charge) with the $c = \frac{1}{2}$ Ising model. While it is not without interest that such a system displays a "hidden" W_3 symmetry, the original hopes that W strings might be new kinds of string theories, maybe even with higher-spin massless states, seem to have evaporated.

In fact, although it is possibly physically less realistic, the two-dimensional W_3 string based on the original two-scalar Miura realisation is mathematically-speaking a much more interesting object. It turns out that the physical states of the two-scalar W_3 string divide into two categories; there are some which generalise to the multi-scalar case if one makes the replacement described above, and then there are other physical states that have no analogues in the multi-scalar case. The first category is in some sense "uninteresting," in that again the physical states are tensor products of $c^{\text{eff}} = \frac{51}{2}$ Virasoro states with Ising-model primary fields. The second category comprises physical states that are not simply tensor products of Virasoro-string states with Ising model states, and as such they could be regarded as states on which the W_3 symmetry is realised in a more non-trivial way.

One could, of course, build a W string theory based on more or less any W algebra, such as W_N. However, since each higher-spin current in a W algebra represents a constraint on the physical states of the associated W string, there is a sense in which one obtains richer W-string theories by considering algebras with fewer higher-spin currents. To this end, recently some examples of other kinds of W strings have been constructed, in which there is just a single current of spin $s > 2$ in addition to the energy-momentum tensor [6]. Curiously, it has been found that one does not necessarily even need an underlying W algebra in order to construct a nilpotent BRST operator. We shall describe some recent work on these spin-2 plus spin-s strings later.

2. W_3 strings

2.1 The physical spectrum

We shall begin by describing the multi-scalar W_3 string. As was discussed in the introduction, the key ingredient for the construction of the W_3 string is its BRST operator. This was found in [7] (see also [8]). It takes the form

$$Q_B = \oint dz \left[c\left(T + \tfrac{1}{2}T_{\text{gh}}\right) + \gamma\left(W + \tfrac{1}{2}W_{\text{gh}}\right) \right], \tag{2.1}$$

and is nilpotent provided that the matter currents T and W generate the W_3 algebra with central charge $c = 100$, and that the ghost currents are chosen to be

$$T_{\text{gh}} = -2b\,\partial c - \partial b\,c - 3\beta\,\partial\gamma - 2\partial\beta\,\gamma \, , \tag{2.2}$$

$$W_{\text{gh}} = -\partial\beta\,c - 3\beta\,\partial c - \tfrac{8}{261}\left[\partial(b\,\gamma\,T) + b\,\partial\gamma\,T\right] + \tfrac{25}{1566}\left(2\gamma\,\partial^3 b + 9\partial\gamma\,\partial^2 b + 15\partial^2\gamma\,\partial b + 10\partial^3\gamma\,b\right), \tag{2.3}$$

where the ghost-antighost pairs $(c,\,b)$ and $(\gamma,\,\beta)$ correspond respectively to the T and W generators. A matter realisation of W_3 with central charge 100 can be given in terms of $n \geq 2$ scalar fields, as follows [4]:

$$T = -\tfrac{1}{2}(\partial\varphi)^2 - \alpha\,\partial^2\varphi + T^{\text{eff}},$$

$$W = -\frac{2i}{\sqrt{261}}\left[\tfrac{1}{3}(\partial\varphi)^3 + \alpha\,\partial\varphi\,\partial^2\varphi + \tfrac{1}{3}\alpha^2\,\partial^3\varphi + 2\,\partial\varphi\,T^{\text{eff}} + \alpha\,\partial T^{\text{eff}}\right], \tag{2.4}$$

where $\alpha^2 = \frac{49}{8}$ and T^{eff} is an energy-momentum tensor with central charge $\frac{51}{2}$ that commutes with φ. Since T^{eff} has a fractional central charge, a background-charge vector a_μ is needed in order to realise it with d scalar fields X^μ:

$$T^{\text{eff}} = -\tfrac{1}{2}\partial X_\mu\partial X^\mu - ia_\mu\,\partial^2 X^\mu, \tag{2.5}$$

with a_μ chosen so that $\frac{51}{2} = d - 12 a_\mu a^\mu$.

It is now in principle a straightforward matter to construct physical states $|\chi\rangle$, by demanding that they be annihilated by the BRST operator and that they be BRST non-trivial:

$$Q_B|\chi\rangle = 0, \qquad |\chi\rangle \neq Q_B|\psi\rangle. \tag{2.6}$$

Rather than follow the historical course of events here, we shall jump ahead to a discovery made in [9], which leads to a considerable simplification of the BRST operator and the form of the physical states. In [9], the the following non-linear redefinition, under which the ghosts and the φ matter field become mixed, was introduced:

$$\begin{aligned}
c &\longrightarrow c + \tfrac{21}{8\sqrt{2}}\,\partial\gamma + \tfrac{9}{8}\,b\,\partial\gamma\,\gamma - \tfrac{3}{2}\,\partial\varphi\,\gamma\,, \\
b &\longrightarrow b\,, \\
\gamma &\longrightarrow \tfrac{9\sqrt{29}i}{8}\gamma\,, \\
\beta &\longrightarrow -\tfrac{8i}{9\sqrt{29}}\Big\{\beta + \tfrac{21}{8\sqrt{2}}\,\partial b + \tfrac{9}{8}\partial b\,b\,\gamma + \tfrac{3}{2}\,\partial\varphi\,b\Big\}, \\
\varphi &\longrightarrow \varphi - \tfrac{3}{2}\,b\,\gamma\,.
\end{aligned} \tag{2.7}$$

These transformations are canonical, in the sense that the redefined fields satisfy the same set of OPEs as the original ones. We have taken the opportunity to rescale away some tiresome numerical coefficients at the same time. In terms of the new fields, which we shall use exclusively from now on, the BRST operator becomes [9]

$$Q_B = Q_0 + Q_1, \tag{2.8}$$

where

$$Q_0 = \oint dz\, c\Big(T^{\text{eff}} + T_\varphi + T_{\gamma,\beta} + \tfrac{1}{2}T_{c,b}\Big), \tag{2.9}$$

$$Q_1 = \oint dz\, \gamma\Big((\partial\varphi)^3 + 3\alpha\,\partial^2\varphi\,\partial\varphi + \tfrac{19}{8}\partial^3\varphi + \tfrac{9}{2}\partial\varphi\,\beta\,\partial\gamma + \tfrac{3}{2}\alpha\,\partial\beta\,\partial\gamma\Big), \tag{2.10}$$

with the energy-momentum tensors given by

$$T_\varphi \equiv -\tfrac{1}{2}(\partial\varphi)^2 - \alpha\,\partial^2\varphi, \tag{2.11}$$

$$T_{\gamma,\beta} \equiv -3\,\beta\,\partial\gamma - 2\,\partial\beta\,\gamma, \tag{2.12}$$

$$T_{c,b} \equiv -2\,b\,\partial c - \partial b\,c, \tag{2.13}$$

$$T^{\text{eff}} \equiv -\tfrac{1}{2}\partial X^\mu\,\partial X^\nu\,\eta_{\mu\nu} - i a_\mu\,\partial^2 X^\mu. \tag{2.14}$$

The BRST operator is graded, with $Q_0^2 = Q_1^2 = \{Q_0, Q_1\} = 0$.

The physical states of the W_3 string have been analysed in considerable detail. The early discussions [5,3,10,11] concentrated on physical states having what one might call "standard" ghost structure. In ordinary bosonic string theory, standard ghost structure means that the physical operators $V(z)$ that create physical states $|\text{phys}\rangle = V(0)|0\rangle$ are of the form $V = cY(X)$, where c is the ghost field and $Y(X)$ is independent of ghosts. ($Y = e^{ip\cdot X}$ for tachyons, $\xi \cdot \partial X^{ip\cdot X}$ at level 1, *etc.* Note that here, and always in this paper, $|0\rangle$ denotes the $SL(2,C)$-invariant vacuum.) For the W_3 string, the analogous notion of states with "standard" ghost structure is when the physical operator V takes the form $V = c\,\partial\gamma\,\gamma\,Y(\varphi, X)$. The physical-state conditions (2.6) imply that such operators are physical if

$$V = c\,\partial\gamma\,\gamma\,e^{\mu\varphi}\,Y_\Delta^{\text{eff}}(X),\tag{2.15}$$

where $Y_\Delta^{\text{eff}}(X)$ is a highest-weight operator under T^{eff} with weight Δ, and

$$\mu = -\tfrac{8}{7}\alpha, \quad \Delta = 1; \quad \text{or} \quad \mu = -\tfrac{6}{7}\alpha, \quad \Delta = 1; \quad \text{or} \quad \mu = -\alpha, \quad \Delta = \tfrac{15}{16}.\tag{2.16}$$

Thus we see that the momentum in the φ direction is frozen to three discrete values. From the point of view of the effective spacetime described by the X^μ coordinates, these physical states look like two sectors of ordinary bosonic strings, one having intercept 1, and the other having intercept $\tfrac{15}{16}$ [5,10].

It turns out that this is by no means the end of the story as far as physical states are concerned. There are also physical states with "non-standard" ghost structure. The first example of such a state was found in [12], in the case of the two-scalar W_3 string. Subsequently, more physical states with non-standard ghost structure were found in [13]. The main focus of [13] was to examine analogues of the "ground-ring" of ghost-number zero physical operators for the two-scalar W_3 string. However, it was also shown that some (but not all) of the non-standard ghost structure physical states of the two-scalar W_3 string generalise to physical states of the multi-scalar W_3 string [13]. For example, there is a physical state described by the operator

$$V = c\gamma\,e^{\mu\varphi}\,Y_\Delta^{\text{eff}}(X)\tag{2.17}$$

with $\mu = -\tfrac{4}{7}\alpha$ and $\Delta = \tfrac{1}{2}$ [13,14,9]. From the effective spacetime point of view, this corresponds to a third sector of effective bosonic string states, with intercept $\tfrac{1}{2}$.

We shall not present a detailed discussion of all the physical states of the W_3 string here, but instead just give a summary of the final conclusions. Further details may be found in [14,9]. Let us first consider what have been called "prime states." These are the physical states of lowest possible ghost number at a given momentum and level. From the prime states, one can build quartets of physical states by normal ordering with either or both of the "ghost boosters" $a_\varphi \equiv [Q_B, \varphi]$ and $a_{X^\mu} \equiv [Q_B, X^\mu]$ [13,14,9]. In addition there are conjugate quartets [13]. Thus it suffices to describe the prime states in order to characterise the entire physical spectrum.

The physical prime states divide into four sectors. The first three of these sectors comprise physical states with continuous on-shell spactime momentum, and are described by operators of the form [9]

$$V = c\,U(\varphi,\beta,\gamma)\,Y^{\mathrm{eff}}_\Delta(X). \tag{2.18}$$

The operator $Y^{\mathrm{eff}}_\Delta(X)$ is highest-weight under T^{eff}, with conformal weight Δ taking one of the three values 1, $\frac{15}{16}$, $\frac{1}{2}$. $U(\varphi,\beta,\gamma)$ has conformal weight $h = 1 - \Delta$ under $T_\varphi + T_{\gamma,\beta}$, and in addition satisfies $[Q_1, U\} = 0$. For simplicity, we may take the operator $Y^{\mathrm{eff}}_\Delta(X)$ to be tachyonic for now, since the process of constructing excited spacetime operators is identical to that in ordinary bosonic string theory. Thus we may classify the physical states by their level number ℓ, which represents the level of ghost and φ excitations only. One finds that as one goes to higher and higher levels, the same set of three Δ values 1, $\frac{15}{16}$, $\frac{1}{2}$ recur repeatedly, with the ghost numbers of the associated operators $U(\varphi,\beta,\gamma)$ becoming more and more negative, and the φ momentum frozen to more and more positive values. The lowest-level examples for $\Delta = 1$ and $\Delta = \frac{15}{16}$ are the standard ghost structure $\ell = 0$ operators (2.16), and the lowest-level example for $\Delta = \frac{1}{2}$ is the level $\ell = 1$ operator (2.17). Many more examples may be found in [14,9].

The fourth sector of physical prime states in the multi-scalar W_3 string comprises states with discrete momentum (in fact $p_\mu = 0$ or $p_\mu = -2a_\mu$) in the effective spacetime described by X^μ. Those with $p_\mu = 0$ take the form

$$V = c\,U_1(\varphi,\beta,\gamma) + U_2(\varphi,\beta,\gamma), \tag{2.19}$$

where the operators $U_i(\varphi,\beta,\gamma)$ satisfy the conditions consequent upon (2.6). The lowest-level examples occur at $\ell = 1$ [13], but a more interesting discrete state is the one at level $\ell = 6$, which is described by the operator [13,9]

$$D = \left[c\beta + \tfrac{1}{4}\left(6\sqrt{2}\,\partial\beta\,\gamma - 3\sqrt{2}\,\beta\,\partial\gamma + 12\partial\varphi\,\beta\,\gamma + 4\sqrt{2}\,\partial\varphi\,\partial\varphi + 2\partial^2\varphi\right)\right]e^{\frac{2}{7}\alpha\varphi}. \tag{2.20}$$

From the pattern of physical states, their frozen φ-momenta and ghost numbers, obtained in [14,9], one can see that the level-6 discrete operator D, and its associated screening current $\oint b\,D$, might be used to build up each entire tower of physical states in each sector from its lowest-level member. Thus the $\Delta = 1$ and $\frac{15}{16}$ level $\ell = 0$ states (2.15) [5,10], and the $\Delta = \frac{1}{2}$ level $\ell = 1$ state (2.17) [13] could be viewed as the basic building blocks for all the continuous-momentum physical states of the multi-scalar W_3 string. This idea was first proposed in [13], where the discrete state D (2.20) was derived as one of the ground-ring generators of the W_3 string. The fermionic screening current $\oint b\,D$ was first explicitly presented in [15]; in the simplified formalism introduced in [9] that we are using in this paper, it takes the form $\beta\,e^{\frac{2}{7}\alpha\varphi}$ [9] (clearly in general, every discrete operator (2.19) has an associated screening current $U_1(\varphi,\beta,\gamma)$). Dealing with the multiple contour integrals that arise is very cumbersome, since only certain powers of D give well-defined products. In practice, it seems that the easiest way to obtain explicit expressions for higher-level physical states is by directly solving the physical-state conditions (2.6). The utility of the screening current $\beta\, e^{\frac{2}{7}\alpha\varphi}$ is largely restricted to a descriptive rôle in organising the higher-level physical spectrum once it is already known by other means, rather than as a tool for *deriving* the full cohomology of the BRST operator. Some further details of its action were presented in [16], where its use in obtaining physical states already found in [14,9] was exhibited. It has recently been shown that there are discrete physical operators at level $\ell = 15$ which are invertible, and are thus guaranteed to give BRST-non-trivial physical states when normal ordered with any physical operators [17]. These, by contrast, can be used to *derive* the complete cohomolgy of the W_3 string, both in the two-scalar and the multi-scalar cases [17].

2.2 Interactions

Constructing BRST-invariant scattering amplitudes for the W_3 string remained an outstanding problem for quite a long time. The difficulty was that with the states of standard ghost structure known at that time, it was simply not possible to build multi-point correlation functions in which the total ghost number of the operators took the correct value ($G_{b,c} = 3$ and $G_{\beta,\gamma} = 5$) and the total φ momentum took the correct value (-2α, where α is the background charge). One can easily see this by looking at the form of the standard ghost structure states (2.16).

The key to building W_3 scattering amplitudes was the discovery [12,13] of physical states in the multi-scalar W_3 string with non-standard ghost structure. Now, it turns out that both of the above difficulties are eliminated at a stroke [14,9]. In fact, one builds BRST-invariant W_3-string scattering amplitudes by *precisely* the same techniques as one uses for the ordinary bosonic string [14,9]. In other words, one simply builds all possible correlation functions of the four sectors of physical operators of the theory. Those which achieve the correct balance of φ momentum and the correct overall ghost structure can be non-zero. (The use of the ghost boosters a_φ and a_{X^μ} is often important in order to achieve the correct overall ghost structure.) For three-point functions, this is the end of the story; since the physical operators have conformal weight zero under the total energy-momentum tensor, the results are simply constants, independent of the locations of the three operators on the worldsheet. For higher N-point functions, it is necessary, as in ordinary bosonic string theory, to make the replacement $V(z_i) \longrightarrow \oint dw\, b(w) V(z_i)$ for $(N-3)$ of the physical operators, so as to make spin-1 currents (screening operators) that can be integrated over the positions z_i to give an invariant amplitude [14,9].[†]

[†] After this procedure for computing scattering amplitudes for the W_3 string was obtained in [14,9], papers appeared [16] making the incorrect claim that the prescription in [14,9] was incomplete. Specifically, the authors of [16] claimed that unless one augmented the prescription of [14,9] by introducing a *new kind* of amplitude in which a certain screening charge was included, then one could not construct all the non-zero

We shall not give any detailed scattering-amplitude calculations here; many examples can be found in [14,9]. The upshot is that the "tower" of physical operators in any one of the four sectors described above all behave equivalently from the effective spacetime point of view, and essentially can be thought of as equivalent representatives of the same effective-spacetime physical state. Thus one effectively has three Virasoro-like string sectors, but with intercepts $\Delta = 1$, $\frac{15}{16}$ and $\frac{1}{2}$, and a fourth sector of discrete operators which, having $\Delta = 0$ in the effective spacetime, all behave like representatives of the identity.

To uncover the pattern of non-vanishing correlation functions, one must choose representatives appropriately from the four sectors (and use ghost-boosters if necessary) so as to achieve the correct φ-momentum balance and overall ghost structure. Note that for some amplitudes the fourth sector, of discrete physical states, can play an essential rôle in achieving the φ-momentum balance. This is the case for example for the scattering of four physical states from the $\Delta = \frac{15}{16}$ sector. In an obvious notation $\langle \frac{15}{16} \ \frac{15}{16} \ \oint b \frac{15}{16} \ \frac{15}{16} \rangle$ itself is zero, but $\langle \frac{15}{16} \ \frac{15}{16} \ \oint b \frac{15}{16} \ \oint bD \ \frac{15}{16} \rangle$ is non-zero, where D is the discrete physical operator (2.20). From the effective-spacetime point of view, the latter is also a four-point function, since D has $p_\mu = 0$. (It happens that this four-point function was not originally evaluated in [14,9]. Possibly the authors of [16] mistakenly thought that this meant that it could not be calculated, giving rise to the incorrect claims in [16] discussed in the footnote on the previous page.)

The full set of tree-level scattering amplitudes for the multi-scalar W_3 string turns out to admit a very simple interpretation: The scattering amplitudes are exactly what one would get for a critical Virasoro string in which one tensors the $c = \frac{51}{2}$ energy-momentum tensor (2.5) with the $c = \frac{1}{2}$ energy-momentum tensor for the Ising model. In fact, if one associates the $\Delta = 1$, $\frac{15}{16}$ and $\frac{1}{2}$ sectors of the theory with the identity operator 1, the spin field σ, and the energy operator ϵ of the Ising model, then the W_3 scattering amplitudes reproduce the fusion rules of the Ising model [14,9].

In a sense, this conclusion is a somewhat unexciting one. What one has learned is that the operators $U(\varphi, \beta, \gamma)$ in (2.18) provide a representation of the primary fields of the Ising model, realised on the (φ, β, γ) system. (From (2.11) and (2.12) one easily sees that T_φ has central charge $\frac{149}{2}$, and $T_{\gamma,\beta}$ has central charge -74, so their total is indeed the central charge $\frac{1}{2}$ of the Ising model.) Some further aspects were discussed in [18].

As mentioned in the introduction, if one solves the physical-state conditions (2.6) for the case of the *two-scalar* W_3 string (*i.e.* just one field X in addition to the field φ), the spectrum turns out to have many more states than simply those corresponding to specialising the

---

amplitudes of the W_3 string. The claim in [16] is obviously wrong, since the authors simply rediscovered the screening current $\oint dw\, b(w)D(z)$ where $D(z)$ is given by (2.20). (As is well known, acting on *any* physical operator with $\oint dw\, b(w)$ gives a screening current.) Thus inclusion of the screening charge is just a particular case of the general procedure of computing N-point correlation functions of physical operators that we described above, where one of the $N > 3$ operators happens to be the discrete physical operator D of (2.20). Apparently the authors of [16] were unaware that the screening current was simply $\oint dw\, b(w)D$, and therefore was already included in the procedure for calculating scattering amplitudes given in [14,9].

physical states discussed in section *2.1* to the case of a single X coordinate [12,13,14,9,19,17]. To put it another way, the two-scalar W_3 string has additional states in its physical spectrum that do not generalise to the multi-scalar case. These states have the property that they cannot be written in a factorised form such as (2.18) or (2.19). For example, at level $\ell = 3$ there is a physical operator $[21bc\gamma - 28\partial\gamma - \frac{16}{3}\alpha c + 12(\alpha\,\partial\varphi + a\,\partial X)\,\gamma]\,e^{-\frac{3}{7}\alpha\varphi + \frac{3}{7}aX}$, where $a = \alpha/\sqrt{3}$ is the background charge for the X coordinate. For such states, there is no effective-spacetime interpretation, and the W_3 symmetry evidently acts in a more non-trivial way than it does on the states of the multi-scalar W_3 string. It may well be, therefore, that the two-scalar W_3 string will provide more insights into the meaning of W_3 geometry.

3. Higher-spin generalisations

A natural generalisation of the above discussion would be to consider a string theory based on a different W algebra, such as the W_N algebra. It is believed that the multi-scalar W_N string would admit an effective spacetime interpretation as a $c = 26 - [1 - \frac{6}{N(N+1)}]$ Virasoro-type string tensored with the N'th unitary Virasoro minimal model [5,20]. More interesting possibilities emerge if one considers a W string with fewer higher-spin currents, since there will be correspondingly fewer constraints and therefore a richer physical spectrum. A case that has been considered recently is where one has just two currents, namely the energy-momentum tensor and a primary current of spin s [6]. The BRST operator for such a string can be obtained by writing an ansatz that generalises (2.8)–(2.14). In fact the only changes are to replace (2.10) by

$$Q_1 = \oint dz\,\gamma\,F(\varphi,\beta,\gamma), \tag{3.1}$$

where $F(\varphi,\beta,\gamma)$ is a spin-s operator to be solved for, and (β,γ) is now a ghost-system for spin s, so β has spin s, γ has spin $(1-s)$, and (2.12) is replaced by

$$T_{\gamma,\beta} = -s\,\beta\,\partial\gamma - (s-1)\,\partial\beta\,\gamma. \tag{3.2}$$

One can now attempt to solve for operators $F(\varphi,\beta,\gamma)$ which give a nilpotent BRST charge.

Solutions for $F(\varphi,\beta,\gamma)$ were found in [6] for the cases of $s = 4$, 5 and 6. It seems likely that solutions exist for arbitrary s, but their complexity grows rapidly with increasing s. In fact for $s = 4$ there are two distinct solutions, for $s = 5$ there is one, and for $s = 6$ there four. Just one solution in each case seems to be associated with a unitary string theory; it corresponds to T^{eff} having the central charge $c^{\mathrm{eff}} = 26 - \frac{2(s-2)}{(s+1)}$ [6].

We shall discuss only the solutions for Q_B that appear to be associated with unitary string theories here. It seems that in the multi-scalar case the physical states can again be interpreted as coming from a Virasoro-like string theory, in this case tensored with a $c = \frac{2(s-2)}{(s+1)}$ minimal model. This is the central charge of the lowest unitary minimal model of the W_{s-1} algebra. The physical states for the models with $s = 4$, 5 and 6 were analysed in some detail in [6], and evidence was found supporting the conjecture that the physical states correspond to those for an effective Virasoro string tensored with the lowest unitary W_{s-1} minimal model. In particular, it was found in [6] that some of the continuous-momentum physical operators, which take the general form (2.18), correspond to effective Virasoro operators $Y^{\text{eff}}(X)$ with intercepts Δ that are conjugate to the conformal weights h of the primary fields of the relevant W_{s-1} minimal model, in the sense that $\Delta = 1 - h$. The remaining continuous-momentum physical operators correspond to intercepts that are conjugate to weights having the form of a positive integer plus a weight of a primary field of the minimal model. This can be easily understood, and was investigated in detail in [21]:

The fields of the W_{s-1} minimal model can be viewed as fields of a Virasoro model with the same central charge. Primary fields of the W_{s-1} model will also be primary under Virasoro. In addition, fields that are W secondaries in the W_{s-1} model (*i.e.* fields that are built by acting with the negative modes of the higher-spin currents of the W_{s-1} algebra) will also be Virasoro primaries, with weights that are increased by some integers relative to those of the W_{s-1} primaries. Thus what we are seeing explicitly in these examples is the way in which the set of primary fields of a given Virasoro model can be more economically described in terms of a smaller set of primaries of a W minimal model of the same central charge. When $s = 4$, the W_3 minimal model has central charge $c = \frac{4}{5}$, so there is just a *finite* set of Virasoro primaries in this case. However, for $s = 5$ and $s = 6$ the corresponding W_4 and W_5 minimal models have central charges $c = 1$ and $c = \frac{8}{7}$ respectively. In these cases, an infinite number of Virasoro primaries are obtained from a finite number of W primaries and their W descendants.

We have seen for the examples of $s = 4$, 5 and 6 that the lowest unitary W_{s-1} minimal model, with central charge $c = \frac{2(s-2)}{(s+1)}$, can be realised in terms of the (φ, β, γ) system, with the primary and W-secondary fields of the model described by the set of operators $U(\varphi, \beta, \gamma)$ appearing in the physical operators (2.18) of the theory. Included amongst the W_{s-1} primary operators are the primary currents of the W_{s-1} algebra themselves, with spins $3, 4, \ldots (s-1)$. In fact they arise as physical operators with zero φ momenta, at levels that can easily be seen to be given by all integers ℓ in the interval $\frac{1}{2}s(s-1) + 3 \leq \ell \leq \frac{1}{2}s(s+1) - 1$. Thus we obtain an explicit realisation of W_{s-1} at central charge $c = \frac{2(s-2)}{(s+1)}$ in terms of a scalar field φ and the (β, γ) ghost system with spins $(s, 1 - s)$. Of course one can bosonise the (β, γ) ghosts, thereby obtaining a two-scalar realisation of W_{s-1} at the specific value of central charge given above. In interesting feature of these rather unusual realisations is that the OPEs of these primary currents close on the relevant W_{s-1} algebra modulo the appearance of certain additional null fields on the right-hand side [21]. For example, the case $s = 4$ gives a two-scalar realisation at $c = \frac{4}{5}$ which closes on W_3 modulo the appearance of an additional spin-4 primary null field in the OPE of the spin-3 generator with itself. Such realisations were studied previously in [22].

4. Conclusions

In this paper we have presented a brief review of the current status of some of the developments in W-string theory. We have concentrated mostly on the W_3 string, since this is the simplest non-trivial example and it already illustrates many of the new features that result from generalising ordinary bosonic string theory. We have also concentrated mostly on the multi-scalar case, since this enables one to have sufficiently many dimensions that one can give the theory a reasonably sensible spacetime interpretation. As we have seen, the spacetime theory that emerges is essentially that of a $c = \frac{51}{2}$ Virasoro string tensored with the Ising model.

Theories that are perhaps more interesting arise if we consider a W string in which the higher-spin local worldsheet symmetries are relatively sparse. In particular, if one considers the case of a W string with just two local symmetries, generated by currents of spins 2 and s, then as we saw in section 3 the multi-scalar theory is essentially equivalent to that of a $c = 26 - \frac{2(s-2)}{(s+1)}$ Virasoro string tensored with the lowest unitary W_{s-1} minimal model. The way in which this W_{s-1} symmetry is realised on the physical states, and the manner in which it enables one to classify the physical states of the theory, is quite intriguing.

In some sense the more subtle aspects of the W symmetry are lost in the multi-scalar W string. This reflects itself in the way in which the physical states factorise, as in (2.18), into the product of Virasoro states times minimal-model fields. If, however, one works instead with the basic "Miura"-type realisations of the W algebra, one finds that there are many more physical states that cannot factorise in this way, suggesting that they correspond to a more non-trivial realisation of the W symmetry. For the W_3 string, and the spin-2 plus spin-s strings discussed in section 3, these basic realisations are in terms of just two scalar fields, φ and X. If one is looking for a non-trivial realisation of the W symmetry, as opposed to a "realistic" spacetime theory, these basic "pure W-gravity" theories are probably of greater interest.

We have not touched on several important topics, notably the construction of a non-critical BRST operator for the W_3 string [15,23], which has two independent realisations of the W_3 symmetry, one for the "Liouville sector" and the other for the "matter sector." This is an extremely interesting area deserving of further study. Another topic that we have not covered is the attempt to give a full and rigorous derivation of the cohomology of the BRST operator. Progress in this direction has recently been achieved in [19,17].

Acknowledgments

I am very grateful to my collaborators Hong Lu, Bengt Nilsson, Stany Schrans, Ergin Sezgin, Kelly Stelle, Kris Thielemans, Xujing Wang and Kaiwen Xu.

REFERENCES

[1] A.B. Zamolodchikov, *Teor. Mat. Fiz.* **65** (1985) 1205; V.A. Fateev and A.B. Zamolodchikov, *Nucl. Phys.* **B280** (1987) 644; V.A. Fateev and S. Lukyanov, *Int. J. Mod. Phys.* **A3** (1988) 507.

[2] A. Bilal and J.-L. Gervais, *Nucl. Phys.* **B314** (1989) 646.

[3] C.N. Pope, L.J. Romans and K.S. Stelle, *Phys. Lett.* **B268** (1991) 167; *Phys. Lett.* **B269** (1991) 287.

[4] L.J. Romans, *Nucl. Phys.* **B352** (1991) 829.

[5] S.R. Das, A. Dhar and S.K. Rama, *Mod. Phys. Lett.* **A6** (1991) 3055; *Int. J. Mod. Phys.* **A7** (1992) 2295.

[6] H. Lu, C.N. Pope and X.J. Wang, "On higher-spin generalisations of string theory," CTP TAMU-22/93, hep-th/9304115.

[7] J. Thierry-Mieg, *Phys. Lett.* **B197** (1987) 368.

[8] F. Bais, P. Bouwknegt, M. Surridge and K. Schoutens, *Nucl. Phys.* **B304** (1988) 348.

[9] H. Lu, C.N. Pope, S. Schrans and X.J. Wang, "On the spectrum and scattering of W_3 strings," CTP TAMU-4/93, hep-th/9301099, to appear in *Nucl. Phys. B.*

[10] C.N. Pope, L.J. Romans, E. Sezgin and K.S. Stelle, *Phys. Lett.* **B274** (1992) 298.

[11] H. Lu, B.E.W. Nilsson, C.N. Pope, K.S. Stelle and P.C. West, *Int. J. Mod. Phys.* **A8** (1993) 4071.

[12] S.K. Rama, *Mod. Phys. Lett.* **A6** (1991) 3531.

[13] C.N. Pope, E. Sezgin, K.S. Stelle and X.J. Wang, *Phys. Lett.* **B299** (1993) 247.

[14] H. Lu, C.N. Pope, S. Schrans and X.J. Wang, *Nucl. Phys.* **B403** (1993) 351.

[15] M. Bershadsky, W. Lerche, D. Nemeschansky and N.P. Warner, *Nucl. Phys.* **B401** (1993) 304.

[16] M.D. Freeman and P.C. West, "The covariant scattering and cohomology of W_3 strings," KCL-TH-93-2, hep-th/9302114; "W_3 strings, parafermions and the Ising model," KCL-93-10, hep-th/9306134.

[17] H. Lu, C.N. Pope, X.J. Wang and K.W. Xu, "The complete cohomology of the W_3 string," CTP TAMU-50/93, hep-th/9309041.

[18] C.M. Hull, "New realisations of minimal models and the structure of W strings," NSF-ITP-93-65, hep-th/9305098.

[19] P. Bouwknegt, J. McCarthy and K. Pilch, "Semi-infinite cohomology of W algebras," USC-93/11, hep-th9302086.

[20] H. Lu, C.N. Pope, S. Schrans and K.W. Xu, *Nucl. Phys.* **B385** (1992) 99.

[21] H. Lu, C.N. Pope, K. Thielemans and X.J. Wang, "Higher-spin strings and W minimal models," CTP TAMU-43/93, KUL-TF-93/34, hep-th/9308114, to appear in *Class. Quantum Grav.*

[22] E. Bergshoeff, H.J. Boonstra and M. de Roo, "Realisations of W_3 symmetry," UG-6/92, hep-th/9209065.

[23] E. Bergshoeff, A. Sevrin and X. Shen, *Phys. Lett* **B296** (1992) 95.

RUNNING COUPLING CONSTANTS IN 2D GRAVITY

Christof Schmidhuber

California Institute of Technology, Pasadeana, CA 91125

The topic of this talk is the renormalization group flow in two–dimensional field theories that are coupled to gravity. I will explain the basic ideas at the example of the sine-Gordon model. I will define in a minute, what is meant by scale–dependent coupling constants in theories with gravity, where the scale itself is a dynamical variable. Before, it is necessary to generalize the theory of David, Distler and Kawai in order to describe two–dimensional quantum gravity coupled to *non*–conformal matter.

1. New Terms in the DDK Action

In David, Distler and Kawai's[1,2] approach, the action for 2D gravity on genus 0, coupled to a scalar matter field x with central charge $c = 1$, is usually assumed as

$$S = \frac{1}{8\pi} \int \sqrt{\hat{g}} \{ (\partial x)^2 + (\partial \phi)^2 + 2\sqrt{2}\hat{R}\phi + \tau^i \, \Phi_i(x) \, e^{\alpha_i \phi}$$
$$+ \text{cosmological constant} \}. \tag{1}$$

Here, $\hat{g}$ is the (fictitious) background metric, ϕ is the Liouville mode and the τ^i are dimensionless coupling constants (powers of a length scale a are omitted here and below). The $\Phi_i(x)$ are scaling operators of the matter theory, such as the sine–Gordon interaction $\cos px$. They are "gravitationally dressed" by exponentials $e^{\alpha_i \phi}$, where α_i is adjusted so that the dimensions of the dressed operators are two. Summation over i is understood in (1). We will ignore all problems associated with the presence of the cosmological constant. It can be argued that they have no qualitative effects on what follows.[3] This will be confirmed below by agreement with matrix model results.

As it stands, theory (1) has a problem: it is not scale invariant for finite τ^i. It should be scale invariant, because in quantum gravity scale invariance is part of the gauge invariance. In particular, all beta functions in the theory should be zero to all orders. But in (1) they aren't: generally, the beta functions are given by[4]

$$\beta^i \equiv \dot{\tau}^i = (\Delta^i_j - 2\delta^i_j)\, \tau^j + \pi c^i_{jk} \tau^j \tau^k + O(\tau^3). \tag{2}$$

Here, Δ^i_j is the dimension matrix computed with S_0, and c^i_{jk} are the operator product coefficients. It is clear that by adjusting α_i in (1) we can make the linear terms in

(2) vanish: $\Delta^i_j = 2\delta^i_j$.[1,2] But whenever there are nontrivial c^i_{jk}'s, the beta functions have quadratic pieces. By "nontrivial c^i_{jk}'s" I mean universal ones, that cannot be eliminated by picking a particular renormalization scheme. In the $c = 1$ model, universal c^i_{jk}'s are present whenever the interactions in (1) are "discrete primary fields", like $\cos px$ near the "special momenta" $p = \sqrt{2}$ or $p = \sqrt{2}/2$.[5]

The only way to cure the problem of nonvanishing beta functions is to add new terms to the action (1), of the form[3]

$$-\pi\, c^k_{ij}\tau^i\tau^j \int X_k(x,\phi) \; + \; \text{higher orders}, \tag{3}$$

where the operators X_k are, up to field redefinitions, uniquely determined by scale invariance at $O(\tau^2)$. Their form is simple, but let me just state the result for the example of the sine–Gordon model coupled to gravity, $\Phi_i(x) = \cos px$ with $p = \sqrt{2}+\epsilon$, ϵ being small. $p = \sqrt{2}$ is the momentum where the Kosterlitz–Thouless transition takes place in the sine–Gordon model without gravity. (1) plus (3) comes out to be:

$$S = \frac{1}{8\pi} \int \sqrt{\hat{g}}\{\partial x^2 + \partial\phi^2 + 2\sqrt{2}\hat{R}\phi + \text{cosmological constant}\}$$
$$+ m \int \cos(\sqrt{2}+\epsilon)x \; e^{\epsilon\phi} - \frac{\pi}{8\sqrt{2}}\, m^2 \int \phi\,\partial x^2. \tag{4}$$

2. Running Coupling Constants

I am now ready to come to my title – running coupling constants. We have just introduced new terms in order to insure that the coupling constants do *not* run – with respect to the background scale $\sqrt{\hat{g}}$. But the physical scale in DDK's approach is $\sqrt{\hat{g}}e^{\alpha\phi}$, $\alpha = -\sqrt{2}$. Therefore, a shift of ϕ by a constant λ,

$$\phi \to \phi + \lambda, \tag{5}$$

is a scale transformation (see refs. 3,6). So we can ask: how do we have to make m and ϵ λ–dependent in (4), so as to absorb the shift (5). $m(\lambda)$ and $\epsilon(\lambda)$ are what I call "running coupling constants." They are easy to find in this example:

$$m(\lambda) = m_0 e^{-\epsilon\lambda}, \quad \epsilon(\lambda) = \epsilon_0 - \frac{\pi^2}{2}\lambda m^2. \tag{7}$$

Here, m_0 and ϵ_0 are constants. In deriving $\epsilon(\lambda)$, the $\lambda m^2 \partial x^2$ term has been absorbed in a redefinition of x and in a shift of ϵ. Defining 'dot' as $\frac{d}{d\lambda}$, we get the lowest order "beta functions"

$$\dot{\epsilon} \sim -m^2, \quad \dot{m} \sim -\epsilon m. \tag{8}$$

The resulting flow diagram is shown in fig.1. It is qualitatively the same as the Kosterlitz–Thouless diagramm of the flat–space sine–Gordon model, although quantitatively it is modified by the effects of gravity. We see that the new term $\phi\,\partial x^2$ in (4) plays a crucial role: Via the shift of ϵ in (7), it causes the flow in the ϵ direction. Ignoring it would be like forgetting field renormalization in the ordinary sine–Gordon model. As a consequence of this term, there is a diagonal critical line at

$$\epsilon > 0, \quad |m| \sim \epsilon,$$

corresponding to a linear phase boundary. This fact agrees to this order in m and ϵ with recent matrix model results by Moore,[7] who found a singularity of the free energy at

$$\epsilon > 0, \quad |m| \sim \epsilon\, e^{\frac{1}{2}\sqrt{2}\epsilon\log\epsilon}. \tag{9}$$

Thus we "see" the $O(m^2)$ term in the Liouville action (4) in the matrix model results. It will be interesting to see if the logarithm in (9) follows from further modifications of (4), needed to keep the interaction near $p = \sqrt{2}$ marginal beyond $O(m^2)$.

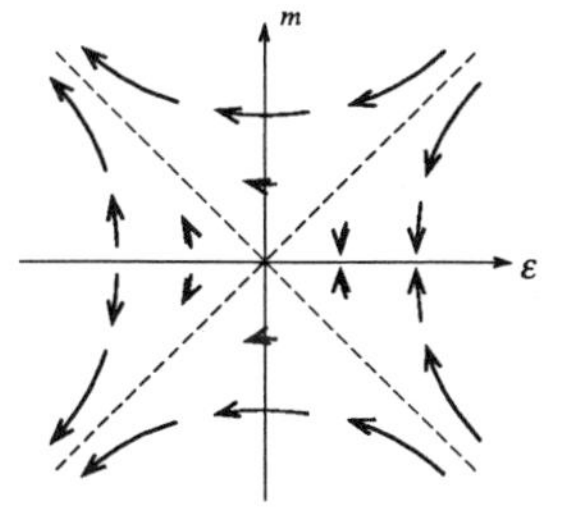

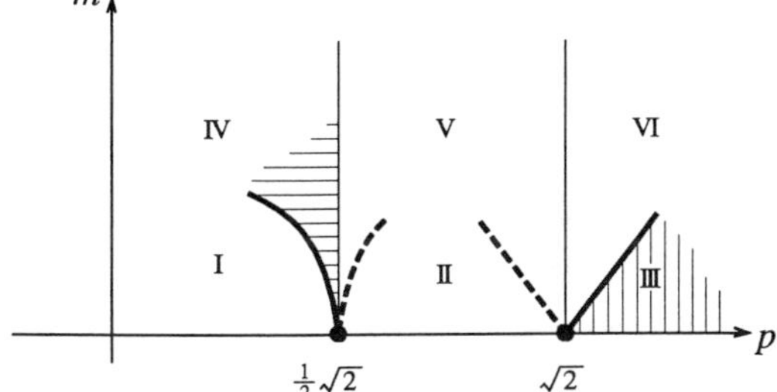

Fig.1: Kosterlitz–Thouless transition with gravity, to leading order; arrows point to the infrared.

Fig.2: Phase diagram for the sine–Gordon model coupled to gravity, to leading order.

3. Phase Diagram of the Sine–Gordon Model with Gravity

We can now interpret the phase diagram of ref. 7(fig. 2) near $m = 0, p = \sqrt{2}$: For $p < \sqrt{2}$ (regions II and V), m grows exponentially in the IR. The x field will thus get stuck in the minima of the sine–Gordon potential and the IR limit will be (infinitely many copies of) the $c = 0$, pure gravity model.[8] For $\epsilon > 0$ but m greater than a critical value $m_c(\epsilon)$ (region VI), the IR limit is again the $c = 0$ model. For $m < m_c(\epsilon)$ (region III), the model flows to the free $c = 1$ model. However, the domain of small ϵ, m is now the IR domain, where the cosmological constant cannot be neglected and further investigation is needed.

As a second example of the correction terms (3), consider the sine–Gordon model in the vicinity of half the Kosterlitz–Thouless momentum, $p = \frac{1}{2}\sqrt{2} + \delta$. To quadratic order, the action is[3]

$$
\begin{aligned}
S = \frac{1}{8\pi} \int & \sqrt{\hat{g}}\{\partial x^2 + \partial \phi^2 + 2\sqrt{2}\hat{R}\phi + \text{ghosts}\} \\
& + m \int \cos(\frac{1}{2}\sqrt{2} + \delta)x \; e^{(-\frac{1}{2}\sqrt{2}+\delta)\phi} + (\frac{\mu}{8\pi} + \frac{m^2\pi}{8\delta}) \int \phi \; e^{-\sqrt{2}\phi},
\end{aligned}
\tag{10}
$$

where we have used the form $\phi e^{-\sqrt{2}\phi}$ for the cosmological constant.[9] A cosmological constant term is induced at order m^2. It becomes comparable with the background cosmological constant at $\delta \sim m^2/\mu$. Although the situation is not entirely clear, we therefore expect some kind of phase transition at

$$
\delta < 0, \quad |m| \sim \sqrt{|\mu\delta|},
\tag{11}
$$

where the last term in (10) becomes negative (fig.2). Indeed, in the matrix model a singularity of the free energy has been found at[7]

$$
\delta < 0, \quad |m| \sim \sqrt{|\mu\delta|} \; e^{\frac{1}{2}\sqrt{2}\delta \log \delta}.
\tag{12}
$$

Let us therefore identify the region where $\mu + \frac{m^2}{\delta}\pi^2$ is negative with region IV of ref. 7. A further interpretation of the phase diagram must be left for the future work.

Let me summarize: There are new terms in the DDK action that describes two–dimensional quantum gravity coupled to nonconformal matter. These terms are required by scale invariance. They are crucial in deriving the renormalization group flow in the continuum theory. The results are consistent with the phase structure observed in the matrix model. Interestingly, we do not seem to make a big error by ignoring the problems associated with the presence of the cosmological constant.

References

1. F. David, *Mod. Phys. Lett.* **A3** (1988), 1651.
2. J. Distler and H. Kawai, *Nucl. Phys.* **B321** (1988), 509.
3. C. Schmidhuber, *Nucl. Phys.* **B404** (1993), 342.
4. J. H. Cardy, *Les Houches summer school, session XLIX* (1988).
5. I. Klebanov and A. M. Polyakov, *Mod. Phys. Lett.* **A6** (1991), 3273.
6. A. Cooper, L. Susskind and L. Thorlacius, *Nucl. Phys.* **B363** (1991), 132.
7. G. Moore, Yale preprint YCTP-P1-92, hepth@xxx/9203061
8. D. Gross and I. Klebanov, *Nucl. Phys.* **B344** (1990), 475.
9. N. Seiberg, *Prog. Theor. Phys.* **S102** (1990), 319.

A SYSTEMATIC APPROACH TO EXTENSIONS OF THE VIRASORO ALGEBRA AND 2D GRAVITIES

ALEXANDER SEVRIN

CERN, CH-1211

Geneva 23, Switzerland

and

KRIS THIELEMANS and WALTER TROOST

Instituut voor Theoretische Fysica, Universiteit Leuven

Celestijnenlaan 200D, B-3001 Leuven, Belgium

ABSTRACT

A very large class of (super)extensions of the Virasoro algebra is constructed from gauged WZW models. The gauge group is determined by an embedding of $sl(2)$ in a (super) Lie algebra. The theory is quantized and the field content of the extended Virasoro algebra is determined. The closure of this algebra is proven and the quantum Miura transformation is constructed. We show that this formulation allows for a simple way to obtain the effective action of the corresponding $2D$ gravity theory. We classify the models for which no coupling constant renormalization beyond 1 loop occurs. We speculate that this indicates the existence of "exotic" $N = 2$ structures which share much of the properties of ordinary $N = 2$ algebras. A concrete example is studied in some detail.

456

1. Introduction

The Virasoro algebra allows for various extensions, for a review see [1], such as the W_n algebras, the supersymmetric Virasoro algebra, etc. They play an important role in the study of conformal field theories, $2D$ gravity and integrable systems. In [2], it was argued that non-trivial embeddings of $sl(2)$ in a semi-simple (super) Lie algebra are closely related to extended Virasoro algebras.

Given an embedding of $sl(2)$ in a (super) Lie algebra $\bar{g}$, any highest weight representation $L(\Lambda)$ of $\bar{g}$ decomposes according to irreducible $sl(2)$ representations [3]:

$$L(\Lambda) = \bigoplus_{j \in \frac{1}{2}\mathbf{N}} n_j(\Lambda)\underline{2j+1}. \tag{1.1}$$

Taking eq. (1.1) for the adjoint representation of $\bar{g}$, one finds the $\bar{g}$ acquires a grading*:

$$\bar{g} = \bigoplus_{m \in \frac{1}{2}\mathbf{Z}} \bar{g}_m \quad \text{where} \quad \bar{g}_m = \{a \in \bar{g} | [e_0, a] = 2ma\}. \tag{1.2}$$

In [2], it was shown that by constraining the affine currents as $J_+ \equiv \frac{\kappa}{2}e_- + T$ where $T \in \ker \mathrm{ad}\, e_+$, the WZW Ward identities reduce to that of an extended Virasoro algebra generated by $n_j(\text{adjoint})$ currents of conformal dimension $j + 1$. Almost [4] all known extended Virasoro algebras can be obtained in this way. The hope exists that all extended Virasoro algebras either fall in this class or can be obtained by twisting, adding dimension 1/2 or 1 currents, further gauging of affine subalgebras, taking subalgebras, etc.

Many aspects of the classical version of these reductions were studied in [5]-[8]. In [10]-[13] important steps towards the understanding of the full quantum theory were made.

2. Gauged WZW Model

The affine Lie algebra $\hat{g}$ is realized by a WZW model† with action $\kappa S^-[g]$. The action $\mathcal{S}_1$

$$\mathcal{S}_1 = \kappa S^-[g] + \frac{1}{\pi x} \int str\, A_- \left(J_+ - \frac{\kappa}{2}e_- - \frac{\kappa}{2}[\tau, e_-] \right) + \frac{\kappa}{4\pi x} \int str[\tau, e_-]\partial_- \tau, \tag{2.1}$$

with the affine currents $J_+ = \frac{\kappa}{2}\partial_+ gg^{-1}$, the gauge fields $A_- \in \Pi_{>0}\bar{g}$ and the "auxiliary" fields $\tau \in \Pi_{1/2}\bar{g}$, is invariant under

$$g \to hg \qquad A_- \to \partial_- hh^{-1} + hA_- h^{-1} \qquad \tau \to \tau + \Pi_{\frac{1}{2}}\eta, \tag{2.2}$$

---

*Throughout the paper we use Π to denote the operator which projects $\bar{g}$ on a subspace determined by a restriction on the grading. E.g. $\Pi_{m \geq n}\bar{g} = \oplus_{m \geq n}\bar{g}_m$, $\Pi_n\bar{g} = \bar{g}_n$, etc.

†We follow the conventions of [11] and [13]

where $h = \exp\eta$, $\eta \in \Pi_{>0}\bar{g}$. The gauge fields A_- appear as Lagrange multipliers, imposing the constraint $\Pi_{<0}J_+ = \frac{\kappa}{2}e_- + \frac{\kappa}{2}[\tau, e_-]$. Using the gauge symmetry we can further reduce this current by choosing $\Pi_{\geq 0}J_+ \in \ker \mathrm{ad}\,e_+$ and $\tau = 0$, thus indeed reproducing the classical constraints of [2].

The generators of the extended Virasoro algebra are the polynomials in the affine currents which are gauge invariant modulo the constraints, *i.e.* the equations of motion of A_-. An explicit way to obtain them is given in [14] and they are of the form $T = CJ_+ + \cdots \in \ker \mathrm{ad}\,e_+$, where C is a normalization constant which will be determined later on. We can couple T to the action S_1:

$$S_2 = S_1 + \frac{1}{\pi x} \int str\,\mu T, \tag{2.3}$$

where $\mu \in \ker \mathrm{ad}\,e_-$, keeping the gauge invariance, provided the transformation rules of the gauge fields, A_-, are suitably modified. The price paid for this is that the gauge algebra now only closes on-shell.

The quantization of such a system calls for the Batalin-Vilkovisky formalism. Introducing ghost fields $c \in \Pi_{>0}\bar{g}$ and anti-fields $J_-^* \in \bar{g}$, $A_+^* \in \Pi_{<0}\bar{g}$, $\tau^* \in \Pi_{-1/2}\bar{g}$ and $c^* \in \Pi_{<0}\bar{g}$, one finds that the solution of the BV master equation is of the form:

$$S_{\mathrm{BV}} = S_2 - \frac{1}{2\pi x} \int str\,c^*cc + \frac{1}{2\pi x} \int str\,J_-^* \left(\frac{\kappa}{2}\partial_+c + [c, J_+]\right) + \frac{1}{2\pi x} \int str\,\tau^*c$$

$$+ \frac{1}{2\pi x} \int str\,A_+^* \left(\partial_-c + [c, A_-] + \text{terms linear in } \mu\right). \tag{2.4}$$

Choosing the gauge $A_- = 0$ amounts to performing a canonical transformation which changes A_+^* into a field, the antighost $b \in \Pi_{<0}\bar{g}$, and A_- into an antifield b^*. The gauge-fixed action reads:

$$S_{\mathrm{gf}} = \kappa S^-[g] + \frac{\kappa}{4\pi x} \int str[\tau, e_-]\partial_-\tau + \frac{1}{2\pi x} \int str\,b\partial_-c + \frac{1}{\pi x} \int str\,\mu\hat{T}, \tag{2.5}$$

and the nilpotent BRST charge is:

$$Q = \frac{1}{4\pi ix} \oint str\left\{c\left(J_+ - \frac{\kappa}{2}e_- - \frac{\kappa}{2}[\tau, e_-] + \frac{1}{2}J_+^{\mathrm{gh}}\right)\right\}, \tag{2.6}$$

where $J_+^{\mathrm{gh}} = \frac{1}{2}\{b, c\}$. The total current $\hat{J}_+ = J_+ + J_+^{\mathrm{gh}}$ has the important property that $\Pi_{\geq 0}\hat{J}_+$ satisfies almost the same operator algebra as $\Pi_{\geq 0}J_+$, the only difference being the form of the central extension. The explicit form of the current $\hat{T}$, except for its classical limit, is unknown. To determine $\hat{T}$, we observe that BRST invariance of the action requires $\hat{T}$ to be BRST invariant. This determines $\hat{T}$ up to BRST exact pieces.

This leads us to the study of the BRST cohomology. Consider the algebra $\mathcal{A}$ generated by the basic fields $\{b, \hat{J}_+, \tau, c\}$, which consists of all regularized products of the basic fields and their derivatives modulo the usual relations between different orderings, derivatives, etc. To every field Φ, we assign a double grading $[\Phi] = (k, l)$, where $k \in \frac{1}{2}\mathbf{Z}$ is the $sl(2)$ grading and $k + l \in \mathbf{Z}$ is the ghost number. The "auxiliary" fields τ are assigned the grading $[0, 0]$. The algebra $\mathcal{A}$ acquires a double grading:

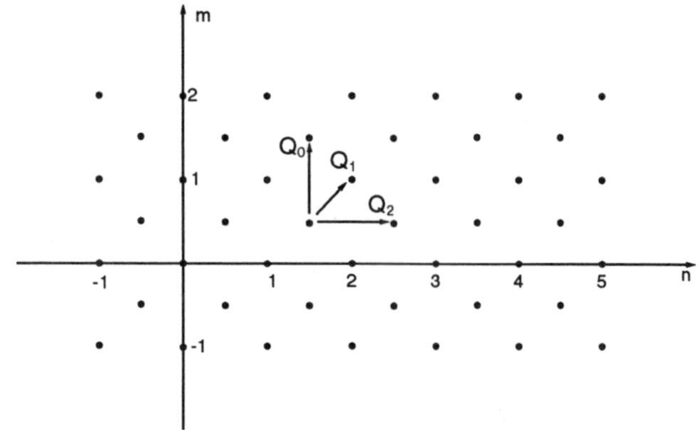

Figure 1: Q_0, Q_1 and Q_2 acting on $\mathcal{A}_{(\frac{1}{2},\frac{3}{2})}$.

$$\mathcal{A} = \bigoplus_{\substack{m,n \in \frac{1}{2}\mathbf{Z} \\ m+n \in \mathbf{Z}}} \mathcal{A}_{(m,n)}, \tag{2.7}$$

and operator product expansions (OPE) preserve the grading. The BRST charge decomposes into three parts of definite grading, $Q = Q_0 + Q_1 + Q_2$, with $[Q_0] = (1,0)$, $[Q_1] = (\frac{1}{2},\frac{1}{2})$ and $[Q_2] = (0,1)$:

$$Q_0 = -\frac{\kappa}{8\pi i x} \oint strce_-$$

$$Q_1 = -\frac{\kappa}{8\pi i x} \oint strc\,[\tau, e_-]. \tag{2.8}$$

As illustrated in Fig. 1, the operators Q_0, Q_1 and Q_2, map $\mathcal{A}_{(m,n)}$ to $\mathcal{A}_{(m+1,n)}$, $\mathcal{A}_{(m+\frac{1}{2},n+\frac{1}{2})}$ and $\mathcal{A}_{(m,n+1)}$ respectively. It follows from $Q^2 = 0$ that $Q_0^2 = Q_2^2 = \{Q_0, Q_1\} = \{Q_1, Q_2\} = Q_1^2 + \{Q_0, Q_2\} = 0$, but

$$Q_1^2 = -\{Q_0, Q_2\} = \frac{\kappa}{32\pi i x} \oint str\left\{c\left[\Pi_{1/2}c, e_-\right]\right\} \tag{2.9}$$

does not vanish. The action of Q_0, Q_1 and Q_2 on the basic fields is given by

$$
\begin{array}{llllllll}
Q_0 : & b & \to & -\frac{\kappa}{2}e_- & Q_1 : & b & \to & -\frac{\kappa}{2}[\tau, e_-] \\
& c & \to & 0 & & c & \to & 0 \\
& \hat{J}_+ & \to & -\frac{\kappa}{4}[e_-, c] & & \hat{J}_+ & \to & -\frac{\kappa}{4}[[\tau, e_-], c] \\
& \tau & \to & 0 & & \tau & \to & \frac{1}{2}\Pi_{+1/2}c
\end{array}
$$

$$
\begin{array}{llll}
Q_2 : & b & \to & \Pi_{<0}\hat{J}_+ \\
& c & \to & \frac{1}{2}cc \\
& \hat{J}_+ & \to & \frac{1}{2}[c, \Pi_{\geq 0}\hat{J}_+] + \frac{\kappa}{4}\partial_+ c \\
& & & \quad -\frac{1}{2}[\Pi_{<0}(t^A), \\
& & & \qquad [\Pi_{>0}(t_A), \partial_+ c]] \\
& \tau & \to & 0,
\end{array}
$$

$$\tag{2.10}$$

where $\vec{[}X, Y] = (-)^{(AB)}\,(X^A Y^B)\,f_{AB}{}^C t_C$, and $(X^A Y^B)$ is a regularized product. [‡] The BRST charge Q acts as a derivation on a regularized product of fields.

[‡]Note that we use for the supertrace a similar formula, *i.e.* without reordering the fields.

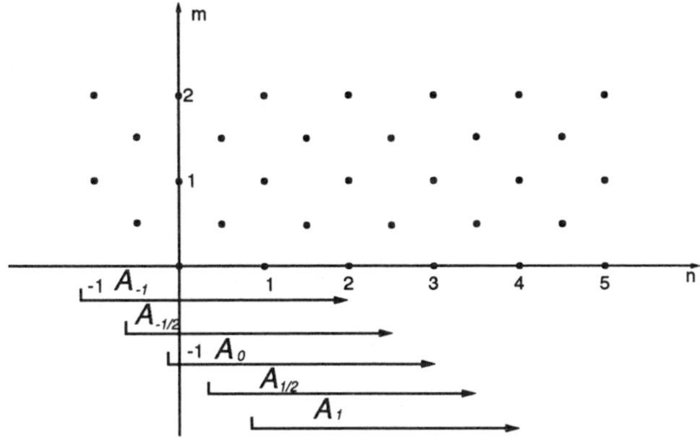

Figure 2: $\widehat{A}$ and its filtration $\widehat{A}^n$.

The subcomplex $\mathcal{A}^{(1)}$, generated by $\{b, \Pi_{<0}\hat{J}_+ - \frac{\kappa}{2}[\tau, e_-]\}$ has a trivial cohomology $H^*(\mathcal{A}^{(1)}; Q) = \mathbf{C}$. From this and eq. (2.10) one shows that one can as well compute the cohomology of the reduced complex $\widehat{A} = \mathcal{A}/\mathcal{A}^{(1)}$, generated by $\{\Pi_{\geq 0}\hat{J}_+, \tau, c\}$ as $H^*(\mathcal{A}) = H^*(\widehat{A})$. The OPEs close on the reduced complex. Obviously the double grading on $\mathcal{A}$ is inherited by $\widehat{A}$. We now apply the theory of spectral sequences [15] to solve the cohomology on $\widehat{A}$. We limit ourselves here to stating the final results. Full details can be found in [13]. The filtration $\widehat{A}^m$, $m \in \frac{1}{2}\mathbf{Z}$ of $\widehat{A}$ (see Fig. 2):

$$\widehat{A}^m \equiv \bigoplus_{k \in \frac{1}{2}\mathbf{Z}} \bigoplus_{l \geq m} \widehat{A}_{(k,l)}. \tag{2.11}$$

leads to a spectral sequence $E_r = H^*(E_{r-1}; d_{r-1})$, $r \geq 1$, converging to $H^*(\widehat{A}; Q)$. One shows $E_1 = H^*(\widehat{A}; Q_0) \simeq \widehat{A}\left[\Pi_{\ker \mathrm{ad}\, e_+} \hat{J}_+\right] \otimes \widehat{A}[\tau] \otimes \widehat{A}\left[\Pi_{\frac{1}{2}}c\right]$, where $\widehat{A}[\Phi]$ denotes the subalgebra of $\widehat{A}$ generated by Φ. The spectral sequence collapses after the next step and we find

$$H^*(\mathcal{A}; Q) \simeq E_2 = H^*(E_1; Q_1) = \widehat{A}\left[\Pi_{\ker \mathrm{ad}\, e_+}\left(\hat{J}_+ + \frac{\kappa}{4}[\tau, [e_-, \tau]]\right)\right]. \tag{2.12}$$

This result means that the dimension of the cohomology equals the number of $sl(2)$ irreps in the branching of the adjoint representation of $\bar{g}$.

We now outline a recursive procedure to obtain explicitly the generators of the cohomology: the tic-tac-toe construction [15] (see Fig. 3). We make a choice for the basis of $\bar{g}$: $\{t_{(jm;\alpha_j)}\}$ where $j \in \frac{1}{2}\mathbf{Z}$, $m = -j, -j+1, \cdots, j$ and $\alpha_j = 1, \cdots n_j$ (adjoint). The generators of the cohomology $\hat{T} \equiv \sum_{j,\alpha_j} \hat{T}^{(j,\alpha_j)} t_{(jj;\alpha_j)} \in \ker \mathrm{ad}\, e_+$ are of the form

$$\hat{T}^{(j,\alpha_j)} = \sum_{r=0}^{2j} \hat{T}_r^{(j,\alpha_j)}, \tag{2.13}$$

where $\hat{T}_r^{(j,\alpha_j)}$ has grading $(j - \frac{r}{2}, -j + \frac{r}{2})$. The leading term $\hat{T}_0^{(j,\alpha_j)}$ is given by

$$\hat{T}_0^{(j,\alpha_j)} = \mathcal{C}\left\{\hat{J}_-^{(jj;\alpha_j)} + \delta_{j,0}\frac{\kappa}{4}[\tau, [e_-, \tau]]^{(00;\alpha_j)}\right\} \tag{2.14}$$

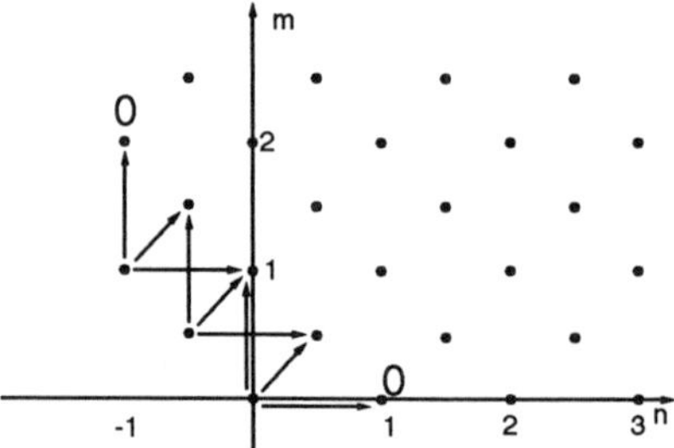

Figure 3: The tic-tac-toe construction for a conformal dimension 2 current.

and the remaining terms are recursively determined by

$$Q_0 \hat{T}_r^{(j,\alpha_j)} = -Q_1 \hat{T}_{r-1}^{(j,\alpha_j)} - Q_2 \hat{T}_{r-2}^{(j,\alpha_j)}, \tag{2.15}$$

where $\hat{T}_r^{(j,\alpha_j)} = 0$ for $r < 0$ or $r > 2j$. Of course, there is a certain ambiguity in this construction as one can always add combinations of the other generators, as this preserves BRS-invariance. The fact that the last term in the expansion Eq. (2.14) has grading $(0,0)$ follows from considering the mirror spectral sequence $E'_r = H^*(E'_{r-1}; d'_{r-1})$, associated to the filtration

$$\hat{\mathcal{A}}'^m \equiv \bigoplus_{l \in \frac{1}{2}\mathbf{Z}} \bigoplus_{k \geq m} \hat{\mathcal{A}}_{(k,l)}. \tag{2.16}$$

Using eq. (2.10), one shows that $E'_1 = H^*(\hat{\mathcal{A}}; Q_2)$ is only non-vanishing at grading $(\frac{m}{2}, \frac{m}{2})$, $m \geq 0$. As we already know that E'_∞ for this spectral sequence vanishes unless the ghostnumber is zero, we find that $\hat{T}_{2j}^{(j,\alpha_j)}$ is indeed the last, non-vanishing term in Eq. (2.14).

For small values of j, we can give explicit expressions for the generators [13], however in general the explicit form of $\hat{T}^{(j,\alpha_j)}$ must be computed on a case by case basis.

The energy-momentum tensor $\hat{T}^{\text{EM}}$ itself can be computed. It is proportional to $\hat{T}^{(1,0)}$. However, we choose to add the "diagonal" quadratic combination of the $\hat{T}^{(0,\alpha_0)}$, which corresponds to the Sugawara tensor of the affine algebra of the $j = 0$ currents. By analogy to the classical case [5], this choice for $\hat{T}^{\text{EM}}$ is expected to give that the other generators are primary (if one doesn't use the freedom of adding additional combinations of the generators there). We end up with

$$\hat{T}^{\text{EM}} = \frac{\kappa}{x(\kappa + \tilde{h})} \left(str\left\{ \hat{J}_+ e_- \right\} + str\left\{ [\tau, e_-] \hat{J}_+ \right\} + \frac{1}{\kappa} str\left\{ \Pi_0(\hat{J}_+) \Pi_0(\hat{J}_+) \right\} + \frac{\kappa + \tilde{h}}{\kappa} str\left\{ e_- \partial_+ \hat{J}'_+ \right\} \right.$$

$$\left. + \frac{1}{\kappa} str\left\{ \left[\Pi_0(t^A), \left[\Pi_0(t_A), \partial_+ \hat{J}'_+ \right] \right] e_- \right\} - \frac{\kappa + \tilde{h}}{4} str\left\{ [\tau, e_-] \partial_+ \tau \right\} \right), \tag{2.17}$$

where $\hat{J}'_+ \in \Pi_{\geq 1} \bar{g}$ is defined by $[e_-, \hat{J}'_+] = \Pi_{\geq 0} \hat{J}_+ - \Pi_{\ker \text{ad } e_+} \hat{J}_+$. The first term is $\hat{T}_0^{(1,0)}$, the second term $\hat{T}_1^{(1,0)}$ and the remainder forms $\hat{T}_2^{(1,0)}$.

It is not hard to verify that $\hat{T}^{\mathrm{IMP}}$[§]

$$\hat{T}^{\mathrm{IMP}} \equiv \frac{1}{x(\kappa+\tilde{h})}str J_+ J_+ - \frac{1}{2x}str e_0 \partial_+ J_+ - \frac{\kappa}{4x}str\left([\tau,e_-]\partial_+\tau\right)$$

$$+\frac{1}{4x}strb[e_0,\partial_+c] - \frac{1}{2x}strb\partial_+c + \frac{1}{4x}str\partial_+b[e_0,c], \tag{2.18}$$

is also BRST invariant. One obtains it from $\hat{T}^{\mathrm{EM}}$ by adding a BRST exact term $Q(\frac{2}{x(\kappa+\tilde{h})}strb(J_+ + \frac{1}{4}\{b,c\}))$ to $\hat{T}^{\mathrm{EM}}$. One finds for the central extension [11]:

$$c = \frac{1}{2}c_{\mathrm{crit}} - \frac{(d_B - d_F)\tilde{h}}{\kappa+\tilde{h}} - 6y(\kappa+\tilde{h}). \tag{2.19}$$

Here d_B and d_F are the number of bosonic and fermionic generators of $\bar{g}$ and c_{crit} is the critical value of the central charge for the extension of the Virasoro algebra under consideration:

$$c_{\mathrm{crit}} = \sum_{j,\alpha_j}(-)^{(\alpha_j)}(12j^2 + 12j + 2). \tag{2.20}$$

Also, y is the index of embedding, which in the case $\tilde{h}\neq 0$ is given by

$$y = \frac{1}{3\tilde{h}}\sum_{j\alpha_j}(-)^{(\alpha_j)}j(j+1)(2j+1), \tag{2.21}$$

and $(-)^{(\alpha_j)} = +1 \ (-1)$ if $t_{(jm,\alpha_j)}$ is bosonic (fermionic). The requirement that $\hat{T}^{\mathrm{EM}}$ generates the Virasoro algebra in the standard normalization fixes the normalization constant $\mathcal{C}$ to

$$\mathcal{C} = \frac{2y\kappa}{\kappa+\tilde{h}}. \tag{2.22}$$

Knowing the leading term of the currents, eq. (2.14), we find that the conformal dimension of $\hat{T}^{(j,\alpha_j)}$ is given by $j+1$.

We now come to the main result: *The generators $\hat{T}$ form an extension of the Virasoro algebra with n_j(adjoint) generators of conformal dimension $j+1$ and with the value of the central charge given by Eq. (2.19).*

To finish the proof of this statement, it only needs to be shown that the currents $\hat{T}$ close under OPEs. From the fact that the OPEs close on $\hat{A}$ and preserve the grading, one deduces that the OPEs of the generators $\hat{T}$ close modulo BRST exact terms. However, there are no states of negative ghostnumber in the reduced complex $\hat{A}$, implying that the OPEs of the generators $\hat{T}$ close.

The quantum Miura transformation is obtained as the map $\hat{T} \to \hat{T}_{(0,0)}$ which, as one can easily verify, is an algebra isomorphism.

3. An example : $OSp(N|2)$

In this section we present an explicit example to make our treatment more concrete (see also [11], especially for the conventions used).

[§] We normalized $\{e_0, e_\pm\}$ such that $[e_0, e_\pm] = \pm 2e_\pm$, $[e_+, e_-] = e_0$, $str(e_0 e_0) = 2str(e_+, e_-) = 4xy$. This formula corrects a misprint in the coefficient of the second term.

The Lie algebra of $OSp(N|2)$ is generated by a set of bosonic generators $\{t_+, t_0, t_=, t_{ab}; t_{ab} = -t_{ba} \text{ and } a, b \in \{1, \cdots, N\}\}$ which form an $Sl(2) \times SO(N)$ Lie algebra and a set of fermionic generators $\{t_{+a}, t_{-a}; a \in \{1, \cdots, N\}\}$. A Lie algebra valued field $A = A^+ t_+ + A^0 t_0 + A^= t_= + \frac{1}{2} A^{ab} t_{ab} + A^{+a} t_{+a} + A^{-a} t_{-a}$ where the representation matrices are in the fundamental representation (index $x = 1/2$), is given by:

$$A \equiv \begin{pmatrix} A^0 & A^+ & A^{+b} \\ A^= & -A^0 & A^{-b} \\ A^{-a} & -A^{+a} & A^{ab} \end{pmatrix}, \tag{3.1}$$

and $A^{ab} = -A^{ba}$. From this, one reads off the generators of $OSp(N|2)$ in the fundamental representations and one can easily compute the (anti)commutation relations. The dual Coxeter number for this algebra is $\tilde{h} = \frac{1}{2}(4 - N)$. The supertrace of two fields is

$$str(A, B) = 2 A^0 B^0 + A^+ B^= + A^= B^+ - A^{ab} B^{ba} + (-1)^B 2 A^{-a} B^{+a} - (-1)^B 2 A^{+a} B^{-a}, \tag{3.2}$$

with $(-1)^B$ a phase factor depending on the parity of the B (-1 for ghosts, $+1$ for other fields).

The constraints implemented by Eq. (2.1) are in our example simply[¶]

$$J^= = \frac{\kappa}{2} \qquad\qquad J^{-a} = -\frac{\kappa}{2}\tau^{+a} . \tag{3.3}$$

The elements of the classical extended Virasoro algebra are polynomials which are invariant (up to equations of motion) under the $\Pi_{>0} OSp(N|2)$ gauge transformations. The generators are found by imposing the unique (finite) $\Pi_{>0} OSp(N|2)$ gauge transformation which yields $J^0 = \tau^{+a} = 0$. They are given by :

$$\begin{aligned}
\tilde{J}^+_+ &= J^+ + \frac{2}{\kappa} J^0 J^0 + 2 J^{+a} \tau^{+a} - \sqrt{2} \lambda_{ab}{}^i J^i \tau^{+a} \tau^{+b} - \partial J^0 - \frac{\kappa}{2} \partial \tau^{+a} \tau^{+a}, \\
\tilde{J}^{+a}_+ &= J^{+a} + \sqrt{2} \lambda_{ab}{}^i J^i \tau^{+b} + J^0 \tau^{+a} - \frac{\kappa}{2} \partial \tau^{+a} \\
\tilde{J}^i_+ &= J^i + \frac{\kappa}{2\sqrt{2}} \lambda_{ab}{}^i \tau^{+a} \tau^{+b} ,
\end{aligned} \tag{3.4}$$

where $\lambda_{ab}{}^{(pq)} \equiv 1/\sqrt{2}(\delta^p_a \delta^q_b - \delta^q_a \delta^p_b)$. We refer to [11] for the variations of these polynomials, and give only one example :

$$\delta \tilde{J}^i_- = -\frac{\pi}{2\sqrt{2}} \lambda_{ab}{}^i \eta^{+a} \frac{\delta S_1}{\delta A^{+b}_+} . \tag{3.5}$$

Computing Poisson brackets of these generators, we can readily identify $G^a = 4i \tilde{J}^{+a}_+$ and $U^i = -2\sqrt{2} \tilde{J}^i_+$ with the generators of the classical N-extended $SO(N)$ superconformal algebra. However, the corresponding Virasoro operator is given by adding to $\tilde{J}^+$ the Sugawara tensor of the $SO(N)$ affine subalgebra (in the $\tilde{J}_-$ basis). We define

$$T_s = 2\tilde{J}^+ + s\frac{4}{\kappa} \tilde{J}^i \tilde{J}^i . \tag{3.6}$$

[¶]Note that τ^{+a} changed sign with respect to [11] because $[\tau_{\text{here}}, e_-] = \tau_{\text{there}},$.

Of course, T_s is gauge-invariant for any s, but it satisfies the Virasoro Poisson brackets only for $s = 0, 1$. The latter choice is in general to be preferred as the other generators are then primary fields [5].

We know couple these generators to the action $\mathcal{S}_1$ (Eq. (2.3)).

$$\mathcal{S}_2 = \mathcal{S}_1 + \frac{1}{\pi} \int hT_s + \psi^a G_a + A^i U_i. \tag{3.7}$$

To preserve the gauge invariance of the resulting action, we have to modify the transformation of the gauge fields such that the equation of motion terms in (3.5) are canceled. This reflects itself in the terms proportional to the antifields A^*_+ in the Batalin-Vilkovisky action. Because the new gauge algebra closes only on-shell, we need terms quadratic in the antifields to find an extended action satisfying the BV master equation. The result is (see Eq. (2.4), and Eq. (4.20) of [11]) $^{\|}$

$$
\begin{aligned}
\mathcal{S}_{\mathrm{BV}} \;=\; & (\text{terms independent of } h, \psi^a, A^i) + \frac{1}{\pi} \int A^{\ddagger *}_+ \left(-c^{\ddagger}\partial h - \frac{4}{\kappa}c^{\ddagger}hJ^0_- - 2ic^{\ddagger}\psi^a\tau^{+a} \right) \\[6pt]
& - \frac{2}{\pi} \int A^{+b*}_+ \left(-\frac{1}{2}\gamma^{+b}\partial h + c^{\ddagger}h\tau^{+b} - \frac{2}{\kappa}\gamma^{+b}hJ^0_- + s\frac{2\sqrt{2}}{\kappa}\lambda_{ab}{}^i\gamma^{+a}J^i_- h \right. \\[6pt]
& \left. - ic^{\ddagger}\psi^b + i\gamma^{+a}\psi^a\tau^{+b} - i\gamma^{+b}\psi^a\tau^{+a} - i\gamma^{+a}\psi^b\tau^{+a} - \lambda_{ab}{}^i\gamma^{+a}A^i \right) \\[6pt]
& - \frac{2}{\pi\kappa} \int A^{\ddagger *}_+ A^{+a*}_+ \gamma^{+a}c^{\ddagger}h + s\frac{1}{\pi\kappa} \int A^{+a*}_+ A^{+a*}_+ \gamma^{+b}\gamma^{+b}h. \tag{3.8}
\end{aligned}
$$

We proceed now by choosing the gauge $A_- = 0$ and rename the antifields into antighosts :

$$A^{+a*}_+ = \beta^{-a}, \qquad A^{\ddagger *}_+ = b^= \tag{3.9}$$

and end up with the gauge-fixed action Eq. (2.5) :

$$\mathcal{S}_{\mathrm{gf}} = \kappa S^-[g] - \frac{\kappa}{\pi} \int \tau^{+a}\bar{\partial}\tau^{+a} + \frac{1}{\pi} \int b^=\bar{\partial}c^{\ddagger} - \frac{2}{\pi} \int \beta^{-a}\bar{\partial}\gamma^{+a} + \frac{1}{\pi} \int \left(h\hat{T}_s + \psi^a\hat{G}_a + A^i\hat{U}_i \right), \tag{3.10}$$

where $\hat{T}_s$, $\hat{G}_a$ and $\hat{U}_i$ have precisely the form given before but with the currents J replaced by their hatted counterparts :

$$
\begin{aligned}
\hat{J}^{\ddagger}_+ \;&=\; J^{\ddagger}_+ \\[6pt]
\hat{J}^{+a}_+ \;&=\; J^{+a}_+ - \frac{1}{2}c^{\ddagger}\beta^{-a} \\[6pt]
\hat{J}^0_+ \;&=\; J^0_+ - \frac{1}{2}b^=c^{\ddagger} + \frac{1}{2}\beta^{-a}\gamma^{+a} \\[6pt]
\hat{J}^i_+ \;&=\; J^i_+ + \frac{1}{\sqrt{2}}\lambda_{ab}{}^i\beta^{-a}\gamma^{+b}. \tag{3.11}
\end{aligned}
$$

$^{\|}$Note that we use a more convenient normalization for the antifields here compared to [11]. We have $A^{+a*}_{+\,here} = -\frac{\pi}{2}A^{+a*}_{z\,there}$ and $A^{\ddagger *}_{+\,here} = \pi A^{\ddagger *}_{z\,there}$.

Note that this substitution rule is valid independent of the value of s one takes. The hatted generators are classically invariant under the action of the BRS-charge Eq. (2.6).

We now go to the quantum case. For this we need the OPEs of all fields, which follow from the action $\mathcal{S}_{\text{gf}}$. They are listed in [11]. We can then use the results for the cohomology of the BRS operator of section 2. $\hat{J}'_+$ is in the $OSp(N|2)$ case given by $-\hat{J}^0_+ t_+$. We get the following representants of the cohomology of Q :

$$\hat{T}^{(0,i)} = \hat{J}^i + \frac{\kappa}{2\sqrt{2}}\lambda_{ab}{}^i \tau^{+a}\tau^{+b}$$

$$\hat{T}^{(1/2,a)} = \hat{J}^{+a} + \sqrt{2}\lambda_{ab}{}^i \hat{J}^i \tau^{+b} + \hat{J}^0 \tau^{+a} - \frac{\kappa+1}{2}\partial\tau^{+a}$$

$$\hat{T}^{(1,0)}_s = \hat{J}^{\ddagger} + \frac{2}{\kappa}\hat{J}^0_+ \hat{J}^0_+ + 2\hat{J}^{+a}_+ \tau^{+a} - \sqrt{2}\lambda_{ab}{}^i \hat{J}^i \tau^{+a}\tau^{+b} - \frac{\kappa+1}{\kappa}\partial\hat{J}^0_+$$

$$\qquad - \frac{2\kappa+3}{4}\partial\tau^{+a}\tau^{+a} + s\frac{2}{\kappa}\hat{T}^{(0,i)}\hat{T}^{(0,i)} . \tag{3.12}$$

These expressions are the same as the classical ones (3.4, 3.6), up to certain corrections coming from ordering ambiguities. Note that in contrast to the classical case, $\hat{T}^{(1,0)}_s$ for $s = 0$ does not satisfy the Virasoro algebra, while it does for $s = 1$.

One can now check that these currents (with $s = 1$) indeed form a representation of the quantum $SO(N)$ superconformal algebra with the appropriate normalization factors and the value for the central charge as given by Eq. (2.19).

4. Application to 2D Gravity

Consider a matter system, where we denote the matter fields collectively by φ, with as action $S[\varphi]$ and with a set of symmetry currents, denoted by $T^i[\varphi]$, which form an extension of the Virasoro algebra. The induced action in the light-cone gauge is defined by

$$e^{-\Gamma[\mu]} = \int [d\varphi] e^{-S[\varphi] - \frac{1}{\pi}\int \mu_i T^i[\varphi]} , \tag{4.1}$$

where μ_i are generalized Beltrami differentials acting as sources. The generating functional of its connected Greens functions, which upon a Legendre transform becomes the effective action, is defined by

$$e^{-W[\tilde{T}]} = \int [d\mu] e^{-\Gamma[\mu] + \frac{1}{\pi}\int \mu_i \tilde{T}^i}$$

$$= \int [d\varphi]\delta(T[\varphi] - \tilde{T})e^{-S[\varphi]}. \tag{4.2}$$

Evaluating this functional integral is in general impossible as it involves the computation of a usually very complicated Jacobian. In our present case, due to the particular choice of matter system, *i.e.* a gauged WZW model, we can actually compute the Jacobian and as a result obtain an all order expression for the effective action.

This program was performed for the case of W_3 gravity in [16], proving the expression for the effective action conjectured in [17, 18]. In [11]-[13] the general case was treated. The effective action is defined by

$$\exp -W[\check{T}] = \int [\delta g g^{-1}][d\tau][dA_-][d\mu] \left(\mathrm{Vol}\left(\Pi_{>0}\bar{g}\right)\right)^{-1} \exp -\left(\mathcal{S}_2 - \frac{1}{\pi x}\int str\mu\check{T}\right). \qquad (4.3)$$

In order to compute the effective action, one chooses the gauge $\Pi_{\geq 0}J_+ \in \ker \mathrm{ad}\ e_+$ and $\tau = 0$ which amounts to a canonical transformation in eq. (2.4) which interchanges fields and anti-fields for $\{\tau, \tau^*\}$ and $\{\Pi_{>0}[e_+, J_+], \Pi_{<0}[e_-, J_+^*]\}$. We find

$$W[\check{T}] = \kappa_c S_-[g] \qquad (4.4)$$

where $\kappa_c = \kappa + 2\tilde{h}$ and we used $[\delta g g^{-1}] = [dJ_+] \exp\left(-2\tilde{h}S^-[g]\right)$. From eq. (2.19) we get the level as a function of the central charge:

$$12 y \kappa_c = 12 y \tilde{h} - \left(c - \frac{1}{2}c_{\mathrm{crit}}\right) - \sqrt{\left(c - \frac{1}{2}c_{\mathrm{crit}}\right)^2 - 24(d_B - d_F)\tilde{h}y} \qquad (4.5)$$

Eq. (4.5) provides an all-order expression for the coupling constant renormalization. The WZW model in eq. (4.4) is constrained by

$$\partial g g^{-1} + \frac{1}{4xy} str\left\{\Pi_{\mathrm{NA}}\left(\partial g g^{-1}\right)\right)\Pi_{\mathrm{NA}}\left(\partial g g^{-1}\right)\right\} e_+ = e_- + \frac{1}{\kappa + \tilde{h}} \sum_{j,\alpha_j} \frac{1}{2^{\frac{3}{2}j-1}y^j} \check{T}^{(j\alpha_j)} t_{(jj,\alpha_j)}, \qquad (4.6)$$

where $\Pi_{\mathrm{NA}}\bar{g}$ is the projection on the centralizer of $sl(2)$ in $\bar{g}$.

5. Discussion

From eq. (4.5), one deduces that for generic values of κ, no renormalization of the coupling constant beyond one loop occurs if and only if either $d_B = d_F$ or $\tilde{h} = 0$ (or both). We get $d_B = d_F$ for $su(m \pm 1|m)$, $osp(m|m)$ and $osp(m+1|m)$ and $\tilde{h} = 0$, for $su(m|m)$, $osp(m+2|m)$ and $D(2,1,\alpha)$. Note that $P(m)$ and $Q(m)$ have not been considered, since the absence of an invariant metric implies that no WZW models exist for them. The nonrenormalization of the couplings is reminiscent of nonrenormalization theorems [7] for extended supersymmetry. These imply that under suitable circumstances at most one loop corrections to the coupling constants are present (the wave function renormalization may have higher order contributions). Comparing our list with the tabulation [8] of super-W algebras obtained from a (classical) reduction of superalgebras, we find that many of them, though not all, have $N = 2$

supersymmetry. For instance, there is an sl_2 embedding in $osp(3|2)$ which gives the super-W_2 algebra of [9], which contains four fields (dimensions 5/2, 2, 2, 3/2). It has no renormalizations, but no $N = 2$ subalgebra. We checked the even stronger statement using a Mathematica^{TM} package [19] that it is not possible to find an associative algebra in $N = 2$ superspace with only a dimension 3/2 superfield with SOPE closing on itself and a central extension. Also, it seems that all superalgebras based on the reduction of the unitary superalgebras $su(m|n)$ contain an $N = 2$ subalgebra, whereas our list contains only the series $|m - n| \leq 1$. Clearly, the structural reason behind the lack of renormalization beyond one loop remains to be clarified.

6. Acknowledgements

We would like to thank Jan de Boer, Alex Deckmyn and Wolfgang Lerche for most enjoyable discussions.

7. References

1 P. Bouwknegt and K. Schoutens, *Phys. Rep.* **223** (1993) 183.

2 F. A. Bais, T. Tjin and P. van Driel, *Nucl. Phys.* **B357** (1991) 632.

3 E. B. Dynkin, *Amer. Math. Soc. Transl.* **6** (1967) 111.

4 H. Kausch, private communication, and W. Eholzer, A. Honecker and R. Hübel, preprint BONN-HE-93-08, hep-th/9302124.

5 L. Feher, L. O' Raifeartaigh, P. Ruelle, I. Tsutsui and A. Wipf, *Phys. Rep.* **222** (1992) 1; preprint DIAS-STP-91-29, hep-th/9112068.

6 L. Feher, L. O'Raifeartaigh, P. Ruelle and I Tsutsui, preprint BONN-HE-93-14, hep-th/9304125.

7 M. T. Grisaru, W. Siegel and M. Roček, Nucl. Phys. **B159** (1979) 429; *Superspace*, S. J. Gates, M. T. Grisatu, W. Siegel and M. Roček, Benjamin/Cummings pub. comp. 1983, p358.

8 L. Frappat, E. Ragoucy and P. Sorba, preprint ENSLAPP-AL-391-92, hep-th/9207102.

9 J. M. Figueroa-O'Farrill and S. Schrans, Phys. Lett. **B257** (1991) 69

10 J. de Boer and T. Tjin, preprint THU-93-05, hep-th/9302006.

11 A. Sevrin, K. Thielemans and W. Troost, *Nucl. Phys.* **B407** (1993) 459.

12 A. Sevrin, K. Thielemans and W. Troost, *Phys. Rev.* **D48** (1993) 1768.

13 A. Sevrin and W. Troost, *Phys. Lett.* **B315** (1993) 304.

14 A. Deckmyn, A. Sevrin, R. Siebelinck and W. Troost, in preparation.

15 R. Bott and L. W. Tu, *Differential Forms in Algebraic Topology*, (Springer Verlag, 1986).

16 J. de Boer and J. Goeree, *Nucl. Phys.* **B401** (1993) 348.

17 H. Ooguri, K. Schoutens, A. Sevrin and P. van Nieuwenhuizen, *Comm. Math.*

Phys. **145** (1992) 515.

18 K. Schoutens, A. Sevrin and P. van Nieuwenhuizen, *Nucl. Phys.* **B371** (1992) 315.

19 S. Krivonos, K. Thielemans, to be published.

Non-Critical String Models as Topological Coset models.

J. Sonnenschein and S. Yankielowicz[*]

School of Physics and Astronomy
Beverly and Raymond Sackler
Faculty of Exact Sciences
Ramat Aviv Tel-Aviv, 69978, Israel

ABSTRACT

The topological coset model appraoch to non-critical string models is summarized. The action of a topological twisted $\frac{G}{H}$ coset model ($rank\ H = rank\ G$) is written down. A "topological coset algebra" is derived and compared with the algebraic structure of the $N = 2$ twisted models. The cohomology on a free field Fock space as well as on the space of irreducible representation of the "matter" affine Lie algebra are extracted. We compare the results of the $A_1^{(N-1)}$ at level $k = \frac{p}{q} - N$ with those of (p, q) W_N strings.

A $\frac{SL(2,R)}{SL(2,R)}$ model which corresponds to the $c = 1$ is written down. A similarity transformation on the BRST charge enables us to extract the full BRST cohomoloy. One to one correspondence between the physical states of the $c = 1$ and the corresponding coset model is found.

---

[*] Work supported in part by the US-Israel Binational Science Foundation and the Israel Academy of Sciences.

.1. INTRODUCTION

Topological quantum field theoreis[1](TQFTs) were recently found to be a very useful tool in the study of string theories. Non-critical string models or 2D gravitational models serve as a laboratory to explore new domains of string theories such as the non-pertubative behaviour. In this talk we summarize the attemp to analayze the $c \leq 1$ models and their W_N generalizations as TQFTs.

A TQFT is a QFT in which all the observables, namely, all correlators of "physical operators", are invariant under any arbitrary deformation of $g_{\alpha\beta}$ the metric of the underlying space-time. Given a set of physcial operators $F_i[\Phi^a(x_i)]$ which are functional of the fields $\Phi^a(x)$, $a = 1, ...p$ of the theory and which are invariant under the symmetries of the theory, then the theory is topological iff

$$\delta_{g_{\alpha\beta}} < \prod_i F_i[\Phi^a(x)] >= 0. \tag{1}$$

In particular this definition implies that all correlators are independent of distances between the operators. In a theory where the energy-momentum tensor $T_{\alpha\beta}$ is exact under a BRST-like symmetry operator Q, namely, $T_{\alpha\beta} = \{Q, G_{\alpha\beta}\}$, all physical operators are in the cohomology of Q property (1) is obeyed. Thus, it is a TQFT.

A two-dimensional theory which is a TQFT as well as conformal, is a topological conformal field theory (TCFT). The algebraic structure of these models is characterized by the fact that T and the Q are both BRST exact. However, there is no unique structure which is common to all TCFTs. In fact it is shown here that the "topological coset algebra" differ from the algebra of the twisted $N = 2$ models.[2]

Every conformal field theroy coupled to 2-D gravity is obviously a TCFT after integrating over all metric degrees of freedom. We now define the concept of a "topological model" which is a TQFT without the introduction and integration over the metric. In more than two dimensions examples of such models are the Chern-Simons theory and the four dimensional theory which corresponds to the Donaldson invariants.[1] In two dimensions an example is the theory of flat gauge connections.[3] We are now ready to introduce the notion of "topological coset models" which are gagued WZW models that are also topological models.[4,5]

The paper is organized as follows. In section 2 we write down the quantum action of a topological $\frac{G}{G}$ and twisted $\frac{G}{H}$ coset models as a sum of "decoupled" matter, gauge and ghost sectors. The algebraic structure is derived in section 3 and it is found that a larger algebra then that of the TCFT[2] algebra. Section 4 is devoted to a brief reminder of the cohomology on a free field Fock space as well as on an irreducible representation of the "matter" affine Lie algebra. In section 5 a comparison between the physical states of topological coset models and the corresponding string models is made. We discuss the relation between gauge holonomies and the twist needed to achieve the correspondence with the gravitational model.

We then write down a topological coset model which describes the $c = 1$ string model. A special treatment of the BRST charge using a double similarity tranformation is used to exactract the space of physical states. The latter are then compared with the ground ring and other states of the two dimensional string theory.

.2. THE QUANTUM ACTION

The topological coset models are constructed by gauging an anomaly free diagonal subgroup $H \in G$ group of a level k WZW model. The $\frac{G}{G}$ models are defined by $H = G$ whereas the general twisted $\frac{G}{H}$ models require that $rank\, H = rank\, G$. In the latter case the gauged action is a twisted supersymmetric G-WZW model, namely, the usual $\frac{G}{H}$ model with an extra set of $(1,0)$ anti-commuting ghosts where the dimension one fields take their values in the positive roots of $\frac{G}{H}$ and the dimension zero fields in the negative ones. A great part of the discussion that follows applies to general compact groups, however, we will mainly concentrate on $G = SU(N)$ and the non-compact $SL(N)$ groups. The latter are needed since we will be interested also in fractional levels. We proceed to derive the quantum action of the $\frac{G}{G}$ and the twisted $\frac{G}{H}$ models.

The $\frac{G}{G}$ model

The classical action of the $\frac{G}{G}$ model takes the form

$$S_k(g, A, \bar{A}) = S_k(g) + i\frac{k}{2\pi} \int_\Sigma d^2z Tr(g^{-1}\partial g\bar{A} + g\bar{\partial}g^{-1}A - \bar{A}g^{-1}Ag + A\bar{A}) \qquad (2)$$

where $g \in G$ and $S_k(g)$ is the WZW action at level k. An essential step in the analysis of this action is a "decoupling" of the matter and gauge degrees of freedom. This can be acheived by rewriting the gauge fields in terms of group elements. In the case of a topologically trivial Σ the gauge field can be parametrized as follows $A = ih^{-1}\partial h, \bar{A} = i\bar{h}\partial\bar{h}^{-1}$ where $h(z) \in G^c$. The action then[6,7,8] reads

$$S_k(g, A) = S_k(g) - S_k(h\bar{h}) \qquad (3)$$

The Jacobian of the change of variables introduces a dimension $(1,0)$ system of anticommuting ghosts χ and ρ in the adjoint representation of the group. The quantum action thus takes the form of

$$S_k(g, h, \rho, \chi) = S_k(g) - S_k(h\bar{h}) - i \int d^2z Tr[\rho\bar{D}\bar{\chi} + c.c] \qquad (4)$$

where $D\chi = \partial\chi - i[A, \chi]$. This action involves an interaction term of the form $Tr_H(\bar{\rho}[h^{-1}\partial h, \bar{\chi}])$ and a similar term for ρ, χ. By performing a chiral rotation $\bar{\rho} \to h^{-1}\bar{\rho}h$ and $\bar{\chi} \to h^{-1}\bar{\chi}h$ with $\rho \to \bar{h}\rho\bar{h}^{-1}$ and $\chi \to \bar{h}\chi\bar{h}^{-1}$, one achieves a decoupling of the whole ghost system. The price of that is an additional $S_{-2C_G}(h\bar{h})$ term in the action resulting form the corresponding anomaly. C_G is the second Casimir of the adjoint representation. This result can be derived by using a non-abelian bosonization of the ghost system.[9] The final quantum action after gauge fixing $\bar{h} = 1$

$$S_k(g, h, \rho, \chi) = S_k(g) - S_{-(k+2C_G)}(h) - i \int d^2 z Tr[\rho \bar{\partial} \chi + c.c], \tag{5}$$

indeed, has the structure of three "decoupled" affine Lie algebra actions.

The twisted $\frac{G}{H}$ model

The classical action of the twisted $\frac{G}{H}$ model is given by

$$S_{(tKS)} = S_k(g, A, \bar{A}) + S_{(gh)}^{\frac{G}{H}}$$

$$S_k(g, A, \bar{A}) = S_k(g) - \frac{k}{2\pi} \int_\Sigma d^2 z Tr_G[g^{-1}\partial g \bar{A}_{\bar{z}} + g\bar{\partial}g^{-1}A - \bar{A}g^{-1}Ag + A\bar{A}] \tag{6}$$

$$S_{gh}^{\frac{G}{H}} = \frac{i}{2\pi} \int d^2 z \sum_{\alpha \in \frac{G}{H}} [\rho^{+\alpha}(\bar{D}\chi)^{-\alpha} + \bar{\rho}^{+\alpha}(D\bar{\chi})^{-\alpha}]$$

Using the same parametrization as for the $\frac{G}{G}$ case, one finds after inserting the Jacobian and gauge fixing the following action

$$S_k(g, A) = S_k(hg\bar{h}) - S_k(h\bar{h}) + \frac{i}{2\pi} \int d^2 z Tr_H[\rho \bar{D}\chi + \bar{\rho}D\bar{\chi}]. \tag{7}$$

The twisted Kazama-Suzuki[10] action is given by

$$\begin{aligned}
S_{(tKS)} = S_k(g) - \sum_{I=1}^{n} S_k(h^{(I)}) - \frac{k}{4\pi} \int d^2 z \sum_{s=1}^{r} \partial \mathcal{H}_s \bar{\partial} \mathcal{H}_s \\
+ \frac{i}{2\pi} \int d^2 z Tr_H[\rho \bar{D}\chi + \bar{\rho}\partial \bar{\chi}] + \frac{i}{2\pi} \int d^2 z \sum_{\alpha \in \frac{G}{H}} [\rho^{+\alpha}(\bar{\partial}\chi)^{-\alpha} + \bar{\rho}^{+\alpha}(D\bar{\chi})^{-\alpha}]
\end{aligned} \tag{8}$$

for $G = SU(N)$ and $H = SU(N_1) \times ... \times SU(N_n) \times U(1)^r$ with $r = N - 1 - \sum_{I=1}^{n} N_I + n$, where the gauge fields A take the form of $A = i \sum_{I=1}^{n} h^{(I)^{(-1)}} \partial h^{(I)} + i \sum_{s=1}^{r} \partial \mathcal{H}_s$.

After chiral rotating the ghost fields the action takes the following form

$$\begin{aligned}
S_k = S_k(g) + \sum_{I=1}^{n} S_{-(k+C_G+C_{H^{(I)}})}(h^{(I)}) + \\
\frac{1}{2\pi} \int d^2 z [\sum_{s=1}^{r} \partial \mathcal{H}_s \bar{\partial} \mathcal{H}_s + i\sqrt{\frac{2}{k+C_G}}(\vec{\rho}_G - \vec{\rho}_H) \cdot \vec{\mathcal{H}}R] \\
+ \frac{i}{2\pi} \int d^2 z Tr_H[\bar{\rho}\partial \chi + \rho \bar{\partial}\chi] + \frac{i}{2\pi} \int d^2 z \sum_{\alpha \in \frac{G}{H}} [\rho^{+\alpha}(\bar{\partial}\chi)^{-\alpha} + \bar{\rho}^{+\alpha}(\partial \bar{\chi})^{-\alpha}]
\end{aligned} \tag{9}$$

where we have normalized the $\vec{\mathcal{H}}$ fields to be free bosons, and $\vec{\rho}_G$ and $\vec{\rho}_H$ are half the sums of the positive roots of G and H respectively. The action is composed of three decoupled sectors: the matter sector, the gauge sector and the ghost sector involving ghosts in H and $\frac{G}{H}$.

.3. THE TOPOLOGICAL COSET ALGEBRA

An important class of TCFT is the twisted $N = 2$ superconformal field theories. These models are characterized by the "TCFT algebra"[2] given by

$$
\begin{aligned}
T(z) &= \{Q, G(z)\} \qquad Q(z) = [Q, j^{\#}(z)] \\
\{Q, Q(z)\} &= 0 \qquad \{G, G(z)\} = 0
\end{aligned}
\tag{10}
$$

where the holomorphic currents, $T(z)$, $j^{\#}(z)$, $Q(z)$ and $G(z)$ are the energy-momentum ternsor, a $U(1)$ current, a BRST-like current and an anti-commuting dimension two current,respectively. Obviously these holomorphic currents have also their anti-holomorphic partners with the same algebra. Let us now examine whether the topological coset model share the same algebraic structure.

The $\frac{G}{G}$ models

Let us first examine the G affine Lie algebra . There are three sets of holomorphic G transformations which leave (7) invariant $\delta_J g = i[\epsilon(z), g]$ $\delta_I h = i[\epsilon(z), h]$ and $\delta_{J(gh)} \chi^a = i f^a_{bc} \epsilon^b \chi^c$; $\delta_{J(gh)} \rho^a = -i f^a_{bc} \epsilon^b \rho^c$ with ϵ in the algebra of G. The corresponding currents J^a, I^a and $J^{(gh)a} = f^a_{bc} \chi_b \rho_c$ satisfy the G affine Lie algebra at levels $k, -(k + 2c_G)$ and $2c_G$ respectively. Out of all possible linear combinations of these cuurent the following one is very useful,

$$
J^{(tot)a} = J^a + I^a + J^{(gh)a} = J^a + I^a + i f^a_{bc} \chi_b \rho_c.
\tag{11}
$$

It obeys an affine Lie algebra at level

$$
k^{(tot)} = k - (k + 2c_G) + 2c_G = 0.
\tag{12}
$$

The energy-momentum tensor $T(z)$ is a sum of Sugawara terms of the J^a and I^a currents and the usual contribution of a $(1, 0)$ ghost system, namely[6,7,8]

$$
T(z) = \frac{1}{k + c_G} : J^a J^a : -\frac{1}{k + c_G} : I^a I^a : +\rho^a \partial \chi^a.
\tag{13}
$$

The corresponding Virasoro central charge vanishes

$$
c^{(tot)} = \frac{k d_G}{k + c_G} - \frac{(k + 2c_G) d_G}{-(k + 2c_G) + c_G} - 2d_G = 0
\tag{14}
$$

This last property is an indication that the $\frac{G}{G}$ model is a TCFT.

The $Q(z)$ cuurent has an obvious realization. It is just the BRST current which emerges from the gauge fixing of the original gauged action. It is then also easy to find its dimension two partner G

$$Q(z) = \chi_a[J^a + I^a + \tfrac{1}{2}J^{(gh)a}]$$

$$G = \frac{1}{k + c_G}\rho_a[J^a - I^a]. \tag{15}$$

It is straightforward to check that T is BRST exact, $T(z) = \{Q, G(z)\}$. The BRST current itself is also BRST exact with respect to the $U(1)$ ghost number current $j^{\#} = \chi^a\rho_a$, $Q(z) = [Q, j^{\#}(z)]$. By its construction the BRST charge is nilpotent so one may conclude that indeed the topological coset model share the TCFT algebra. In fact the algebraic structure is different since G defined above is not a nilpotent operator (apart from the case of $U(1)$). Instead one finds $\{G, G(z)\} \equiv W(z) = \frac{1}{4C_G}f_{abc}J^{(tot)a}\rho^a\rho^b + \partial\rho^a\rho_a$. To close the algebra[11] one has to introduce one additional anticommuting current of dimemsion three $U = \frac{1}{12C_G}f_{abc}J^{(tot)}\rho^a\rho^b$ such that the full "topological coset algebra" is given by

$$\begin{aligned}
\tilde{T}(z) &= \{Q, G(z)\} \quad & Q(z) &= \{Q, j^{\#}(z)\} \\
\{Q, Q(z)\} &= 0 \quad & W(z) &\equiv \{G, G(z)\} \\
\tilde{W}(z) &= \{Q, U(z)\} \quad & [W, W(z)] &= 0.
\end{aligned} \tag{16}$$

The twisted $\frac{G}{H}$ models

There are two twisted $N = 2$ symmetry algebras in the twisted $\frac{G}{H}$ models. One of them which emerges from the H gauge fixing part, is of the "topological coset algebra" type whereas the other, which corresponds to the fact that the origianal action is a twisted $N = 2$ WZW model, is a "TCFT algebra". For the derivaition of the "physical states" the relevant algebra is a "topological coset algebra" which is a direct sum of the the two. Let us start with the H sector. The following set of currents obey the "topological coset model" of eqn. (16):

$$\begin{aligned}
T^H(z) = {}& \frac{1}{2(k + c_G)}g_{ab} : (J^a + J^a_{\frac{G}{H}})(J^b + J^b_{\frac{G}{H}}) : - \frac{1}{2(k + c_G)}g_{ab} : I^aI^b : \\
& - \frac{\sqrt{2}}{k + C_G}(\vec{\rho}_G - \vec{\rho}_H) \cdot \partial(\vec{J} + \vec{I} + \vec{J}_{\frac{G}{H}}) + g_{ab}\rho^a\partial\chi^b \\
J^{\#}(z) = {}& \rho^a\chi_a \\
Q(z) = {}& g_{ab}\chi^a[J^b + I^b + J^b_{\frac{G}{H}} + \tfrac{1}{2}J^{(gh)b}_H] \\
= {}& g_{ab}\chi^a[J^b + I^b + \tfrac{i}{2}f^b_{cd}\rho^c\chi^d + if^b_{\gamma,-\beta}\rho^\gamma\chi^{-\beta}] \\
G^H = {}& \frac{g_{ab}}{2(k + c_G)}\rho^a[J^b - I^b + if^b_{\gamma,-\beta}\rho^\gamma\chi^{-\beta}] - \frac{\sqrt{2}}{k + C_G}(\vec{\rho}_G - \vec{\rho}_H) \cdot \partial\vec{\rho}
\end{aligned} \tag{17}$$

474

where $\vec{J}$, $\vec{I}$ and $\vec{J}_{\frac{G}{H}}$ are the Cartan-subalgebra currents given in the basis in which $[J_n^i, J_m^j] = kn\delta^{ij}\delta_{m+n}$.

The $\frac{G}{H}$ algebra on the other hand involves the following currnts

$$T^{\frac{G}{H}}(z) = \frac{1}{2(k+c_G)}g_{\tilde{a}\tilde{b}} : J^{\tilde{a}}J^{\tilde{b}} : -\frac{1}{2(k+c_G)}g_{ab} : (J^a + J_{\frac{G}{H}}^a)(J^b + J_{\frac{G}{H}}^b) :$$

$$+ \frac{\sqrt{2}}{k+C_G}(\vec{\rho}_G - \vec{\rho}_H)\cdot\partial(\vec{J} + \vec{J}_{\frac{G}{H}}) + \sum_{\alpha\in\frac{G}{H}}\rho^{+\alpha}(\partial\chi)^{-\alpha}$$

$$J^{\#}(z) = \rho^{+\alpha}\chi_{-\alpha} \tag{18}$$

$$Q^{\frac{G}{H}} = \sum_{\alpha\beta\gamma\in\frac{G}{H}} \chi^{-\alpha}(J^\alpha + \frac{i}{2}f_{\gamma,-\beta}^\alpha\rho^\gamma\chi^{-\beta})$$

$$G^{\frac{G}{H}} = \frac{1}{k+C_G}\sum_{\alpha\beta\gamma\in\frac{G}{H}}\rho^\alpha(J^{-\alpha} + \frac{i}{2}f_{\gamma-\beta}^{-\alpha}\rho^\gamma\chi^{-\beta}).$$

where $\tilde{a}$ and $\tilde{b}$ go over the adjoint of G. The combined algebra is based on the set of generators which have the form $A(z) = A^H(z) + A^{\frac{G}{H}}(z)$. The combined energy momentum tensor aquires now the simplified form

$$T(z) = \frac{1}{2(k+c_G)}g_{\tilde{a}\tilde{b}} : J^{\tilde{a}}J^{\tilde{b}} : -\frac{1}{2(k+c_G)}g_{ab} : I^a I^b : +g_{ab}\rho^a\partial\chi^b$$

$$- \frac{\sqrt{2}}{k+C_G}(\vec{\rho}_G - \vec{\rho}_H)\partial\vec{I} + \sum_{\alpha\in\frac{G}{H}}\rho^{+\alpha}(\partial\chi)^{-\alpha} \tag{19}$$

with a total Virasoro central charge is found to be

$$c = \frac{kd_G}{k+C_G} + \sum_{I=1}^{n}\frac{(k+C_G+C_{H^I})d_{H^I}}{k+C_G} + r - 2d_H - (d_G - d_H) + 6[\sqrt{\frac{2}{k+C_G}}(\vec{\rho}_G - \vec{\rho}_H)]^2 = 0 \tag{20}$$

where we have used, assuming G is a simply laced group, the relations $12\rho_G^2 = d_G C_G$, and $\vec{\rho}_H \cdot (\vec{\rho}_G - \vec{\rho}_H) = 0$.

.4. BRST COHOMOLOGY AND PHYSICAL STATES

Next we proceed to extract the space of physical states of the model. We take as our definition of a physical state a state in the cohomology of $Q = Q^{(BRST)} + Q^{\frac{G}{H}}$, namely, $|phys> \in H^*(Q)$. The computation of the cohomology is based on a spectral sequence decomposition approach.[12] The extraction of the physical states was worked out in detail in refs. [9,13,14]. Here we will mention one feature of this method and the final results. The method is based on " Wakimoto bosonization" of the matter (J) and "gauge" (I) currents.[13] There are two possible bosonizations which are related by an automorphism, for instance in the $SL(2,R)$ case, $J^+ \leftrightarrow J^-, J^0 \leftrightarrow -J^0$. Denoting the two parametrizations by $+$ and $-$ one has the two options $(+,+)$ and $(+,-)$ for the (J,I) system. In [9,13,14] we have used a $(+,-)$ scheme since only in this way a convenient grading decomposition is possible. Eventhough this bosonization lacks an $SL(2,R)$ invariant vacuum, after projecting to the space of irreducible representations of affine Lie algebra[15] the appropriate vacuum invariance is restored. The full cohomology on the Fock space of free fields was found to be

$$H(Q) \simeq H^{rel}(Q) \oplus \sum_{\{k_1,...,k_l\}} \chi_0^{k_1} \cdots \chi_0^{k_l} H^{rel}(Q) \quad . \tag{21}$$

where the sum is over $k_1, ..., k_l$ namely all possible subsets of $1, ..., N-1$.

The relative cohomology , $H^{rel}(Q)$ is the cohomology on the space of vanishing zero modes of the components of ρ in the Cartan sub-alegebra of H which is given by

$$H^{rel}(Q) = \{ \prod_{\alpha \in H, \alpha > 0} \chi_0^\alpha |\vec{J}, \vec{I} >; \quad \vec{J} + \vec{I} + 2\vec{\rho}_H = 0\}. \tag{22}$$

where $\vec{\rho}_H$ is half the sum of the positive roots of H.

The topological coset models, as will be shown below, are intimately realted to gravitational (and W_N) models. This correspondence occurs once the cohomology of the theory is taken in the space of irreducible representations of the J sector affine Lie algebra . To simplify the picture we present here only the results the of the $A_1^{(1)} \frac{G}{G}$ model.[13] The results for the general topological coset model cases are given in refs. [14,9]. The space of physical states is composed of states built on $J = J_{r,s}$ where r and s are integers with either $r, s \geq 1$ or $r < 0, s \leq 0$ and $2J_{r,s} + 1 = r - (s-1)(k+2)$. For each such $J_{r,s}$ there is an infinite set of states with $I = I_{-r-2lp,s}$ $G = -2l$ and $I = I_{r-2lp,s}$ $G = 1 - 2l$ for every positive integer l. For integer k we have $J = 0, .., \frac{k}{2}$. Let us now examine the index interpretation of the torus partition function.[7] One has to insert the values of $\hat{L}_0 = L_0 - \frac{1}{k+2}[J(J+1) - I(I+1)]$ and $\hat{J}^0_{(tot)} = J^{(tot)0}_0 - (J + I + 1+)$ into $Tr[(-)^G q^{\hat{L}_0} e^{i\pi\theta \hat{J}^0_{(tot)}}]$. The end result[13] is

$$Tr[(-)^G q^{\hat{L}_0} e^{i\pi\theta \hat{J}^0_{(tot)}}] = 2iq^{\frac{-1}{4(k+2)}} e^{-i\pi\frac{\theta}{2}} M_{k,J}(\tau,\theta). \tag{23}$$

where

$$M_{k,j}(\tau,\theta) = \sum_{l=-\infty}^{\infty} q^{(k+2)(l+\frac{j+\frac{1}{2}}{(k+2)})^2} \sin\{\pi\theta[(k+2)l + j + \tfrac{1}{2}]\} \tag{24}$$

is the numerator of the character which corresponds to the highest weight state J. We have, thus, rederived using the BRST cohomology the path integral results of ref. [7], for the torus partition function.

.5. COMPARISON WITH STRING MODELS

The main motivation to study the topological coset model is the idea that they are closely related to non-critical string models. More specifically, we expect a correspondence between the $A_{N-1}^{(1)}$ twisted $\frac{G}{H}$ models and W_N strings and, in particular, between the case of $G = SL(2)$ and minimal models coupled to gravity. Therefore, we would like to examine now whether one can map the topological coset models into string models. In fact, for reasons that will be clarified shortly, the comparison with the gravitational models should be done with the topological coset models only after twisting their energy-momentum tensor. For $G = SL(N, R)$ the latter is given by

$$T(z) \to \tilde{T}(z) = T(z) + \sum_{i=1}^{N-1} \partial J^{(tot)^i}(z) \tag{25}$$

For the $N = 2$ case this type of twist in $T(z)$ corresponds to an addition of the term proportional to $\omega \bar{J}_0^{tot} + \bar{\omega} J^{(tot)}{}_0$ to the action of eqn. (7), where we use the following expression for the curvature $R^2 = \bar{\partial}\omega + \partial\bar{\omega}$. It is easy to realize that a similar modification of the action arises when one introduces an holonomy θ^0 in the parametrization of the gauge fields namely $A^0 = Tr[T^0 h^{-1}\partial h] + \theta^0$ and identifies θ^0 with ω. In the general $SL(N, R)$ case the holonomies are in the Kartan sub-algebra and they give rise to a $\theta^i J^{(tot)}{}_i + \bar{\theta}^i J^{(tot)}{}_i$ term.

Obviously, since $T(z)$ and $\partial J^{(tot)^i}(z)$ are BRST exact so is $\tilde{T}(z)$. Thus, the total Virasoro anomaly is unchanged. However, the contribution of each sector to c is modified as follows

$$c_J \to \tilde{c}_J = c_J - d_G C_G k \qquad c_{H(I)} \to \tilde{c}_{H(I)} = c_{H(I)} + d_{H(I)} C_{H(I)}(k + C_G + C_{H(I)}), \tag{26}$$

and the shift in the ghost contribution is given by a similar expression which can be found from the fact that the sum of the shifts vanishes. In what follows we consider, for simplicity, the case of $G = SL(N, R)$. The twisted ghost sector includes the ghosts of a W_N gravity, namely, a sequence of ghosts with dimensions $(i, 1 - i)$ for $i = 2, ..., N$ contributing $\tilde{c}_{Wgh} = -2(N - 1)[(N + 1)^2 + N^2]$ to $\tilde{c}$. The rest of the ghosts are paired with commuting fields of the same conformal structure coming from the J and I sectors. For $N = 2$ one finds $\tilde{c}_{Wgh} = -26 - 2\#_{pairs}$ where there are two pairs in the $\frac{SL(2,R)}{SL(2,R)}$ model and one in the $\frac{SL(2)}{U(1)}$ case. The net matter degrees of freedom have the following Virasoro anomaly $c = \tilde{c}_J - \frac{1}{2}[\tilde{c}_{(gh)} - \tilde{c}_{Wgh}] = (N - 1)[(2N^2 + 2N + 1) - N(N + 1)(t + \frac{1}{t})]$ which is exactly that of a (p, q) minimal W_N matter sector[16] provided $t \equiv k + N = \frac{p}{q}$. This was explicitly verified by analyzing the dimensions and contributions to $\tilde{c}$ of the various free fields in the J sector[14]. The expression for c reduces to that of the p, q minimal model for $N = 2$.

Next we want to compare the partition function of the topological coset model to that of the correspondig (W) string models. For simplicity we concentrate on the relation between the (p,q) minimal models and the $\frac{G}{G}$ for $G = SL(2)$ at level $k = \frac{p}{q} - 2$. The character of the minimal models coupled to gravity is given by the numerator of the matter character of the minimal models[18]Comparing eqn. (23) to the numerator of the character of the minimal model, it is clear that a correspondence may be achieved only provided one takes $\tau = -\frac{1}{2}\theta$. Recall that in the topological coset models we integrate in the path-integral only over θ (and not over τ) and the result is τ independent.[7]In this case the numerator of the character in the minimal model which is proportional to $Tr[(-1)^G q^{\hat{L}_0}]$ is mapped into $Tr[(-1)^G u^{\hat{L}_0 - \widehat{J^{(tot)}}{}^0_0}]$ where $u = e^{2i\pi\theta}$ in the $\frac{G}{G}$ model. The integration over the moduli parameter of the torus is therefore replaced by the integration over the moduli of flat gauge connection. To establish this mapping we compare the number of states at a given level and ghost number in the minimal models with the corresponding numbers at the same ghost number and "twisted level" of the $\frac{SL(2,R)}{SL(2,R)}$ model. The latter are given for $J = J_{r,s}$ by

$$
\begin{aligned}
I &= I_{-r-2lp,s} \quad G = -2l \quad &\hat{L}_0 - \widehat{J^{(tot)}}{}^0_0 &= l^2 pq + l(qr - sp) \\
I &= I_{r-2lp,s} \quad G = 1 - 2l \quad &\hat{L}_0 - \widehat{J^{(tot)}}{}^0_0 &= l^2 pq - l(qr + sp) + rs
\end{aligned}
\tag{27}
$$

In the minimal models we have states built on vacua labeled by the pair r, s with $1 \le r \le p-1$ and $1 \le s \le q - 1$ with $ps > qr$ which have dimension $h_{r,s} = \frac{(qr-sp)^2 - (p-q)^2}{4pq}$. The levels of the excitations are $\hat{L}_0 = \Delta - h_{r,s}$. For $G = 2l + 1$ one has $\Delta = A(l) = \frac{[(2pql+qr+sp)^2 - (q-p)^2]}{4pq}$ and for $G = 2l$ $\Delta = B(l) = \frac{[(2pql-qr+sp)^2 - (q-p)^2]}{4pq}$[17,18] Hence, the the contribution of the various levels to the partition function are identical to those of $\hat{L}_0 - \widehat{J^{(tot)}}{}^0_0$ in eqn. (27) for the same ghost numbers. The respective vacua satisfy $J = \sqrt{\frac{p}{2q}}p_m$ and $I = -\sqrt{\frac{p}{2q}}p_L$ where p_m and p_L are the matter and Liouville momenta respectively. It is thus clear that for a given r, s we get the same number of states with the same ghost number parity in the two models and the two partition functions on the torus are in fact identical.

.6.   C=1 STRING AS A TOPOLOGICAL COSET MODEL

Now that the connection between the fractional level twisted $\frac{G}{H}$ models and the W_N string models has been established, we would like to describe the two dimensional string theory, the $c = 1$ model, as a topological coset model. It is straightforward to realize that $\frac{SL(2,R)}{SL(2,R)}$ model with $k = -1$ or $\frac{SL(2,R)}{U(1)}$ with $k = -3$ conatins a matter sector with $c = 1$. There is however an important difference between the $c < 1$ and $c = 1$ cases. In the latter there is no background charge for the matter sector and thus no double complex in the BRST structure. Recall that the latter enabled us to use the $(+, -)$ bosonization inspite of the lack of an explicit $SL(2, R)$ invariant vacuum. In the $(+, +)$ scheme there is no aparent way to define degrees such that the spectral sequence machinery can be applied. A direct computation of the BRST cohomology seems also intracktable due to the cubic and quartic

478

terms in Q. The new idea with which enables us to bypass all these obstacles, is to similarity transforam Q into an operator with an isomorphic cohomology , The new operator is a sum of terms acting on different sectors so that its cohomology is a direct sum of simpler cohomologies. A detailed analysis of this appraoch is presented in ref.[19]. Here we give a brief description of the method and its results. In the $(+,+)$ parametrization T and $J^{(tot)}$ are given by

$$T^{(total)} = -\partial\phi^+\partial\phi^- - i\partial^2\phi^+ - \beta^+\partial\gamma^+ - \beta^-\partial\gamma^- - \rho^+\partial\chi^- - \rho^-\partial\chi^+ - 2\rho_0^2\chi_0 \tag{28}$$

$$\begin{aligned}
J^{(tot)+} &= \beta^+ + \chi^+\rho^0 - \chi^0\rho^+ \\
J^{0(total)} &= \beta^+\gamma^+ + \beta^-\gamma^- + i\partial\phi^- + \chi^+\rho^- - \chi^-\rho^+ \\
J^{(tot)-} &= \beta^+(\gamma^{+2} + \gamma^{-2}) + \beta^-\gamma^+\gamma^- - 2i(\gamma^+\phi^- + t\gamma^-\phi^+) + 2(2\partial\gamma^+ - t\partial\gamma^-) + \chi^-\rho^0 - \chi^0\rho^-.
\end{aligned} \tag{29}$$

where $t = k + 2$. The (β^+,γ^+) and (β^-,γ^-) are $(1,0)$ commuting system and ϕ^+ and ϕ^- are scalars.

Let us now define the dimension $(0,0)$ operators of zero ghost number

$$\begin{aligned}
R &= \oint \frac{dz}{2\pi i}(\chi^+\rho^+\gamma^-\gamma^- + 2\chi^+\rho^0\gamma^+ - \chi^+\rho^+\gamma^+\gamma^+) \\
P &= -\oint \frac{dz}{2\pi i}{}' (i\phi^+(\beta^+\gamma^+ + \beta^-\gamma^- + \chi^+\rho^- - \chi^-\rho^+))',
\end{aligned} \tag{30}$$

where $\oint \frac{dz}{2\pi i}{}'$ means that the zero modes of ϕ^+ are excluded. We then use these operators to transform $Q_{BRST}{}^{(rel)}$ to the desired form in the following way

$$e^{-P}e^R Q_{BRST}{}^{(rel)}e^{-R}e^P = Q_{tr}^{(rel)} \tag{31}$$

with

$$\begin{aligned}
Q_{tr}^{(rel)} &= \oint \frac{dz}{2\pi i}[\chi^-\beta^+ + 2i\chi^0\partial\phi^- - 2t\chi^+\partial\gamma^- - 2t\phi_0^+\chi^+\gamma^-] \\
&= 2\sum_{n\neq 0}\chi^0_{-n}\phi^-_n + \sum_n(\chi^-_{-n}\beta^+_n - 2t(\phi_0^+ - n - 1)\chi^+_{-n}\gamma^-_n).
\end{aligned} \tag{32}$$

The mode expansions are relative to the vacuum of the twisted theory (i.e. $\gamma(z) = \sum_n \gamma_n z^{-(n+1)}$). From (31) it follows that the cohomologies of $Q_{BRST}{}^{(rel)}$ and of $Q_{tr}^{(rel)}$ are isomorphic, namely, for every state $|\Phi_0>$ in the cohomology of $Q_{tr}^{(rel)}$, the state $|\Psi> = e^{-R}e^P|\Phi_0>$ is in the cohomology of $Q_{BRST}{}^{(rel)}$ and vice versa.

In the following direct sum of Fock spaces

$$\bigoplus_{n\neq 0} F(\chi^0_{-n}, \rho^0_n, \phi^-_{-n}, \phi^+_n) \bigoplus_n F(\chi^-_{-n}, \rho^+_n, \gamma^+_{-n}, \beta^+_n) \bigoplus_n F(\chi^+_{-n}, \rho^-_n, \beta^-_{-n}, \gamma^-_n) \qquad (33)$$

the first term is subjected to the action of the first term in eqn. (32), and similarly for the second and third terms. It is thus apparent that $Q^{(rel)}_{tr}$ indeed decomposes into a sum of anti-commuting terms which act on separate Fock spaces and, therefore, that the cohomology ring is a direct sum of smaller ones.

The nontrivial $Q^{(rel)}_{tr}$-cohomology states are spanned by

$$|\Phi_0> = \rho^-_{-n}\gamma^{-r}_{-n}|\phi^+_0 = -n+1,\ n > 0, \phi^-_0 = r\rangle$$

$$|\Phi_0> = \gamma^{-r}_{-n}|\phi^+_0 = -n+1, n > 0, \phi^-_0 = r-1\rangle$$

$$|\Phi_0> = \chi^+_{-n}\beta^{-r}_{-n}|\phi^+_0 = n+1 > 0, \phi^-_0 = -r-2\rangle$$

$$|\Phi_0> = \beta^{-r}_{-n}|\phi^+_0 = n+1 > 0, \phi^-_0 = -r-1\rangle$$

for $r = 0, 1, 2, ...$, and

$$|\Phi_0> = |any\phi^+_0, \phi^-_0 = -1> .$$

We can now insert $|\Phi_0>$ into the expressions for the states in the cohomology of $Q_{BRST}^{(rel)}$ as follows

$$|\Psi> = \sum_{n=0}^{\infty} \frac{(-1)^n}{n!} R^n \sum_{m=0}^{\infty} \frac{P_0{}^m}{m!}|\Phi_0> = e^{-R}e^{P_0}|\Phi_0>$$

.7. **PHYSICAL STATES OF THE $\frac{SL(2,R)}{SL(2,R)}$ MODELS VERSUS THOSE OF THE $c \leq 1$ MODELS.**

The use of the similarity transforamtion method was motivated by the $k = -1$ $\frac{SL(2,R)}{SL(2,R)}$ case which corresponds to the $c = 1$ string model. In fact this appraoch is also adequate for any rational vaule of t and thus produces a unified description of the topological coset models which are the counterparts of the $c \leq 1$ Liouville models. At ghost number $N_G = -1$, we expect that the discrete states found above would correspond to elements of the ground ring[20] (recall the shift in the ghost number when moving from states to operators because $|0>_{phys} = \chi^+_1|0>_{SL(2,C)}$). The lowest level state is simply $\rho^-_{-1}|\phi^+_0 = \phi^-_0 = 0>$ which corresponds to the identity operator. The next two states of the cohomology of $Q_{tr}^{(rel)}$ which are at level 2 translate into operators in the cohomology of $Q_{BRST}^{(rel)}$ as follows :

$$\rho_{-1}^{-}\gamma_{-1}^{-}|\phi_0^{+} = 0, \phi_0^{-} = 1 > \;\to\; \tilde{x} = \gamma^{-}e^{i\phi^{+}}$$

$$\rho_{-2}^{-}|\phi_0^{+} = -1, \phi_0^{-} = 0 > \;\to\; \tilde{y} = [-i\partial\phi^{+} + \chi^{+}(\rho^{-} + 2\rho^0\gamma^{+} + \rho^{+}[(\gamma^{-})^2 - (\gamma^{+})^2])]e^{-i\phi^{-}}$$

$$(34)$$

These states are (with the appropriate identification) at the same momenta as those of the ground ring generators in the $c \leq 1$ models. In fact $\tilde{y}$ is equal to y of ref. [20] with some additions from the "topological sectors". One can also change the form of $\tilde{x}$ so it resembles that of the ground ring x by adding a $Q_{BRST}^{(rel)}$ exact term as follows

$$\tilde{x} = \{Q_{BRST}^{(rel)}, \tfrac{1}{2}\rho^0(\beta^{-})^{-1}e^{i\phi^{+}}\} + (\beta^{-})^{-1}(\chi^{+}\rho^{-} + i\partial\phi^{-} + \beta^{+}\gamma^{+} - \chi^{-}\rho^{+})e^{i\phi^{+}} \quad (35)$$

The ground ring cohomology is now generated by $\tilde{x}^n\tilde{y}^m$. As in the ground ring of ref. [20], it is easy to realize that area preserving diffeomorphisms leave the ground ring invariant. These W_∞ transformations are generated by currents constructed by acting on the $N_G = 1$ cohomology operators with G_{-1}. Recall that $G = \rho^{-}(J^{+} - I^{+}) + 2\rho^0(J^0 - I^0) + \rho^{+}(J^{-} - I^{-}) + \partial\rho^0$. For instance the generators $\partial_{\tilde{x}}$ and $\partial_{\tilde{y}}$ take the following form

$$\partial_{\tilde{x}} = G_{-1}(\chi^{+}e^{-i\phi^{+}}) = \beta^{-}e^{-i\phi^{+}}$$

$$\partial_{\tilde{y}} = G_{-1}(\chi^{+}(\beta^{-})^{-1}e^{i\phi^{-}}) = e^{i\phi^{-}}$$

$$(36)$$

It is easy to check that indeed, as is hinted by the notations,

$$\partial_{\tilde{x}}\tilde{x} = \partial_{\tilde{y}}\tilde{y} = 1 \qquad \partial_{\tilde{x}}\tilde{y} = \partial_{\tilde{y}}\tilde{x} = 0. \quad (37)$$

One may wonder about the operator $(\beta^{-})^{-1}$ which does not seem to be an appropriate operator to use since $\beta^{-} = J^{+} - I^{+}$. Without the inclusion of arbitrary powers of β^{-} the space of physical states of the $\frac{SL(2,R)}{SL(2,R)}$ model does not recover that of the $c \leq 1$ models. A similar situation is facing us also in the tachyonic sector. One possible prescription for regaining a full equivalence in the states is to implement an idea of ref. [21] where a further bosonization is invoked for the (β^{-}, γ^{-}) system. In this bosonization $\beta^{-} \equiv e^{u-iv}$ and $\gamma^{-} \equiv -i\partial v e^{-u+iv}$, where u, v are free bosons with a background charge of $-\frac{1}{2}$ and $\frac{i}{2}$ respectively. In terms of the latter bosons, one is entitled to take any arbitrary power of β^{-} and hence we complete the missing states in the comparison with the gravitational models. For another prescription see ref. [19]. One branch of the tachyons of the $c \leq 1$ model can be easily identified with a sector of the cohomology of the $\frac{SL(2,R)}{SL(2,R)}$ model: this is the vacuum of the latter, $|\phi_0^{+} = p^{+}, \phi_0^{-} = -1 >$ which corresponds to the operator $\chi^{+}e^{ip^{+}\phi^{-} - i\phi^{+}}$. If one identifies ϕ_J with the matter field X, ϕ_I with the Liouville field ϕ and χ^{+} with c, the tachyonic states of one branch are indeed found. However, the other branch $\chi^{+}e^{ip^{-}\phi^{+} + i\phi^{-}}$, is missing in the cohomology of $Q_{BRST}^{(rel)}$. There are, however, additional states with no excitations at $N_G = 0$. These are the states $(\beta_0^-)^r |\phi_0^+ = 1, \phi_0^- = -r - 1 >$ corresponding to the operators $\chi^+ (\beta^-)^r e^{i\phi_0^- \phi^+ + i\phi^-}$. Apart from the appearence of the operator β^- these states are identical to a discrete series of the other branch of the tachyons. If again we bosonize β^- then r can take any real number and thus one finds states which correspond to the full missing branch. For $k = -1$, restricting the values of r to the integers would correspond to the $c = 1$ model at the self-dual radius.

The states of other ghost number are also in one to one correspondence with those of the $c \leq 1$ Fock space relative cohomology. The only exception is that our second branch of the tachyon appears in both $N_G = 1$ and $N_G = 2$ whereas in the Liouville model it appears only in the former. A similar situation is revealed in the $\frac{SL(2,R)}{U(1)}$ analysis of ref. [21].

REFERENCES

1. E. Witten, *Comm. Math. Phys.* **117** (1988) 353.

2. R. Dijkgraaf, E. Verlinde, and H. Verlinde, *Nucl. Phys.* **B352** (1991) 59.

3. D. Montano, J. Sonnenschein , *Nucl. Phys.* **B324** (1989) 348, J. Sonnenschein *Phys. Rev.* **D42** (1990) 2080.

4. E. Witten, "On Holomorphic Factorization of WZW and Coset Models" IASSNS-91-25.

5. M. Spiegelglas and S. Yankielowicz *Nucl. Phys.* **B393**,(1993) 301.

6. K. Bardacki, E. Rabinovici, and B. Serin *Nucl. Phys.* **B299** (1988) 151.

7. K. Gawedzki and A. Kupianen , *Phys. Lett.* **215B** (1988) 119, *Nucl. Phys.* **B320** (1989)649.

8. D. Karabali and H. J. Schnitzer, *Nucl. Phys.* **B329** (1990) 625.

9. O. Aharony,O. Ganor J. Sonnenschein and S. Yankielowicz , *Nucl. Phys.* **B399** (1993) 560.

10. Y. Kazama and H. Suzuki, *Nucl. Phys.* **B321** (1989).

11. J. Isidro and A. V. Ramallo " gl(N,N) current algebras and TFTs" US-FT-3/93.

12. P. Bouwknegt, J. McCarthy and K. Pilch *Phys. Lett.* **234B** (1990) 297, *Comm. Math. Phys.* **131** (1990) 125;Cern Preprint TH-6162/91

13. O. Aharony,O. Ganor N. Sochen J. Sonnenschein and S. Yankielowicz ,*Nucl. Phys.* **B399** (1993) 527.

14. O. Aharony, J. Sonnenschein and S. Yankielowicz , *Phys. Lett.* **289B** (1992) 309.

15. D. Bernard and G. Felder *Comm. Math. Phys.* **127** (1991) 145.

16. Fateev and Lukeanov *Int. Jour. of Mod. Phys.* **A31** (1988) 507.

17. M. Bershadsky and I. Klebanov *Nucl. Phys.* **B360** (1991) 559.

18. B. Lian and G. Zuckerman *Phys. Lett.* **254B** (1991) 417.

19. O. Aharony,O. Ganor J. Sonnenschein and S. Yankielowicz , *Phys. Lett.* **305B**,(1993) 35-42.

20. E. Witten *Nucl. Phys.* **B377** (1992) 55.

21. S. Mukhi and C. Vafa " Two Dimensional Black hole as a Topological Coset model of $c = 1$ String Theory",HUTP-93/A002, TIFR/TH/93-01.

6. Appendix

NEGATIVE PHYSICS

V. Gates, M. Roachcock, Empty Kangaroo, and W.C. Gall *

Institute for Tired Palindromes,
State Univ. of New York at Stony Brook, Stony Brook, New York, Unit. States of A.

ABSTRACT

Physics is getting more negative: black HOLES, NOT theory, NO-go theorems, DARK matter, etc.

Four dimensions good, two dimensions bad better
           - - - - - - - - - Geor Georwell, *Conformal Farm*
If time and space are relative, then why didn't they come to my birthday party?
           - - - - - - - - - Alexei Sayle
A sweet potato, therefore a yam - René AlaCartes
Proof? You mean you actually expect me to give you the proof, too?
    That's the last theorem you're ever going to get from me! - - Pierre de Fermat

Black holes are even harder to write papers on than an empty stomach [1], because they swallow up all your results. Black holes also cause a loss of information [2], since after you start working on them people stop reading your papers. This information gets lost even if you put it on supercomputers, since it gets radiated away as crayons. Although they have no hair, they do have an electric charge, of about $1.3/kwatt-hr. This is represented by the following diagram (drawn with a Penrose):

---

* Made you look!

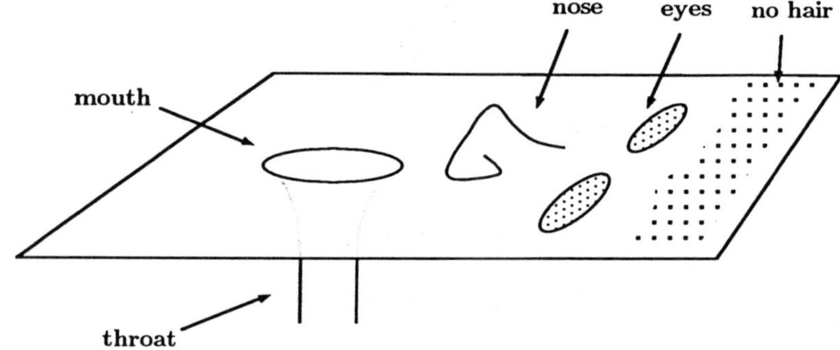

Figaro. Black hole [3]

No-go theorems are the hardest to find proofs for [4], since physicists tend to take the seminar titles literally*.

N=2 strings are the same as N=4 strings, since 2+2=4. This means that space-time has 2 space and 2 time dimensions, so the Lorentz group is SO(Tutu) [5]. In N=2 strings the coordinates are complexified, appearing as

$$\bar{X} \Box X$$

where X=Christ and $\bar{X}$=antiChrist [6]. The background metric for these strings has self-dual curvature, but this curvature can be replaced with torsion. This method is known as "paratelellelism." In this formalallalism, the cohomolollology of the BEST operator [7] describes spaces with torsion-bar suspension $\overline{T}_{abc}$, where the vectors carry indices, except in biology, where vectors carry in disease. We use the Dirack notation

$$\text{ket:} \quad |\rangle$$
$$\text{bra:} \quad \langle|$$

We also choose hypergolic coordinates

$$t = r \ cough \ \theta, \quad x = r \ sigh \ \theta$$

*

(This is known as putting Deshorses besfores Descartes.) The resulting field equation is the Witten-Gordon equation [8,9]:

$$\Box\phi + \phi = 0$$

where we have used the Croissant bracket:

$$\{P,Q\} = 8$$

$\Box$ *is a marginal operator.* This may be related to WZW$_N$-gravity. In the random lattice approach to string theory, the string is treated as a connected collection of points, as a string of beads [10]. This foolows fram an application of the uncretinty principle for matter at high dentistries [11]. Since string theory has no devirgins, all infancies are canceled*. As a consequence, this theory can describe neither oriented or unoriented strings, but only disoriented ones.

All these ideas can be applied to cosmology, since anything can be applied to cosmology. The main problem in cosmology is missing dark matter. The reason this matter is missing is because it isn't really dark: It's transparent, *space* is dark. (The confusion arose from the erroneous argument that the missing matter must contribute a lot of mass, and therefore must be heavy, and thus not light, and hence dark. This argument was invented by some astronomers who got fat masses from eating too many Higgses.) Most of this missing matter is in the form of albinos, which are particles that have no color (like neutrinos, except lighter). By properly taking these albinos into account we can shed some light on the true nature of dark matter, or at least make light of the whole situation.

These ideas can also be applied to more important issues: For example, strings can replace the ozone layer, since they don't suffer from ultraviolet divergences.

ACKNOWLEDGMENT

We would like to thank the secretaries for checking our speling. Some of the calculations in an earlier version were performed by various colleagues, so we dropped those sections.

---

* If Siamese twins apply for a position, how many would they fill (and do they file a joined tax return)?

REFERENCES

[1] D. Letterman, Stupid metrics, NBC preprint

[2] Bugsy Beatty, Insanitons in Jelly-Roll Relativity, Stony Brook preprint RA-GG-MOPP-93-87 (1993.61427)

[3] D. Seuss and A. Hoponpopolous, Grand unified cohosmology and the lunar neutrino problem, Texas M&M preprint

[4] e-mail Martinec, Do quantum mechanics use Lagrange multi-pliers?, Chicanogo preprint

[5] Light Cohn, *in* Proc. XXth Int. Conf. on The Usual Stuff, Le Whoosh!, July 3 - June 27, last year ─OOO
                                    XOX

[6] X. Pope, Malcolm X, and Alexander 7, WWW strings and 2D-frutti gravity

[7] Bart Zweabart, BVD quantiztion is the BEST quantization for closed string field theory

[8] Edward Physicshands, If it's knot theory, thekn it must be experimeknt, Princetokn prepriknt

[9] K. Gordon and S. Gordon, Sophomoric scalars, *Phys. Rev.* **D11** (1975) 2088

[10] Ben and Aretha Franklin, Chain of fools, Philadelphia preprint

[11] Philadelphia Nelson, There ain't no such thing as a free lagrangian, Benjamin book

[12] Tropicana, Twistor in an 8 oz. can, 7-11 preprint

[13] Askok Das and Häagen Dasz, Sugar-cone string field theory

[14] V. Horowitz, Off-shell strings break violin variance

[15] E. Verlinde and H. Verlinde, Calabi-Yau and Ginsparg-Lanyau are different faces of the same theory

[16] J-pole Chinski, Exact aspirin solution for a 4-d black hole throat

[17] Von Jones and Von Nieuwenhuizen, Loops and conformal field theory

[18] Author One and Author Two, Recent seminal results, EFI-92-41

CONFERENCE PHOTO

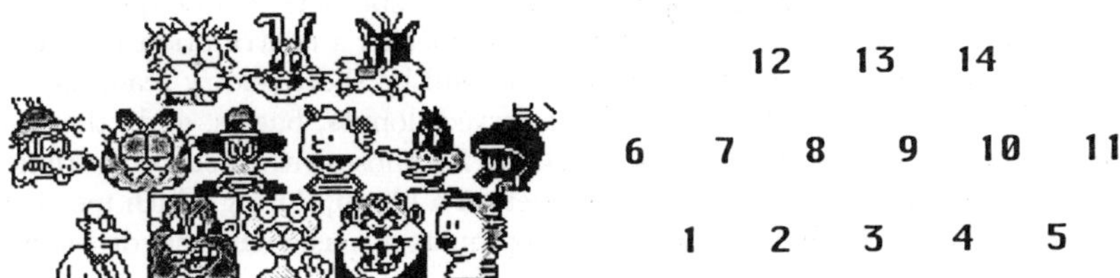

First row: [1] Unidentified (This guy doesn't even look like a physicict), [2] J.C. Maxwell, [3] W.C. Gall, [4] W. Pauli, [5] T. Hobbes.
Second row: [6] A. Volta, [7] E. Schrödinger, [8] D. Hilbert, [9] J. Calvin, [10] A.S. Eddington, [11] W. Heisenberg.
Third row: [12] A. Einstein, [13] R.P. Feynman, [14] V. Gates.

PRELIMINARY ANNOUNCEMENT FOR NEXT YEAR'S CONFERENCE

Next year our conference will be held at the beautiful scenic resort of Stony Brook:

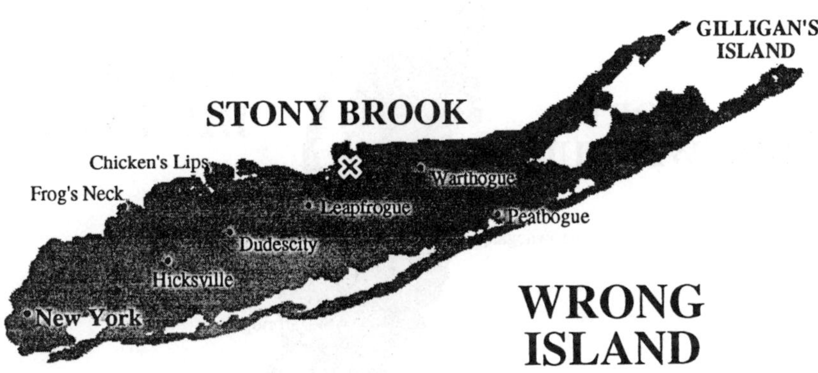

We have already received generous funding from NSF, DOE, KGB, BHT, LSD, BVD, and TLC. The topics will be the same as this year, but with names changed to protect the innocent. Already 42 speakers have been invited. (Unfortunately, all the attendees are so far unconfirmed.) The proceedings will again be published. There will also be additional 5-minute talks, which will be published in the margins. The proceedings will also contain photographs of all the speakers, whose quality will be inversely proportional to the time it takes each speaker's contribution to reach the conference organizers. For entertainment, there will be short poetry readings during the coffee breaks, during which you will be served donuts, bagels, and other 1-loop corrections. (Tree graphs will also be available for those with dietary restrictions.) In addition, there will be a conference banquet on a blimp, during which members of the University's Glee Club will sing barbershop quartet music while skydiving off the observation deck.

The following year the conference will be held in Wales: